Nutrition, Choice and Health-Related Claims

Nutrition, Choice and Health-Related Claims

Special Issue Editor

Tiziana de-Magistris

MDPI • Basel • Beijing • Wuhan • Barcelona • Belgrade • Manchester • Tokyo • Cluj • Tianjin

Special Issue Editor
Tiziana de-Magistris
Centro de Investigación y
Tecnologia Agroalimentaria de
Aragón (CITA)
Spain

Editorial Office
MDPI
St. Alban-Anlage 66
4052 Basel, Switzerland

This is a reprint of articles from the Special Issue published online in the open access journal *Nutrients* (ISSN 2072-6643) (available at: https://www.mdpi.com/journal/nutrients/special_issues/Nutrition_Health_Claims).

For citation purposes, cite each article independently as indicated on the article page online and as indicated below:

LastName, A.A.; LastName, B.B.; LastName, C.C. Article Title. *Journal Name* **Year**, *Article Number*, Page Range.

ISBN 978-3-03928-648-5 (Pbk)
ISBN 978-3-03928-649-2 (PDF)

Contents

About the Special Issue Editor

Tiziana de Magistris (Ph.D) obtained her doctoral in Agricultural and Natural Resource Economy (2004) at the Universitá degli Studi "Parthenope" of Naples, Italy. She has conducted research in agricultural economy at the Agro-Food and Natural Resources Economy unit at the Agro-Food Research and Technology Centre of Aragon (CITA) since 2006. Her research activity focuses on consumer preferences for food labelling, hypothetical and non-hypothetical valuation methods and, psychological and behavioral mechanisms that drive better consumer decision-making. During her scientific career, she has participated in 6 European projects Fellows (leading two of them as Marie Curie Fellow: Obesclaim, FOODLABEL), 4 national competitive projects (leading two of them), and 15 private contracts/services with companies. She is co-author of more than 50 scientific publications and more than 10 book chapters, and she has contributed with more than 40 communications at scientific conferences. She is also invited professor at the Universita degli Studi "Parthenope" of Naples, Italy, teaching about consumers' preferences for food products.

nutrients

MDPI

Editorial

Nutrition, Choice and Health-Related Claims

Tiziana de-Magistris [1,2]

[1] Unidad de Economía Agroalimentaria y de los Recursos Naturales—Centro de Investigación y Tecnología Agroalimentaria de Aragón (CITA), Gobierno de Aragón, 50004 Zaragoza, Spain; tmagistris@gmail.com

[2] Instituto Agroalimentario de Aragón, IA2 (CITA-Universidad de Zaragoza), 50009 Zaragoza, Spain

Received: 20 February 2020; Accepted: 26 February 2020; Published: 28 February 2020

Scientific evidence shows that food consumption is one of the main causes that increases the risk of developing a non-communicable disease (NCD). One of the mechanisms introduced to ensure more informed food purchases that lead to healthier diets is the introduction in the marketplace of functional food products to provide information on the nutritional and health properties that certain foods possess. This information is transmitted to consumers via different nutritional and health claims.

Two studies investigated the prevalence of front-of-package (FoP) claims in Brazilian packaged food. Duran et al. [1] found that nutritional claims (NCs) were the most prevalent, followed by health claims (HCs), especially in breakfast cereals and dairy beverages. Zancheta Ricardo et al. [2] examined the presence of trans fat information on the nutrition facts panel. The authors reported that 81.3% of the 11,434 products analyzed, did not present a source of trans fats in the list of ingredients. However, bakery products, cookies and crackers, candies and desserts, snacks, and convenience foods had the highest percentages of trans fat claims.

Two studies explored the context of foods and drinks with healthy and nutritious attributes in the United Kingdom (UK). Cesar Revoredo-Giha et al. [3] indicated that trading down in quality occurs in most of the studied categories and countries, and when households trade down, they moved to products with worse nutritional qualities. Likewise, Monserrat and Revoredo-Giha [4] assessed to what extent health and nutrition claims made by breakfast cereals had an impact on their market success.

Four studies focused on consumers' preferences and willingness to pay (WTP) for functional food products, while one study examined the role of functional food in disease prevention. Vischeccia et al. [5] analyzed a mozzarella cheese carrying reduced fat and enriched in omega-3 claims. The authors found that consumers´ willingness to pay for health claims was higher than nutrition claims and that naturally enriched omega-3 was the most preferred claim. Castellari et al [6] indicated that providing new nutritional information significantly increased the WTP for a jam-like fruit compote enriched with aloe vera gel. Panea and Ripoll [7] investigated the perception of pork quality that consumers attached to the functional combination of the addition of extracts derived from plants (pork-derived extracts added to pork feed) and the meat conservation conditions (packaging and time exposure). Likewise, Verneau et al. [8] analyzed the effect of information about the health benefit produced by lycopene on the WTP for canned crushed tomatoes enriched with lycopene. The results showed a relevant impact of information on WTP in the case of lycopene-enriched products. Finally, Plasek et al. [9] examined the role of functional foods in disease prevention in Hungary on about 13 diseases with four prevention methods. The results reported that functional foods prevented digestive problems, a weakened immune system and a high cholesterol level.

Two studies investigated consumers´ preferences for food products carrying sustainable and nutritional labels simultaneously. Almli et al. [10] observed a broad heterogeneity in health attitudes among Norwegian, Romanian, and Turkish organic consumers. Akaichi et al. [11] showed that consumers did not view desirable food attributes (such as organic, local, fair trade and high animal welfare) as unrelated to health-related food labels.

Two lines of investigation analyzed how consumers ranked different nutritional claims. Gracia and Barreiro-Hurlé [12] reported that the ranking of claims differs between biscuits and pastries and across consumers. Moreover, the results indicated that, for the average consumer, the most important nutritional claims for the two cereal products were "reduced saturated fat" and "with no added sugar". On the other hand, the least important claim was "low salt". Annunziata and Mariani [13] examined six specific claims on the basis of a web survey carried out on a sample of 504 consumers. Findings revealed that there is little attention paid to nutritional health claims and their use is not widespread.

Two studies explored consumers' choices for health information. Sogari et al. [14] tested the impact of labeling wholegrain pasta with a health message descriptor displayed at the point-of-purchase (POP) on consumer choice in a campus dining setting. Findings indicated that only the message about vitamin benefits had a significant effect on this choice, with a higher probability of selecting this pasta than the no-message condition and also a higher probability than the fiber message condition. Sajdakowska and Tekień [15] determined different segments of consumers based on their preferences towards some statements related to nutrition presented on a yoghurt label with a precise focus on aspects of the increased and decreased content of some ingredient.

Four studies focused on psychological factors affecting consumers' preferences for NCs. Guzek et al. [16] determined the influence of food neophobia (FN) about allergens on the food product choices. The respondents characterized by a high level of FN less commonly chose dishes characterized by neophobic potential as a starter (carpaccio), main course (risotto ai frutti di mare) and dessert (zabaglione). Benson et al. [17] identified knowledge as the key factor influencing how much individuals believe nutritional claims and their perceptions. López-Galán and de-Magistris [18], assessed whether an emotional eating style influenced the purchase of food products carrying these claims. Findings of this study suggested that emotional eating negatively impacts purchasing behavior related to nutritional claims. Finally, Bazzani et al. [19] showed that health consciousness was an important driver in the use of wine labels, such as clean labels and alcohol content.

The last two studies investigated new determinants influencing purchase intention for functional foods. Berhaupt-Glickstein et al. [20], investigated the effects of health claims carried by green tea on purchase intentions among adults 55 years of age and older living in the US. Factors that mitigated the claim's effects on purchase intentions were: race/ethnicity, age, importance of health claims, supplement use, health, worry about health/becoming sick with cancer, worry that led to dietary change, green tea consumption and the relationship between green tea and cancer. Finally, Temesi et al. [21] revealed that perceived fit of the carrier and the ingredient is a major determinant of purchase intention, together with health concerns and attitudes to functional foods.

The present Special Issue focused on the role of nutritional properties and/or health-related claims of food products and functional food products on choice preferences, choice behavior, healthy eating/healthy diet and the willingness to pay for certain foods.

Funding: This research received no external funding.

Conflicts of Interest: The author declares no conflict of interest.

References

1. Duran, A.C.; Zancheta Ricardo, C.; Mais, A.; Bortoletto Martins, A.P.; Smith Taillie, L. Conflicting Messages on Food and Beverage Packages: Front-of-Package Nutritional Labeling, Health and Nutrition Claims in Brazil. *Nutrients* **2019**, *11*, 2967. [CrossRef]
2. Zancheta Ricardo, C.; Mateus Peroseni, I.; Amaral Mais, L.; Bortoletto Martins, A.P.; Duran, A.C. Trans Fat Labeling Information on Brazilian Packaged Foods. *Nutrients* **2019**, *11*, 2130. [CrossRef] [PubMed]
3. Revoredo-Giha, C.; Akaichi, F.; Chalmers, N. Trading on Food Quality due to Changes in Prices: Are There Any Nutritional Effects? *Nutrients* **2020**, *12*, 23. [CrossRef] [PubMed]
4. Costa-Font, M.; Revoredo-Giha, C. Importance of Health Claims in the Adoption of New Breakfast Cereal Products in the UK. *Nutrients* **2019**, *11*, 3076. [CrossRef] [PubMed]

5. Viscecchia, R.; Nocella, G.; De Devitiis, B.; Bimbo, F.; Carlucci, D.; Seccia, A.; Nardone, G. Consumers' Trade-Off between Nutrition and Health Claims under Regulation 1924/2006: Insights from a Choice Experiment Analysis. *Nutrients* **2019**, *11*, 2881. [CrossRef] [PubMed]
6. Castellari, E.; Ricci, E.C.; Stranieri, S.; Marette, S.; Sarnataro, M.; Soregaroli, C. Relationships Between Health and Environmental Information on the Willingness to Pay for Functional Foods: The Case of a New Aloe Vera Based Product. *Nutrients* **2019**, *11*, 2781. [CrossRef] [PubMed]
7. Panea, B.; Ripoll, G. Plant-Derived Extracts Feed-Addition and Packaging Type Influence Consumer Sensory Perception of Pork. *Nutrients* **2019**, *11*, 2652. [CrossRef]
8. Verneau, F.; La Barbera, F.; Furno, M. The Role of Health Information in Consumers' Willingness to Pay for Canned Crushed Tomatoes Enriched with Lycopene. *Nutrients* **2019**, *11*, 2173. [CrossRef]
9. Plasek, B.; Lakner, Z.; Kasza, G.; Temesi, A. Consumer Evaluation of the Role of Functional Food Products in Disease Prevention and the Characteristics of Target Groups. *Nutrients* **2020**, *12*, 69. [CrossRef]
10. Almli, V.L.; Asioli, D.; Rocha, C. Organic Consumer Choices for Nutrient Labels on Dried Strawberries among Different Health Attitude Segments in Norway, Romania, and Turkey. *Nutrients* **2019**, *11*, 2951. [CrossRef]
11. Akaichi, F.; Revoredo Giha, C.; Glenk, C.; Gil, J.M. How Consumers in the UK and Spain Value the Coexistence of the Claims Low Fat, Local, Organic and Low Greenhouse Gas Emissions. *Nutrients* **2020**, *12*, 120 [CrossRef] [PubMed]
12. Azucena Gracia, A.; Barreiro-Hurlé, J. Making Sense of Information Overload: Consumer Ranking of Nutritional Claims in Cereal Based Products. *Nutrients* **2019**, *11*, 2858. [CrossRef]
13. Annunziata, A.; Mariani, A. Do Consumers Care about Nutrition and Health Claims? Some Evidence from Italy. *Nutrients* **2019**, *11*, 2735. [CrossRef] [PubMed]
14. Sogari, G.; Li, L.; Lefebvre, M.; Menozzi, D.; Pellegrini, N.; Cirelli, M.; Gómez, M.I.; Mora, C. The Influence of Health Messages in Nudging Consumption of Whole Grain Pasta. *Nutrients* **2019**, *11*, 2993. [CrossRef] [PubMed]
15. Sajdakowska, M.; Tekień, A. To Raise or Not to Raise the Level of Ingredients in Yoghurts: Polish Consumer Preferences Regarding Dairy Products. *Nutrients* **2019**, *11*, 2526. [CrossRef] [PubMed]
16. Guzek, D.; Pęska, J.; Głąbska, D. Role of Food Neophobia and Allergen Content in Food Choices for a Polish Cohort of Young Women. *Nutrients* **2019**, *11*, 2622. [CrossRef]
17. Benson, T.; Lavelle, F.; McCloat, A.; Mooney, E.; Bucher, T.; Egan, B.; Dean, M. Are the Claims to Blame? A Qualitative Study to Understand the Effects of Nutrition and Health Claims on Perceptions and Consumption of Food. *Nutrients* **2019**, *11*, 2058. [CrossRef]
18. López-Galán, B.; de-Magistris, T. Testing Emotional Eating Style in Relation to Willingness to Pay for Nutritional Claims. *Nutrients* **2019**, *11*, 1773. [CrossRef]
19. Bazzani, C.; Capitello, R.; Ricci, E.C.; Scarpa, R.; Begalli, D. Nutritional Knowledge and Health Consciousness: Do They Affect Consumer Wine Choices? Evidence from a Survey in Italy. *Nutrients* **2020**, 12. [CrossRef]
20. Berhaupt-Glickstein, A.; Hooker, N.C.; Hallman, W.K. Qualified Health Claim Language affects Purchase Intentions for Green Tea Products in the United States. *Nutrients* **2019**, *11*, 921. [CrossRef]
21. Temesi, Á.; Bacsó, Á.; Grunert, K.G.; Lakner, Z. Perceived Correspondence of Health Effects as a New Determinant Influencing Purchase Intention for Functional Food. *Nutrients* **2019**, *11*, 740. [CrossRef] [PubMed]

nutrients

MDPI

Article

How Consumers in the UK and Spain Value the Coexistence of the Claims Low Fat, Local, Organic and Low Greenhouse Gas Emissions

Faical Akaichi [1,*], Cesar Revoredo Giha [1], Klaus Glenk [1] and Jose Maria Gil [2]

1 Department of Rural Economy, Environment and Society, Scotland's Rural College, Edinburgh EH9 3JG, UK; cesar.revoredo@sruc.ac.uk (C.R.G.); Klaus.glenk@sruc.ac.uk (K.G.)

2 CREDA-UPC-IRTA, 08860 Barcelona, Spain; chema.gil@creda.es

* Correspondence: Faical.akaichi@sruc.ac.uk; Tel.: +44-131-535-4217

Received: 31 October 2019; Accepted: 11 December 2019; Published: 1 January 2020

check for updates

Abstract: This study investigates the substitution and complementary effects for beef mince attributes drawing on data from large choice experiments conducted in the UK and Spain. In both countries, consumers were found to be willing to pay a price premium for the individual use of the labels "Low Fat" (UK: €3.41, Spain: €1.94), "Moderate Fat" (UK: €2.23, Spain: €1.57), "Local" (UK: €1.54, Spain: €1.61), "National" (UK: €1.33, Spain: €1.37), "Organic" (UK: €1.02, Spain: €1.09) and "Low Greenhouse Gas Emissions (GHG)" (UK: €2.05, Spain: €0.96). The results showed that consumers in both countries do not treat desirable food attributes as unrelated. In particular, consumers in Spain are willing to pay a price premium for the use of the labels "Local", "Organic" and "Low GHG" on beef mince that is also labelled as having low or moderate fat content. By contrast, consumers in the UK were found to discount the coexistence of the labels "Low Fat" and "Organic", "Low Fat" and "Low GHG" and "Moderate Fat" and "Low GHG". The results, however, suggest that in the UK the demand for beef mince with moderate (low) fat content can be increased if it is also labelled as "Organic" or "Low GHG" ("Local").

Keywords: health; local; organic; greenhouse gas emissions; consumer; choice experiment; willingness to pay; trade-offs

1. Introduction

The prevalence of overweight and obesity is increasing at an alarming rate. It is estimated that approximately 2 billion adults are overweight and over 600 million are obese globally [1,2]. The increasing prevalence of overweight and obesity is placing a considerable burden on the economy and public health, including increases in the risk of developing serious health conditions, with direct healthcare costs amounting to billions [3–5].

Fortunately, obesity is preventable due to its strong, although not exclusive, link to diet. In fact, there is strong evidence that the prevalence of overweight and obesity is linked to the growing consumption of energy-dense foods and sugar-loaded beverages that are generally inexpensive, palatable and convenient [6–8]. As a result, it has been recognised that changing dietary habits and lifestyle would contribute to the reduction in overnutrition and its serious health and economic consequences [9–12].

In response, a whole raft of policy approaches have been designed and implemented to promote healthy diets and make the food selection environment more conducive to healthy choices. These approaches include mandates, restrictions, economic incentives, marketing limits, information provision and health campaigns [13,14]. Among these policy approaches, nutrition labelling is probably the most studied population-based health approach [15–20]. In general, these studies found that getting consumers to choose and eat healthier foods is not a trivial task. On the one hand, nutrition labels and health claims were found to have the potential to increase consumers' demand for a healthier diet and help them to make more informed food choices. On the other hand, difficulty in understanding traditional nutrient declarations, especially those provided on the back of product packaging, was the most cited barrier to the use of nutrition labels.

Another aspect that may reduce the effectiveness of nutrition labels is the fact that this type of label is competing with other food attributes for consumer awareness. In fact, in addition to traditional food attributes such as price, income, taste and convenience, consumers are increasingly showing interest in less tangible food attributes, such as the sustainability, local origin, animal friendliness and social fairness of the production and processing of food products [21–25]. As a result, the strong interest in nutrition information exhibited by consumers in research studies may not translate into actual purchases of healthier food products. For example, a lamb consumer who is willing to pay a price premium for the labels "Local" and "Low Fat", with the premium being higher for the former label, is likely to end up buying lamb labelled "Local" if the lamb carrying the label "Low Fat" is offered at the same or a higher price than the local lamb.

While extensive research has been devoted to assessing consumer understanding and use of nutrition labels and health claims, relatively little research has assessed how consumers weigh health-related food labels in comparison to labels for other desirable food attributes (such as organic, local, fair trade and high animal welfare) [21,24–29]. In general, it was found that despite the high interest shown by consumers in nutrition labels and health claims, it is possible that this interest does not translate into actual purchase, partly due to the trade-offs they make when choosing between food products with different desirable attributes.

Another factor that can affect (positively or negatively) the effectiveness of nutrition labels and health claims in increasing the demand for healthier foods is attribute bundling. In fact, because of consumers' increasing interest in desirable attributes, such as organic, local and animal welfare, producers and marketers may bundle these attributes to increase their products' differentiation, satisfy the needs of a larger number of ethically minded consumers and increase their sales. However, bundling desirable attributes is only a plausible strategy if consumers perceive them as independent or complementary. In other words, their value for the bundle of attributes is equal to or greater than the sum of their values for each individual attribute. If the desirable attributes are perceived as substituting or overlapping each other, bundling them will decrease consumers' total marginal willingness to pay (WTP) for the bundle (A product is said to be complementary if it is used or consumed jointly with another product. Such a product usually has more value when paired with its complement than when used separately. A product is said to be a substitute for another product if it satisfies the same (or at least some of the) basic wants as the other product. Substitute products usually have more value when used separately than when used together.). For example, consumers can perceive the labels "Low Fat" and "Organic" as complementary if they think that the two labels refer to two complementary dimensions of food sustainability: health and environment. However, if consumers are expecting organic meat to have lower fat content, they may see the two labels as communicating partially overlapping information and, hence, discount the coexistence of both.

Most of the papers that investigated how consumers trade off different food attributes against each other assumed zero interactions between the attributes. Relaxing this assumption allows: (a) testing the effect of attribute bundling; and (b) correctly computing total consumers' WTP for bundles of attributes. A food product is generally a bundle of different attributes. Consumers' total WTP for the product is equal to the sum of their WTP for the individual attributes, forming the bundle, plus the value of the interactions between the bundled attributes. It is noteworthy that estimating the interactions between the attributes considered, for example, in a choice experiment, requires a larger number of observations and generally makes the estimation of a choice model with a high number (e.g., more than 10) of random main and interaction effects very challenging. Studies by Nilsson [27] and

Bond et al. [28] were among the first to explicitly consider interaction effects between health-related attributes and other desirable attributes (e.g., organic, local).

Nilsson [27] found that the value of pork labelled as "Environmentally Certified" is enhanced if it also carries the label "Certified Free of Antibiotics". However, they found that US pork consumers perceive the labels "Environmentally Certified" and "Certified for Animal Well-being", and the labels "Certified for Animal Well-being" and "Certified Free of Antibiotics" as unrelated. Bond et al. [28] found that the coexistence of different health claims reduces the total marginal WTP for the bundles (i.e., perceived as substitutes). They found, however, that on top of price premiums for the labels "Organic" and "Excellent Source of Vitamin C", consumers were willing to pay an additional premium for the coexistence of these two labels on the same product.

In addition to the policy and market implications of the findings of Nilsson [27] and Bond et al. [28], both studies provided evidence that focusing on the main effects of health-related attributes and ignoring their interactions with other desirable attributes may lead to biased and misleading results. This conclusion was also confirmed by other studies that looked at the interaction between non-health-related attributes [21,24,30].

This study contributes to the literature on how consumers trade off health-related attributes against other food attributes in three ways. First, we assess consumer preference and WTP for various labels of nonconventional attributes on beef mince products, with a focus on the labels "Low Fat", "Moderate Fat", "High Fat", "Local", "National", "Imported", "Organic", "Low Greenhouse Gas Emissions (GHG)", "Moderate GHG" and "High GHG". To the best of our knowledge, this is the first study to investigate how consumers trade off the attribute fat content against the attributes origin, type of production and level of greenhouse gas emissions from production. Second, we investigate potential competition and complementarities between the labels "Low Fat", "Moderate Fat", "Local", "Organic" and "Low GHG" to reveal much of the potential marketing information that could be used to promote healthier meat products. Third, we analyse how consumers' preferences and WTP for individual and bundles of desirable food attributes vary by country (the UK versus Spain) and across consumer groups (gender and age group).

Therefore, this study aims to answer three empirical questions: (a) how do consumers perceive and value alternative health and "Sustainable" labels; (b) do they perceive these labels as unrelated or do they consider them as substitutes or complementary; (c) what degree of heterogeneity is there among consumers' preferences and WTP?

2. Materials and Methods

The data were collected in the UK and Spain through a national web-based choice experiment. A choice experiment is a quantitative research technique that involves asking individuals to state their preference over hypothetical alternative scenarios, products or services. Each alternative is described by several attributes. Individuals' responses are used to determine whether their preferences are significantly influenced by the attributes. The responses are also used to determine the relative importance of the attributes. Choice experiment has been used extensively in different research disciplines (e.g., marketing, health economics, environmental economics, the economics of transport) due to close resemblance to the real-world decision [31,32].

The initial design of the choice experiment was developed and revised based on input from a small sample of 110 respondents in each country. These respondents were not included in the dataset used for the econometric analysis. The final version of the survey was administered by a market research company. A total of 1211 and 1206 primary grocery shoppers in the UK and Spain, respectively, completed the survey. All subjects gave their informed consent for inclusion before they participated in the study. In both countries, the sample was required to be representative of the population in terms of gender, age, employment status and geographical area of the country. These hard quotas were achieved in both samples, except the age quota in the Spanish sample. In Spain, consumers aged 18 to 54 years old were slightly oversampled and consumers over the age of 54 were undersampled ((18–24)

11% vs. 8%; (25–34) 11% vs. 14%; (35–44) 28% vs. 19%; (45–54) 25% vs. 19%; (55+) 26% vs. 39%). Please note also that female consumers in both countries were slightly oversampled. This is because in EU, the majority of food buyers are female buyers). Because the product considered in this study is beef mince, only meat consumers were allowed to take part in the survey. The quality of the data was checked, and all the ineligible observations were discarded and replaced by eligible ones from new respondents. The socio-demographic characteristics of the two samples are provided in Table 1.

Table 1. Respondents' sociodemographic characteristics.

Characteristic	UK	Spain
Gender		
Female	60%	60%
Male	40%	40%
Age		
18–24	11%	10%
25–34	16%	10%
35–44	19%	27%
45–54	17%	25%
55+	36%	27%
Employment status		
Employed	61%	44%
Self-employed	8%	13%
Retired	5%	13%
Homemaker	5%	8%
Student	7%	7%
Other	7%	1%
Unemployed	7%	14%
Sub-country (UK)/Region (Spain)		
Scotland	11%	–
England	80%	–
Wales	5%	–
Northern Ireland	3%	–
Northwest (Galicia, Principado de Asturias)	–	8%
Castilla-León	–	5%
North (Cantabria, País Vasco, La Rioja, C. Foral de Navarra)	–	8%
Northeast (Aragón, Islas Baleares, Cataluña)	–	20%
Levante (Comunidad Valenciana, Región de Murcia)	–	14%
Centre-south (Castilla La Mancha, Extremadura, Madrid)	–	21%
Andalucía y Canarias	–	24%

In each country, respondents were successively shown nine choice sets. Each choice set consists of three hypothetical beef mince alternatives and an opt-out alternative. An example of one of the choice sets used in the study is displayed in Figure 1. Each alternative of beef mince is described in terms of five attributes: fat content, type of production, origin, level of GHG emissions and price. The attributes and their corresponding levels were chosen based on the literature and the outcome of a shelf audit that was carried out in the major supermarkets in both countries. The attributes and their levels are described in Table 2.

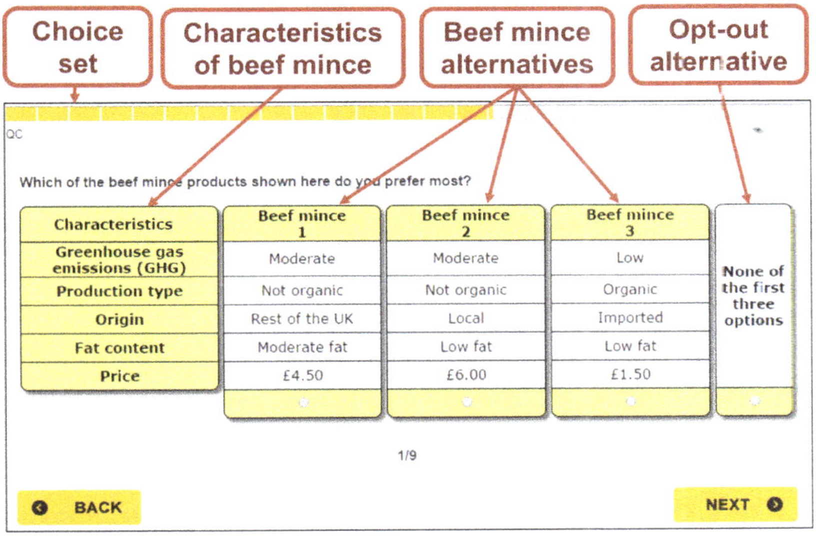

Figure 1. An example of a choice set used in the choice experiment conducted in the UK.

Table 2. Attribute levels of beef mince.

Attribute	Levels (UK)	Levels (Spain)
Fat content	**Low**: 3 g per 100 g serving of beef mince **Moderate**: 12 g per 100 g serving of beef mince **High**: 21 g per 100 g serving of beef mince	**Low**: 3 g per 100 g serving of beef mince **Moderate**: 12 g per 100 g serving of beef mince **High**: 21 g per 100 g serving of beef mince
Origin	**Local**: the beef cattle were raised and the beef mince was produced in the UK sub-country (Scotland, England, Wales or Northern Ireland) where the respondent lives	**Local**: the beef cattle were raised and the beef mince was produced in the autonomous region ("Comunidad Autónoma") where you the respondent lives
	Rest of the UK/National: if the beef cattle were raised and the beef mince was produced in the UK but not in the UK sub-country where the respondent lives	**Rest of Spain/National**: if the beef cattle were raised and the beef mince was produced in Spain but not in the autonomous region ("Comunidad Autónoma") where the respondent lives
	Imported: if beef mince was not produced in the UK, but has its origin in an EU (90% of beef imports) or non-EU country (10% of beef imports)	**Imported**: if beef mince was not produced in Spain, but has its origin in an EU (85% of beef imports) or non-EU country (15% of beef imports)
Type of production	**No label**: Beef mince is not labelled as "Organic"	**No label**: Beef mince is not labelled as "Organic"
	Organic: if the beef cattle was born and had been raised on organic pasture, had never received antibiotics and growth hormones; was fed only organic feed; and had unrestricted outdoor access	**Organic**: if the beef cattle was born and had been raised on organic pasture, had never received antibiotics and growth hormones; was fed only organic feed; and had unrestricted outdoor access
Greenhouse gas emissions *	**Low**: 5.9 kg of CO_2e per 500 g of beef mince **Moderate**: 19.1 kg of CO_2e per 500 g of beef **High**: 32.2 kg of CO_2e per 500 g of beef mince	**Low**: 5.9 kg of CO_2e per 500 g of beef mince **Moderate**: 19.1 kg of CO_2e per 500 g of beef **High**: 32.2 kg of CO_2e per 500 g of beef mince
Price	£1.50 £3.00 £4.50 £6.00	2.30 € 3.10 € 3.90 € 4.70 €

* Note that this does not include the emissions resulting from the processing and transportation of the meat.

Given all the attribute levels, a full factorial design of 216 (i.e., 3 × 2 × 3 × 3 × 4 = 216) profiles can be generated. Since presenting participants with 216 profiles would be time-consuming and cognitively very challenging, Ngene Software was used to generate a Bayesian D-optimal design that allows robust estimation of all main and two-way interaction effects [33]. The Bayesian D-optimal design was obtained after 25,000 iterations with 500 Halton draws per iteration, achieving a Db-error of 0.11 and 0.15 for the designs used in the UK and Spain, respectively. Each of the final designs of the choice experiment consisted of 36 choice sets of four alternatives each (i.e., three beef mince alternatives plus the opt-out alternative). To make the choice task cognitively easier for respondents, the design was blocked in four blocks (i.e., nine choice sets per respondent). In the choice task, respondents were successively shown nine different choice sets and were repeatedly asked to choose the alternative they prefer most.

To reduce the effect of hypothetical bias, we followed the approach used by Ladenburg, and Olsen [34]. This approach consists in including a cheap talk script [35] right before the choice task. Then the cheap talk script is augmented with a repeated opt-out reminder, showed to participants before each choice set. The cheap talk script and the opt-out reminder used in this study are displayed in Figures 2 and 3, respectively.

> From previous similar studies on consumer preferences and willingness to pay (WTP), we know that people often respond in one way but act differently. For instance, there is strong research evidence that in studies where people do not actually have to pay money for a product when indicating a particular preference or WTP, people overestimate their preferences and WTP.
>
> A possible reason for this is that people do not really consider how big the impact of this extra cost is on the available family budget. It is easy to be generous when you do not really have to pay for it. In the store, people might think in a different way; the amount of money spent on one product cannot be spent on other things.
>
> So please respond to each of the following nine choice questions just exactly as you would if you were in a real store and had to pay for your choice.

Figure 2. The cheap talk script used in this study.

Figure 3. The opt-out reminder that was shown to respondents before each choice set.

Note that the order of showing the nine choice sets was randomised for each respondent. The choice task was followed by a questionnaire. The questionnaire was used to collect information on respondents' socio-demographic characteristics as well as their purchasing habits and attitudes toward several food attributes (e.g., health, environmental sustainability, origin, price, labelling). The questionnaire was also used to collect information on various aspects of respondents' choice behaviour such as attribute nonattendance, certainty about choice responses, and respondents' level of altruism and free riding.

The data collected were analysed within a random utility framework [36]. Thus, an individual n presented with j alternatives at a choice occasion t is expected to choose the alternative that maximises his/her utility. Following Lancaster's [37] concept that any product is a bundle of attributes, the utility that an individual n derives from the consumption of a product is assumed to be equal to the sum of his/her marginal utility for each of the product's attributes. Consequently, if we assume a sample of N respondents who are presented with T choice occasions of J alternatives each, individual n's utility (U_{njt}) from choosing the jth alternative at a tth choice occasion takes the form

$$U_{njt} = V_{njt} + \varepsilon_{njt} \quad (1)$$

where V_{njt} is the deterministic (observed) component and ε_{njt} is the random (unobserved) component. ε_{njt} is assumed to be independent and identically distributed. Assuming that the deterministic component of the utility is linear-in-parameter, Equation (1) can be written as

$$U_{njt} = \beta X_{njt} + \varepsilon_{njt} \quad (2)$$

where β denotes the K × 1 vector of unknown utility parameters. As described in more detail further below in this study, X_{njt} represent the following level of attributes "Low Fat", "Moderate Fat", "Local", "Rest of the country", "Organic", "Low GHG", "Moderate GHG" and "Price" as well as the six two-way interactions ("Low Fat and Organic", "Low Fat and Local", "Low Fat and Low GHG", "Moderate Fat and Organic", "Moderate Fat and Local", "Moderate Fat and Low GHG"). (We were not able to estimate the choice model with all possible two-way interactions due to problems of convergence we had faced during the estimation of the choice model). The levels "High Fat", "Imported", "Not Organic (No label)" and "High GHG" were dropped from the estimation to avoid the problem of perfect multicolinearity. They are also used as the baseline levels when interpreting the estimated effects.

Conditional logit (CL) [36] is the workhorse model for analysing discrete choice data. However, its assumptions (i.e., homogeneity of respondents' preferences and the independence of the alternatives included in any choice set) do not generally hold [32]. Revelt and Train [38] proposed a less restrictive model (Random Parameter Logit (RPL)) that allows individuals' preferences to be heterogeneous and the assumption of the independence of alternatives to be relaxed. In the RPL, at least one parameter is specified as random. In other words, each individual is considered to have a unique set of preferences, reflected in the individual parameters β_i. Since the unconditional choice probability does not have a closed-form solution, simulation methods are used to estimate the parameters (see Revelt and Train [38] for details).

In this study, the parameters for all the non-price attributes as well as the six two-way interactions were assumed to be normally distributed. Theoretically, the estimated coefficient for the price is expected to be negative. Therefore, to avoid obtaining unrealistic positive values for the parameter price, we first multiplied the variable price by −1. Then, a lognormal distribution was imposed on the variable price instead of a normal distribution [39].

In addition to obtaining information on consumers' preferences, the use of discrete choice models allows the derivation of measures designed to determine the amount of money individuals are willing to give up in order to obtain some benefit from the non-price attributes of the product (e.g., low fat, organic, local). Such measures are referred to as measures of WTP. The most used approach to calculate consumers' WTP consists of computing the ratio of two estimated parameters, holding all else constant.

In particular, WTP is commonly expressed as the negative ratio of the non-price attribute coefficient (e.g., the coefficient for the level organic) to the price coefficient:

$$WTP_{non\text{-}price\ attribute} = -\frac{\beta_{non\text{-}price\ attribute}}{\beta_{price}} \tag{3}$$

The calculated value represents the respondents' marginal WTP. In this study, the attributes' levels considered in the estimation of the RPL model were all coded as dummies. Therefore, the calculated WTP value represents respondents' marginal WTP for the attribute level considered in the estimation (e.g., "Low fat") relative to the baseline level (e.g., "High fat"). Note that it is more robust to estimate consumers' marginal WTP following the approach proposed by Train and Weeks [40]. Train and Weeks [40] proposed to estimate the RPL model in WTP space instead. This involves estimating the distribution of willingness to pay directly by re-formulating the model in such a way that the coefficients to be estimated represent the WTP measures. The estimation of the RPL model in WTP space is very time consuming for large sample size and high number of random parameters. We tried to estimate the RPL in WTP space but we had to abort the estimation after nine days running without reaching convergence.

In some cases, the estimated coefficient was found to have a statistically insignificant mean but significant standard deviation. A statistically significant standard deviation suggests that consumers' preferences for the estimated effect are heterogeneous. This heterogeneity can be the cause of the nonsignificance of the estimated mean of the effect if consumers are equally split into both positive and negative sides of the preference scale. If this is the case, the positive and negative effects can cancel each other out, resulting in a statistically insignificant mean of the investigated effect. To test this hypothesis, we followed the approach mentioned in Train [41] to compute the percentage of respondents who placed a positive (or negative) value on the estimated effect using the following formula:

$$100 \times \Phi\left(\frac{-\beta_k}{S_k}\right) \tag{4}$$

where Φ is the cumulative standard normal distribution and β_k and S_k are the mean and the standard deviation of the kth interaction parameter, respectively. Note that this formula is only applicable if the random parameter of interest has a symmetric normal distribution.

After estimating the RPL model for the whole sample, we found that consumers' preferences in both countries are strongly heterogeneous. To evidence the heterogeneity of consumers' preferences and how they vary across different consumer groups, we estimated the RPL model for five consumer segments: "Women", "Men", "Youth (<30 years old)", "Adults (30–60 years old)", and "Elderly (>60 years old)". So, in total, six RPL models were estimated. Furthermore, the standard error of consumers' WTP was computed using the delta method [42].

To test whether the differences in respondents' WTP in the different segments are statistically significant, the Complete Combinatorial Test, proposed by Poe et al. [43], was used. The test, first, requires the generation of distribution of 3000 WTP estimates using the parametric bootstrapping method proposed by Krinsky and Robb [44]. Then, the complete combinatorial test is used to compare the bootstrapped WTP values in the different segments.

3. Results

The RPL models were estimated using Stata 15, with 2000 Halton draws to simulate the 15 random parameters (i.e., nine main effects and six interaction effects). The results of the estimated marginal utilities (i.e., preferences) and their standard deviations, as well as the marginal WTP of the sampled respondents in the UK (Model 1) and Spain (Model 2), are presented in Table 3. The results show that the estimated RPL models for panel data fit the data better when tested against the basic conditional logit: $\chi^2 = 2214.35$, $p < 0.01$ in the case of the UK data and $\chi^2 = 1718.10$, $p < 0.01$ in the case of the data collected in Spain.

3.1. Main Effects

The results displayed in Table 3 show that the main effect coefficients are all significant at 1% and of the expected sign. Another general result worth mentioning is that the significance of the standard deviations of all the main effects (except for the label "Moderate Fat" in the UK model) shows that heterogeneity is indeed an issue to be considered when investigating consumers' preferences for desirable attributes of beef mince. A binary variable "No Buy Option" was included in the estimation of the RPL models to capture respondents' preferences for the opt-out alternative. In both models, the estimated coefficient is highly significant. The estimated coefficient was found to be positive in Model 1 and negative in Model 2. The coefficient is negative when more than two-thirds of respondents (77% in Spain versus 69.7% in the UK) have a higher preference for the beef mince alternative than for the opt-out alternative, and vice versa if the sign of the estimated coefficient is positive. The significant and negative value of the price coefficient in both models indicates that consumers in both countries prefer cheaper beef mince, all else being equal.

In both models, the positive coefficient value for the labels "Low Fat", "Moderate Fat", "Local", "Rest of the Country", "Organic", "Low GHG" and "Moderate GHG" indicates that the utility associated with beef mince packages carrying these labels is higher than the utility for the labels set as baseline (i.e., "High Fat", "Imported", "Not Organic" and "High GHG"). In particular, the results suggest that consumers in the UK and Spain favour beef mince labelled as "Low Fat" or "Moderate Fat", as opposed to beef mince carrying the label "High Fat". Note that respondents in both countries prefer beef mince with low fat content over beef mince with moderate fat content.

Regarding respondents' preferences for the labels communicating information about the origin of the beef mince, in both models consumer utility was found to be the highest for local beef mince, followed by national and imported beef mince in second and third place, respectively. The majority of sampled respondents in the UK and Spain were found to prefer organic beef mince as opposed to its conventional counterpart. Consumers in both countries were found to be more likely to choose beef mince that carries the labels "Low GHG" or "Moderate GHG" as opposed to beef mince that carries the label "High GHG". While respondents in the UK have a higher preference for beef mince labelled as "Low GHG" than beef mince that carries the label "Moderate GHG", respondents in Spain showed higher preferences for the label "Moderate GHG" than the label "Low GHG".

Regarding respondents' preferences for the labels communicating information about the origin of the beef mince, in both models consumer utility was found to be the highest for local beef mince, followed by national and imported beef mince in second and third place, respectively. The majority of sampled respondents in the UK and Spain were found to prefer organic beef mince as opposed to its conventional counterpart. Consumers in both countries were found to be more likely to choose beef mince that carries the labels "Low GHG" or "Moderate GHG" as opposed to beef mince that carries the label "High GHG". While respondents in the UK have a higher preference for beef mince labelled as "Low GHG" than beef mince that carries the label "Moderate GHG", respondents in Spain showed higher preference for the label "Moderate GHG" than the label "Low GHG".

Table 3. Estimated marginal utilities and willingness to pay.

Variables	Preferences				Marginal Willingness to Pay		
	Mean (UK)	Std. Deviation (UK)	Mean (Spain)	Std. Deviation (Spain)	Mean (UK)	Mean (Spain)	*p*-Value (Poe Test)
Low Fat	2.700 ***	1.606 ***	1.819 ***	1.252 ***	3.41 ***	1.94 ***	0.00
Moderate Fat	1.769 ***	−0.323	1.469 ***	0.715 ***	2.23 ***	1.57 ***	0.00
Local	1.222 ***	0.784 ***	1.512 ***	0.973 ***	1.54 ***	1.61 ***	0.31
Rest of the Country/National	1.051 ***	0.431 ***	1.283 ***	0.832 ***	1.33 ***	1.37 ***	0.37
Organic	0.807 ***	1.172 ***	1.019 ***	1.051 ***	1.02 ***	1.09 ***	0.31
Low GHG	1.621 ***	−0.410 ***	0.901 ***	−0.033	2.05 ***	0.96 ***	0.00
Moderate GHG	0.697 ***	0.696 ***	1.163 ***	0.543 ***	0.88 ***	1.24 ***	0.00
Low Fat and Local	0.197 **	−0.052	0.229 **	1.221 ***	0.25 **	0.24 **	0.49
Low Fat and Organic	−0.304 ***	−0.728 ***	0.270 ***	−0.095	−0.38 ***	0.29 ***	0.00
Low Fat and Low GHG	−0.866 ***	0.110	0.499 ***	0.012	−1.09 ***	0.53 ***	0.00
Moderate Fat and Local	−0.067	0.465 *	0.417 ***	−0.065	0.00	0.44 ***	0.00
Moderate Fat and Organic	0.099	−0.457 **	0.211 **	0.073	0.00	0.22 **	0.28
Moderate Fat and Low GHG	−0.777 ***	0.527 ***	0.270 **	−0.003	−0.98 ***	0.29 **	0.00
Price	−0.456 ***	0.828 ***	−0.245 ***	0.599 ***	–	–	–
NoBuyOption	0.372 ***	2.282 ***	−0.649 ***	2.610 ***	–	–	–
Number of observations	10,899		10,854		–	–	–
Log likelihood	−11,036.67		−11,671.099		–	–	–
Chi square (χ^2)	3307.94		3323.69		–	–	–
p-value	<0.01		<0.01		–	–	–

Note that (***), (**) and (*) indicate that the corresponding value is statistically significant at (1%), (5%) and (10%) level, respectively. GHG stands for Greenhouse gas emissions. Furthermore, the number of observation is equal to the number of respondents multiplied by the number of choice sets (9 per respondent).

Despite their usefulness in investigating how consumers weigh different food attributes and in predicting the probability of possible future choices, estimated utility coefficients do not provide direct information for welfare and other policy analyses. Instead, they are commonly converted into money values (e.g., WTP). Respondents' marginal WTP for the different labels considered in this study and their interactions are displayed in the last three columns of Table 3. The results show that the labels "Low Fat" and "Moderate Fat" are highly valued by consumers in both countries. Nonetheless, UK consumers' price premiums for the labels "Low Fat" and "Moderate Fat" are respectively 76% (€3.41 versus €1.94) and 42% (€2.23 versus €1.57) higher than the price premiums of Spanish consumers. Regarding the value consumers give the origin-related labels, the sampled consumers in both countries have statistically similar marginal WTPs for the beef mince labelled as "Local" or "Rest of the UK" / "Rest of Spain". However, note that respondents in the UK and Spain are willing to pay 16% (€1.54 versus €1.33) and 18% (€1.61 versus €1.37) more, respectively, for the label "Local" than for the label "Rest of the UK" / "Rest of Spain".

Consumers in the UK and Spain were found to be willing to pay a comparable price premium for beef mince labelled "Organic". In both countries, consumers' price premium for the label "Organic" is significantly lower than their price premiums for five of the labels considered in the analysis. For example, in the UK, consumers' price premium for organic beef mince is 234%, 119%, 101%, 51% and 30% less than their price premiums for the labels "Low Fat", "Moderate Fat", "Low GHG", "Local" and "Rest of the UK", respectively. In both countries, beef mince carrying the labels "Low GHG" or "Moderate GHG" received a significantly higher value than beef mince labelled as "High GHG". Nonetheless, in the UK, consumers value the label "Low GHG" more (+133%) than the label "Moderate GHG", as opposed to consumers in Spain, who value beef mince labelled as "Moderate GHG" more highly (29%) than beef mince that carries the label "Low GHG". Note that, when compared with the label "Organic", consumers in the UK value the label "Low GHG" more highly, while consumers in Spain value "Moderate GHG" more highly. Furthermore, the value UK consumers give the label "Low GHG" is 114% higher than the value consumers in Spain give it. However, UK consumers' price premium for beef mince labelled as "Moderate GHG" is 41% lower than Spanish consumers' price premium for the same label.

3.2. Interaction Effects

In line with the findings of previous studies [21,24,25,28–30], the results from the estimation of the main effects show that consumers in the UK and Spain value the labels "Low Fat" and "Moderate Fat" highly and they are also willing to pay an economically significant premium for the labels "Local", "Organic" and "Low GHG". As explained in the introduction, a key contribution of this study to the literature on consumers' preferences and WTP for healthier food products is its assessment of whether bundling health-related labels with other positively valued labels (e.g., "Local", "Organic" and "Low GHG") can boost the demand for healthier beef mince. To answer this question, six two-way interactions were estimated ("Low Fat and Organic", "Low Fat and Local", "Low Fat and Low GHG", "Moderate Fat and Organic", "Moderate Fat and Local" and "Moderate Fat and Low GHG"). The estimated marginal utilities and WTP for the six interactions are presented in Table 3.

Contrary to the results of the estimated main effects, the estimated interaction parameters do not have the same sign across countries. In the case of Spain, all the estimated interactions are positive and statistically significant. This indicates that consumers in Spain perceive the labels constituting the six bundles as complementary. This, in turn, implies that they are willing to pay a significant price premium for the coexistence of the bundled labels in addition to the price premium they are willing to pay for the individual labels. For example, if the beef mince that carries the label "Low Fat" is also labelled as "Local", consumers' total willingness to pay for the bundle is equal to their price premium for labels "Low Fat" (€1.94) and "Local" (€1.61) plus their price premium for the coexistence of the two labels on the same product (€0.24). Therefore, it can be deduced that Spanish consumers' total marginal WTP for beef mince that carries the bundle of labels "Low Fat and Local" equals €3.79,

"Low Fat and Organic" equals €3.32, "Low Fat and Low GHG" equals €3.71, "Moderate Fat and Local" equals €3.62, "Moderate Fat and Organic" equals €2.88 and "Moderate Fat and Low GHG" equals €3.10. It is noteworthy that with the exception of the interaction "Low Fat and Local", the estimated standard deviations for the other five interactions are statistically insignificant. This implies that consumers' positive valuation of the coexistence of the labels constituting the bundles "Low Fat and Local", "Low Fat and Organic", "Low Fat and Low GHG", "Moderate Fat and Local", "Moderate Fat and Organic" and "Moderate Fat and Low GHG" are highly homogeneous. Overall the results suggest that, in the case of Spain, adding the labels "Local", "Organic" or "Low GHG" to beef mince that is labelled "Low Fat" or "Moderate Fat" can boost the demand for it.

In the case of the UK, the results show that only the coexistence of the labels "Low Fat" and "Local" is valued positively (i.e., the two labels are perceived as complementary). In particular, UK consumers are willing to pay an additional premium of €0.25 for beef mince that is labelled as "Low Fat" and "Local" at the same time. UK consumers were also found to perceive the labels constituting the bundles "Low Fat and Organic", "Low Fat and Low GHG" and "Moderate Fat and Low GHG" as (partial) substitutes. In particular, consumers discount their total marginal WTP for beef mince that is simultaneously labelled as "Low Fat" and "Organic" by €0.38. This is equivalent to offsetting 37% of consumers' price premium for the label "Organic" (i.e., €0.38/€1.02) *100). Similarly, labelling beef mince that carries the label "Low Fat" or the label "Moderate Fat" as "Low GHG" is expected to reduce consumers' total price premium for the bundles "Low Fat and Low GHG" and "Moderate Fat and Low GHG" by €1.09 and €0.98, respectively. This is equivalent to offsetting the positive impact of the label "Low GHG" by 53% when bundled with the label "Low Fat" and by 48% when bundled with the label "Moderate Fat".

The estimated coefficients of the interactions "Moderate Fat and Local" and "Moderate Fat and Organic" are statistically insignificant. This result seems to suggest that UK consumers perceive labels constituting the bundles "Moderate Fat and Local" and "Moderate Fat and Organic" as unrelated. Nonetheless, the estimated standard deviations of the two interactions are statistically significant, suggesting that consumers' preferences for these bundles are heterogeneous. As noted above, this heterogeneity can be the cause of the nonsignificance of the estimated interaction parameters if consumers are equally split into both positive and negative sides of the preference scale. To compute the percentage of respondents who gave a positive (or negative) value to the bundles "Moderate Fat and Local" and "Moderate Fat and Organic", we used the formula in Equation (4).

We found that, in the case of the bundle "Moderate Fat and Local", 56% of UK consumers perceived the labels "Moderate Fat" and "Local" as substitutes, while 44% of them perceived them as complementary. In the case of the bundle "Moderate Fat and Organic", we found that while 59% of UK consumers positively valued the coexistence of the labels "Moderate Fat" and "Organic", 41% of them valued it negatively. These results suggest that the heterogeneity of consumers' preferences is possibly behind the nonsignificance of the estimated interaction parameters for the bundles "Moderate Fat and Local" and "Moderate Fat and Organic". However, we recommend using these results with caution due to the strong assumptions and low robustness of the approach proposed by Train [41] used here.

3.3. Heterogeneity of Consumers' WTP

As noted above, the significance of the estimated standard deviations (Table 3) indicates that consumers' preferences and marginal WTP are heterogeneous, especially for the main effects. The results from the estimation of the RPL models for the segments "Women", "Men, "Youth (<30 years old)", "Adult (30–60 years old)" and "Elderly (>60 years old)" are displayed in Table 4 (for the UK) and Table 5 (for Spain).

Table 4. Heterogeneity of consumers' WTP–UK.

Claims	Marginal Willingness to Pay					p-Value			
	Women	Men	Young	Adults	Elderly	Women vs. Men	Young vs. Adult	Young vs. Elderly	Adult vs. Elderly
Low Fat	3.81 ***	3.11 ***	3.28 ***	3.20 ***	4.49 ***	0.02	0.43	0.04	0.00
Moderate Fat	2.34 ***	2.19 ***	2.72 ***	1.90 ***	3.24 ***	0.33	0.06	0.24	0.01
Local	1.75 ***	1.40 ***	1.42 ***	1.35 ***	2.56 ***	0.05	0.41	0.01	0.01
Rest of the Country/National	1.46 ***	1.23 ***	1.10 ***	1.09 ***	2.65 ***	0.11	0.49	0.00	0.00
Organic	1.18 ***	0.90 ***	1.23 ***	1.06 ***	0.73 ***	0.10	0.30	0.12	0.00
Low GHG	2.22 ***	1.95 ***	2.01 ***	2.02 ***	2.16 ***	0.17	0.50	0.40	0.16
Moderate GHG	1.10 ***	0.65 ***	1.09 ***	0.75 ***	1.25 ***	0.00	0.06	0.30	0.37
Low Fat and Local	0.33 *	0.16	0.16	0.23	0.32 *	0.25	0.43	0.38	0.01
Low Fat and Organic	−0.52 ***	−0.27	0.23	−0.52 ***	−0.57	0.17	0.02	0.07	0.41
Low Fat and Low GHG	−1.11 ***	−1.13 ***	−0.90 *	−1.12 ***	−1.16 ***	0.47	0.35	0.36	0.45
Moderate Fat and Local	−0.13	−0.05	−0.51	0.02	−0.18	0.39	0.13	0.30	0.46
Moderate Fat and Organic	0.40 **	−0.15	0.46	0.04	0.04	0.02	0.15	0.22	0.33
Moderate Fat and Low GHG	−0.86 ***	−1.14 ***	−0.91 *	−0.89 ***	−1.42 ***	0.21	0.48	0.25	0.50

Note that (***), (**) and (*) indicate that the corresponding value is statistically significant at (1%), (5%) and (10%) level, respectively. GHG stands for Greenhouse gas emissions.

Table 5. Heterogeneity of consumers' WTP–Spain.

Claims	Marginal Willingness to Pay										*p*-Value			
	Women		Men		Young		Adults		Elderly		Women vs. Men	Young vs. Adult	Young vs. Elderly	Adult vs. Elderly
Low Fat	1.96	***	1.87	***	2.40	***	1.77	***	2.19	***	0.34	0.04	0.34	0.12
Moderate Fat	1.62	***	1.47	***	1.78	***	1.52	***	1.43	***	0.26	0.26	0.24	0.38
Local	1.62	***	1.58	***	2.07	***	1.52	***	1.48	***	0.41	0.04	0.07	0.44
Rest of the Country/National	1.25	***	1.45	***	1.64	***	1.36	***	0.96	***	0.10	0.14	0.02	0.04
Organic	1.03	***	1.13	***	1.12	***	1.00	***	1.45	***	0.28	0.34	0.17	0.04
Low GHG	0.80	***	1.12	***	1.52	***	0.80	***	1.21	***	0.06	0.02	0.24	0.10
Moderate GHG	1.26	***	1.17	***	1.65	***	1.13	***	1.39	***	0.25	0.01	0.18	0.09
Low Fat and Local	0.11		0.38	**	−0.12		0.28	**	0.28		0.10	0.14	0.19	0.49
Low Fat and Organic	0.24	**	0.30	**	0.27		0.37	***	−0.23		0.37	0.37	0.10	0.02
Low Fat and Low GHG	0.75	***	0.19		0.50		0.56	***	0.42		0.02	0.43	0.45	0.35
Moderate Fat and Local	0.34	**	0.46	***	0.19		0.45	***	0.51		0.31	0.25	0.25	0.43
Moderate Fat and Organic	0.31	**	0.10		0.27		0.30	***	−0.30		0.14	0.46	0.08	0.02
Moderate Fat and Low GHG	0.36	**	0.16		0.22		0.19		0.84	**	0.21	0.47	0.11	0.00

Note that (***) and (**) indicate that the corresponding value is statistically significant at (1%) and (5%) level, respectively. GHG stands for Greenhouse gas emissions.

The results show that gender and age can partially explain the heterogeneity of consumers' preferences and WTP. The results in Table 4 show that, in the UK, female consumers are willing to pay a significantly higher price premium for the labels "Low Fat", "Local" and "Moderate GHG". Both female and male consumers valued the label bundles considered in this study similarly, with the exception of the bundle "Moderate Fat and Organic". While male respondents perceived the labels "Moderate Fat" and "Organic" as independent, female consumers considered the two labels to be complementary and were willing to pay an additional premium of €0.40 for beef mince that is simultaneously labelled as "Moderate Fat" and "Organic". Interestingly, in the case of Spain, the results in Table 5 show that female and male consumers have similar preferences and WTP for the main effects (except for the label "Low Fat"). The results for the interaction effects show that while female consumers are willing to pay an additional premium for all the bundles (except "Low Fat and Local"), male respondents in Spain seem to perceive the labels constituting the bundles "Low Fat and Low GHG", "Moderate Fat and Organic" and "Moderate Fat and Low GHG" as independent.

With regard to the differences across age segments, the results in Table 4 show that UK respondents over the age of 60 are willing to pay the highest premium for the labels "Low Fat", "Moderate Fat", "Local", "Rest of the Country", "Low GHG" and "Moderate GHG". Young UK consumers' price premium for the beef mince labelled as "Organic" is the highest among the three age segments. For the interaction effects, the results from the data collected in the UK show that consumers' preferences and WTP are similar among the three age segments, except for the bundles "Low Fat and Local" and "Low Fat and Low GHG". While older respondents perceived the labels "Low Fat" and "Local" as complementary, young and adult consumers perceived them as unrelated. Furthermore, while young and elderly respondents considered the labels "Low Fat" and "Organic" to be independent, adult consumers would discount the total marginal WTP for beef mince simultaneously labelled as "Low Fat" and "Organic" by €0.52.

As opposed to UK consumers, young consumers in Spain (Table 5) were found to have the highest price premium for the labels "Low Fat", "Moderate Fat", "Local", "National", "Low GHG" and "Moderate GHG". While elderly UK consumers were willing to pay the lowest price premium for organic beef mince, elderly respondents sampled in Spain were found to be willing to pay the highest price premium for beef mince labelled as "Organic". The results in Table 5 show that adult consumers in Spain perceive the labels forming the bundles "Low Fat and Organic", "Low Fat and Local", "Low Fat and Low GHG", "Moderate Fat and Organic" and "Moderate Fat and Local" as complementary, as opposed to the young and elderly consumers who consider these labels to be unrelated.

4. Discussion and Conclusions

The results suggest that labels indicating that the beef mince has a low or moderate fat content are of significant importance for consumers in the UK and Spain. This finding is in line with the dietary recommendations of the UK and Spanish governments [45–47], as well as the World Health Organisation [48], aiming to curb the epidemic prevalence of obesity through encouraging consumers to eat fewer food products that are high in fat, sugar or salt. Previous studies [25,49,50] similarly found that consumers in different European countries prefer their food to have a low fat content.

Despite the high interest in beef mince carrying the label "Low Fat", producing and selling low-fat beef mince seems to be challenging. In fact, the authors carried out an online shelf audit of the major supermarkets in the UK (Asda, Tesco, Sainsbury's and Waitrose) and Spain (Carrefour, Alcampo, Eroski Caprabo, Mercadona and Corte Inglés) to check the availability and the retail prices of beef mince with a low or moderate fat content. We found that in both countries, none of the beef mince products sold online can be technically labelled as low in fat (i.e., containing 3 g or less of fat per 100 g of beef mince). Nonetheless, in both countries, most of the beef mince products sold online have a moderate level of fat. We also noticed that in Spain, fat-content-related labels are only occasionally displayed on the front of packages of beef mince, as opposed to the UK where the use of front-of-package health labels and claims is more common. The reasons (e.g., taste, cost, technical and processing factors) for this very limited availability of beef mince with a fat content of less than 3% in grocery stores, as well as the uncommon use of front-of-package nutrition labels in Spain still need to be studied in future research. Uncovering these reasons should help to improve the supply and marketing of low-fat beef mince, thus satisfying the needs of an increasing number of consumers interested in health.

We also learned from the online shelf audit that beef mince with a moderate fat content is sold at a retail premium that ranges between €3.40 and €6.20 per kg in the UK and it does not exceed €3 per kilogramme in Spain. These retail premiums are close to the estimated price premiums that the sampled consumers in the UK (€4.46) and Spain (€3.14) are willing to pay for beef mince labelled as "Moderate Fat". However, a more rigorous analysis that makes use of data with a good temporal and spatial coverage of beef mince's retail prices is needed to find out whether healthier beef mince products are currently sold at affordable prices. Our results clearly show that consumers in the UK are willing to pay a significantly higher price premium for healthier beef mince than consumers interviewed in Spain. This suggests that marketers who are interested in selling beef mince in both countries should consider the substantial gap between consumers' price premiums for healthier beef mince in the two countries (note that consumers' WTP in both countries was not adjusted for the differences between the two countries in terms of food prices, consumers' income, etc.).

In line with the results of several previous studies [21–25], we found that the majority of sampled consumers in the UK and Spain also positively value the labels "Local", "Rest of the Country", "Organic", "Low GHG" and "Moderate GHG". Interestingly, in both countries, consumers' price premium for beef labelled as "Organic" is significantly lower than their price premium for beef mince that carriers the label "Local" or "Rest of the Country". Onozaka and McFadden [21] and Meas et al. [24] similarly found that US consumers have a higher preference for local food than organic food. While our paper does not investigate the reasons for the higher preferences for local beef mince (see [51,52]), it suggests that labelling locally (nationally) produced and sold beef mince as "Local" ("National") can increase its competitive power vis-à-vis organic and imported beef mince.

The results also suggest that there is a potential market in the UK and Spain for beef mince produced with low or moderate greenhouse gas emissions. Carbon labels are not currently used in the UK and Spain to label meat and meat products, especially those produced with low GHG emissions. This is probably due to the high carbon footprint of most of the meat produced and sold in both countries, as well as the difficulty of continuously collating precise data on GHG emissions from meat production [53,54]. As a result of the increasing calls to reduce meat consumption [55], meat producers and processors are under unprecedented pressure to reduce the environmental impact of their products. In this context, our results are sending a positive signal back to producers and marketers of beef mince that consumers are willing to bear a considerable increase in beef mince price (up to €2.05 in the UK and €1.24 in Spain) if the product is produced with a significantly lower environmental impact.

Despite consumers' high price premium for beef mince labelled as "Low Fat" or "Moderate Fat", the results showed that consumers in both countries may trade off this beef mince against a counterpart that is carrying other desirable labels, such as "Local", "Organic" and "Low GHG". This competition is further increased if the non-health-related labels are bundled. For example, the results suggest that if beef mince products labelled "Moderate Fat", "Local" and "Organic" are offered at the same price, consumers in the UK and Spain are more likely to buy the beef mince with a moderate fat content because of their higher price premium for this type of beef mince. Nonetheless, if the organic beef mince is also labelled as "Local" when sold in the local market, consumers in both countries may purchase the locally produced organic beef mince instead (UK: €2.56 versus €2.23, Spain: €2.70 versus €1.57). To sum up, when pricing and marketing their products, producers and marketers of healthier beef mince should take into account the trade-offs that consumers may make when they are faced with products carrying other desirable food labels such as "Local" and "Organic".

Bundling desirable attributes can also boost the demand for healthy food products if consumers perceive the attributes bundled to be complementary or independent. In this study, consumers were found to positively value the labels "Low Fat", "Moderate Fat", "Local", "Organic" and "Low GHG".

This implies that bundling these labels can improve the demand for beef mince if consumers perceive them as complementary or independent. Nonetheless, if they are perceived as substitutes, it implies that consumers negatively value the coexistence of the bundled labels. The results show that sampled consumers in Spain think that the labels "Local", "Organic" and "Low GHG" complement the labels "Low Fat" and "Moderate Fat". This suggests that the demand for low- or moderate-fat beef mince can be improved if it is also labelled as "Local", "Organic" or "Low GHG". Therefore, it is recommended that Spanish producers and marketers of beef mince with low or moderate fat content also label their product as local when it is sold in local markets. The findings of this study suggest that producers of organic beef mince can also benefit from reducing the fat content of their product and promote this fat reduction by labelling the organic product as "Low Fat" or "Moderate Fat".

The results from the data collected in Spain show that the coexistence of the labels "Low Fat" and "Low GHG" received the highest price premium. Therefore, the demand for low-fat beef mince in Spain can be increased if the product is also labelled as produced with low GHG emissions. However, it remains to be seen whether producing low-fat beef mince, which may necessitate the use of a higher quantity of beef meat, is feasible with a low environmental impact. The coexistence of the labels "Moderate Fat" and "Low GHG" was also positively valued by the sampled consumers in Spain. This is further encouraging evidence of the existence of potential demand for beef mince with a low carbon footprint. Most importantly, this also suggests that the producers and marketers of beef mince with moderate fat content, which is currently the most marketed type of beef mince in Spain, can receive higher price premiums if they manage to significantly reduce the total environmental impact of their product.

Sampled consumers in the UK were found to value the coexistence of multiple labels differently in comparison with consumers in Spain. The only exception is the coexistence of the labels "Low Fat" and "Local". UK consumers were found to positively value the coexistence of the two labels on the same beef mince product, suggesting that UK producers and marketers of low-fat beef mince can increase their sales in the local market if they also label their product as "Local". In contrast with the preferences of the sampled consumers in Spain, UK consumers were found to discount the simultaneous use of the label "Low GHG" or the label "Organic" with the labels "Low Fat" and "Moderate Fat". Whilst producers and marketers of low- or moderate-fat beef mince should still label their products "Organic" or "Low GHG", it is recommended that they take into account the value of the discount when computing the total return of the discounted bundles (e.g., "Low Fat and Organic").

Furthermore, consumers in the UK were found to perceive the labels constituting the bundles "Moderate Fat and Local" and "Moderate Fat and Organic" as unrelated. This implies that the total price premium for the bundle is equal to the sum of the price premiums for the individual labels. This is noteworthy for at least two reasons. First, producers and marketers of beef mince with moderate fat content can increase the demand for their product if they label it as "Local" when it is sold in local markets. The results also suggest that the demand for organic beef mince can be boosted if it is labelled as having a moderate fat content. Second, moderately reducing the fat content of beef mince (to be between 3.1% and 20% of fat) is technically and financially more feasible than producing beef mince with less than 3% fat content. In fact, most of the beef mince that is currently sold in the UK and Spain is eligible for the label "Moderate Fat" without any further processing.

As mentioned in the introduction, most of the research studies that have investigated individuals' preferences and willingness to pay for food products with multiple attributes assumed that respondents treat food attributes as unrelated. Our findings clearly show that this assumption is inappropriate and can lead to biased and misleading results, which in turn is likely to bias the output from subsequent analyses such as cost–benefit analysis. This is on top of ignoring important information on how consumers trade off food products with multiple attributes. In this study, we found that only two out of 12 estimated interaction effects were insignificant. Therefore, we join Onozaka and McFadden [22] and Meas et al. [24], among others, in encouraging researchers who want to investigate individuals' preferences and choices in multiple attribute settings to design the choice experiment in a way that they can later at least estimate the two-way interaction effects.

Previous studies (e.g., [23–25]) investigated the heterogeneity of consumers' preferences for food products with desirable attributes. They found that consumers' preferences and WTP for food products labelled as healthier, local or organic vary by age, income level, education level and gender. In line with those studies, we found that consumers' preferences are heterogeneous across countries and consumers within the same country. For example, we found that while elderly people in the UK constitute the group of consumers with the highest price premium for the desirable attribute levels, young consumers in Spain seem to be the segment with the highest demand for beef mince carrying desirable labels such as "Low Fat", "Moderate Fat" and "Local". The results also highlight the importance of assessing the heterogeneity of consumers' preferences, especially if the outcome of the assessment is intended to help in designing effective policy and marketing strategies that are tailored to the needs of different consumer groups.

Like any other empirical study, the research work described in this paper has some limitations. For example, the data was collected using web-based survey. The use of web-based surveys has many advantages such as ease of data gathering, minimal costs, automation in data input and handling, increase in response rates, flexibility of design, potentially better targeting and convenience to participants. However, one should be aware of the drawbacks of online surveys. Perhaps the heaviest disadvantage of online surveys is the difficulty to avoid getting answers from respondents who answer online surveys mainly for getting the monetary incentive and not with a desire to contribute to the advancement of the study by reporting honest answers. As noted in the methods' section, we used some incentives (e.g., cheap talk, opt-out reminder and required minimum time for evaluating each choice card) to reduce the occurrence of this problem.

These incentives were also used to reduce the effect of hypothetical bias, which is another typical limitation of the use of web-based surveys. Hypothetical bias occurs because respondents tend to respond differently to hypothetical scenarios than they do to the same scenarios in the real world. In this study, only one product (beef mince) and one type of consumer (meat eaters) were considered due to the high cost of conducting large web-based surveys. Therefore, the finding of this study should not be generalised to all food products, nor can they help to understand non-meat eaters' preferences for heathier food products.

In this study, we focused on assessing consumers' preferences and WTP for nutrition labels and how they are affected by the presence of other desirable labels such as "Organic" and "Local". However, not only food attributes and their corresponding labels influence how consumers perceive the usefulness of nutrition labels and health claims. We believe that our knowledge on the use of nutrition labels and health claims would benefit from further research on whether and how factors such as type of diet, social norms, perception of body image and culture affect consumers' understanding and use of health-related labels and claims. Furthermore, due to the importance of investigating the heterogeneity of consumer preferences and behaviour, we believe that the effect of those factors (e.g., diet, body image and cultural factors) should also be assessed for different consumer segments (e.g., consumers segmented by age group, income level, household composition and purchasing habits).

Author Contributions: Conceptualization, F.A. and C.R.G.; methodology, F.A. and K.G.; software, F.A.; validation, F.A.; formal analysis, F.A.; investigation, F.A.; resources, F.A., C.R.G., and J.M.G.; data curation, F.A. and K.G.; writing—original draft preparation, F.A.; writing—review and editing, F.A., C.R.G., K.G., and J.M.G.; project administration, F.A.; funding acquisition, C.R.G. and J.M.G. All authors have read and agreed to the published version of the manuscript.

Funding: This research was funded by the Scottish Government as part of the 2016–2021 Research Program on Food Security (Workpackage 3.3).

Acknowledgments: This study derives from work commissioned by the Scottish Government as part of the 2016–2021 Research Program on Food Security (Workpackage 3.3). All the opinions are the sole responsibility of the authors.

Conflicts of Interest: The authors declare no conflict of interest.

References

1. Food and Agriculture Organization of the United Nations; International Fund for Agricultural Development; The United Nations Children's Fund; World Food Programme; World Health Organization. The State of Food Security and Nutrition in the World 2019. In *Safeguarding against Economic Slowdowns and Downturns*; FAO: Rome, Italy, 2019.
2. Abarca-Gómez, L.; Abdeen, Z.A.; Hamid, Z.A.; Abu-Rmeileh, N.M.; Acosta-Cazares, B.; Acuin, C.; Agyemang, C. Worldwide trends in body-mass index, underweight, overweight, and obesity from 1975 to 2016: A pooled analysis of 2416 population-based measurement studies in 1289 million children, adolescents, and adults. *Lancet* **2017**, *390*, 2627–2642. [CrossRef]
3. Wang, Y.C.; McPherson, K.; Marsh, T.; Gortmaker, S.L.; Brown, M. Health and economic burden of the projected obesity trends in the USA and the UK. *Lancet* **2011**, *378*, 815–825. [CrossRef]
4. Withrow, D.; Alter, D.A. The economic burden of obesity worldwide: A systematic review of the direct costs of obesity. *Obes. Rev.* **2011**, *12*, 131–141. [CrossRef] [PubMed]
5. Tremmel, M.; Gerdtham, U.G.; Nilsson, P.M.; Saha, S. Economic burden of obesity: A systematic literature review. *Int. J. Environ. Res. Public Health* **2017**, *14*, 435. [CrossRef]
6. Darmon, N.; Briend, A.; Drewnowski, A. Energy dense diets are associated with lower diet costs: A community study of French adults. *Public Health Nutr.* **2004**, *7*, 21–27. [CrossRef]
7. Rouhani, M.H.; Haghighatdoost, F.; Surkan, P.J.; Azadbakht, L. Associations between dietary energy density and obesity: A systematic review and meta-analysis of observational studies. *Nutrition* **2016**, *32*, 1037–1047. [CrossRef]
8. Soliman, G.A.; Huang, T.T. Causes of childhood obesity. In *Obesity in Childhood and Adolescence*; ABC-CLIO: California, CA, USA, 2018; Volume 2, p. 25.
9. World Health Organization. *Obesity: Preventing and Managing the Global Epidemic*; Number 894; World Health Organization: Geneva, Switzerland, 2000.
10. Lean, M.; Lara, J.; Hill, J.O. Strategies for preventing obesity. *BMJ* **2006**, *333*, 959–962. [CrossRef]
11. Wansink, B. *Mindless Eating: Why We Eat More Than We Think*; Bantam; Hay House: London, UK, 2007.
12. Waters, E.; de Silva-Sanigorski, A.; Burford, B.J.; Brown, T.; Campbell, K.J.; Gao, Y.; Armstrong, R.; Prosser, L.; Summerbell, C.D. Interventions for preventing obesity in children. *Cochrane Database Syst. Rev.* **2011**, *12*, 1–210. [CrossRef]
13. Gorski, M.T.; Roberto, C.A. Public health policies to encourage healthy eating habits: Recent perspectives. *J. Healthc. Leadersh.* **2015**, *7*, 81.
14. Reisch, L.A.; Sunstein, C.R.; Gwozdz, W. Beyond carrots and sticks: Europeans support health nudges. *Food Policy* **2017**, *69*, 1–10. [CrossRef]
15. Cowburn, G.; Stockley, L. Consumer understanding and use of nutrition labelling: A systematic review. *Public Health Nutr.* **2005**, *8*, 21–28. [CrossRef] [PubMed]
16. Grunert, K.G.; Wills, J.M. A review of European research on consumer response to nutrition information on food labels. *J. Public Health* **2007**, *15*, 385–399. [CrossRef]
17. Cecchini, M.; Warin, L. Impact of food labelling systems on food choices and eating behaviours: A systematic review and meta-analysis of randomized studies. *Obes. Rev.* **2016**, *17*, 201–210. [CrossRef] [PubMed]
18. Ares, G.; Varela, F.; Machin, L.; Antúnez, L.; Giménez, A.; Curutchet, M.R.; Aschemann-Witzel, J. Comparative performance of three interpretative front-of-pack nutrition labelling schemes: Insights for policy making. *Food Q. Prefer.* **2018**, *68*, 215–225. [CrossRef]
19. Vandevijvere, S.; Vanderlee, L. Effect of formulation, labelling, and taxation policies on the nutritional quality of the food supply. *Cur. Nutr. Rep.* **2019**, *8*, 240–249. [CrossRef] [PubMed]
20. Gao, Z.; Schroeder, T.C. Effects of label information on consumer willingness-to-pay for food attributes. *Am. J. Agric. Econ.* **2009**, *91*, 795–809. [CrossRef]
21. Onozaka, Y.; McFadden, D.T. Does local labeling complement or compete with other sustainable labels? A conjoint analysis of direct and joint values for fresh produce claim. *Am. J. Agric. Econ.* **2011**, *93*, 693–706. [CrossRef]
22. Wezemael, L.V.; Caputo, V.; Nayga, R.M.; Chryssochoidis, G.; Verbeke, W. European consumer preferences for beef with nutrition and health claims: A multi-country investigation using discrete choice experiments. *Food Policy* **2014**, *44*, 167–176. [CrossRef]

23. Meas, T.; Hu, W.; Batte, M.T.; Woods, T.A.; Ernst, S. Substitutes or complements? Consumer preference for local and organic food attributes. *Am. J. Agric. Econ.* **2014**, *97*, 1044–1071. [CrossRef]
24. Grunert, K.G.; Sonntag, W.I.; Glanz-Chanos, V.; Forum, S. Consumer interest in environmental impact, safety, health and animal welfare aspects of modern pig production: Results of a cross-national choice experiment. *Meat Sci.* **2018**, *137*, 123–129. [CrossRef]
25. Hu, W.; Hünnemeyer, A.; Veeman, M.; Adamowicz, W.; Srivastava, L. Trading off health, environmental and genetic modification attributes in food. *Eur. Rev. Agric. Econ.* **2004**, *31*, 389–408. [CrossRef]
26. Nilsson, T.; Foster, K.; Lusk, J.L. Marketing opportunities for certified pork chops. *Can. J. Agric. Econ. Rev. Can. Agroecon.* **2006**, *54*, 567–583. [CrossRef]
27. Bond, C.A.; Thilmany, D.D.; Bond, J.K. What to choose? The value of label claims to fresh produce consumers. *J. Agric. Res. Econ.* **2008**, *33*, 402–427.
28. Kamphuis, C.B.; de Bekker-Grob, E.W.; van Lenthe, F.J. Factors affecting food choices of older adults from high and low socioeconomic groups: A discrete choice experiment. *Am. J. Clin. Nutr.* **2015**, *101*, 768–774. [CrossRef]
29. Gracia, A.; Barreiro-Hurlé, J.; Galán, B.L. Are local and organic claims complements or substitutes? A consumer preferences study for eggs. *J. Agric. Econ.* **2014**, *65*, 49–67. [CrossRef]
30. Louviere, J.J.; Hensher, D.A.; Swait, J.D. *Stated Choice Methods: Analysis and Applications*; Cambridge University Press: Cambridge, UK, 2000.
31. Hensher, D.A.; Rose, J.M.; Greene, W.H. *Applied Choice Analysis: A Primer*, 2nd ed.; Cambridge University Press: Cambridge, UK, 2015.
32. ChoiceMetrics. Ngene User Manual. 2018. Available online: http://www.choice-metrics.com/NgeneManual120.pdf (accessed on 20 November 2019).
33. Ladenburg, J.; Olsen, S.B. Augmenting short cheap talk scripts with a repeated opt-out reminder in choice experiment surveys. *Res. Energy Econ.* **2014**, *37*, 39–63. [CrossRef]
34. Cummings, R.G.; Taylor, L.O. Unbiased value estimates for environmental goods: A cheap talk design for the contingent valuation method. *Am. Econ. Rev.* **1999**, *89*, 649–665. [CrossRef]
35. McFadden. Conditional Logit Analysis of Qualitative Choice Behaviour. University of California at Berkeley, 1974. Available online: https://eml.berkeley.edu/reprints/mcfadden/zarembka.pdf (accessed on 20 November 2019).
36. Lancaster, K.J. A new approach to consumer theory. *J. Political Econ.* **1966**, *74*, 132–157. [CrossRef]
37. Revelt, D.; Train, K. Mixed logit with repeated choices: households' choices of appliance efficiency level. *Rev. Econ. Stat.* **1998**, *80*, 647–657. [CrossRef]
38. Hensher, D.A.; Greene, W.H. The mixed logit model: The state of practice. *Transportation* **2003**, *30*, 133–176. [CrossRef]
39. Train, K.; Weeks, M. *Discrete Choice Models in Preference Space and Willingness-to-Pay Space*; Springer Netherlands: Dordrecht, The Netherlands, 2005; pp. 1–16.
40. Train, K. *Discrete Choice Methods with Simulation*; Cambridge: Cambridge, UK, 2003.
41. Cox, C. Delta method. *Encycl. Biostat.* **2005**, *2*. [CrossRef]
42. Poe, G.L.; Giraud, K.L.; Loomis, J.B. Computational methods for measuring the difference of empirical distributions. *Am. J. Agric. Econ.* **2005**, *87*, 353–365. [CrossRef]
43. Krinsky, I.; Robb, A.L. On approximating the statistical properties of elasticities. *Rev. Econ. Stat.* **1986**, *68*, 715–719. [CrossRef]
44. Public Health of England. A Quick Guide to the Government's Healthy Eating Recommendations. 2018. Available online: https://assets.publishing.service.gov.uk/government/uploads/system/uploads/attachment_data/file/742746/A_quick_guide_to_govt_healthy_eating_update.pdf (accessed on 20 November 2019).
45. Scottish Government. Preventing Overweight and Obesity in Scotland: A Route Map towards Healthy Weight. 2010. Available online: https://www.gov.scot/publications/preventing-overweight-obesity-scotland-route-map-towards-healthy-weight/pages/0/ (accessed on 20 November 2019).
46. Gobierno de España. PLAN de Colaboración para la Mejora de la Composición de los Alimentos y Bebidas y otras Medidas 2020. 2017. Available online: http://www.aecosan.msssi.gob.es/AECOSAN/docs/documentos/nutricion/PLAN_COLABORACION_2020.pdf (accessed on 20 November 2019).
47. World Health Organization. Healthy Diet. 2015. Available online: https://www.who.int/nutrition/publications/nutrientrequirements/healthydiet_factsheet394.pdf (accessed on 20 November 2019).

48. Balcombe, K.; Fraser, I.; Di Falco, S. Traffic lights and food choice: A choice experiment examining the relationship between nutritional food labels and price. *Food Policy* **2010**, *35*, 211–220. [CrossRef]
49. Koistinen, L.; Pouta, E.; Heikkilä, J.; Forsman-Hugg, S.; Kotro, J.; Mäkelä, J.; Niva, M. The impact of fat content, production methods and carbon footprint information on consumer preferences for minced meat. *Food Q. Prefer.* **2013**, *29*, 126–136. [CrossRef]
50. Adams, D.C.; Adams, A.E. De-placing local at the farmers' market: Consumer conceptions of local foods. *J. Rural Soc. Sci.* **2011**, *26*, 74.
51. Clark, M.; Tilman, D. Comparative analysis of environmental impacts of agricultural production systems, agricultural input efficiency, and food choice. *Environ. Res. Lett.* **2017**, *12*, 064016. [CrossRef]
52. Röös, E.; Mie, A.; Wivstad, M.; Salomon, E.; Johansson, B.; Gunnarsson, S.; Wallenbeck, A.; Hoffmann, R.; Nilsson, U.; Sundberg, C.; et al. Risks and opportunities of increasing yields in organic farming. A review. *Agron. Sustain. Dev.* **2018**, *38*, 14.
53. Nijdam, D.; Rood, T.; Westhoek, H. The price of protein: Review of land use and carbon footprints from life cycle assessments of animal food products and their substitutes. *Food Policy* **2012**, *37*, 760–770. [CrossRef]
54. Poore, J.; Nemecek, T. Reducing food's environmental impacts through producers and consumers. *Science* **2018**, *360*, 987–992. [CrossRef]
55. Springmann, M.; Clark, M.; Mason-D'Croz, D.; Wiebe, K.; Bodirsky, B.L.; Lassaletta, L.; de Vries, W.; Vermeulen, S.J.; Herrero, M.; Carlson, K.M.; et al. Options for keeping the food system within environmental limits. *Nature* **2018**, *562*, 519. [CrossRef] [PubMed]

nutrients

MDPI

Article

Nutritional Knowledge and Health Consciousness: Do They Affect Consumer Wine Choices? Evidence from a Survey in Italy

Claudia Bazzani, Roberta Capitello *, Elena Claire Ricci, Riccardo Scarpa and Diego Begalli

Department of Business Administration, University of Verona, 37129 Verona, Italy; claudia.bazzani@univr.it (C.B.); elenaclaire.ricci@univr.it (E.C.R.); riccardo.scarpa@univr.it (R.S.); diego.begalli@univr.it (D.B.)

* Correspondence: roberta.capitello@univr.it; Tel.: +39-045-802-8488

Received: 11 December 2019; Accepted: 24 December 2019; Published: 27 December 2019

Abstract: Wine is one of the few food products not subject to mandatory nutritional labelling, except for alcohol content. As such, health-related characteristics might be inferred by attributes related to production methods and alcohol content. This research focuses on the set of information currently reported on wine bottle labels, investigates the consumer's use of such labels, and their preferences for information associated with 'naturalness' such as clean labels and alcohol content. We conducted a survey on Italian consumers of red wine, which included a choice experiment. Results showed that health consciousness is an important driver in the use of wine labels. Estimates from a latent class model suggest that health consciousness, along with age, plays a significant role in defining consumer preference segments: the majority of our sample tended to prefer red wine characterized by 'clean labels', but younger and more health-conscious consumers showed a significant disutility for higher alcohol content. More traditional consumers revealed disutility for more unconventional 'clean labels', which were instead appreciated by a third group of consumers, called here 'new clean trend lovers'. Preference for nutritional information such as lower alcohol content and clean labels distinguished the more health-conscious consumers, who belonged to the most likely preference class. Together, the results may suggest that nutritional information currently not mandatory for wine would be appreciated by a significant share of wine consumers.

Keywords: food choices; health consciousness; red wine; choice experiment; clean labels; alcohol content; nutritional knowledge

1. Introduction

Consumer preferences for food products are evolving. An increasingly wider set of attributes is being considered by consumers when making food choices [1–4]. In the last few decades, more relevance has been given to sustainability attributes linked to, for example, nutritional, health-related properties, and environmental aspects of production processes [5]. Eating habits are indeed important from a societal point of view as they may have relevant consequences on citizen welfare and health. On the health side, the incidence of food-related illnesses and non-communicable chronic diseases such as, for example, obesity, diabetes, and cardiovascular diseases, has been increasing and is becoming a major problem for developed and developing countries [6]. This has increased medical spending and has also gradually amplified public awareness [7]. Thus, the nutritional aspects of food products and their relation to the individual's future health have become particularly important for both policy makers and consumers [8].

In the context of wine, choices are particularly difficult because of the large number of different intrinsic and extrinsic product attributes [9]. Indeed, wine supply has been undergoing a strong process of differentiation, not only based on sensorial characteristics related to grape varieties and production methods, origin, and brand [10], but also on sustainability issues [11–13]. Some wineries are focusing on environmental impacts of wine production such as the type of agricultural practices (including, but not limited to, the use of energy-intensive pesticides and fertilizers), water and carbon footprints, the impacts on biodiversity, and sustainable waste management [14–20]. Indeed, following the environmental impacts of wine production, in the last decades, several codes of practices have emerged in the wine industries of California, New Zealand, Australia, South Africa, and Chile [15,21]. Other firms have directed their differentiation efforts on country of origin and on traditional production methods or health-related issues [22]. Some brands have started to promote sensible drinking campaigns to signal their care for consumers and to align with health-related public policies [23]. However, a debate in the literature exists regarding the appropriateness of such actions and their related intent [23–26].

Even if there are still issues of understanding and comprehension of the information contained in labels [27,28], labelling may constitute an important instrument to help fill the information gap between consumers and producers [29]. In particular, in relation to the health dimension of sustainability, the nutritional information in labels is viewed as one of the most effective ways to promote healthier food choices among consumers [30–32]. In the last 10–15 years, some of the issues related to nutritional properties and the health-effects of food products have been addressed by regulation. Wine remains one of the few food products not subject to regulation on nutritional labelling (for Europe see Regulation (EU) 1169/2011) [33,34]), even though discussion on how to change this is ongoing [35,36]. Indeed, the frequency and amount of alcohol consumption are strongly related to health. Although some benefits have been found for moderate wine consumption (especially for red) [37–39], in several western countries, wine constitutes a large share of the energy intake of individuals [40–42] and this may be associated with weight gain and non-communicable diseases [43,44]. More specifically, flavonoids, phenolics, and other phytochemicals contained in wine have been found to have antioxidant properties that may have a positive health effect in relation to chronic diseases [45,46]. Indeed, in relation to coronary heart diseases, the so-called "French paradox" suggests that wine consumption (especially red wine) may prevent artery damage by cholesterol and other circulatory issues [47,48]. Other studies suggest that moderate wine consumption can, in some cases, reduce the risk of cancer [49,50]. Moderate wine drinking seems to also positively affect other types of illnesses like bone density, ulcers, gall and kidney stones [50,51]. On the other hand, excessive wine consumption has been associated with an increase in health risks and mortality, in relation, for example, to alcoholism and its associated risks (liver cirrhosis, injuries related to drunk driving, and social behavior) and some types of cancer [39,48,51]. Alcohol poses further risks if consumed during pregnancy [52,53]. All of these effects are associated with high social costs [54]. Moreover, alcohol is an energy intense drink with low density nutritional value. Thus, drinking wine increases the calorie intake, and has been associated with food-related chronic health conditions [40,55,56].

Even if currently there are no mandatory labelling requirements concerning ingredient list and nutritional information [57,58], the wine economics literature has turned its attention toward consumer interest in nutritional labelling and health warning messages [59–61]. These studies highlight the growth in consumer interest in such indications and the contemporary low levels of knowledge about the nutritional aspects of wine consumption [62,63]. It was also observed that attitudes toward these labels depend on several factors including wine involvement, health consciousness, and attitudes toward nutritional information [33,63]. This is indeed consistent with the food-related literature, which highlights how health conscious and health-oriented consumers as well as those with higher than average levels of nutritional knowledge tend to generally read food labels more [31,32,64,65].

Food-related studies have also investigated consumer interest for the so-called 'clean labels' [66]. As Asioli et al. [66] emphasized, there is no clear definition of what a 'clean' label is; however, this concept is often associated with 'natural' labels [67] such as those indicating organic, sustainable, 'free from' artificial additives/ingredients, or traditionally produced with minimal processing. Moreover, some studies have shown that "natural" products tend to be perceived by consumers as a healthier choice [68].

Specifically, the literature on consumer wine choices has focused on 'clean' labels related to aspects such as organic, biodiversity friendly, and biodynamic wine production processes [69–72], harvesting techniques, and the absence of additives [73–76]. This literature confirms the link between naturalness and health [77,78].

However, to the best of our knowledge, the existing literature that has explored consumer trade-offs for wine attributes that can be directly or indirectly linked to health issues is quite limited. This study focuses on the information set currently reported on (some) wine bottle labels and investigates consumer preferences for such wine attributes to evaluate if these may provide cues to promote healthier wine choices (e.g., lower alcohol content, sustainable agricultural practices) and, if so, among which types of consumers. In particular, we focused on 'clean' labels identified by Asioli et al. [36] as 'organic' and 'natural' labels (i.e., we did not focus on 'free from artificial additives/ingredients' labels) and labelled information directly linked to nutritional aspects such as alcohol content. We also aimed to explore whether attitudinal (e.g., health consciousness, wine interest, nutritional knowledge) and socio-demographic variables influenced consumer preferences for wine labels potentially linked to health [79].

Specifically, our research is directed at shedding light on three main questions:

- Is health orientation a driver of the consumer use of wine labels as in the case of food labels?
- What kind of labelled information (e.g., clean labels and/or alcohol content) directs wine choices toward more health-oriented consumers?
- What profiles characterize consumers interested in clean labels on wine and/or alcohol content?

Finally, we focused on at-home consumption as we aimed to investigate wine choices and eating habits. From a public policy point of view, this type of every-day behavior is interesting for studying and promoting healthier and more sustainable eating. Indeed, food consumption patterns and preferences for different types of product attributes may also differ depending on the eating occasion, the choice context, and the purchasing occasion [80–83], and some literature has highlighted that being in non-designated eating places, or more generally, eating-out, are associated with less healthy choices [84].

2. Materials and Methods

To achieve the objectives of this study, we conducted an online survey on a sample of Italian red wine consumers. Respondents were recruited in Italy by an online panel provider, were aged 18 years or older (i.e., the legal drinking age in Italy), and self-declared consuming red wine at home more than once a month. The questionnaire was administrated online in November 2018. The sample was stratified by gender, age, and region of residence, and consisted of 278 valid respondents.

The survey was primarily devoted to the collection of information, on one side, on consumer attitudes toward wine and health-related aspects; on the other, on consumer preferences for red wine consumed at home. Consumer preferences were elicited through a choice experiment (CE), the design of which is described in Section 2.1.

Consumer attitudes were surveyed through a set of questions making use of previously validated scales, which were aimed at evaluating consumer behavior and attitudes. These included the following:

- health consciousness [85,86];
- use of food labels and use of wine labels [87];
- perception of own weight status [88];
- food nutritional knowledge (adapted from Drichoutis et al. [32]);

- wine interest (adapted from Bruwer and Huang [89]); and
- wine consumption consciousness [90].

Appendix A presents the list of items included in the questionnaire according to the chosen scales. A seven-point Likert scale was used to measure the respondents' levels of agreement on statements.

Furthermore, the questionnaire investigated wine consumption habits (e.g., frequency, occasions, preferred types of wine) and wine purchasing behavior (e.g., purchasing factors, preferred wine shops) [9,91–93]. We also evaluated if the respondent was aware that the number of calories in wine was proportional to the alcohol percentage in volume.

The final part of the questionnaire included questions about the respondents' socio-demographic characteristics.

2.1. Choice Experiment Design

CEs are one of the most popular stated preference methods used in food marketing to elicit individuals' preferences and willingness-to-pay (WTP) for a good or a service. Their popularity is due to their ability to simultaneously estimate the utility of different attributes and attribute levels. Moreover, CEs are consistent with well-established theories, namely Lancaster's economic theory of consumer behavior [94], which assumes that the total utility of a good can be segregated in partial utilities given by the different attributes of the good, and with random utility theory, which assumes that individuals make choices to maximize their utility under a budget constraint [95].

In this study, the product item was red wine in a 750 mL bottle. We selected a red wine because the latter presents a wider range of alcohol content than white wine. Moreover, red wine is believed to have a more positive impact on health than other types of wine [51,96].

To avoid any kind of bias derived from the provenance and origin of the brands, we referred to a red wine produced in Italy without any protected designation of origin or geographical indication. No additional information on brand, grape variety, or sensory characteristic was indicated on the label.

To achieve the study objective, the attributes and levels were selected in order to propose several types of red wine characterized by product cues and production practices potentially affecting the healthiness and naturalness of wine. Here, the intent was to focus the design on some of the most relevant cues considered by the contemporary wine consumer and used as new product differentiation options by wineries [97].

Two pilot tests on a sample of 81 (pilot 1) and 80 (pilot 2) respondents, respectively, were performed to test the CE and refine the implemented attributes and their levels.

Table 1 presents the selected attributes and their levels as used in the CE. Specifically, five attributes were used:

- sustainability certification;
- hand-picked grapes;
- unfiltered wine;
- alcohol content; and
- price.

The attribute related to 'sustainability certification' was described by three levels: (1) organic certification, represented through the 'European Leaf' logo; (2) biodynamic certification, using the logo of the private protocol 'Demeter®'; and (3) a certification aimed at communicating the adoption of agricultural practices that protect or enhance biodiversity (i.e., the private protocol 'Biodiversity Friend®', supported by the WBA World Biodiversity Association). We used these product information features in our analysis, since attributes related to sustainable agricultural practices might be linked to healthier food [66,98]. To illustrate, organic, biodynamic, and environmentally friendly products are often perceived as healthier and more natural [97–102]. The literature suggests that an important motivation for purchasing organic products is related to health considerations [103–106] and that health-conscious consumers are more in favor of organic food [107]. Indeed, organic buyers are found to also care highly for healthy diets [103,105].

Table 1. Attributes and attribute levels used in the choice experiment.

Attribute	Attribute Levels	Indication on The Back Label
Sustainability certification	Organic	
	Biodynamic	demeter
	Biodiversity friend®	biodiversity friend
	None	-
		-
Hand-picked grapes	-Present -Absent	Uva raccolta a mano
Unfiltered wine	-Present -Absent	Vino non filtrato
Alcohol content (% ABV)	11	11% vol
	12	12% vol
	13	13% vol
	14	14% vol
Price (€/750 mL bottle)	2.10	-
	3.60	-
	5.10	-
	8.10	-
	11.10	-
	14.10	-

We considered two further claims, 'hand-picked grapes' and 'unfiltered wine', which we indicated, when present, graphically through a logo that we conceived anew for the purpose of this study. Hand harvested and unfiltered claims where selected given that they tend to recall traditional production methods and indicate low levels of automation and processing, which as Asioli et al. [66] explained, can be considered as aspects of 'clean labelling'. Indeed, some of the literature has highlighted how food processing and 'machine-made' (vs. hand-made) processes tend to reduce the level of 'naturalness' perceived by consumers [108–110], and how natural products tend to be perceived as healthier by consumers [68].

Furthermore, we investigated the role of alcohol content in wine choices This indication is particularly interesting as it is already available on all wine bottles and it is linked to calorie and energy content [111]. Four levels of alcohol by volume (ABV) were used: 11%, 12%, 13%, and 14%, which allowed us to include a widespread alcohol content percentage of red wine in the Italian wine market.

The price attribute had six levels (€2.10, €3.60, €5.10, €8.10, €11.10, and €14.10) to take into account the extended price differentiation characterizing the wine on offer in Italy 'off-premise' stores (e.g., supermarkets, wine shops, or by the producers) [112].

Based on Thiene, Franceschinis, and Scarpa [88], the allocation of attributes and attribute levels to product alternatives was designed by employing a WTP_b efficient design that used as Bayes priors the WTP values obtained from the estimation of a multinomial logit model (MNL) on data from the second pilot (Choice Metrics, Ngene v1.0.1, 2011). The final design consisted of sixty choice tasks divided in five blocks of twelve choice tasks each. Thus, each respondent was asked to face twelve choice tasks.

Each choice task presented three alternatives: two different bottles and a 'neither of these' alternative in order to let respondents opt out of purchase in case they did not prefer either bottle on offer. The two bottles in the choice task were described by their respective back label. Prices were indicated under the pictures displaying the back labels. The order of appearance of the choice tasks was also randomized as the order of the two product alternatives for each choice task.

To provide a homogenous context, respondents were asked to imagine buying a 750 mL bottle of red wine for a home consumption occasion in their usual shop. For each of the 12 choice tasks, respondents were asked to indicate the wine they preferred or to opt out. To limit the hypothetical bias, the choice experiment was introduced by a 'cheap talk' script [113]. Specifically, respondents were warned about the potential problem of hypothetical bias and were asked not to overstate their true WTP and focus on their responses as if they were in a real-life setting [114].

2.2. Models Specification

2.2.1. Ordinary Least Square (OLS) regression

In order to answer our first research question (i.e., whether health-oriented individuals tend to be more prone to use wine labels), we performed an Ordinary Least Square regression specified as follows:

$$y_n = \alpha_0 + \sum_{k=1}^{K} \alpha_k s_{kn} + \epsilon_n \quad (1)$$

where y_n is the dependent variable; α_0, is the intercept of the model; s_{kn} corresponds to the kth explanatory variable of the model (k = 1 to K); α_k is the average increment in y_n associated with a unitary variation of s_{kn}; ϵ_n is the independent and identically distributed (i.i.d.) random error with zero mean. In our case, we specified the dependent variable (y_n) as the mean value of the five items capturing the use of wine labels, while we included the following as explanatory variables (s_{kn}): (1) health consciousness; (2) attitudinal information, which in the literature regarding food choices, have been determined as drivers in food label use (nutritional knowledge and being on diet or not); (3) information related to the respondents' interest/knowledge of the wine product (wine interest, wine consumption consciousness, knowledge of the relationship between energy value, and alcohol degree); and (4) socio-demographic information.

Finally, as a robustness check, we also tested a similar model on food label use in general to test if these behavioral variables affected wine label use in a similar way to food label use.

2.2.2. Discrete Choice Models

Regarding our research questions related to consumer preferences for a set of labels that could potentially be related to health issues, we used discrete choice models (DCMs) to estimate the utility function determining the CE responses.

In DCMs, the utility for individual n of choosing alternative i in the choice situation t can be specified as follows:

$$U_{nit} = \beta_n' x_{nit} + \varepsilon_{nit} \quad (2)$$

where x_{nit} is the vector of the observed variables relating to alternative i and individual n in choice task t; β_n is a vector of individual structural taste parameters characterizing choices; ε_{nit} is the unobserved utility term, assumed to be an independently and identically distributed Gumbel (Extreme Value Type I).

In an MNL model, the probability that consumer n chooses alternative i in choice task t is defined as follows:

$$\Pr_{nit} \frac{\exp(\beta_n' x_{nit})}{\sum_{j=1}^{J} \exp\left(\beta_n' x_{njt}\right)} \quad (3)$$

In this specification, it is assumed that the consumers' preferences are heterogeneous as described in studies for food or wine products [115]. However, studies have shown that consumers have preferences that may cluster into classes. This heterogeneity specification allows for a more parsimonious model than other random parameter logit (RPL) models accounting for preference heterogeneity [116]. RPL models assume heterogeneity to occur with a continuous distribution, meaning that each individual is assumed to have his/her own preferences. However, heterogeneity can be described by a discrete mixture, namely, different groups of consumers might have different preferences across groups, but preferences are identical within the group [117]. Given our interest in defining different preference clusters of consumers, in this study, we chose to adopt a latent class model (LCM) which takes into account that, within each segment, consumer preferences are homogeneous, but preferences vary between these segments [116,118].

In LCMs, respondents can be probabilistically assigned to a finite set of C classes on the basis of their observed pattern of choices. Each class is characterized by a unique class-specific vector of utility parameters β_c for each of the attributes in the choice task. Given membership to a class c, the probability of the sequence of choices y_n over the T choice occasions can be described as:

$$\mathrm{Pr}_n(y|c, x_{nit}) = \prod_{t=1}^{T} \frac{\exp(\beta_c' x_{nit})}{\sum_{j=1}^{J} \exp\left(\beta_c' x_{njt}\right)} \tag{4}$$

Membership probabilities for each latent class c are also defined according to a logit probability as:

$$\pi_c \; \frac{\exp\left(a_c + \gamma_c' Z_n\right)}{\sum_{c=1}^{C} \exp(a_c + \gamma_c' \, Z_n)} \tag{5}$$

where Z_n is the vector of co-variates characterizing respondent n, and γ is the vector of associated parameters to be estimated, while a_c is a class-specific constant. In estimation, for identification purposes, only C−1 sets of coefficients can be identified. For one arbitrary class c the vector γ_c and a_c are both set to zero for identification purposes. In our analysis, we included as co-variates the same variables we used in the linear regression to explain wine/food label use. If these co-variates are significant, then we can use this information to explain segment membership.

In order to facilitate the inclusion of attitudinal and socio-demographic information in our model specifications (both in the OLS and the LCM), these variables were specified as follows:

- Age = dummy variable with value equal to 1 if the respondent has the median age or is older, 0 otherwise;
- Gender = dummy variable with value equal to 1 if the respondent was a woman, 0 otherwise;
- Education level = dummy variable with value equal to 1 if the respondent has the median education level or a higher one, 0 otherwise;
- Health consciousness = mean value of the health consciousness scale;
- Nutritional knowledge = mean value of the correct answers (if answer is correct = 1, 0 otherwise) to the questions related to food nutritional knowledge;
- Weight status control = dummy variable taking value of 1 for individuals who stated to be on a diet or to control their weight status, 0 otherwise;
- Knowledge calories/alcohol in wine = dummy variable taking value equal to 1 if the respondent answered correctly to the question related to wine nutritional knowledge, 0 otherwise
- Wine consumption consciousness = dummy with value equal to 1 if the respondent has a score equal to the median or higher, 0 otherwise;
- Wine interest = mean of the scores from the wine interest scale;
- Food labels use = mean of the scores from the use of food label scale; and
- Wine labels use = mean of the scores from the use of wine label scale.

3. Results

3.1. Sample Description, Attitudes, and Knowledge on Health, Food, and Wine Label Use and Wine Drinking

Table 2 illustrates the respondents' socio-demographic characteristics. The proportion of women was slightly higher than that of men, and all age groups from 25–34 to 64–74 years were almost equally represented. The level of education was high (51.44% of respondents completed high school), while the level of stated household income was medium to low range.

Table 2. Socio-demographic characteristics of respondents (n = 278).

Variables	Sample
Age group (%)	
18–24 years	13.31
25–34 years	18.71
35–44 years	17.63
45–54 years	15.47
55–64 years	16.55
65–74 years	17.63
74+ years	0.72
Gender (%)	
Woman	52.52
Education (%)	
Comprehensive school	0.36
Intermediate school	12.23
High school	51.44
Bachelor degree	11.87
Master degree	18.35
PhD degree or post-doc	5.76
Total annual household net income (%)	
Less than €10,000	14.75
€10,000–€20,000	21.22
€20,001–€30,000	23.74
€30,001–€40,000	14.39
€40,001–€50,000	4.68
€50,001–€60,000	2.52
€60,001–€70,000	2.88
€70,001–€80,000	1.80
€80,001–€90,000	2.88
More than €90,000	2.52
I prefer not to answer this question.	8.63

Most respondents were interested in wine: they tended to like it and considered it as an important consumption good in their life; however, they were less involved in wine-related activities such as reading books or magazines on the subject or attending wine events (Table 3). They were mostly aware about the link between food and heart disease prevention while they are less knowledgeable about the nutritional compounds of food and wine. Most respondents paid attention to their weight and were quite interested in using labels when choosing food and wine products, and were conscious about the constant monitoring of their health status.

Table 3. Respondents' attitudes and knowledge on health, food, and wine label use and wine drinking (n = 278).

Variables	Sample
Weight status control (%)	
I'm trying to lose weight.	35.25
I'm trying to maintain my weight.	44.24
I don't do anything to regulate my weight.	20.50
Wine consumption consciousness (mean)	
It is important to limit the amount of alcohol you consume	5.88
Wine Interest (mean)	
Wine is an important product to me.	5.38
I like wine.	5.86
I read magazines and books related to wine.	3.79
I take part in events which are related to wine (wine tourism, wine festivals, wine shows)	4.23
Food nutritional knowledge (%)	
Eating more fiber helps to prevent heart disease.	
Yes	71.58
No	10.43
Don't know	17.99
Eating less salt helps to prevent heart disease.	
Yes	84.17
No	7.91
Don't know	7.91
Eating more fruit and vegetables helps to prevent heart disease.	
Yes	83.81
No	8.27
Don't know	7.91
Eating less saturated fats helps to prevent heart disease.	
Yes	73.74
No	7.19
Don't know	19.06
Which has more calories: 100 grams of durum wheat pasta or 100 grams of white rice?	
Pasta	38.85
White rice	19.42
The same	13.31
Don't know	28.42
Which fat do you think experts are saying is most important to cut down?	
Monosaturated fat	16.19
Saturated fat	55.40
Not sure	28.42
Knowledge calories/alcohol in wine (%)	
The amount of calories in wine is proportional to the alcohol percentage.	
True	48.56
False	20.86
Don't know	30.58
Use of wine labels (mean)	
Reading wine labels takes more time that I can spend. (R)	4.50
Reading wine labels makes it easier to choose wines.	5.22
When I use wine labels, I make better wine choices.	5.10
Using wine labels to choose wines is better than just relying on my own knowledge about what is in them.	5.17
The information on wine labels is hard to interpret. (R)	4.56
Use of food labels (mean)	
Reading food labels takes more time that I can spend. (R)	4.12
Reading food labels makes it easier to choose foods.	5.38
When I use food labels, I make better food choices.	5.42
Using food labels to choose foods is better than just relying on my own knowledge about what is in them.	5.34
The nutritional information on food labels is hard to interpret. (R)	4.03

Table 3. *Cont.*

Variables	Sample
Health consciousness (mean)	
I reflect about my health a lot.	5.32
I am very self-conscious about my health.	5.40
I am alert to changes in my health.	5.46
I am usually aware of my health.	5.38
I take responsibility for the state of my health.	5.62
I am aware of the state of my health as I go through the day.	5.28

3.2. Ordinary Least Square Estimation

In relation to the first of our research questions, we first investigated the association between a set of variables related to health consciousness, nutritional knowledge, and socio-demographics with label use. With regard to food labels, our results confirm previous research findings that associate health consciousness and nutritional knowledge to higher food label use [32,64,65]. Specifically, results indicate that—ceteris paribus—more health-conscious respondents and those with a higher nutritional knowledge tended to have more favorable attitudes toward food label use (Table 4). Consumers who were conscious of wine consumption tended to use food labels more; however, knowledge about the link between alcohol consumption and calorie intake does not seem to have had an effect. Unlike what is usually found in the literature, the socio-demographic variables failed to play a significant role in our sample. Other variables such as the desire to control bodyweight, and other more wine-related variables were also found to be insignificant with food label use.

Table 4. Ordinary Least Square regression model results concerning reading food and wine labels.

	Food Label Use (Mean)	Wine Label Use (Mean)
Age	0.114 (0.104)	0.195 ** (0.097)
Gender	0.054 (0.101)	−0.111 (0.094)
Education level	−0.065 (0.108)	0.027 (0.101)
Health consciousness	0.364 *** (0.060)	0.247 *** (0.056)
Weight status control	0.126 (0.129)	−0.001(0.120)
Wine interest	−0.012 (0.046)	0.088 ** (0.043)
Knowledge calories/alcohol in wine	0.089 (0.104)	−0.036 (0.097)
Food nutritional knowledge	0.398 ** (0.191)	0.243 (0.179)
Wine consumption consciousness	0.223 ** (0.109)	0.212 ** (0.101)
Intercept	2.234 *** (0.330)	2.830 *** (0.309)
Model statistics		
R-square	0.240	0.175
No. of observations	278	278

***, ** denote significance at the 1%, 5%, and 10% level, respectively; 1 = Number in brackets are standard errors.

Table 4 also depicts the results of the same model applied to wine labels. With regard to wine label use, the results concerning health consciousness and wine consumption consciousness were confirmed. In this model, another variable more strictly related to wine becomes relevant. Indeed, as may be expected, wine interest seems to play a significant role on wine label reading. Moreover, age was statistically significant, indicating that older consumers tended to use wine labels more than younger consumers.

3.3. Latent Class Model

Different steps were developed in order to select the best number of classes for our data. At first, we used model fit statistics criteria, which led us to identify seven classes. However, as Ruto, Garrod, and Scarpa [119] explain, parsimony in the number of classes and the identification of sufficient explanatory classes are important criteria to take into account when selecting the optimal number of groups. Accordingly, in our study, we selected a Latent Class Model with three classes (Table 5).

Table 5. Latent class model estimates.

	Parameters		
	Class 1	**Class 2**	**Class 3**
Attributes Variable	Coefficients	Coefficients	Coefficients
Hand-picked grapes	0.553 *** (0.076)	0.766 *** (0.168)	0.902 *** (0.322)
Unfiltered wine	0.201 *** (0.057)	−0.038 (0.147)	0.339 (0.208)
Organic	0.792 *** (0.085)	0.516 *** (0.170)	0.353 (0.308)
Biodynamic	0.130 (0.108)	−0.589 ** (0.233)	0.514 * (0.279)
Biodiversity	0.589 *** (0.087)	−0.542 *** (0.193)	0.830 *** (0.247)
Alcohol content	−0.095 *** (0.033)	−0.015 (0.078)	0.147 (0.102)
Price	−0.021 *** (0.006)	−0.050 ** (0.021)	−0.605 *** (0.041)
No-buy	−3.438 *** (0.423)	0.819 (0.950)	−1.264 (1.259)
Co-variates			
Constant	−1.203 (1.485)	−0.870 (1.735)	0
Weight status control	0.177 (0.483)	1.107 (0.751)	0
Wine interest	0.158 (0.186)	−0.136 (0.244)	0
Knowledge calories/alcohol in wine	−0.535 (0.386)	0.065 (0.520)	0
Food nutritional knowledge	−1.141 (0.826)	−1.047 (0.996)	0
Health consciousness	0.683 *** (0.264)	0.195 (0.335)	0
Wine consumption consciousness	−0.665 (0.442)	0.017 (0.584)	0
Education level	0.314 (0.465)	0.122 (0.595)	0
Age (dummy)	−1.674 *** (0.494)	−0.705 (0.605)	0
Gender	0.258 (0.422)	0.995 (0.607)	0
Size	63%	18%	19%
Model Statistics			
No. of observations		3336	
Log likelihood		−2695.389	
Akaike Information Criterion		5478.8	

***, **, * denote significance at the 1%, 5% and 10% level, respectively; 1 = Number in brackets are standard errors.

In our econometric analysis, all the 'clean labels' were specified as dummy variables, taking a value of 1 if the label was present on the bottle of wine, or 0 otherwise. On the other hand, alcohol content and price were treated as continuous variables describing the experimentally designed levels. As Table 5 shows, Class 3 was considered as the reference class for the determination of the effect of individual characteristics.

On the basis of the LCM model utility parameter estimates, the three classes can be described as follows:

Class 1. The health-conscious wine consumer: This class represents the highest share of respondents (63% of probability). These are the consumers who are the most prone to purchasing a bottle of wine for home consumption, given the negative and statistically significant coefficient of the no-buy option. Their probability of purchasing a bottle of wine significantly increases when one of the clean labels is present, with the exception of the biodynamic certification. What distinguishes this segment from the others is the negative and statistically significant parameter of the alcohol variable, indicating that an increase in alcohol degree in wine decreases the respondents' utility from choosing a bottle of wine. Moreover, we observe that this was the only segment of consumers who have a strong significant preference for nearly all clean labels including the unfiltered wine. Finally, as expected, with an increase in price, individuals tend to have a higher disutility for buying the bottle. Belonging to this class is determined by an increase in health consciousness and by being of a younger age.

Class 2. The opponents: This is the segment with the lowest probability of membership (18%). Given the insignificant and positive value of the opt-out coefficient, this group of individuals generally do not tend to choose to buy a bottle of red wine for home consumption. Their probability of purchasing the bottle increases only with more 'traditional-processing' (i.e., hand-picked grapes) and more widespread organic certifications, while it decreases in the case of unconventional claims (e.g., biodiversity friendly or biodynamic wine). Alcohol content is not statistically significant, while we observe that an increase in price further decreases the utility of the opponents to purchase a bottle of wine.

Class 3. The new clean trend lovers: These consumers are not generally interested in buying a bottle of wine, but they have a significant preference for clean labels, except for organic certification, which represents the most widespread clean label in the wine market. However, it is important to specify that the coefficient of the unfiltered claim is borderline significant ($p = 0.103$). Even in this case, we did not observe any significant effect of alcohol content on the consumers' utility formation, while price has a negative effect.

4. Discussion

Our first research question was directed at investigating whether health-oriented consumers were more prone to reading wine labels. The OLS estimation results showed that attitude toward reading food and wine labels was mainly favored by some attitudinal antecedents of personal status consciousness (i.e., health and wine consumption). These findings are in line with several previous studies carried out in different countries, for example, Drichoutis et al. [32], Aschemann-Witzel et al. [5], and Cavaliere et al. [64].

With regard to nutritional knowledge, we observed that this information explained food label use as confirmed in the literature review by Drichoutis et al. [65], but not wine label use. Instead, the opposite occurred with the wine interest variable. Namely, the latter attitude significantly affected only wine labels use, as Annunziata et al. [33] and Vecchio et al. [63] also found. However, if we consider food nutritional knowledge as a proxy for involvement with food product, this result might indicate that interest in product characteristics is an important factor for explaining the use of labels, as, indeed, is suggested by the literature (see Cavaliere et al., [64] in the case of food products or Annunziata et al. [59] and Bruwer and Huang [89] in the case of wine).

On the other hand, the significant effect of wine interest might be interpreted as the attitude of wine consumers to also gather product information from additional sources such as reading wine magazines or books, and participating in wine events [9,22,23,51,81,89]. Therefore, from a marketing prospective, these findings would suggest that the communication process toward consumers is quite challenging for wine companies and the label is only one component of such process, although it is of relevance during the purchasing choice, especially at off-premise points of sale. This is particularly important in relation to the age of consumers, as highlighted by our results and by other previous studies [60,70,82].

In conclusion, on the basis of our OLS regression, in relation to our research question of whether health-oriented consumers tend to use wine labels more, as in the case of food labels, our data suggest that this is indeed the case. In particular, our results suggest that wine and food label use is driven by similar attitudes.

As for the second and third research questions, 'which kind of labelled information directs wine choices for health-oriented consumers' and 'who are the consumers interested in clean labels and/or alcohol content', the latent class analysis suggests the existence of a large cluster of consumers (Class 1) who are driven by their health consciousness when choosing a wine. They prefer hand-made products (hand-picked grapes and unfiltered wine) and sustainable product certifications. This cluster may potentially represent an important market target, estimated around 60% of our sample that corresponds to the contemporary young consumer, who is particularly keen on quality assurance information. The above results are in line with previous work on quality attribute certifications [17,19] and on young consumers [38,70].

Importantly, what differentiates these health-conscious consumers from the other two clusters is the aversion toward a higher degree of alcohol content. This was actually expected since the literature shows that more health-oriented consumers tend to quite negatively perceive those wines with high alcohol content (see Yoo et al. [90]; Drichoutis et al. [32]; Martin-Moreno et al [35]; Casini et al. [93]).

The cluster of the 'opponents' includes consumers who are not price-sensitive. They prefer organic certification and hand-picked grapes to other more recent claims. These findings suggest that this group of consumers tends to be driven by conventional product cues during their choice while health-related awareness does not seem to influence their preferences. Therefore, organic production and traditional practices potentially act as the principal preference drivers of choice for about one fifth of our respondents. This group could be an interesting potential market to be targeted and developed by organic producers. This aligns with previous findings from Schäufele and Hamm [12]; Schleenbecker and Hamm [106]; Fotopoulos et al [103]; de Magistris and Gracia [79]; and Castellini et al. [22].

The third class is characterized by consumers attracted to more innovative sustainability-related labels. For these consumers, price represents a constraint, but they demonstrate an interest in new label claims. They prefer hand-made to industrial wines (highlighted by the hand-picked grapes claim) and innovative ways to produce eco-friendly wine, like biodynamic and biodiversity friendly protocols [11,36,101]. What seems to emerge is an interest only for innovative sustainability-related labels. This is also the case among consumers who have a strong disutility for higher priced products, which is interesting given that a common finding in the literature is that richer people—who can afford pricier wines—have the strongest preferences for such attributes [12]. Our results suggest that this type of consumer potentially represents one fifth of the market target and are therefore of interest to producers proposing new types of natural or sustainable production schemes. Moreover, we did not observe health consciousness as a determinant of membership to this group.

5. Conclusions

Given the interest of health-conscious consumers toward alcohol content information and wine label information in general, we believe that our results promote the use of nutritional information, even in the case of wine products. Nutritional information on wine may indeed add value to the nutritional aspects of wine in comparison with other types of alcoholic drinks. Nutritional information on wine labels could also represent a way of diffusing more awareness and of favoring the consumption of less-alcoholic wine, and therefore, a more responsible drinking behavior. Specifically, our findings suggest that the inclusion of nutritional information on wine labels could incentivize a more conscious use of the product in the home context. Indeed, within a daily (Mediterranean) diet, wine—as it also happens with extra virgin olive oil—can be considered as a 'functional' food, since a moderated quantity of the product is supposed to have positive effects for health [63,90]. This policy could be particularly oriented toward Italian mature adults who are more prone to using wine labels than younger consumers. Moreover, the latter are shifting their wine drinking patterns from regular consumption to occasional consumption during weekends, preferring out-of-home contexts, outside meals, and high alcohol beverages [51,60,70,81,82]. Thus, new tools should be developed to provide information in this context [120,121].

Before closing, it is important to point out that our results need to be interpreted with caution due to the size and geographical scope of the sample of respondents. Moreover, it is important to point out that the results may suffer from the hypothetical nature of the experiment. However, wine is a well-known food product with which Italian consumers are familiar in the context of making purchase choices. Thus, at least the novelty bias should be reduced [1,122]. However, some of the labelled indications might be new and unknown to some respondents.

Future research could explore attribute non-attendance (ANA) in order to corroborate our conjecture that health conscious individuals tend to give more attention to nutritional information such as alcohol over clean labels. Moreover, we believe that future research would advance the literature by testing the effect of providing nutritional information on consumer preference formation for different wine cues such as clean labels or information regarding brand, origin, and variety.

Author Contributions: Conceptualization, C.B. and R.C.; Methodology, C.B. and R.S.; Supervision, D.B.; Writing—original draft, C.B., R.C. and E.C.R.; Writing—review & editing, R.S. and D.B. All authors have read and agreed to the published version of the manuscript.

Funding: This research received no external funding.

Acknowledgments: We thank Davide Quinzi for assistance with the data collection, and Demeter Associazione Italia and WBA World Biodiversity Association for allowing us to use the certification logos.

Conflicts of Interest: The authors declare no conflicts of interest.

Appendix A

Table A1. Scales used to assess consumer attitudes and knowledge on health, food, and wine label use and wine drinking used in this study.

Health consciousness [85,86]
(seven-point Likert from strongly disagree to strongly agree)
I reflect about my health a lot.
I am very self-conscious about my health.
I am alert to changes in my health.
I am usually aware of my health.
I take responsibility for the state of my health.
I am aware of the state of my health as I go through the day.

Use of wine labels (Adapted from Žeželj et al. [87])
(seven-point Likert from strongly disagree to strongly agree)
Reading wine labels takes more time that I can spend.
Reading wine labels makes it easier to choose wines.
When I use wine labels, I make better wine choices.
Using wine labels to choose wines is better than just relying on my own knowledge about what is in them.
The information on wine labels is hard to interpret.

Use of food labels [87]
(seven-point Likert from strongly disagree to strongly agree)
Reading food labels takes more time that I can spend. R
Reading food labels makes it easier to choose foods.
When I use food labels, I make better food choices.
Using food labels to choose foods is better than just relying on my own knowledge about what is in them.
The nutritional information on food labels is hard to interpret. R

Weight status control (Adapted from Thiene et al. [88])
Which of the following statements best describe your actual status?
- I'm trying to lose weight.
- I'm trying to maintain my weight.
- I don't do anything to regulate my weight.

Food Nutritional Knowledge [32]
Which has more calories: 100 grams of durum wheat pasta or 100 grams of white rice?
- Pasta
- White rice
- The same
- Don't know

Which fat do you think experts are saying is most important to cut down?
- Monosaturated fat
- Saturated fat
- Not sure

Table A1. *Cont.*

Which fat do you think experts are saying is most important to cut down?
- Monosaturated fat
- Saturated fat
- Not sure
Do you think these behaviors help to prevent heart disease? *(Yes, No, don't know)*
- Eating more fiber
- Eating more fiber
- Eating less salt
- Eating more fruit and vegetables
- Eating less saturated fat
Knowledge calories/alcohol in wine (adapted from Drichoutis et al.) [32] The amount of calories in wine is proportional to the alcohol percentage.
- True
- False
- Don't know
Wine interest (adapted from Bruwer and Huang [89]) *(seven-point Likert from strongly disagree to strongly agree)*
- Wine is an important product to me.
- I like wine.
- I read magazines related to wine.
- I take part in events which are related to wine (wine tourism, wine festivals, wine shows).
Wine consumption consciousness (mean) [90] *(seven-point Likert from strongly disagree to strongly agree)*
- It is important to limit the amount of alcohol you consume

References

1. Rousseau, S. The role of organic and fair trade labels when choosing chocolate. *Food Qual. Prefer.* **2015**, *44*, 92–100. [CrossRef]
2. Huang, C.L. Consumer preferences and attitudes towards organically grown produce. *Eur. Rev. Agric. Econ.* **1996**, *23*, 331–342. [CrossRef]
3. Bazzani, C.; Gustavsen, G.W.; Nayga, R.M.; Rickertsen, K. A comparative study of food values between the United States and Norway. *Eur. Rev. Agric. Econ.* **2018**, *45*, 239–272. [CrossRef]
4. Caputo, V.; Scarpa, R.; Nayga, R.M. Cue versus independent food attributes: The effect of adding attributes in choice experiments. *Eur. Rev. Agric. Econ.* **2017**, *44*, 211–230. [CrossRef]
5. Aschemann-Witzel, J.; Grunert, K.G.; van Trijp, H.C.; Bialkova, S.; Raats, M.M.; Hodgkins, C.; Wasowicz-Kirylo, G.; Koenigstorfer, J. Effects of nutrition label format and product assortment on the healthfulness of food choice. *Appetite* **2013**, *71*, 63–74. [CrossRef]
6. WHO. *Obesity and Overweight*; WHO: Geneva, Switzerland, 2017.
7. Cavaliere, A.; Ricci, E.C.; Solesin, M.; Banterle, A. Can health and environmental concerns meet in food choices? *Sustainability* **2014**, *6*, 9494–9509. [CrossRef]
8. Grunert, K.G.; Wills, J.M. A review of European research on consumer response to nutrition information on food labels. *J. Public Health* **2007**, *15*, 385–399. [CrossRef]
9. Lockshin, L.; Jarvis, W.; d'Hauteville, F.; Perrouty, J.-P. Using simulations from discrete choice experiments to measure consumer sensitivity to brand, region, price, and awards in wine choice. *Food Qual. Prefer.* **2006**, *17*, 166–178. [CrossRef]
10. Capitello, R.; Agnoli, L.; Begalli, D. Drivers of high-involvement consumers' intention to buy PDO wines: Valpolicella PDO case study. *J. Sci. Food Agric.* **2016**, *96*, 3407–3417. [CrossRef]

11. Gilinsky, A.; Newton, S.K.; Vega, R.F. Sustainability in the Global Wine Industry: Concepts and Cases. *Agric. Agric. Sci. Procedia* **2016**, *8*, 37–49. [CrossRef]
12. Schäufele, I.; Hamm, U. Consumers' perceptions, preferences and willingness-to-pay for wine with sustainability characteristics: A review. *J. Clean. Prod.* **2017**, *147*, 379–394. [CrossRef]
13. Szolnoki, G. A cross-national comparison of sustainability in the wine industry. *J. Clean. Prod.* **2013**, *53*, 243–251. [CrossRef]
14. Bisson, L.F.; Waterhouse, A.L.; Ebeler, S.E.; Walker, M.A.; Lapsley, J.T. The present and future of the international wine industry. *Nature* **2002**, *418*, 696. [CrossRef] [PubMed]
15. Marshall, R.S.; Cordano, M.; Silverman, M. Exploring individual and institutional drivers of proactive environmentalism in the US Wine industry. *Bus. Strat. Environ.* **2005**, *14*, 92–109. [CrossRef]
16. Knowles, L.; Hill, R. Environmental initiatives in South African wineries: A comparison between small and large wineries. *Eco-Manag. Audit.* **2001**, *8*, 210–228. [CrossRef]
17. Santini, C.; Cavicchi, A.; Casini, L. Sustainability in the wine industry: Key questions and research trendsa. *Agric. Food Econ.* **2013**, *1*, 9. [CrossRef]
18. Warner, K.D. The quality of sustainability: Agroecological partnerships and the geographic branding of California winegrapes. *J. Rural Stud.* **2007**, *23*, 142–155. [CrossRef]
19. Forbes, S.L.; De Silva, T.-A. Analysis of environmental management systems in New Zealand wineries. *Int. J. Wine Bus. Res.* **2012**, *24*, 98–114. [CrossRef]
20. Moscovici, D.; Reed, A. Comparing wine sustainability certifications around the world: History, status and opportunity. *J. Wine Res.* **2018**, *29*, 1–25. [CrossRef]
21. Gabzdylova, B.; Raffensperger, J.F.; Castka, P. Sustainability in the New Zealand wine industry: Drivers, stakeholders and practices. *J. Clean. Prod.* **2009**, *17*, 992–998. [CrossRef]
22. Castellini, A.; Mauracher, C.; Procidano, I.; Sacchi, G. Italian market of organic wine: A survey on production system characteristics and marketing strategies. *Wine Econ. Pol.* **2014**, *3*, 71–80. [CrossRef]
23. Barry, A.E.; Goodson, P. Use (and Misuse) of the Responsible Drinking Message in Public Health and Alcohol Advertising: A Review. *Health Educ. Behav.* **2010**, *37*, 288–303. [CrossRef] [PubMed]
24. Foster, S.E.; Vaughan, R.D.; Foster, W.H.; Califano, J.A. Estimate of the commercial value of underage drinking and adult abusive and dependent drinking to the alcohol industry. *Arch. Pediatr. Adolesc. Med.* **2006**, *160*, 473–478. [CrossRef] [PubMed]
25. Munro, G. An addiction agency's collaboration with the drinks industry: Moo Joose as a case study. *Addiction* **2004**, *99*, 1370–1374. [CrossRef] [PubMed]
26. McCreanor, T.; Casswell, S.; Hill, L. ICAP and the perils of partnership. *Addiction* **2000**, *95*, 179–185. [CrossRef] [PubMed]
27. Grunert, K.G.; Hieke, S.; Wills, J. Sustainability labels on food products: Consumer motivation, understanding and use. *Food Pol.* **2014**, *44*, 177–189. [CrossRef]
28. Grunert, K.G. Sustainability in the Food Sector: A Consumer Behaviour Perspective. *Int. J. Food Syst. Dyn.* **2011**, *2*, 207–218.
29. Banterle, A.; Ricci, E.C.; Cavaliere, A. Environmental sustainability and the food system. In *Regulating Food Safety Law in the EU—A Legal-Economic Perspective*; Bremmers, H.J., Purnhagen, K., Eds.; Springer: Berlin/Heidelberg, Germany, 2018.
30. Capacci, S.; Mazzocchi, M.; Shankar, B.; Brambila Macias, J.; Verbeke, W.; Pérez-Cueto, F.J.; Kozioł-Kozakowska, A.; Piórecka, B.; Niedzwiedzka, B.; D'Addesa, D. Policies to promote healthy eating in Europe: A structured review of policies and their effectiveness. *Nutr. Rev.* **2012**, *70*, 188–200. [CrossRef]
31. Nayga Jr, R.M. Nutrition knowledge, gender, and food label use. *J. Consum. Aff.* **2000**, *34*, 97–112. [CrossRef]
32. Drichoutis, A.C.; Lazaridis, P.; Nayga, R.M. Nutrition knowledge and consumer use of nutritional food labels. *Eur. Rev. Agric. Econ.* **2005**, *32*, 93–118. [CrossRef]
33. Annunziata, A.; Pomarici, E.; Vecchio, R.; Mariani, A. Do Consumers Want More Nutritional and Health Information on Wine Labels? Insights from the EU and USA. *Nutrients* **2016**, *8*, 416. [CrossRef] [PubMed]
34. Petticrew, M.; Douglas, N.; Knai, C.; Durand, M.A.; Eastmure, E.; Mays, N. Health information on alcoholic beverage containers: Has the alcohol industry's pledge in England to improve labelling been met? *Addiction* **2016**, *111*, 51–55. [CrossRef] [PubMed]

35. Martin-Moreno, J.M.; Harris, M.E.; Breda, J.; Møller, L.; Alfonso-Sanchez, J.L.; Gorgojo, L. Enhanced labelling on alcoholic drinks: Reviewing the evidence to guide alcohol policy. *Eur. J. Public Health* **2013**, *23*, 1082–1087. [CrossRef] [PubMed]
36. Pabst, E.; Szolnoki, G.; Mueller Loose, S. How will mandatory nutrition and ingredient labelling affect the wine industry? A quantitative study of producers' perspectives. *Wine Econ. Pol.* **2019**, *8*, 103–113. [CrossRef]
37. Rosenzweig, T.; Skalka, N.; Rozenberg, K.; Elyasiyan, U.; Pinkus, A.; Green, B.; Stanevsky, M.; Drori, E. Red wine and wine pomace reduced the development of insulin resistance and liver steatosis in HFD-fed mice. *J. Funct. Foods* **2017**, *34*, 379–389. [CrossRef]
38. Chang, K.J.; Liz Thach, M.; Olsen, J. Wine and health perceptions: Exploring the impact of gender, age and ethnicity on consumer perceptions of wine and health. *Wine Econ. Pol.* **2016**, *5*, 105–113 [CrossRef]
39. Plunk, A.D.; Syed-Mohammed, H.; Cavazos-Rehg, P.; Bierut, L.J.; Grucza, R.A. Alcohol consumption, heavy drinking, and mortality: Rethinking the J-Shaped curve. *Alcohol. Clin. Exp. Res.* **2014**, *38*, 471–478. [CrossRef]
40. Barry, A.E.; Whiteman, S.D.; Cremeens-Matthews, J. Alcohol-Attributable Calories Consumed as a Result of Binge Drinking: A National Survey of Drinkers in the United States. *Subst. Use Misuse* **2016**, *51*, 932–936. [CrossRef]
41. Nielsen, S.J.; Kit, B.K.; Fakhouri, T.; Ogden, C.L. Calories Consumed from Alcoholic Beverages by US Adults, 2007–2010. *NCHS Data Br.* **2012**, 1–8.
42. Barry, A.E.; Merianos, A.L. Alcohol as Food/Calories. *Curr. Addict. Rep.* **2016**, *3*, 85–90. [CrossRef]
43. Sayon-Orea, C.; Martinez-Gonzalez, M.A.; Bes-Rastrollo, M. Alcohol consumption and body weight: A systematic review. *Nutr. Res.* **2011**, *69*, 419–431. [CrossRef] [PubMed]
44. Traversy, G.; Chaput, J.-P. Alcohol consumption and obesity: An update. *Curr. Obes.* **2015**, *4*, 122–130. [CrossRef] [PubMed]
45. Nijveldt, R.J.; van Nood, E.; van Hoorn, D.E.; Boelens, P.G.; van Norren, K.; van Leeuwen, P.A. Flavonoids: A review of probable mechanisms of action and potential applications. *Am. J. Clin. Nutr.* **2001**, *74*, 418–425. [CrossRef] [PubMed]
46. Kinsella, J.E.; Frankel, E.; German, B.; Kanner, J. Possible mechanisms for the protective role of antioxidants in wine and plant foods. *Food Technol.* **1993**, *47*, 85–89.
47. Renaud, S.; de Lorgeril, M. Wine, alcohol, platelets, and the French paradox for coronary heart disease. *Lancet* **1992**, *339*, 1523–1526. [CrossRef]
48. Flesch, M.; Rosenkranz, S.; Erdmann, E.; Böhm, M. Alcohol and the risk of myocardial infarction. *Basic Res. Cardiol.* **2001**, *96*, 128–135. [CrossRef]
49. Grønbæk, M. Type of Alcohol Consumed and Mortality from All Causes, Coronary Heart Disease, and Cancer. *Ann. Intern. Med.* **2000**, *133*, 411. [CrossRef]
50. De Lorimier, A.A. Alcohol, wine, and health. *Am. J. Surg.* **2000**, *180*, 357–361. [CrossRef]
51. Vecchio, R.; Decordi, G.; Grésillon, L.; Gugenberger, C.; Mahéo, M.; Jourjon, F. European consumers' perception of moderate wine consumption on health. *Wine Econ. Pol.* **2017**, *6*, 14–22. [CrossRef]
52. Albertsen, K.; Andersen, A.-M.N.; Olsen, J.; Grønbæk, M. Alcohol Consumption during Pregnancy and the Risk of Preterm Delivery. *Am. J. Epidemiol.* **2004**, *159*, 155–161. [CrossRef]
53. Kaminski, M.; Rumeau, C.; Schwartz, D. Alcohol Consumption in Pregnant Women and the Outcome of Pregnancy. *Alcohol. Clin. Exp. Res.* **1978**, *2*, 155–163. [CrossRef] [PubMed]
54. Nutt, D.J.; Rehm, J. Doing it by numbers: A simple approach to reducing the harms of alcohol. *J. Psychopharmacol.* **2014**, *28*, 3–7. [CrossRef] [PubMed]
55. Wang, H.; Fang, H.; Yu, X.; Liang, S. How real time pricing modifies Chinese households' electricity consumption. *J. Clean. Prod.* **2018**, *178*, 776–790. [CrossRef]
56. Farhat, T.; Iannotti, R.J.; Simons-Morton, B.G. Overweight, obesity, youth, and health-risk behaviors. *Am. J. Prev. Med.* **2010**, *38*, 258–267. [CrossRef]
57. WHO. *Alcohol Labelling. A Discussion Document on Policy Options*; WHO: Geneva, Switzerland, 2017.
58. Botterman, S.; De Cuyper, K.; Tresignie, C. *State of Play in the Use of Alcoholic Beverage Labels to Inform Consumers about Health Aspects Action to Prevent and Reduce Harm from Alcohol*; EU Health Programme (2008–2013); GfK Belgium, Rand Europe, European Union: Leuven, Belgium, 2014.
59. Annunziata, A.; Pomarici, E.; Vecchio, R.; Mariani, A. Nutritional information and health warnings on wine labels: Exploring consumer interest and preferences. *Appetite* **2016**, *106*, 58–69. [CrossRef]

60. Annunziata, A.; Agnoli, L.; Vecchio, R.; Charters, S.; Mariani, A. Health warnings on wine labels: A discrete choice analysis of Italian and French Generation Y consumers. *Wine Econ. Pol.* **2019**, *8*, 81–90. [CrossRef]
61. Wigg, S.; Stafford, L.D. Health Warnings on Alcoholic Beverages: Perceptions of the Health Risks and Intentions towards Alcohol Consumption. *PLoS ONE* **2016**, *11*, e0153027. [CrossRef]
62. Annunziata, A.; Pomarici, E.; Vecchio, R.; Mariani, A. European consumers' interest toward nutritional information on wine labeling: A cross-country analysis. *BIO Web Conf.* **2015**, *5*, 04003. [CrossRef]
63. Vecchio, R.; Annunziata, A.; Mariani, A. Is More Better? Insights on Consumers' Preferences for Nutritional Information on Wine Labelling. *Nutrients* **2018**, *10*, 1667. [CrossRef]
64. Cavaliere, A.; De Marchi, E.; Banterle, A. Investigation on the role of consumer health orientation in the use of food labels. *Public Health* **2017**, *147*, 119–127. [CrossRef]
65. Drichoutis, A.C.; Lazaridis, P.; Nayga Jr, R.M. Consumers' use of nutritional labels: A review of research studies and issues. *Acad. Mark. Sci. Rev.* **2006**, *2006*, 1.
66. Asioli, D.; Aschemann-Witzel, J.; Caputo, V.; Vecchio, R.; Annunziata, A.; Næs, T.; Varela, P. Making sense of the "clean label" trends: A review of consumer food choice behavior and discussion of industry implications. *Food Res. Int.* **2017**, *99*, 58–71. [CrossRef]
67. Osborn, S. 12—Labelling relating to natural ingredients and additives. In *Advances in Food and Beverage Labelling*; Berryman, P., Ed.; Woodhead Publishing: Oxford, UK, 2015; pp. 207–221. ISBN 978-1-78242-085-9.
68. Ricci, E.C.; Banterle, A.; Stranieri, S. Trust to Go Green: An Exploration of Consumer Intentions for Eco-friendly Convenience Food. *Ecol. Econ.* **2018**, *148*, 54–65. [CrossRef]
69. Sogari, G.; Corbo, C.; Macconi, M.; Menozzi, D.; Mora, C. Consumer attitude towards sustainable-labelled wine: An exploratory approach. *Int. J. Wine Bus. Res.* **2015**, *27*, 312–328. [CrossRef]
70. Pomarici, E.; Vecchio, R. Millennial generation attitudes to sustainable wine: An exploratory study on Italian consumers. *J. Clean. Prod.* **2014**, *66*, 537–545. [CrossRef]
71. Mazzocchi, C.; Ruggeri, G.; Corsi, S. Consumers' preferences for biodiversity in vineyards: A choice experiment on wine. *Wine Econ. Pol.* **2019**, *8*, 155–164. [CrossRef]
72. Delmas, M.A.; Grant, L.E. Eco-Labeling Strategies and Price-Premium: The Wine Industry Puzzle. *Bus. Soc.* **2014**, *53*, 6–44. [CrossRef]
73. Amato, M.; Ballco, P.; López-Galán, B.; De Magistris, T.; Verneau, F. Exploring consumers' perception and willingness to pay for "Non-Added Sulphite" wines through experimental auctions: A case study in Italy and Spain. *Wine Econ. Pol.* **2017**, *6*, 146–154. [CrossRef]
74. Kubota, S.; Sawano, H.; Kono, H. Japanese consumer preferences for additive-free wine labeling. *Agric. Food Econ.* **2017**, *5*, 4. [CrossRef]
75. Saltman, Y.; Johnson, T.; Wilkinson, K.; Bastian, S. Australian wine consumers' acceptance of and attitudes toward the use of additives in wine and food production. *Int. J. Wine Bus. Res.* **2015**, *7*, 83–92.
76. Dominici, A.; Boncinelli, F.; Gerini, F.; Marone, E. Consumer preference for wine from hand-harvested grapes. *Br. Food J.* **2019**, in press. [CrossRef]
77. Galati, A.; Schifani, G.; Crescimanno, M.; Migliore, G. "Natural wine" consumers and interest in label information: An analysis of willingness to pay in a new Italian wine market segment. *J. Clean. Prod.* **2019**, *227*, 405–413. [CrossRef]
78. D'Amico, M.; Di Vita, G.; Monaco, L. Exploring environmental consciousness and consumer preferences for organic wines without sulfites. *J. Clean. Prod.* **2016**, *120*, 64–71. [CrossRef]
79. De Magistris, T.; Gracia, A. The decision to buy organic food products in Southern Italy. *Br. Food J.* **2008**, *110*, 929–947. [CrossRef]
80. Fayet-Moore, F.; McConnell, A.; Kim, J.; Mathias, K.C. Identifying Eating Occasion-Based Opportunities to Improve the Overall Diets of Australian Adolescents. *Nutrients* **2017**, *9*, 608. [CrossRef] [PubMed]
81. Capitello, R.; Bazzani, C.; Begalli, D. Consumer personality, attitudes and preferences in out-of-home contexts: The case of rosé wine in Italy. *Int. J. Wine Bus. Res.* **2019**, *31*, 48–67. [CrossRef]
82. Agnoli, L.; Boeri, M.; Scarpa, R.; Capitello, R.; Begalli, D. Behavioural patterns in Mediterranean-style drinking: Generation Y preferences in alcoholic beverage consumption. *J. Behav. Exp. Econ.* **2018**, *75*, 117–125. [CrossRef]
83. Boncinelli, F.; Dominici, A.; Gerini, F.; Marone, E. Consumers wine preferences according to purchase occasion: Personal consumption and gift-giving. *Food Qual. Prefer.* **2019**, *71*, 270–278. [CrossRef]

84. Liu, J.L.; Han, B.; Cohen, D.A. Associations between eating occasions and places of consumption among adults. *Appetite* **2015**, *87*, 199–204. [CrossRef]
85. Gould, S.J. Consumer attitudes toward health and health care: A differential perspective. *J. Consum. Aff.* **1988**, *22*, 96–118. [CrossRef]
86. Michaelidou, N.; Hassan, L.M. The role of health consciousness, food safety concern and ethical identity on attitudes and intentions towards organic food. *Int. J. Consum. Stud.* **2008**, *32*, 163–170. [CrossRef]
87. Žeželj, I.; Milošević, J.; Stojanović, Ž.; Ognjanov, G. The motivational and informational basis of attitudes toward foods with health claims. *Appetite* **2012**, *59*, 960–967. [CrossRef] [PubMed]
88. Thiene, M.; Scarpa, R.; Longo, A.; Hutchinson, W.G. Types of front of pack food labels: Do obese consumers care? Evidence from Northern Ireland. *Food Pol.* **2018**, *80*, 84–102. [CrossRef]
89. Bruwer, J.; Huang, J. Wine product involvement and consumers' BYOB behaviour in the South Australian on-premise market. *Asia Pac. J. Mark. Logist.* **2012**, *24*, 461–481. [CrossRef]
90. Yoo, Y.J.; Saliba, A.J.; MacDonald, J.B.; Prenzler, P.D.; Ryan, D. A cross-cultural study of wine consumers with respect to health benefits of wine. *Food Qual. Prefer.* **2013**, *28*, 531–538. [CrossRef]
91. Thach, L. Time for wine? Identifying differences in wine-drinking occasions for male and female wine consumers. *J. Wine Res.* **2012**, *23*, 134–154. [CrossRef]
92. Van Loo, E.J.; Caputo, V.; Nayga, R.M., Jr.; Meullenet, J.-F.; Ricke, S.C. Consumers' willingness to pay for organic chicken breast: Evidence from choice experiment. *Food Qual. Prefer.* **2011**, *22*, 603–613. [CrossRef]
93. Casini, L.; Rungie, C.; Corsi, A.M. How loyal are Italian consumers to wine attributes? *J Wine Res.* **2009**, *20*, 125–142. [CrossRef]
94. Lancaster, K.J. A new approach to consumer theory. *J. Political Econ.* **1966**, *74*, 132–157. [CrossRef]
95. Manski, C.F. The structure of random utility models. *Decis* **1977**, *8*, 229–254. [CrossRef]
96. Di Castelnuovo, A.; Rotondo, S.; Iacoviello, L.; Donati, M.B.; De Gaetano, G. Meta-analysis of wine and beer consumption in relation to vascular risk. *Circulation* **2002**, *105*, 2836–2844. [CrossRef] [PubMed]
97. Waldrop, M.E.; McCluskey, J.J.; Mittelhammer, R.C. Products with multiple certifications: Insights from the US wine market. *Eur. Rev. Agric. Econ.* **2017**, *44*, 658–682. [CrossRef]
98. Brennan, C.; Gallagher, K.; McEachern, M. A review of the 'consumer interest' in organic meat. *Int. J. Consum. Stud.* **2003**, *27*, 381–394. [CrossRef]
99. Loebnitz, N.; Aschemann-Witzel, J. Communicating organic food quality in China: Consumer perceptions of organic products and the effect of environmental value priming. *Food Qual. Prefer.* **2016**, *50*, 102–108. [CrossRef]
100. Demartini, E.; Ricci, E.C.; Mattavelli, S.; Stranieri, S.; Gaviglio, A.; Banterle, A.; Richetin, J.; Perugini, M. Exploring Consumer Biased Evaluations: Halos Effects of Local Food and of Related Attributes. *Int. J. Food Syst. Dyn.* **2018**, *9*, 375–389.
101. Capitello, R.; Sirieix, L. Consumers' Perceptions of Sustainable Wine: An Exploratory Study in France and Italy. *Economies* **2019**, *7*, 33. [CrossRef]
102. Hemmerling, S.; Canavari, M.; Spiller, A. Preference for naturalness of European organic consumers. *Br. Food J.* **2016**, *118*, 2287–2307. [CrossRef]
103. Fotopoulos, C.; Krystallis, A.; Ness, M. Wine produced by organic grapes in Greece: Using means—End chains analysis to reveal organic buyers' purchasing motives in comparison to the non-buyers. *Food Qual. Prefer.* **2003**, *14*, 549–566. [CrossRef]
104. Magnusson, M.K.; Sjödén, P.; Åberg, L.; Koivisto Hursti, U.; Arvola, A. Attitudes towards organic foods among Swedish consumers. *Br. Food J.* **2001**, *103*, 209–227. [CrossRef]
105. Zanoli, R.; Naspetti, S. Consumer motivations in the purchase of organic food. *Br. Food J.* **2002**, *104*, 643–653. [CrossRef]
106. Schleenbecker, R.; Hamm, U. Consumers' perception of organic product characteristics. A review. *Appetite* **2013**, *71*, 420–429. [CrossRef] [PubMed]
107. Govindasamy, R.; DeCongelio, M.; Bhuyan, S. An Evaluation of Consumer Willingness to Pay for Organic Produce in the Northeastern U.S. *J. Food Prod. Mark.* **2006**, *11*, 3–20. [CrossRef]
108. Abouab, N.; Gomez, P. Human contact imagined during the production process increases food naturalness perceptions. *Appetite* **2015**, *91*, 273–277. [CrossRef] [PubMed]
109. Evans, G.; de Challemaison, B.; Cox, D.N. Consumers' ratings of the natural and unnatural qualities of foods. *Appetite* **2010**, *54*, 557–563. [CrossRef]

110. Cheung, T.T.L.; Junghans, A.F.; Dijksterhuis, G.B.; Kroese, F.; Johansson, P.; Hall, L.; De Ridder, D.T.D. Consumers' choice-blindness to ingredient information. *Appetite* **2016**, *106*, 2–12. [CrossRef]
111. Yeomans, M.R. Alcohol, appetite and energy balance: Is alcohol intake a risk factor for obesity? *Physiol. Behav.* **2010**, *100*, 82–89. [CrossRef]
112. Contini, C.; Boncinelli, F.; Romano, C.; Scozzafava, G.; Casini, L. Price vector issue in a choice experiment: A methodological proposal. *Food Qual. Prefer.* **2019**, *75*, 23–27. [CrossRef]
113. Cummings, R.G.; Taylor, L.O. Unbiased value estimates for environmental goods: A cheap talk design for the contingent valuation method. *Am. Econ. Rev.* **1999**, *89*, 649–665. [CrossRef]
114. Murphy, J.J.; Stevens, T.; Weatherhead, D. Is Cheap Talk Effective at Eliminating Hypothetical Bias in a Provision Point Mechanism? *Environ. Resour. Econ.* **2005**, *30*, 327–343. [CrossRef]
115. Mueller, S.; Osidacz, P.; Francis, I.L.; Lockshin, L. Combining discrete choice and informed sensory testing in a two-stage process: Can it predict wine market share? *Food Qual. Prefer.* **2010**, *21*, 741–754. [CrossRef]
116. Train, K.E. *Discrete Choice Methods with Simulation*; Cambridge University Press: Cambridge, UK, 2009.
117. Scarpa, R.; Drucker, A.G.; Anderson, S.; Ferraes-Ehuan, N.; Gomez, V.; Risopatron, C.R.; Rubio-Leonel, O. Valuing genetic resources in peasant economies: The case of 'hairless' creole pigs in Yucatan. *Ecol. Econ.* **2003**, *45*, 427–443. [CrossRef]
118. Peschel, A.O.; Grebitus, C.; Steiner, B.; Veeman, M. How does consumer knowledge affect environmentally sustainable choices? Evidence from a cross-country latent class analysis of food labels. *Appetite* **2016**, *106*, 78–91. [CrossRef] [PubMed]
119. Ruto, E.; Garrod, G.; Scarpa, R. Valuing animal genetic resources: A choice modeling application to indigenous cattle in Kenya. *Agric. Econ.* **2008**, *38*, 89–98. [CrossRef]
120. Sogari, G.; Pucci, T.; Aquilani, B.; Zanni, L. Millennial Generation and Environmental Sustainability: The Role of Social Media in the Consumer Purchasing Behavior for Wine. *Sustainability* **2017**, *9*, 1911. [CrossRef]
121. Vandenbroele, J.; Vermeir, I.; Geuens, M.; Slabbinck, H.; Kerckhove, A.V. Nudging to get our food choices on a sustainable track. *Proc. Nutr. Soc.* **2019**, *78*, 1–14. [CrossRef]
122. List, J.A.; Shogren, J.F. Price information and bidding behavior in repeated second-price auctions. *Am. J. Agric. Econ.* **1999**, *81*, 942–949. [CrossRef]

nutrients

MDPI

Article

Consumer Evaluation of the Role of Functional Food Products in Disease Prevention and the Characteristics of Target Groups

Brigitta Plasek [1], Zoltán Lakner [1], Gyula Kasza [2] and Ágoston Temesi [1,*]

1 Department of Food Economics, Faculty of Food Science, Szent István University, Villányi str. 29-43, 1118 Budapest, Hungary; plasek.brigitta@etk.szie.hu (B.P.); lakner.zoltan@etk.szie.hu (Z.L.)

2 Directorate for Food Safety Risk Assessment, National Food Chain Safety Office, Keleti Károly str. 24, 1024 Budapest, Hungary; kaszagy@nebih.gov.hu

* Correspondence: temesi.agoston@etk.szie.hu; Tel.: +36-1-305-7100 (ext. 6178)

Received: 2 December 2019; Accepted: 23 December 2019; Published: 26 December 2019

check for updates

Abstract: Our research explores the methods consumers would consider using in order to prevent non-communicable diseases, with consuming functional food products being one of these methods. Previous research has pointed out the importance of missing information such as which diseases worry consumers and what they would use to avoid them. We conducted a personal survey with 1027 people in Hungary about 13 diseases and four prevention methods. We analyzed the results with descriptive statistical methods, binary logistic regression, and random forest. According to our results, the highest proportion of worried respondents think it is justified to consume functional foods for the prevention of digestive problems, a weakened immune system, and high cholesterol level. Our results help to characterize the target group for these three diseases. Completed education plays a key role in choosing functional foods to prevent a weakened immune system. Those with tertiary education are the most likely to choose this prevention method. With the other two diseases, age played a crucial role. All age groups over 36 would be more likely to choose functional foods to prevent digestive problems, whereas in case of high cholesterol level, the 36–50 age group would be more likely to apply this method.

Keywords: consumer research; willingness to purchase; non-communicable diseases; binary logistic regression; random forest

1. Introduction

Changes in lifestyle, less physical activity, and improper nutrition have resulted in the spread of non-communicable diseases to the extent that today they pose serious problems, and a significant ratio of deaths are connected to them worldwide [1]. Globally, 41 million people die from non-communicable diseases annually, which constitutes 71% of all deaths. Of these, about 15 million people are in the 30–69 age group. Cardiovascular diseases pose the biggest risk (17.9 million deaths) followed by cancer-related diseases (9.0 million), respiratory diseases (3.9 million), and diabetes (1.6 million) [2].

In Hungary, non-communicable diseases present a host of problems for society. Based on the statistical data of World Health Organization (WHO), it was concluded that in Hungary the average number of years spent in good health was 67.4 years in 2015, and the average life expectancy was 75.9 years [3]. This means that the average Hungarian spends 8–9 years suffering from some disease. Statistical data show that in Hungary, cardiovascular diseases, malignant tumors, chronic respiratory diseases, diabetes, and some other non-communicable diseases accounted for 94% of all deaths in 2016 [4].

Medicaments offer a classic solution to treat diseases. Recently, as a result of regulations, medicaments have become more readily available for consumers, so today they have an array of non-prescription products at their disposal that they can apply to cure themselves [5]. Dietary supplements also offer a widely-used solution; however, these pills play a role in the prevention of diseases already. A survey has shown that elderly consumers consider dietary supplements at least as important as prescription medicine. The same research also concluded that the most common reasons for taking dietary supplements were health protection, disease prevention, and the advice of a health expert [6]. Over the last decades, consumption of dietary supplements has grown exponentially; as Starr [7] demonstrated, in the United States, 4000 products were sold in 1994, and this number grew to more than 90,000 by 2014.

Apart from lifestyle changes and the consumption of dietary supplements, the consumption of functional foods can also reduce the risk of developing diseases. "A food can be regarded as 'functional' if it is satisfactorily demonstrated to affect beneficially one or more target functions in the body, beyond adequate nutritional effects in a way that is relevant to either an improved state of health and well-being and/or reduction of risk of disease." (p. S26, [8]). The different health claims manufacturers often use to communicate the positive effects of functional foods on health to consumers fall under European Food Safety Authority regulations. The specific regulation is included in regulation (EC) no. 1924/2006 [9]; the EFSA database makes all applicable health claims and nutrition claims available. The EFSA distinguishes different types of claims in the category of health claims, such as "Function health claims" (Article 13 claims), "Risk reduction claims" (Article 14(1)(a)), and "Claims referring to children's development" (Article 14(1)(b)) [10]. For the purposes of our research, one type of health claim, namely, risk reduction claims, are worth highlighting as they communicate the effect on the reduction of the risk of diseases [10].

In their research in Hungary, Szakály et al. came to the straightforward conclusion that "In the current situation, there is no other choice than to bring the public's attention to food products that possess an extra nutritional advantage" (p. 16, [11]). Owing to their ingredients, functional food products can help to prevent a variety of chronic diseases [12]. Encouragingly, several studies have indicated that due to increasing consumer concerns related to health, consumers would be willing to change their dietary habits and include products that would have a positive effect on their health in one way or another [13,14]. From the results of their focus group interviews, Szakály et al. concluded that if participants "have to choose between medication and functional food products, all of them would vote for the latter" (p. 23 [15]).

Due to the complexities of developing functional foods, it is important to assess which diseases consumers are concerned about [16]. However, apart from identifying these diseases, it is also important to examine how consumers would prevent them. Without this information we cannot talk about functional foods developed for, and communicated to, a wide range of consumers, and we cannot expect functional foods to have a bigger role in diets. However, if we are aware of the specific diseases that a wide circle of consumers would choose to prevent with the help of functional foods, then the affected food manufacturers could concentrate on these diseases, and a more remarkable public health effect could be achieved.

At the moment it is not yet known which diseases consumers would seek to reduce the risk of through the consumption of health-protecting food products [17,18]. As Siró et al. put it, "It is more interesting, however, to identify those diseases that consumers would make much account of preventing them by nutrition" (p. 462 [17]). We believe that this missing piece of information is what is needed to get a more complete picture of consumer behavior regarding functional foods. Kraus [19] came closest to answering the question with his research where the potential health effects of functional food products were ranked. Baba et al. [20] maintained that health-related consumer concerns may be a crucial factor in food consumption, and thus also in developing new products. Consumer attitudes to functional food products in light of worrisome diseases have already been investigated in general terms. In accordance with previous research [21,22], Chen's [23] research results showed that those who were concerned about modern lifestyle diseases displayed a more positive attitude towards functional food products and had a larger likelihood of accepting and using them in order to prevent these diseases.

The spread of non-communicable diseases poses numerous societal and economic risks. Nevertheless, it is still an unanswered question as to which prevention methods consumers would use to avoid/treat specific diseases, so the aim of the research was to answer this question by asking them.

2. Materials and Methods

In our research, we conducted a survey with personal interviews (PPI) with 1027 participants between 4–19 March 2013, at busy transport hubs of five big cities in Hungary, namely in Budapest, Pécs, Sopron, Nyíregyháza, and Szeged. The respondents were rewarded with a small non-food gift for participating in the survey. At the beginning of the survey, participants were informed about the aim of the research; at the same time, the responses were anonymous. Before completing the survey, respondents gave implicit consent to their answers being recorded and later analyzed; the study was conducted in accordance with the Declaration of Helsinki. Respondents had the option to refuse to answer any question or stop answering the survey at any point.

We employed quota sampling, as opposed to probability sampling, so our sample significantly follows the segmentation of society. Among our respondents, compared to the population, the ratio of people between ages 36–50 and >50 is smaller, and the ratio of those living in Budapest and having tertiary education is higher.

In the survey, we asked participants questions in relation to 13 diseases: which are the ones they are not concerned about; the ones they are concerned about but would not make a financial sacrifice to prevent them; the ones they are concerned about and would make a financial sacrifice to prevent them. Moreover, supposing these diseases can be avoided/prevented through various methods, we wanted to find out exactly how much respondents would spend. Demographic variables included gender, age, place of living, highest education, estimated income status of the household, and the presence of underage children in the household. Finally, our questions included whether it was important for the respondents to do something for their health, whether they paid attention to their diet, and who did the grocery shopping in the household.

After collapsing the results, we used the statistics software SPSS 25.0 to analyze the answers. The collected data were first analyzed through descriptive statistical methods, followed by binary logistic regression to find the differences based on demographic variables.

In the course of the regression analysis, we considered only respondents who were concerned and would make a financial sacrifice to avoid the disease under investigation. In the models employed, the dependent variable was the financial expense of the prevention method (purchasing functional food (yes/no or Y/N)) chosen in order to avoid the specified disease, and the independent variables were the characteristics of the respondents, presented in Table 1. In our analyses, we applied the 5% significance level ($p < 0.05$) and the forward Wald method.

Table 1. Grouping of respondents based on variables (n = 1027).

Characteristics of Survey Respondents		%
Gender	Female	53.5%
	Male	46.5%
Age brackets	<25	47.7%
	25–35	18.8%
	36–50	14.2%
	50<	19.3%
Place of living	Budapest	39%
	Another city	48.6%
	Village	12.4%

Table 1. *Cont.*

Characteristics of Survey Respondents		%
Highest accomplished qualification	Primary school/vocational school	15.1%
	Secondary technical school/secondary school diploma	40.0%
	Tertiary education	44.9%
Subjective estimation of income position	Below average	35.5%
	Average	53.6%
	Above average	10.9%
Presence of a child under 18 in the household	Yes	79.5%
	No	20.5%
Overweight or underweight status of the respondent (based on BMI) *	Undernourished	6.8%
	Normal	61.3%
	Obese/overweight	31.9%
It is important for me to do something for my health	Important	88.9%
	Less important/not important	11.1%
I pay attention to healthy eating	Important	72.8%
	Less important/not important	27.2%
Grocery shopper	Respondent	25.7%
	Another person	22.7%
	Shared	51.6%

* Based on World Health Organization (WHO) body mass index (BMI) [24].

The principle of triangulation provided another perspective from which our study approached the research questions related to the identification of diseases that consumers would prevent with the help of functional food products, and the socio-demographic features that have a stronger influence on this choice.

As Meerza et al. (p. 29, [25]) summarized it: "binomial logistic regression generally works well as a classifier when the relationship between dependent and independent variables is linear and the data are relatively balanced between classes [26], the random forest (RF), a machine learning method, outperforms the logistic regression when the non-linear characteristics and complex interactions among predictor variables exist [27]".

With the aim of forecasting consumer choice, we ran additional statistical analysis on the data. Random forest is a method widely used for making predictions, with the advantages of being wieldy, robust, and fast [28]. We used the random forest package of the R statistical software to examine the extent to which consumer choice can be predicted. The analysis was conducted by including all the examined variables. "Random forests are a combination of tree predictors such that each tree depends on the values of a random vector sampled independently and with the same distribution for all trees in the forest" (p. 5, [29]). During the analysis, we used the "OOB Estimate of Error Rate" value to examine accuracy, and the "MeanDecreaseGini" values to examine the importance of the analyzed variables. With each analysis, 70% of the total sample was used for training and 30% was used to validate the models.

3. Results

3.1. Identifying the Diseases That Worry Respondents

One of the questions we searched an answer for in our research was to identify the non-communicable diseases that concern consumers the most. Table 2 shows the ratios of the respondents concerned about a specific disease and their propensity to make a monetary sacrifice to avoid it.

Table 2. The attitude of respondents towards different health problems and their propensity to spend money to decrease the probability of their occurrence.

Disease	Concerns, and the Respondent Is Ready to Spend Money to Prevent It	Concerns, but the Respondent Does Not Want to Spend Money to Prevent It	Does Not Concern Respondents
Cancer	57%	21%	22%
Cardiovascular diseases	49%	28%	23%
Weakened immune system	48%	24%	28%
Diabetes	42%	24%	34%
Osteoporosis	38%	26%	36%
Joint diseases	37%	31%	32%
Skin diseases	33%	28%	39%
Digestive problems	33%	33%	34%
High level of cholesterol	32%	33%	35%
Respiratory diseases	32%	32%	36%
Memory lapse/concentration problems	32%	30%	38%
Migraine	26%	27%	47%
Unbalanced mood/apnoea	21%	35%	44%

According to our results, surveyed consumers are most concerned about cancer-related diseases (78%). This same disease is the one with the highest ratio of respondents willing to make a financial sacrifice. A high ratio of respondents would also make a monetary sacrifice to prevent cardiovascular diseases, a weakened immune system, and diabetes. Surveyed consumers are least concerned about migraines and unbalanced mood.

3.2. *The Form of Financial Sacrifice Made Towards Preventing Diseases*

Our research also aimed at finding out what exactly, if anything, worried respondents would spend on. The four factors we examined were taking medicine, using dietary and nutritional supplements, consuming functional foods, and lifestyle changes. Table 3 summarizes the results of the question of what would worried respondents, who are willing to make a financial sacrifice, spend to avoid a given disease. Respondents were free to indicate more than one answer. In the table we included only the answers of respondents who were worried and willing to make a financial sacrifice to avoid the disease. Thus, the displayed percentages for the prevention methods do not emerge from the whole sample, but from a smaller part of it, which we consider to be 100% here.

Our results show that in order to prevent diseases, most people say that they would implement lifestyle changes. Respondents ascribe the biggest importance to it in the case of cardiovascular diseases, which means that consumers would apply this method to prevent the disease which carries the biggest noncommunicable disease (NCD) risk in Hungary, based on data by WHO [4]. Thus, we can also conclude that consumers are aware of the connection between lifestyle and the emergence of cardiovascular diseases.

Table 3. The form of financial sacrifice worried respondents are willing to make.

	Concerned, Ready to Make a Financial Sacrifice to Prevent It		Form of Financial Sacrifice			
			Medicine	Dietary Supplement	Functional Food	Lifestyle Changes
	n	%	%	%	%	%
Digestive problems	327	100%	13%	29%	43%	55%
Weakened immune system	473	100%	22%	36%	39%	56%
High cholesterol level	322	100%	12%	19%	37%	64%
Diabetes	418	100%	36%	17%	35%	59%
Cancer	552	100%	47%	19%	31%	48%
Osteoporosis	376	100%	38%	24%	30%	40%
Cardiovascular disease	480	100%	25%	15%	29%	70%
Joint diseases	368	100%	30%	26%	27%	48%
Skin diseases/eczema	325	100%	47%	14%	26%	37%
Memory lapses/concentration	317	100%	25%	26%	26%	46%
Mood disorders/apnoea	211	100%	12%	21%	24%	66%
Respiratory diseases	311	100%	39%	15%	21%	47%
Migraine	253	100%	43%	11%	21%	44%

The main aim of our research, however, was to identify the diseases that most people are likely to prevent with the consumption of functional food products. To the question raised by Siró et al. [17], namely, what are the diseases that consumers would prevent by choosing functional foods, our results yielded the following answer: digestive problems, a weakened immune system, and high cholesterol level. These are the three diseases for which consumers would choose functional foods in the highest proportion. If we combine this result with the disease that worries respondents the most, we can conclude that functional food products can play the biggest role in the prevention of a weakened immune system for a wide circle of consumers. It should also be added that functional foods can play an important role in preventing many other diseases apart from the three highlighted in our study, in the case of which a significant reluctance of consumers has to be overcome.

In his research, Kraus [19] ranked the potential health effects of functional food products according to their importance. Consistent with our results, strengthening the immune system and lowering the risk of cancer-related diseases also ranked on top of his list, and improving memory ranked lower.

3.3. Disease Prevention with Functional Foods—Consumers' Characteristics

We considered it important to utilize the characteristics of respondents to identify the groups that can be most associated with specific prevention methods. Table 4 displays the results derived from the binary logistic regression. The table shows only the diseases and characteristics of respondents for which significant differences emerged between the groups within a certain variable. The analyses included all the variables from Table 1. The odds ratios received from the regression analysis show the values compared to the first group of the variable, that is, we can see the results compared to females in the case of gender; compared to the 18–25 age group in the case of age; to the primary/vocational school educated in the case of education; and a different town or village compared to Budapest in the case of place of living. Regarding the question about the person who does the shopping, the basis of comparison is the case when the respondent does the shopping; to this we compare the case when someone else does it or when it is a shared responsibility. In the case of the statement "I pay attention to healthy eating", we compare those for whom it is important to those for whom it is not.

Table 4. Results of the binary logistic regression.

Number of Cases	Characteristics of the Respondent (Independent Variable)	Level of Attribute	Would Buy Functional Food to Avoid the Specified Disease (Dependent Variable)						
			B	SE	Wald	Sig.	Exp(B)	95% CI for EXP (B)	
								Lower	Upper
			High cholesterol level						
289	Constant		−0.752	0.189	15.773	0.000	0.471		
	Age	36–50	0.919	0.346	7.056	0.008	2.508	1.273	4.942
			Cancer-related diseases						
498	Constant		−0.745	0.371	4.024	0.045	0.475		
	Gender	Men	−0.469	0.208	5.108	0.024	0.626	0.416	0.940
	Level of education	College/university	0.992	0.356	7.749	0.005	2.697	1.341	5.423
	Grocery shopper	Another person	−0.862	0.296	8.498	0.004	0.422	0.236	0.754
			Skin diseases						
293	Constant		−0.455	0.251	3.301	0.069	0.634		
	Grocery shopper	Another person	−0.743	0.362	4.205	0.040	0.476	0.234	0.968
		Shared	−0.797	0.321	6.170	0.013	0.451	0.240	0.845
			Memory lapse/concentration problems						
284	Constant		−1.871	0.483	15.006	0.000	0.154		
	Age	25–35	0.875	0.359	5.933	0.015	2.399	1.186	4.852
	Level of education	College/university	1.004	0.494	4.132	0.042	2.728	1.037	7.181
			Digestion problems						
291	Constant		−1.103	0.396	7.758	0.005	0.332		
	Age	36–50	1.055	0.362	8.478	0.004	2.872	1.412	5.844
		50+	0.940	0.332	8.013	0.005	2.559	1.335	4.906

Table 4. *Cont.*

Number of Cases	Characteristics of the Respondent (Independent Variable)	Level of Attribute	Would Buy Functional Food to Avoid the Specified Disease (Dependent Variable)						
			B	SE	Wald	Sig.	Exp(B)	95% CI for EXP (B)	
								Lower	Upper
			Migraine						
224	Constant		−0.927	0.197	22.183	0.000	0.396		
	Gender	Men	−1.129	0.376	9.022	0.003	0.323	0.155	0.675
			Diabetes						
375	Constant		−1.247	0.240	26.923	0.000	0.287		
	Age	36–50	0.997	0.330	9.121	0.003	2.710	1.419	5.177
	Place of living	Another city	0.509	0.252	4.074	0.044	1.664	1.015	2.729
		Village	0.984	0.357	7.588	0.006	2.674	1.328	5.385
			Joint diseases						
329	Constant		0.118	0.402	0.086	0.770	1.125		
	Level of education	Secondary technical school/secondary school diploma	−0.981	0.408	5.783	0.016	0.375	0.168	0.834
	Grocery shopper	Another person	−0.956	0.379	6.351	0.012	0.384	0.183	0.809
			Osteoporosis						
333	Constant		−1.495	0.345	18.827	0.000	0.224		
	Gender	Men	−0.518	0.254	4.161	0.041	0.596	0.362	0.980
	Age	36–50	1.087	0.347	9.816	0.002	2.966	1.502	5.856
	I pay attention to healthy eating	Important	0.875	0.345	18.827	0.012	2.398	1.209	4.758
			Weakened immune system						
431	Constant		−1.881	0.439	18.325	0.000	0.152		
	Gender	Men	−0.472	0.213	4.925	0.026	0.624	0.411	0.946
	Age	36–50	0.938	0.314	8.916	0.003	2.554	1.380	4.727
	Level of education	College/university	1.355	0.381	12.656	0.000	3.875	1.837	8.174
	I pay attention to healthy eating	Important	0.616	0.266	5.358	0.021	1.852	1.099	3.121

The table in Appendix A includes the goodness of fit test of models and the indicators that characterize the explanatory power of the models. According to the results of the Hosmer–Lemeshow test, the goodness of fit of the models is appropriate, and apart from one instance, none of the values is significant. However, based on the two other indicators (Cox–Snell R^2 and Nagelkerke R^2), the considered independent demographic variables only partially explain the dependent variables. The Cox–Snell R^2 values were around 0.05 on average, and the Nagelkerke R^2 values were around 0.07. From the small values we can conclude that apart from the demographic variables, other factors also influence whether a consumer chooses the prevention method under consideration in order to avoid diseases.

Based on the results displayed in Table 4, males consider the consumption of functional food as relevant prevention methods to a lesser extent in the prevention of cancer, migraine, osteoporosis, and a weakened immune system. Thus, we can emphasize that in the case of these diseases, it is important that companies address their prevention-related communication primarily at women. This result is in line with previous literature [30–34], which state that women are more open to functional food; at the same time, our results provide a more nuanced picture in that they identify the diseases for the prevention of which functional food development should be targeted to women.

For six of the 13 examined diseases, we obtained a significant result related to the effect of age. The results show that compared to the 18–25 age group, it was the 36–50 age group that yielded significant differences in the highest number. This age group considered functional food products more important in the prevention of high cholesterol level, diabetes, osteoporosis, a weakened immune system, and digestion problems. For this last problem, buying and consuming functional foods can be an important prevention method also for the 50+ age group, whereas for memory and concentration problems, the 25–35 age group may consider this method. This result of ours clarifies the statements of previous literature [17,31,32,34,35], according to which it is the middle-aged and the elderly rather than the young who would consume such products.

It is also important to highlight the aspect of the person who does the grocery shopping, as for three diseases, we obtained a significant result in this respect. For all of the three diseases, namely, cancer-related diseases, skin diseases, and joint problems, it is a shared finding that respondents in the household of whom someone else does the shopping consider the consumption of functional foods a less relevant prevention method than respondents who themselves do the shopping.

Table 4 also includes significant results related to the highest level of completed education of respondents. The responses of those with a higher level of education were compared to those that were primary/vocational school educated. In accordance with results of previous research literature [17,30–33], those with a college/university education consider the consumption of functional foods more suitable for the prevention of certain diseases. Those with a college/university education consider the consumption of functional foods a suitable solution for the prevention of cancer-related diseases, memory and concentration problems, and a weakened immune system, whereas for the prevention of joint problems, it is rather those with a secondary technical school/secondary school diploma who would choose this method.

For certain diseases, place of living or the emphasis on a healthy diet proved to be an important factor. Place of living gained importance with diabetes. Respondents living in the capital consider the potential functional foods less suitable for its prevention. The importance of a healthy diet may play a main role in the choice of products aimed at preventing a weakened immune system or osteoporosis.

Of all the diseases, we obtained the highest exp (B) value related to a weakened immune system. Our results show that for the prevention of this disease, it is respondents with a higher level of education who consider the consumption of functional foods the most suitable method.

We can make sense of the results in Tables 2–4 collectively. Comparing Tables 2 and 3, we can highlight a weakened immune system, diabetes, and digestive problems as diseases that worry a significant number of respondents (Table 2), for the prevention of which a higher ratio of the worried would also choose functional foods (Table 3). Adding in the results of the binary logistic regression

(Table 4) with the characteristics of respondents, the consumer segment which is more likely to consume functional foods in order to prevent specific diseases can be properly outlined.

Thus, for the prevention of a weakened immune system, functional foods are considered the most suitable by women, by those with a higher level of education, by the 36–50 age group, and by those for whom a healthy diet is important. If a product manufacturer would like to communicate the prevention of a weakened immune system—an aim towards which, according to our results, 39% of those concerned and willing to spend would purchase functional food—to its future consumers, then it is important for them to assess how to reach consumers with these characteristics. For the prevention of diabetes, it is the 36–50 age group and those living in a place other than the capital who would give an important role to functional foods. In connection with digestive problems, the over 36 age group can be highlighted, as they consider the consumption of functional foods a relevant prevention method for this disease.

3.4. Random Forest with the Aim of Identifying Factors that Influence Consumer Decision

The results were also analyzed with the random forest method. Our aim was to identify the demographic variables that most influence the decisions of consumers who would choose functional food products to prevent diseases. The analysis was conducted by including all 10 variables. In Table 5, only three variables are presented, namely those with the highest MeanDecreaseGini value. The importance of the other variables is summarized in Appendix B.

The method worked with the lowest error rate in the case of mood disorders/apnoea, migraine, joint diseases, and respiratory diseases for the grouping of respondents and the prediction of their choice of functional food products to prevent diseases. During the statistical analysis, we examined the importance of the 10 variables summarized in Table 1 with the help of MeanDecreaseGini. The age and education of the consumer have a stronger influence during prediction. Age was more important than the other factors for 12 of the 13 examined diseases, whereas in one case, education was more important. Consumers' place of living emerged among the first three most important influencing factors for nine diseases, whereas in two cases, consumers' BMI rating and perception of income, and in one case, the identity of the primary grocery shopper, were among the most important influencing factors. Contrary to previous results [30–33], which found that an important factor in the consumption of functional foods is the gender of the consumer, in our analysis this factor was not among the top three factors influencing prediction for any of the diseases.

During the triangulation process, we also compared the results obtained via the random forest method with those of the regression analysis in order to arrive at a more accurate and justified result in terms of the most relevant target group for disease prevention through functional foods. The ExpB values of the regression analysis and the MeanDecreaseGini values of the random forest analysis complete each other such that the value of importance indicates the most influential factor in the prediction of consumer choice, while the odds ratio makes it more precise.

This information helps to identify the factors worth emphasizing for individual diseases. In the case of weakened immune system, education is the most important factor for making predictions, and based on the regression analysis, those with tertiary education can be an important target group.

In terms of the age of the consumer, five diseases can be highlighted among those for which age was the most important factor when making predictions. During product development and the preparation of marketing communication, companies should focus on the 36–50 age group for the prevention of high cholesterol level, digestive problems, diabetes, and osteoporosis, the 25–35 age group for the prevention of memory lapses and concentration problems and the 36 or older age groups for digestive problems.

Table 5. Results obtained with the help of random forest.

Disease	OOB Estimate of Error Rate	Relative Importance	
			MeanDecreaseGini
Cardiovascular disease	33.66%	age	23.52
		grocery shopper	19.92
		education	19.40
High cholesterol level	40.89%	age	27.12
		education	22.02
		subjective estimation of income position	21.54
Cancer	38.11%	age	21.88
		education	20.85
		place of living	19.60
Mood disorders/Apnoea	26.15%	age	20.15
		place of living	17.57
		education	17.19
Respiratory diseases	31.15%	age	20.74
		education	18.90
		place of living	18.22
Skin diseases, eczema	33.38%	age	24.67
		education	19.89
		place of living	19.89
Memory lapses, concentration	34.08%	age	22.18
		education	19.78
		place of living	18.56
Digestive problems	46.73%	age	28.10
		place of living	24.55
		education	23.90
Migraine	28.09%	age	20.34
		education	16.70
		BMI category	16.57
Diabetes	45.48%	age	24.66
		place of living	22.16
		subjective estimation of income position	21.48
Joint diseases	30.38%	age	23.96
		education	21.71
		BMI category	20.18
Osteoporosis	36.68%	age	24.53
		education	19.66
		place of living	19.32
Weakened immune system	39.64%	education	23.12
		age	23.00
		place of living	21.40

Notes: 70% of the sample was used to train, 30% was used to validate the models; accuracy, sensitivity and specificity values of models can be found in Appendix C.

4. Discussion

In our study we investigated which of the most dangerous non-communicable diseases worry consumers the most and also what exactly people would spend on if they are concerned about a disease and would make a monetary sacrifice to avoid it. We examined 13 diseases and four prevention methods.

The conclusion of our research is that most consumers are aware of the risks of non-communicable diseases and they are worried about them, and, depending on the disease, only a small group of the concerned is not willing to make a financial sacrifice to avoid them.

Our results show that with most diseases, lifestyle change is the most favored prevention method of consumers, followed by, depending on the disease, the consumption of functional foods or dietary/nutritional supplements. Except for a few cases, medication as a treatment method is the least sought-after, and the most divisive option for consumers.

Lifestyle changes are the most preferred methods for most of the diseases, such as for the prevention of cardiovascular diseases and unbalanced mood/apnoea and a high level of cholesterol, while dietary supplements are chosen to prevent a weakened immune system. It is important to point out that with several diseases, the smallest ratio of respondents chose dietary supplements. Further results indicate that consumers are the most concerned about cancer-related diseases. This is the only disease where, besides lifestyle changes, consumers consider taking medicaments as the best option.

Consumers did not consider the consumption of health-protecting food products the best solution in any of the cases. Even if not the best option when trying to prevent/treat diseases, it is still a relevant choice when trying to avoid digestion problems, a weakened immune system, high level of cholesterol, diabetes, and cancer. In his research, Kraus [19] drew up an order of importance regarding the health effects of a product. Some of his results coincide with the order of importance we arrived at. Strengthening the immune system and the prevention of cancer-related diseases as motivating factors for the consumption of functional foods also appeared at the top position in that research, whereas improving memory was ranked lower.

In order to utilize the characteristics of respondents to identify the groups that can be most associated with the specific prevention methods, we conducted binary logistic regression. In the course of the regression analysis, we investigated the 10 examined variables to identify the groups we can emphasize within each variable for each disease. Altogether we obtained multiple significant results in relation to age (the 36–50 age group), completed education (college/university), place of living, and the primary grocery shopper of the household.

Based on the analysis done with the random forest statistical method, we can state that the choice made by consumers participating in the survey can be predicted with an error rate of 26%–47%. Among the examined variables, age and education showed greater importance; also, with some diseases, place of living, BMI rating, and the identity of the primary grocery shopper in the household played an important role in the prediction of decisions.

In accordance with the principle of triangulation, we examined our aim from several perspectives. Comparing the results of the regression and random forest, we can more firmly state that the completed education of respondents plays a bigger role in the prediction of the choice for functional food products in the prevention of a weakened immune system; more specifically, this prevention method is relevant for those with tertiary education. Age plays a more important role in predicting whether respondents purchase functional foods or not to prevent high cholesterol level, digestive problems, diabetes, osteoporosis, memory lapses and concentration problems. For the first four diseases, the 36–50 age group is the relevant target, for memory lapses and concentration problems the 25–35 age group is the relevant target and for digestive problems, the 36 or older age group is the relevant target.

Numerous literature sources [17,30–35] emphasize the important role of gender, age, and completed education as key demographic factors influencing the consumption of functional foods. However, by looking at such products from a different angle, apart from age and completed education, the results of our research also confirm the effects of other influencing factors, such as place of living, BMI rating, and the perception of income. Moreover, our results also show that in relation to disease prevention, the gender of the consumer does not have a significant effect on choosing functional foods.

These new pieces of information are filling a long-standing information gap. Our results help functional food manufacturers with the identification of diseases for which the effects of their products are worth emphasizing, and with the identification of consumer target groups The results of the research are useful for experts working in the field of product development and for marketing communication professionals as well. We recommend that both functional food developmental research and manufacturing companies' research and development (R&D) departments, as well as their marketing experts devote their resources to developing and communicating about products helping to prevent diseases that have the highest consumer acceptance. It would also help to facilitate all of this if this viewpoint—that the planned functional food development should aim at preventing a disease that consumers would like to prevent by consuming functional foods—was included as a basis for the assessment of international or national grants. This is how we can expect functional food consumption to increase significantly and play a more important role in improving society's health.

Limitations

This research is the very first step towards answering a complex question. Due to the limited scope of the analysis, some of the steps in this research were aimed to be simple. Future research should include the analysis of the threat of multi-collinearity as well.

Although the current health conditions of the participants and their family can influence their answer to the given questions, we did not address this problem. Future research could focus on this aspect as well.

The answer we have received is not a general one, as Hungary is a small country in Europe. For a firmer, more widely accepted answer, international research might be useful.

Author Contributions: Conceptualization, Á.T.; methodology, Z.L., B.P., and Á.T.; formal analysis, B.P.; investigation, G.K. and Á.T.; writing—original draft preparation, B.P. and Á.T.; visualization, B.P.; supervision, Á.T. and Z.L.; funding acquisition, G.K. and Z.L. All authors have read and agreed to the published version of the manuscript.

Funding: The project is supported by the European Union and co-financed by the European Social Fund (grant agreement no. EFOP-3.6.3-VEKOP-16-2017-00005).

Conflicts of Interest: The authors declare no conflicts of interest.

Appendix A

Table A1. Goodness of fit tests of model.

Would Buy Functional Food to Avoid the Specific Disease	Hosmer–Lemeshow Test (p-Value)	Cox–Snell R^2	Nagelkerke R^2
High cholesterol level	1.000	0.027	0.036
Cancer-related diseases	0.880	0.059	0.082
Skin diseases	1.000	0.023	0.033
Memory lapse/concentration problems	0.892	0.064	0.093
Digestion problems	0.951	0.074	0.099
Migraine	-	0.044	0.069
Diabetes	0.014	0.052	0.071
Joint disease	0.991	0.040	0.057
Osteoporosis	0.997	0.073	0.102
Weakened immune system	0.913	0.102	0.138

Appendix B

Table A2. Further results obtained with the help of random forest.

Disease	OOB Estimate of Error Rate	Relative Importance	
			MeanDecreaseGini
Cardiovascular disease		Place of living	18.98
		Gender	13.41
		Presence of a child under 18 in the household	9.98
		Subjective estimation of income position	18.95
		It is important for me to do something for my health	7.73
		I pay attention to healthy eating	10.20
		BMI category	19.34
High cholesterol level		Place of living	20.95
		Gender	14.04
		Presence of a child under 18 in the household	11.63
		Grocery shopper	18.62
		It is important for me to do something for my health	8.50
		I pay attention to healthy eating	12.51
		BMI category	20.21
Cancer		Gender	11.85
		Presence of a child under 18 in the household	11.74
		Grocery shopper	18.59
		Subjective estimation of income position	19.83
		It is important for me to do something for my health	7.71
		I pay attention to healthy eating	11.61
		BMI category	18.90
Mood disorders/apnoea		Gender	9.29
		Presence of a child under 18 in the household	9.91
		Grocery shopper	15.55
		Subjective estimation of income position	14.61
		It is important for me to do something for my health	7.06
		I pay attention to healthy eating	9.45
		BMI category	16.66

Table A2. *Cont.*

Disease	OOB Estimate of Error Rate	Relative Importance	
			MeanDecreaseGini
Respiratory diseases		Gender	12.54
		Presence of a child under 18 in the household	11.13
		Grocery shopper	17.81
		Subjective estimation of income position	17.54
		It is important for me to do something for my health	8.09
		I pay attention to healthy eating	11.02
		BMI category	16.42
Skin diseases/eczema		Gender	13.10
		Presence of a child under 18 in the household	11.26
		Grocery shopper	17.96
		Subjective estimation of income position	16.37
		It is important for me to do something for my health	7.68
		I pay attention to healthy eating	11.05
		BMI category	19.29
Memory lapses/ concentration		Gender	12.15
		Presence of a child under 18 in the household	11.65
		Grocery shopper	17.54
		Subjective estimation of income position	17.64
		It is important for me to do something for my health	8.31
		I pay attention to healthy eating	10.98
		BMI category	17.87
Digestive problems		Gender	15.24
		Presence of a child under 18 in the household	11.90
		Grocery shopper	22.28
		Subjective estimation of income position	22.65
		It is important for me to do something for my health	9.65
		I pay attention to healthy eating	11.22
		BMI category	21.93

Table A2. *Cont.*

Disease	OOB Estimate of Error Rate	Relative Importance	
			MeanDecreaseGini
Migraine		Place of living	16.51
		Gender	12.04
		Presence of a child under 18 in the household	9.34
		Grocery shopper	15.10
		Subjective estimation of income position	15.43
		It is important for me to do something for my health	7.33
		I pay attention to healthy eating	10.01
Diabetes		Gender	13.82
		Presence of a child under 18 in the household	12.48
		Grocery shopper	20.31
		Subjective estimation of income position	21.48
		It is important for me to do something for my health	9.21
		I pay attention to healthy eating	12.44
		BMI category	19.70
Joint diseases		Place of living	18.70
		Gender	13.39
		Presence of a child under 18 in the household	11.81
		Grocery shopper	17.88
		Subjective estimation of income position	17.87
		It is important for me to do something for my health	8.88
		I pay attention to healthy eating	11.68
Osteoporosis		Gender	12.90
		Presence of a child under 18 in the household	11.75
		Grocery shopper	18.58
		Subjective estimation of income position	17.93
		It is important for me to do something for my health	8.80
		I pay attention to healthy eating	10.15
		BMI category	17.36
Weakened immune system		Gender	13.69
		Presence of a child under 18 in the household	12.29
		Grocery shopper	20.62
		Subjective estimation of income position	20.45
		It is important for me to do something for my health	9.98
		I pay attention to healthy eating	12.27
		BMI category	19.13

Appendix C

Table A3. Accuracy, sensitivity, and specificity of random forest models.

Disease	Accuracy	Sensitivity	Specificity
Cardiovascular disease	0.6861	0.9227	0.1011
High cholesterol level	0.6246	0.8557	0.1944
Cancer	0.6214	0.88835	0.08738
Mood disorders/apnoea	0.7184	0.97778	0.02381
Respiratory diseases	0.7152	0.93191	0.02703
Skin diseases/eczema	0.7023	0.94619	0.06977
Memory lapses/concentration	0.6634	0.8711	0.1071
Digestive problems	0.5372	0.7596	0.2143
Migraine	0.7379	0.9619	0.0137
Diabetes	0.6375	0.8702	0.1584
Joint diseases	0.6828	0.9178	0.1111
Osteoporosis	0.6204	0.8820	0.1354
Weakened immune system	0.5761	0.8939	0.1385

References

1. Diepeveen, S.; Ling, T.; Suhrcke, M.; Roland, M.; Marteau, T.M. Public acceptability of government intervention to change health-related behaviours: A systematic review and narrative synthesis. *BMC Public Health* **2013**, *13*, 756. [CrossRef] [PubMed]
2. WHO. Noncommunicable Diseases. 2018. Available online: https://www.who.int/news-room/fact-sheets/detail/noncommunicable-diseases (accessed on 14 November 2019).
3. WHO. *World Health Statistics 2016: Monitoring Health for the SDGs, Sustainable Development Goals*; WHO Press: Geneva, Switzerland, 2016.
4. WHO. World Health Organization—Noncommunicable Diseases (NCD) Country Profiles 2018: Hungary. Available online: https://www.who.int/nmh/countries/2018/hun_en.pdf?ua=1 (accessed on 14 November 2019).
5. Smith, F.; Newbould, J. Perspectives of community groups on the reclassification of medicines from prescription to non-prescription. *Int. J. Health Promot. Educ.* **2002**, *40*, 21–30. [CrossRef]
6. Albright, C.L.; Schembre, S.M.; Steffen, A.D.; Wilkens, L.R.; Monroe, K.R.; Yonemori, K.M.; Murphy, S.P. Differences by race/ethnicity in older adults' beliefs about the relative importance of dietary supplements vs prescription medications: Results from the SURE Study. *J. Acad. Nutr. Diet.* **2012**, *112*, 1223–1229. [CrossRef] [PubMed]
7. Starr, R.R. Too little, too late: Ineffective regulation of dietary supplements in the United States. *Am. J. Public Health* **2015**, *105*, 478–485. [CrossRef] [PubMed]
8. Diplock, A.T.; Aggett, P.J.; Ashwell, M.; Bornet, F.; Fern, E.B.; Roberfroid, M.B. Scientific concepts of functional foods in Europe: Concensus document. *Br. J. Nutr.* **1999**, *81* (Suppl. 1), S1–S27.
9. Regulation (EC) No 1924/2006 of the European Parliament and of the Council. Available online: https://eur-lex.europa.eu/legal-content/EN/TXT/PDF/?uri=CELEX:02006R1924-20141213 (accessed on 18 November 2019).
10. European Commission. Health Claims. Available online: https://ec.europa.eu/food/safety/labelling_nutrition/claims/health_claims_en (accessed on 18 November 2019).
11. Szakály, Z.; Kiss, M.; Jasák, H. Funkcionális Élelmiszerek, Fogyasztói Attitűdök És Személyre Szabott Táplálkozás (Functional Foods, Consumer Attitudes and Personalized Nutrition). *Táplálkozásmarketing* **2014**, *1*, 3–17.
12. Alkhatib, A.; Tsang, C.; Tiss, A.; Bahorun, T.; Arefanian, H.; Barake, R.; Khadir, A.; Tuomilehto, J. Functional foods and lifestyle approaches for diabetes prevention and management. *Nutrients* **2017**, *9*, 1310. [CrossRef]
13. Prättälä, R. Dietary changes in Finland—Success stories and future challenges. *Appetite* **2003**, *41*, 245–249. [CrossRef]

14. Niva, M. 'All foods affect health': Understandings of functional foods and healthy eating among health-oriented Finns. *Appetite* **2007**, *48*, 384–393. [CrossRef]
15. Szakály, Z.; Berke, S.; Biacs, P.; Szakály, S.; Széles, G.; Szente, V.; Szigeti, O. Funkcionális élelmiszerek piaci helyzetének vizsgálata= Market Analysis of Functional Foods. *OTKA Kutatási Jelentések|OTKA Res. Rep.* **2010**. Available online: http://real.mtak.hu/12130/1/69336_ZJ1.pdf (accessed on 23 December 2019).
16. Van Kleef, E.; Van Trijp, H.C.M.; Luning, P. Functional foods: Health claim-food product compatibility and the impact of health claim framing on consumer evaluation. *Appetite* **2005**, *44*, 299–308. [CrossRef]
17. Siró, I.; Kápolna, E.; Kápolna, B.; Lugasi, A. Functional food. Product development, marketing and consumer acceptance—A review. *Appetite* **2008**, *51*, 456–467. [CrossRef]
18. Bagchi, D.; Nair, S. (Eds.) *Developing New Functional Food and Nutraceutical Products*; Academic Press: Cambridge, MA, USA, 2016.
19. Kraus, A. Development of functional food with the participation of the consumer. Motivators for consumption of functional products. *Int. J. Consum. Stud.* **2015**, *39*, 2–11. [CrossRef]
20. Baba, Y.; Kallas, Z.; Costa-Font, M.; Gil, J.M.; Realini, C.E. Impact of hedonic evaluation on consumers' preferences for beef attributes including its enrichment with n-3 and CLA fatty acids. *Meat Sci.* **2016**, *111*, 9–17. [CrossRef] [PubMed]
21. Devcich, D.A.; Pedersen, I.K.; Petrie, K.J. You eat what you are: Modern health worries and the acceptance of natural and synthetic additives in functional foods. *Appetite* **2007**, *48*, 333–337. [CrossRef] [PubMed]
22. Petrie, K.J.; Sivertsen, B.; Hysing, M.; Broadbent, E.; Moss-Morris, R.; Eriksen, H.R.; Ursin, H. Thoroughly modern worries: The relationship of worries about modernity to reported symptoms, health and medical care utilization. *J. Psychosom. Res.* **2001**, *51*, 395–401. [CrossRef]
23. Chen, M.F. The mediating role of subjective health complaints on willingness to use selected functional foods. *Food Qual. Prefer.* **2011**, *22*, 110–118. [CrossRef]
24. Body Mass Index—BMI. Available online: http://www.euro.who.int/en/health-topics/disease-prevention/nutrition/a-healthy-lifestyle/body-mass-index-bmi (accessed on 11 January 2019).
25. Meerza, S.I.A.; Yiannaka, A.; Brooks, K.R.; Gustafson, C.R. Information avoidance behavior: Does ignorance keep us uninformed? In Proceedings of the Annual Meeting—Agricultural and Applied Economics Association, Atlanta, Georgia, 21–23 July 2019; Available online: https://ageconsearch.umn.edu/record/290757 (accessed on 23 December 2019). [CrossRef]
26. Muchlinski, D.; Siroky, D.; He, J.; Kocher, M. Comparing random forest with logistic regression for predicting class-imbalanced civil war onset data. *Political Anal.* **2016**, *24*, 87–103. [CrossRef]
27. Cutler, D.R.; Edwards, T.C., Jr.; Beard, K.H.; Cutler, A.; Hess, K.T.; Gibson, J.; Lawler, J.J. Random forests for classification in ecology. *Ecology* **2007**, *88*, 2783–2792. [CrossRef]
28. Gupta, A.; Gupta, G. Comparative study of random forest and neural network for prediction in direct marketing. In *Applications of Artificial Intelligence Techniques in Engineering*; Springer: Singapore, 2019; pp. 401–410. [CrossRef]
29. Breiman, L. Random forests. *Mach. Learn.* **2001**, *45*, 5–32. [CrossRef]
30. De Jong, N.; Ocke, M.C.; Branderhorst, H.A.; Friele, R. Demographic and lifestyle characteristics of functional food consumers and dietary supplement users. *Br. J. Nutr.* **2003**, *89*, 273–281. [CrossRef]
31. Urala, N. *Functional Foods in Finland: Consumers' Views, Attitudes and Willingness to Use*; VTT Publications: Espoo, Finland, 2005.
32. Kraus, A.; Annunziata, A.; Vecchio, R. Sociodemographic factors differentiating the consumer and the motivations for functional food consumption. *J. Am. Coll. Nutr.* **2017**, *36*, 116–126. [CrossRef] [PubMed]
33. Meyerding, S.; Kürzdörfer, A.; Gassler, B. Consumer preferences for superfood ingredients—The case of bread in Germany. *Sustainability* **2018**, *10*, 4667. [CrossRef]
34. Verneau, F.; La Barbera, F.; Furno, M. The role of health information in consumers' willingness to pay for canned crushed tomatoes enriched with Lycopene. *Nutrients* **2019**, *11*, 2173. [CrossRef] [PubMed]
35. Bimbo, F.; Bonanno, A.; Nocella, G.; Viscecchia, R.; Nardone, G.; De Devitiis, B.; Carlucci, D. Consumers' acceptance and preferences for nutrition-modified and functional dairy products: A systematic review. *Appetite* **2017**, *113*, 141–154. [CrossRef]

nutrients

MDPI

Article

Trading on Food Quality due to Changes in Prices: Are There Any Nutritional Effects?

Cesar Revoredo-Giha [1,*], Faical Akaichi [1] and Neil Chalmers [2]

1 Food Marketing Research, Rural Economy, Environment and Society Department, Scotland's Rural College, West Mains Road, Edinburgh EH9 3JG, UK; faical.akaichi@sruc.ac.uk
2 Rowett Institute of Health and Nutrition, University of Aberdeen, Ashgrove Rd W, Aberdeen AB25 2ZD, UK; neil.chalmers@abdn.ac.uk
* Correspondence: cesar.revoredo@sruc.ac.uk

Received: 13 November 2019; Accepted: 18 December 2019; Published: 20 December 2019

check for updates

Abstract: The analysis of changes in prices is not only important because they directly affect households' affordability and, therefore, their food security but also because they may trigger changes in the composition of their food and drink choices. Thus, an increase in prices may force a household with limited resources to choose a bundle of goods with lower prices that substitute their original choices and are probably of lower quality. This paper considers the situation of each UK country and the implications that trading down in quality within a food and drink category has on nutrition. Two motivations to pursue these analyses are to explore the sort of substitutions that households do within a category due to an increase in prices and, in the UK leaving the European Union (Brexit) context, the impact that an increase in food prices may have on nutrition. After computing estimates for trading down for each country for the period 2007–2014, we regress the annual rate of change by nutrient with respect to the annual trading in quality for six food qualities that are major contributors of fat, sugar and salt to the diet. The results indicate that trading down in quality occurs in most of the studied categories and countries, and when households trade down, they move to products with worse nutritional quality. This points out the need to keep improving the quality of products through reformulation, ensure that consumers are well informed of nutritional quality of products and monitor the effect of changes in prices.

Keywords: food choice; changes in quality; changes in prices; nutrition; United Kingdom; Brexit

1. Introduction

Product quality can be described as a bundle of characteristics (attributes) that determine a product's performance. Major categories of food product quality attributes include food safety (e.g., levels of microbial pathogens and residues), nutritional, value (e.g., compositional integrity and taste), package and process (e.g., animal welfare and environmental impact) attributes [1].

Changes in food prices have the possibility to modify not only the quantity consumed of a particular product but also the quality of the products that are chosen, for instance, within a food category [2,3].

The United Kingdom (UK) Department of Food, Environment and Rural Affairs (Defra) produces, as part of its regular publication "Family Food" [4,5], a calculation of the changes in quality due to changes in prices (i.e., trading up/down in quality). In a study considering the years 2007 to 2014, it showed that when considering several food and drink categories, trading down in quality was a very common pattern in UK households together and first income decile households. In fact, their calculation showed that households traded down significantly in 15 out of 22 food and drink categories in the case of all the UK households and 14 out of 22 food and drink categories for the poorest income decile.

It is important to note that Defra's 'trading down in quality' definition considers that it "means switching to purchases of cheaper products within a food grouping [i.e., a category]. Cheaper is equivalent to lower quality in some way. The quality reduction could occur in any quality attribute of the product, e.g., packaging, brand name, provenance, nutrient content, taste. "Trading down into a completely different food type is not captured" [4,5]. This, however, does not need to be the case as it depends on the attributes behind the idea of quality. For instance, if cheaper products are less tasteful than the more expensive ones but they are also healthier due to a better nutritional composition (e.g., lower sugar content), then consumers probably are not worse off by the substitution from a more expensive product to a cheaper one. In other terms, the relationship between trading down in quality and nutrition is an empirical question, as the change in the composition of the basket might not have any nutritional implication.

The contribution of this paper and its aim is to explore what the implications of increases in food and drink prices are in terms of nutrition; more specifically, when consumers trade-down in quality, do they choose products that are worse off in nutritional quality? Moreover, the analysis of this question is done considering the four UK countries (i.e., England, Northern Ireland, Scotland and Wales), which may have different reactions to changes in prices.

There are several reasons for exploring the aforementioned question; one is to explore the effect of changes in purchases made after recession, which has radically changed consumers' purchase behaviour to the point that the new market situation has been described as "age of thrift" [6]. Another reason to explore the nutritional effect of price increases is that it has become a constant concern in the context of the UK leaving the European Union (Brexit), e.g., due to a depreciation of the British Pound (GBP) or new trading conditions imply higher import costs for the country. Thus, potential price rises due to Brexit have been mentioned by several sources [7–9]. Johnson et al. [10] pointed out that the Bank of England have estimated that 60 per cent of the fall in the GBP was passed through higher prices to firms and consumers and that a higher percentage was expected in the long term. In this context, where healthy foods have been portrayed as relatively expensive [11,12] and the UK self-sufficiency rate is 58 per cent in vegetables and 11 per cent in fruit-price increases will certainly have an impact on consumers' healthy diet.

As mentioned, Defra's calculations of trading in quality were done for all the UK households and those on the first income decile; however, the availability of disaggregated data behind the Family Food publication from the UK Data Archive allows extending Defra's analysis by carrying it out by UK country (i.e., England, Northern Ireland, Scotland and Wales). These additional computations are interesting in the Brexit context given the potential differences by country.

The structure of the paper is as follows: it starts with a literature review that briefly summarises the literature of consumers' quality choice and prices. The next section presents the methodology and the data used for the analysis. It is then followed by the discussion of the results, i.e., the estimation of the trading in quality by UK country and their impact on nutrition. Finally, conclusions from the analysis are stated.

2. Literature Review

The analysis of price changes is important not only due to the fact that they directly affect households' affordability and, therefore, their food security, but also because they may trigger changes in the composition of food and drink choices, which have an impact on the quality of the products they buy.

The discussion of changes in prices and their effect on quality changes has been extensively discussed in the demand analysis literature as regards work done with family budget data related to the estimation of price elasticities [2,13–15].

As pointed out by Theil [15], most "commodities" distinguished in budget enquiries are heterogeneous; often they can be described as "aggregates" of more perfectly homogeneous "qualities." A consequence of this well-known fact is that an increase in the expenditures on a certain commodity brought about by, e.g., an increase in income may be caused partly by an increase in the quantity bought (measured in physical units), partly by a transition to a better quality.

In household surveys reporting both expenditures and physical quantities, it is possible to divide one by the other to obtain unit values. These unit values suggest that there is substantial spatial variation in prices; however, it is not possible to use unit values as direct substitutes for true market prices in the analysis of demand patterns. Consumers choose the quality of their purchases, and unit values reflect this choice. Moreover, quality choice may itself reflect the influence of prices as consumers respond to price changes by altering both quantity and quality [2].

McKelvey [3] reported that when estimating demand for a heterogeneous commodity group that is comprised of many different individual goods, commonly used demand system estimation techniques may produce erroneous estimates of the price elasticity of quantity demanded. If quality response to price is ignored, estimated price elasticities of quantity demand will conflate responses on the quantity and quality margins. Policy makers in many countries are considering, or have imposed, taxes to cut intake of unhealthy items like fats and sugar-sweetened beverages. Proponents of taxes assume that quantity demand is fairly responsive to price. The empirical estimates of demand elasticities that applied economists have published over several decades undoubtedly contribute to this view. Yet evidence in many of these countries is from household surveys, with budget shares that vary with the choice of quantity and quality. If quality response to price is ignored, estimated price elasticities of quantity demand include quality responses and overstating likely effects of taxes.

The source of this problem is that it has become increasingly common to use commodity group budget share as the dependent variable when estimating demand systems. When regressing budget share on price, there are two pathways through which price can impact budget share. Again, taking meat as our example, one possibility is that when the price of meat increases, the household might decide to decrease the quantity of meat purchased, thus reducing the budget share devoted to meat. Another possibility is that the household may substitute to cheaper cuts of meat, which would also reduce the budget share devoted to meat. Either way, one would observe a negative correlation between meat price and meat budget share, and this regression provides no way to determine whether this correlation is due to a response on the quantity margin or a response on the quality margin. Moreover, confusing these two responses may result in perverse policy decisions, since taxing a commodity group whose expenditure is responsive to price changes could decrease the quantity consumed or it could instead cause consumers to substitute to lower quality versions of the commodity.

As regards the relationship between prices and nutrition, Huang [16] linked food choice with nutritional status in the context of the classical demand framework by developing a procedure to measure nutrient availability by way of demand elasticities for food items from a traditional demand analysis. Although it is possible to find the relationship between changes in prices and the change in nutrients using Huang's procedure, it does not provide an aggregated index of quality such as in the Deaton and McKelvey's method [2,3]. In what remains of the paper, their index will be used to explore whether changes in prices had any impact on the nutritional quality of the diet.

3. Materials and Methods

3.1. Methods

The starting point of the methodology is based on McKelvey [3], who studied the relationships between unit value and the quality demanded on commodity groups. He follows Deaton [2] in assuming that commodities are divided into groups, each consisting of a number of consumer goods;

for instance, if the group is confectionery, it will comprise chocolates, candies. The expenditure for group c is given by:

$$E_c = p_c{\cdot}q_c \tag{1}$$

where p_c and q_c are vectors containing the price and quantity of each consumption good in the group. The price vector is then decomposed as:

$$p_c = \overline{p}_c{\cdot}p_c^* \tag{2}$$

where p_c^* is a vector capturing the relative price of commodities within the group and $\overline{p}_c$ is a scalar capturing the component of price that is common to all the commodities in the group (e.g., a Laspeyres price index). In this analysis $\overline{p}_c$ is understood as a price index for the category c.

Replacing (2) into (1), one obtains:

$$E_c = \overline{p}_c{\cdot}p_c^*{\cdot}q_c \tag{3}$$

Multiplying and dividing the right hand side by the total quantity purchased, i.e., $Q_c = 1{\cdot}q_c$ and replacing $v_c = \frac{p_c^*{\cdot}q_c}{Q_c}$, (3) becomes:

$$E_c = \overline{p}_c{\cdot}Q_c{\cdot}v_c \tag{4}$$

As explained by McKelvey [3], v_c is a measure of the quality of group c purchases. If the relative prices are constant, a higher v_c means that the chosen q_c is more expensive per unit of group c consumption and this is interpreted as buying a higher quality of group c goods. However, note that the term quality is used in vague way, as it is not associated with any particular attribute.

Expressing (4) as discrete changes from period t-1 to t we obtain (5):

$$\left(1+\hat{E}_{ct}\right) = \left(1+\hat{\overline{p}}_{ct}\right){\cdot}\left(1+\hat{Q}_{ct}\right){\cdot}\left(1+\hat{v}_{ct}\right) \tag{5}$$

where '^' indicate the relative change with respect to the previous period. From (5), it is clear that to keep the expenditure constant after an increase in the category price, a consumer can reduce its total quantity purchase or trade down the quality of the purchases or a combination of both. If $\hat{v}_{ct}$ is negative, the consumer is trading down in quality.

The next step once $\hat{v}_{ct}$ is computed from (5) is to explore to what extent the change in quality as explained above is associated with a change of nutrients. This is an empirical question, as the cheaper products in a category might not be least nutritional ones. To do this, we considered the following pooled regression based on Suits [17]:

$$\hat{N}_{ijt} = \alpha + \alpha_{Eng} + \alpha_{NI} + \alpha_{Sc} + \alpha_{W} + \beta{\cdot}\hat{v}_{ct} \tag{6}$$

where $\hat{N}_{ijt}$ is the relative change of nutrient i in country j in time period t; α and β are the intercept and slope of the regression, respectively, and α_{Eng}, α_{NI}, α_{Sc} and α_W are parameters based on dummies (i.e., that take the value of one if the observation is for the country and zero otherwise) that measure the deviations with respect to α for the cases of the four UK countries, namely England (Eng), Northern Ireland (NI), Scotland (Sc) and Wales (W).

The most important parameter of regression (6) is the slope β, which indicates the reaction of the nutrient (i.e., how it changes in relative terms) to a trading in quality. Note that if the coefficient of the slope is negative and statistically significant (using a t test to check whether the slope is different than zero), it means that trading down in quality (which is a negative relative change in the real unit value) increases the nutrients. In contrast, if the slope is positive, trading down in quality decreases the demand for nutrients. Of course, if the slope is not statistically significant, it means that trading in quality is unrelated to nutrition.

3.2. Data Used on the Analysis

The data available for the computation of the trading up/down in quality are disaggregated information from the Defra's "Family Food" publication for each of the four UK countries. These data provide information about weekly per capita expenditures and purchases [4,5].

The prices used for the categories were from the Office for National Statistics (CNS) Retail Price Indices for the studied groups (base 1987 = 100), which match the groups of the Family Food statistics.

The information about nutrients per food category was taken from disaggregated data underpinning the "Family Food" publication and can be found in the UK Data Archive as part of the Living Costs and Food Surveys. These nutrient intakes are calculated from food purchases using nutrient composition data supplied by Public Health England (PHE) [4,5]. The majority of the data are from PHE's nutrient analysis programme, supplemented by values from manufacturers and retailers. The nutrient conversion excludes inedible parts of purchased foods, such as fish heads, banana peels; it assumes all food is eaten. Intakes from dietary supplements are not included in any of the tables. The nutrients categories considered in the analysis were energy (i.e., calories), macronutrients (i.e., fat, carbohydrates and proteins), saturated fats, Non-milk extrinsic (NME) sugars, sodium, total Sugars and cholesterol.

In terms of the categories used for the analysis, these included the following Defra's Family Food categories. Note that the aggregation of the groups is due to the fact that the retail price indices are only available on an aggregated way:

Food and drink: which was made of: milk and milk products excluding cheese, cheese, carcase meat, non-carcase meat and meat products, fish, eggs, fats, sugar and preserves, fresh and processed fruit and vegetables, including potatoes, bread, flour, cakes, buns and pastries, biscuits and crispbreads, other cereals and cereal products, beverages, other food and drink, soft drinks, confectionery and alcoholic drinks.

Biscuits and cakes: chocolate biscuits, sweet biscuits (not chocolate) and cereal bars, cream crackers and other unsweetened biscuits and crispbread.

Confectionery: chocolate bars—solid, chocolate bars—filled, chewing gum, mints and boiled sweets, fudges, toffees, caramels and takeaway confectionery.

Processed potato products: chips—frozen or not frozen, takeaway chips, instant potato, canned potatoes, crisps and potato snacks and other potato products, frozen or not frozen.

Processed vegetables (excluding processed potatoes): tomatoes, canned or bottled, peas, canned, beans canned, baked beans in sauce, other canned beans and pulses, other canned vegetables, dried pulses other than air-dried, air-dried vegetables, vegetable juices and purees, tomato puree and vegetable purees, vegetable juices, e.g., tomato juice, carrot juice, peas, frozen, beans, frozen, ready meals and other vegetable products (including takeaways), ready meals and other vegetable products, frozen or not frozen, all vegetable takeaway products and other frozen vegetables.

Soft drinks: concentrated (not low calorie), not concentrated (not low calorie), concentrated (low calorie) and not concentrated (low calorie).

Note that the cases of biscuits and cakes, confectionery and soft drinks are part of the discretionary food categories as defined by the Food Standard Scotland (FSS); these are categories not necessary for nutrition and which are contributors of sugar, saturated fats and salts [18]. Thus, according to the FSS, these foods have a significant impact on the diet, accounting for, on average, about one-fifth of total calories, total fat and saturated fats and over half of daily free sugars consumption. Moreover, the FSS concluded that a key step towards meeting dietary goals would be to reduce the intake of these foods.

4. Results and Discussion

This section presents two sets of results from the analysis: first, the trading in quality considering for all the selected categories, the changes in prices between 2007 and 2014 and the reaction in terms of change in expenditure, change in quantities purchases and change in the composition of the purchases (i.e., trading in quality). Second, the results of the regressions between the change in nutrients per year between 2007 and 2014 (i.e., 28 observations, this is seven growth changes multiplied by four UK countries) and the trading in quality. It is important to mention that the reason why this discussion does not compare the results with other results from the UK or elsewhere is due to the fact that to our knowledge of this type of analysis (i.e., the relationship between nutrients and trading down in quality) has not been studied before.

4.1. Trading in Quality by Food Category

Figures 1–6 present the trading in quality analyses by UK country by category. For the aggregated food and drink category (Figure 1), the reaction to the increase in prices is very similar in all countries. It increases in expenditure, decreases in quantities (except in Northern Ireland) and all of them show trading down in quality.

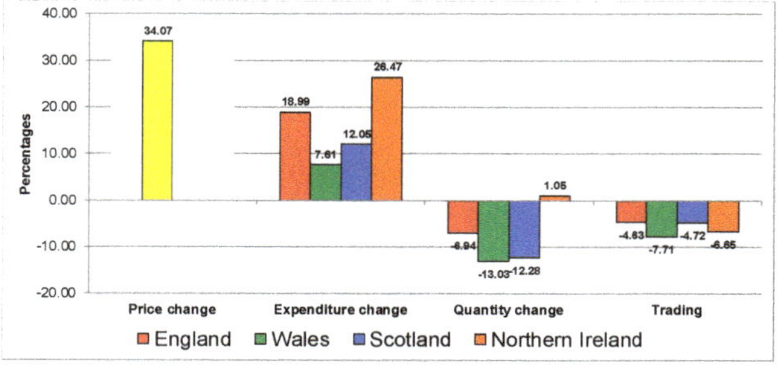

Figure 1. Trading in quality on the food and drink category in UK 2007–2014.

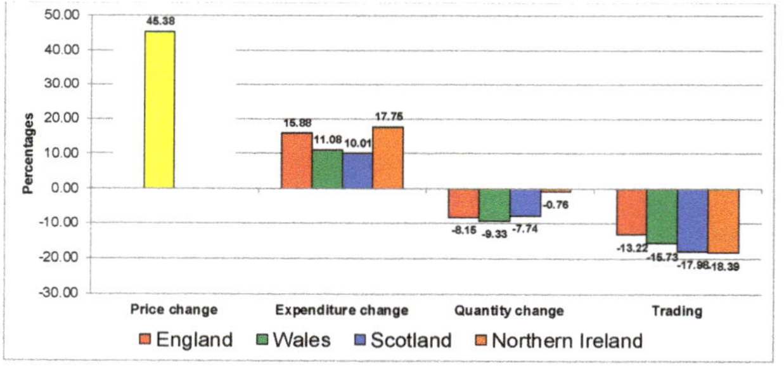

Figure 2. Trading in quality on the biscuits and cake category in UK 2007–2014.

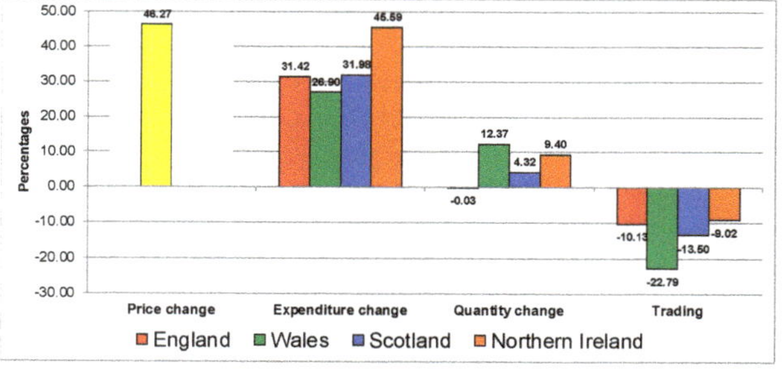

Figure 3. Trading in quality on the confectionery category in UK 2007–2014.

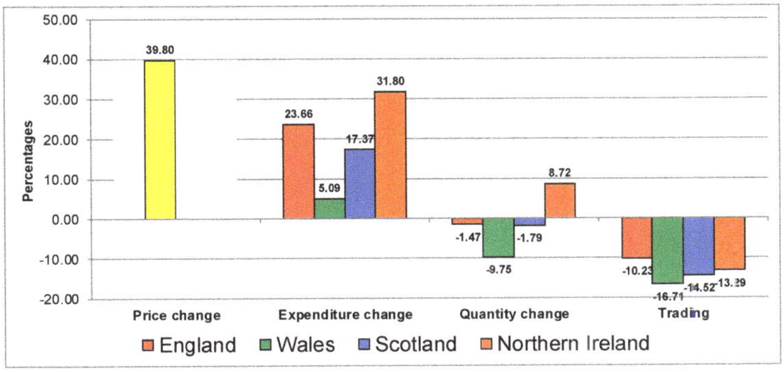

Figure 4. Trading in quality on the processed potatoes category in UK 2007–2014.

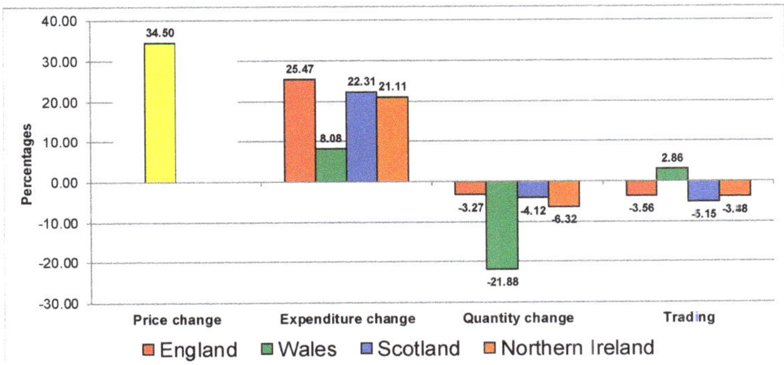

Figure 5. Trading in quality on the processed vegetables category in UK 2007–2014.

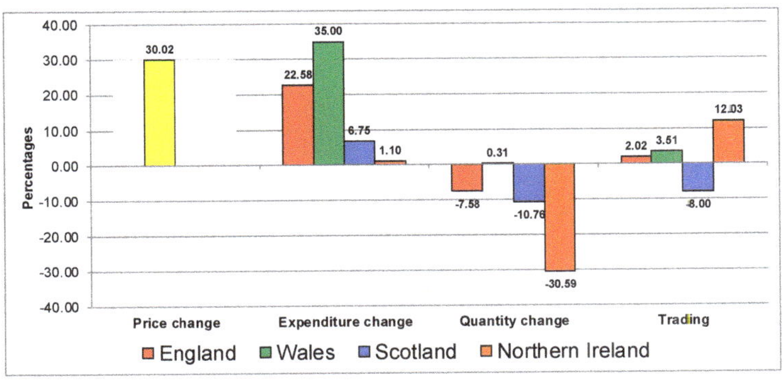

Figure 6. Trading in quality on the soft drinks category in UK 2007–2014.

Figure 2 shows the response for biscuits and cakes category. The response to the increase in prices on the category was similar in all the countries, characterised by increases in expenditure, decreases in quantities and by trading down in quality.

Figure 3 shows the results for the confectionery category, which is similar to the biscuits and cakes category. The response to the increase in prices on the category was similar in all countries, characterised by increases in expenditure, decreases in quantities purchases (England's quantity remained the same) and all countries trading down in quality.

The increase in the price of processed potatoes (Figure 4) was followed by an increase in expenditure, decrease in quantity in all countries except in Northern Ireland and trading down in all the UK countries.

Figure 5 shows that increases in prices of processed vegetables was followed up on average by an increase in expenditure, a decrease in quantity (particularly in Wales) and trading down in most countries except in Wales.

The results to an increase in the price of soft drinks was more varied (Figure 6). It showed an increase in in expenditure; quantities decreased in all countries except Wales, which remained almost the same. All countries except Scotland traded up in quality.

4.2. Nutritional Effect of Trading in Quality

Given the prevalence of trading down in quality in all the UK countries and in all the studied food and drink categories, it is important to analyse whether by purchasing cheaper products within a category, households are moving to a poorer nutritional quality basket—this is done in the next section.

This section presents the results of the regressions between nutrients and trading in quality. The selected nutritional items were energy (calories); macronutrients: fat, carbohydrates and proteins; nutrients that are harmful in excess: saturated fatty acids, non-milk extrinsic sugars and sodium; total sugars and cholesterol.

The relative change of each nutrient in each country was regressed with respect to the relative trading in quality. Since both the dependent and independent variable are in relative change over time, the coefficient can be interpreted as an elasticity (i.e., the reaction of the quantity consumed of a particular nutrient due to a change in the category trading in quality index).

Tables 1–6 present the results for all the studied food and drink categories. Note that each line of the Tables is one regression, i.e., each nutrient with respect to the trading in quality. In addition, recall that the most important coefficient is β, which indicates the percentage change in a nutrient due to a percentage change in trade in quality index. If the value of β is negative, it means that trading down in quality generates an increase in the nutrient consumption.

As shown in Table 1, the UK country dummies representing the deviations to the mean coefficients were not statistically significant. However, the value of β was statistically significant in all the regressions fluctuating from −0.7 (sodium) to −1.37 (NME sugars). The results indicate that trading down in quality increases all the nutrients on the food and drink categories. On the negative side, the results show that increases in the prices of the category encourage consumers to reduce their outlay by substituting within the category towards the uptake of products that are higher in saturated fats, sugar and salt.

The above results are of importance because the latest information National Diet and Nutrition Survey (NDNS) [19] collected from 2014 to 2016 show that sugar makes up 13.5 per cent of 4–10 year-olds and 14.1% of 11–18 year-olds daily calorie intake respectively when the official recommendation is to limit sugar to no more than 5%. Moreover, the survey confirms the UK population continues to consume too much saturated fat and not enough fruit, vegetables and fibre. Thus, the average saturated fat intake for adults (i.e., 19- to 64-year-olds) is 12.5 per cent of daily calorie intake, above the 11 per cent recommended maximum. The results from Table 1 indicate that an increase in food and drink prices may bring product substitutions within the food and drink category that may deteriorate even more the nutritional situation of the UK population.

Table 1. Food and drink—regressions between nutrients and trading in quality.

Nutrient	Regression Coefficients						Intercept Dummies by Country												R^2
	α	*t*-Stat.	Sig	β	*t*-Stat.	Sig	England	*t*-Stat.	Sig	Wales	*t*-Stat.	Sig	Scotland	*t*-Stat.	Sig	*n.* Ireland	*t*-Stat.	Sig	
Energy	−1.82	−2.58	*	−0.90	−4.81	*	0.37	0.46		−1.17	−0.99		−0.11	−0.07		0.91	0.78		0.37
Fat	−1.82	−2.31	*	−0.93	−4.70	*	0.55	0.55		−1.40	−1.08		−0.17	−0.11		1.02	0.63		0.32
Carbohydrates	−1.81	−2.36	*	−0.84	−4.28	*	0.30	0.37		−0.95	−0.75		−0.06	−0.03		0.71	0.70		0.31
Proteins	−1.81	−3.06	*	−0.71	−2.82	*	0.42	0.54		−1.18	−0.84		−0.42	−0.37		1.18	1.09		0.30
Saturated	−2.13	−2.74	*	−1.04	−5.15	*	0.43	0.47		−1.17	−0.92		−0.38	−0.26		1.11	0.75		0.39
NME Sugars	−2.51	−2.07		−1.37	−3.82	*	−0.01	−0.01		0.03	0.01		−0.43	−0.21		0.41	0.19		0.30
Sodium	−1.91	−2.70	*	−0.70	−4.36	*	0.53	0.61		−1.17	−1.06		0.07	0.04		0.57	0.50		0.26
Total Sugars	−2.17	−2.16	*	−1.09	−4.25	*	0.05	0.04		−0.29	−0.17		−0.50	−0.28		0.74	0.40		0.30
Cholesterol	−2.58	−3.33	*	−1.00	−3.21	*	0.46	0.43		−1.25	−0.69		−0.43	−0.28		1.22	0.95		0.32

Note: * stands for statistically significant at 95 per cent. Each regression was run with 28 observations.

Table 2. Biscuits and cakes—regressions between nutrients and trading in quality.

Nutrient	Regression Coefficients						R^2
	α	t-stat.	Sig	β	t-Stat.	Sig	
Energy	−1.85	−2.18	*	−0.33	−1.88		0.17
Fat	−1.70	−1.54		−0.27	−1.46		0.10
Carbohydrates	−1.90	−2.32	*	−0.33	−1.90		0.18
Proteins	−1.93	−3.27	*	−0.30	−2.02		0.17
Saturated	−2.21	−2.53	*	−0.40	−2.77	*	0.20
NME Sugars	−3.15	−3.10	*	−0.75	−3.70	*	0.35
Sodium	−1.96	−2.43	*	−0.27	−2.01		0.13
Total Sugars	−2.56	−2.98	*	−0.55	−3.14	*	0.28
Cholesterol	−2.55	−3.02	*	−0.34	−1.66		0.13

Note: * stands for statistically significant at 95 per cent significant. Each regression was run with 28 observations.

Table 3. Confectionery—regressions between nutrients and trading in quality.

Nutrient	Regression Coefficients						R^2
	α	t-Stat.	Sig	β	t-Stat.	Sig	
Energy	−1.52	−1.68		−0.23	−2.12	*	0.08
Fat	−1.53	−1.59		−0.24	−1.98		0.07
Carbohydrates	−1.62	−1.75		−0.26	−2.52	*	0.10
Proteins	−1.53	−1.90		−0.15	−1.65		0.04
Saturated	−1.75	−1.87		−0.24	−1.75		0.07
NME Sugars	−2.40	−1.78		−0.52	−2.74	*	0.16
Sodium	−1.82	−2.18	*	−0.25	−2.67	*	0.11
Total Sugars	−2.07	−1.93		−0.41	−2.57	*	0.15
Cholesterol	−2.03	−1.90		−0.14	−0.93		0.02

Note: * stands for statistically significant at 95 per cent significant. Each regression was run with 28 observations.

Table 4. Processed potatoes—regressions between nutrients and trading in quality.

Nutrient	Regression Coefficients						R^2
	α	t-Stat.	Sig	β	t-Stat.	Sig	
Energy	−1.33	−1.39		−0.14	−0.80		0.05
Fat	−1.35	−1.35		−0.16	−0.94		0.05
Carbohydrates	−1.31	−1.35		−0.11	−0.63		0.03
Proteins	−1.40	−1.69		−0.09	−0.53		0.03
Saturated	−1.57	−1.51		−0.16	−0.94		0.06
NME Sugars	−1.63	−0.96		−0.14	−0.49		0.02
Sodium	−1.50	−1.76		−0.09	−0.67		0.02
Total Sugars	−1.41	−1.02		−0.08	−0.35		0.01
Cholesterol	−2.07	−1.93		−0.17	−0.77		0.06

Table 5. Processed vegetables—regressions between nutrients and trading in quality.

Nutrient	Regression Coefficients						R^2
	α	t-Stat.	Sig	β	t-Stat.	Sig	
Energy	−1.13	−1.44		−0.30	−2.09	*	0.15
Fat	−1.13	−1.29		−0.37	−2.65	*	0.18
Carbohydrates	−1.16	−1.39		−0.21	−1.45		0.07
Proteins	−1.28	−1.84		−0.30	−2.05		0.17
Saturated	−1.37	−1.78		−0.52	−3.53	*	0.34
NME Sugars	−1.48	−1.16		−0.53	−2.33	*	0.17
Sodium	−1.40	−1.97		−0.34	−2.91	*	0.21
Total Sugars	−1.35	−1.32		−0.44	−2.27	*	0.18
Cholesterol	−1.81	−1.89		−0.30	−1.44		0.10

Note: * stands for statistically significant at 95 per cent significant. Each regression was run with 28 observations.

Table 6. Soft drinks—regressions between nutrients and trading in quality.

Nutrient	Regression Coefficients						R^2
	α	*t*-Stat.	Sig	β	*t*-Stat.	Sig	
Energy	−1.07	−1.23		−0.02	−0.13		0.00
Fat	−1.01	−1.01		−0.08	−0.40		0.01
Carbohydrates	−1.09	−1.27		−0.04	−0.26		0.00
Proteins	−1.25	−1.62		0.04	0.33		0.00
Saturated	−1.30	−1.33		0.05	0.26		0.00
NME Sugars	−1.45	−1.05		0.15	0.58		0.02
Sodium	−1.33	−1.66		−0.01	−0.07		0.00
Total Sugars	−1.31	−1.18		0.08	0.40		0.01
Cholesterol	−1.73	−1.69		−0.05	−0.29		0.00

As none of the remaining regressions showed that the UK country dummies were significant, they were eliminated from the remaining regressions. Table 2 presents the regressions for biscuits and cakes. The impact of trading down indicates that consumers substitute for products that are cheaper and higher in NME sugar and saturated fat. This result is important because biscuits and cakes are classified by the FSS [18] as a discretionary category and are top contributors of saturated fat and sugar on the diet. Therefore, if the price of the category increases (as shown in Figure 2), consumers would substitute towards products within the category of poorer nutritional quality going against the FSS [18] recommendations to meet dietary goals.

Note that a potential solution to avoid the worsening of nutritional quality is through reformulation of the products within a category, so that when consumers make their substitutions, their nutrition does not suffer. However, it is important to highlight that the biscuits and cakes category is one where the goals of the PHE reformulation programme (which is focused on sugar and calories) is not being achieved [20]. The information to date indicates that the sugar content has not reached the 5 per cent reduction set for 2017—in fact the average sugar of the category has not changed at all, which means that the goal of 20 per cent reduction for 2020 is highly unlikely to be reached.

Table 3 shows that in the case of confectionery, trading down has an impact on calories and carbohydrates but particularly important is that it increases sugars (particularly NME sugars, with the highest β coefficient equal to −0.52). This is particularly important because confectionery is the greatest contributor of sugar on the diet and another discretionary food category [18].

According to information from PHE [20], similar to the case of biscuits and cakes, the goal of sugar reduction for 2017 was not achieved (i.e., chocolate confectionery content of sugar did not change at all and in the case of sugar confectionery, it decreased by 1 per cent).

Table 4 present the case of processed potatoes. This is a significant category because some of its products (e.g., chips) are associated with takeaway food in the UK. Moreover, as shown by Defra's Family Food, chips are still the UK's preferred form of potato [21].

As shown in the Table, no significant relationship was found between trading down/up on processed potatoes purchases and the nutrients. It is important to remember that these results do not indicate that processed potatoes, such as chips, do not contribute, for instance, with, e.g., fat to the diet; they indicate that trading in quality does not have nutritional effects.

Table 5 presents the results for processed vegetables. As shown above this is a heterogeneous category with several subcategories known for the presence of NME sugar (e.g., baked beans) or fat (e.g., takeaways).

The regressions show that trading down in quality has a significant impact on energy, fat, fatty acids and NME sugars. This may indicate the need to monitoring the category for the so called "ultraprocessed" processed food, which are typically energy dense; have a high glycaemic load; are low in dietary fibre, micronutrients and phytochemicals; and are high in unhealthy types of dietary fat, free sugars and sodium [22].

Table 6 shows the regressions for soft drinks. Similar to the case of processed potatoes, no statistically significant relationship was found between trading in quality in soft drink purchases and the nutrients. As shown in Figure 6, only Scotland traded down in this category; however, the results indicated that the change was not significant in terms of change in NME sugar, which is the most important nutritional issue within the category.

Recent information from the UK Food and Drink Federation [23] indicates that soft drinks included within the Soft Drinks Industry Levy had reduced sugars by 11 per cent and calories in single serve products by 6 per cent. Moreover, even before coming into effect, the levy was already working as about 50 per cent of manufacturers have reformulated their drinks. Although announce in March 2016, the levy started to operate in April 2018 and companies are supposed to pay 24 pennies per litre of drink if it contains 8 g of sugar per 100 millilitres or 18 pennies per litre of drink if it contains between 5–8 g of sugar per 100 millilitres.

5. Conclusions

The UK Department of Environment, Food and Rural Affairs (Defra) publishes an estimation of trading in quality; i.e., the change in the composition of the food basket or in a particular category when the food prices or the prices of a category changes. When consumers trade down in quality, it means that they choose a cheaper basket, which is assumed to be of poorer quality.

The term quality in the aforementioned context is very vague as attributes behind quality, e.g., flavour and colour, are not mentioned. Therefore, the contribution of this paper has been to consider the degree of association between trading down in quality and nutrition (i.e., we selected nutrition as the quality attribute). This, of course, is an empirical question, as the change in the composition of the basket might not have any nutritional implication.

The topic is important because changes in prices may force consumers to not only change the quantities consumed but also the composition of their basket, which may worsen nutritional constraints. The analysis was carried out for six food and drink categories and for pooled data of all the UK countries.

The results indicate that changes in category prices affect not only expenditures and the quantities purchased but also the choice of products within a category, which have implications for the quality of the diet.

In general, UK countries responded similarly to increases in prices (though there were differences in some categories). Trading down in quality was a common reaction to the increase in prices in all countries. In addition, it was found that trading down in quality, i.e., substituting toward cheaper products within a category, depending on the category, was found to be associated with increases in energy, fats, fatty acids, NME sugars and carbohydrates.

Trading down in quality in cakes, biscuits and confectionary was found to increase NME sugars and fat in the diet. Moreover, in the case of processed vegetable products, trading down in quality triggers an increase in foods higher in energy, fats, fatty acids and NME sugars. All these products enter into the description of ultra-processed products and should be monitored to avoid detrimental effects on the diet.

Given the results of the analysis, there is a clear need to not only monitor food prices and ensure that consumers understand the nutritional information in labels, as households might worsen their consumption when they choose cheaper products, but also from the supply side, to continue with the reformulation campaign, as it will stop consumers worsening their diets when they make substitutions. These results are important in the context of exiting the European Union, because this means that if prices increase due to a depreciation of the UK sterling pound, different tariffs, or higher international prices, then experiencing negative changes in nutritional quality will be possible.

A caveat of the analysis is its aggregated nature. Nevertheless, the results are not unreasonable, and they are statistically significant. Further work should consider more disaggregated datasets, i.e., at the level of households, for longer time spans, and explore the substitutions that consumers make. This would require the construction of price indices at a more disaggregated level.

Author Contributions: Conceptualization, C.R.-G.; methodology, C.R.-G.; formal analysis, C.R.-G., F.A. and N.C.; investigation, C.R.-G., F.A. and N.C.; writing–original draft preparation, C.R.-G.; writing—review and editing, F.A. and N.C. All authors have read and agreed to the published version of the manuscript.

Funding: This research received no external funding.

Acknowledgments: This paper derives from work commissioned by the Scottish Government as part of the 2016–2021 Research Programme on Food, Health and Wellbeing (Theme 3) and research undertaken as part of the ERANET-SUSFOOD Project SUSDIET "Implementing sustainable diets in Europe" funded by Defra 2015–2017.

Conflicts of Interest: The authors declare no conflict of interest.

References

1. Caswell, J.A.; Mojduszka, E. Using information labelling to influence the market for quality in food products. *Am. J. Agric. Econ.* **1996**, *78*, 1248–1253. [CrossRef]
2. Deaton, A. Quality, quantity, and spatial variation of price. *Am. Econ. Rev.* **1988**, *78*, 418–430.
3. McKelvey, C. Price, unit value, and quality demanded. *J. Dev. Econ.* **2011**, *95*, 157–169. [CrossRef]
4. Department for Environment, Food and Rural Affairs (Defra) Family Food 2015. Available online: https://www.gov.uk/government/statistics/family-food-2015 (accessed on 12 November 2019).
5. Department for Environment, Food and Rural Affairs (Defra) (2017) Family food datasets. Available online: https://www.gov.uk/government/statistical-data-sets/family-food-datasets (accessed on 12 November 2019).
6. Piercy, N.F.; Cravens, D.W.; Lane, N. Marketing out of the recession: Recovery is coming, but things will never be the same again. *Mark. Rev.* **2010**, *10*, 3–23. [CrossRef]
7. Agricultural and Horticultural Development Board (AHDB). What Might Brexit Mean for UK Trade in Agricultural Products? 2016. Available online: http://www.ahdb.org.uk/documents/Horizon_Brexit_Analysis_Report-Oct2016.pdf (accessed on 8 October 2019).
8. HM Government (HMG). The United Kingdom's Exit from and New Partnership with the European Union. 2017. Available online: https://www.gov.uk/government/uploads/system/uploads/attachment_data/file/589191/The_United_Kingdoms_exit_from_and_partnership_with_the_EU_Web.pdf (accessed on 8 October 2019).
9. Office for National Statistics (ONS). Prices Economic Commentary: Feb 2017. 2017. Available online: https://www.ons.gov.uk/economy/inflationandpriceindices/articles/priceseconomiccommentary/feb2017#although-food-prices-are-generally-lower-than-2-years-ago-meat-and-vegetables-are-getting-more-expensive (accessed on 8 October 2019).
10. Johnson, P.; Levell, P.; Waters, T. The Fall in Sterling: Who is Hit by the Rise in Inflation? The Institute for Fiscal Studies. 2016. Available online: https://www.ifs.org.uk/publications/8730 (accessed on 8 October 2019).
11. BBC News. Is the 10-a-day Diet only for the Wealthy? 2017. Available online: www.bbc.co.uk/news/health-39077242 (accessed on 1 December 2019).
12. Jones, N.R.; Conklin, A.I.; Suhrcke, M.; Monsivais, P. The growing price gap between more and less healthy foods: Analysis of a novel longitudinal UK dataset. *PLoS ONE* **2014**, *9*, e109343. [CrossRef] [PubMed]
13. Stone, R. The demand for food in the United Kingdom before the war. *Metroeconomica* **1951**, *3*, 8–27. [CrossRef]
14. Houthakker, H.S. Compensated changes in quantities and qualities consumed. *Rev. Econ. Stud.* **1952**, *19*, 155–164. [CrossRef]
15. Theil, H. Qualities, prices and budget enquiries. *Rev. Econ. Stud.* **1952**, *19*, 129–147. [CrossRef]
16. Huang, K.S. Nutrient elasticities in a complete food demand system. *Am. J. Agric. Econ.* **1996**, *78*, 21–29. [CrossRef]
17. Suits, D.B. Dummy variables: Mechanics v. interpretation. *Rev. Econ. Stat.* **1984**, 177–180. [CrossRef]
18. Food Standards Scotland. Monitoring Retail Purchase and Price Promotions in Scotland (2010–2016). 2017. Available online: https://www.foodstandards.gov.scot/publications-and-research/publications/monitoring-retail-purchase-and-price-promotions-in-scotland-2010--2016 (accessed on 1 December 2019).
19. Public Health England. PHE Publishes Latest Data on Nation's Diet: Press Release. 2018. Available online: https://www.gov.uk/government/news/phe-publishes-latest-data-on-nations-diet (accessed on 16 March 2018).
20. Bhatia, T. PHE Sugar Reduction Programme. 2019. Available online: https://ec.europa.eu/health/sites/health/files/non_communicable_diseases/docs/ev_201906201_co021_en.pdf (accessed on 1 December 2019).

21. Kelly, J.; Bates, C. 10 Ways the UK's Eating Habits Have Changed. 2016. Available online: https://www.bbc.co.uk/news/magazine-35595530 (accessed on 10 December 2019).
22. Monteiro, C.A.; Moubarac, J.C.; Levy, R.B.; Canella, D.S.; Louzada, M.L.C.; Cannon, G. Household availability of ultra-processed foods and obesity in nineteen European countries. *Public Health Nutr.* **2017**, *21*, 18–26. [CrossRef] [PubMed]
23. UK Food and Drink Federation. Reformulation. 2019. Available online: https://www.fdf.org.uk/sme-reformulation.aspx (accessed on 10 December 2019).

nutrients

MDPI

Article

Importance of Health Claims in the Adoption of New Breakfast Cereal Products in the UK

Montserrat Costa-Font * and Cesar Revoredo-Giha

Rural Economy, Environment and Society Department, Scotland's Rural College (SRUC), King's Buildings, West Mains Road, Edinburgh EH9 3JG, UK; Cesar.Revoredo@sruc.ac.uk

* Correspondence: Montse.CostaFont@sruc.ac.uk; Tel.: +44-(0)131-535-4225

Received: 1 November 2019; Accepted: 12 December 2019; Published: 17 December 2019

check for updates

Abstract: Regular breakfast consumption has the potential to prevent the prevalence of NCDs and to improve the nutritional profile of diets. Given consumers' interest in improving their diets, food suppliers are interested in introducing new cereal products making different health claims to capture consumers' attention. The purpose of this study is threefold: first, it aims to understand whether UK food suppliers are working to increase the availability of breakfast cereals with healthy and nutritious attributes; second, it explores which companies are leading the launch of these products; and third, it assesses to what extent health and nutrition claims made by breakfast cereals have an impact on their market success. The study employs an assembled database combining data from Mintel Global New Products Database (GNPD) and Kantar Worldpanel Dataset (KWDS) for the UK. A hazard-based duration model was used to analyse the success of the new products launched in the UK market in 2011 following them up to 2015. Our results reveal that UK suppliers broadened the number of breakfast cereals on offer in the period 2000 to 2018, with a particular focus on multigrain cereals, porridge and granola. Health and nutrition claims were added to 27% of these products. Although consumers welcome healthy alternatives such as muesli, the impact of positional claims on the success of newly developed breakfast cereals is claim-specific. No clear pattern regarding the impact of health and nutrition claims is identified. However, other elements such as celiac-friendly ingredients and UK origin do have a positive impact on the success of breakfast cereals.

Keywords: health claims; new product development; nutrition claims; market success

1. Introduction

Diet quality has been reported to be an important risk factor for non-communicative chronic diseases (e.g., diabetes, stroke, cancer) [1,2]. Poor-quality diets are associated with an inadequate intake of key micronutrients, as well as a lack of consumption of whole grains, fruit, nuts, seeds and omega-3 fatty acids, among other elements [1]. Micronutrient deficiencies such as in iron, vitamin A, D, magnesium and zinc, among others, are common in many developing countries and named as "hidden hunger". An important step towards achieving a healthy and rich diet is a regular consumption of breakfast, which provides a higher percentage of micronutrients than other meals [3–6]. In 1960, Adelle Davis stressed the status of breakfast as the most important meal of the day [7] and later studies detailed that ideally between 15% and 25% of our daily energy intake should be consumed at breakfast [3,7,8].

Gaal [9] and Reeves [10] revealed that the majority of the UK population (around 95%) are regular breakfast consumers. Breakfast cereals such as porridge and muesli are also reported to contribute to more than 50% of the breakfast energy intake for the UK population [9–12]. When consumed regularly, breakfast cereals have been linked to an increase in the nutritional profile of diets, only with the exception of a high level of simple sugars associated with those ranges with a high added sugars content [13,14]. Furthermore, evidence suggests an inverse association between breakfast cereal consumption and body weight [15,16]. Paradoxically, despite the interest of consumers in healthy products and their health effects, the trend of breakfast cereal consumption in the UK exhibited a 4% decline in the volume of sales from 2014 to 2019 [17]. The reduction in breakfast cereals sales and consumption has been related to social and governmental pressure to reduce sugar, negative media coverage regarding ultra-processed foods, and changing breakfast habits with movement towards more convenient, healthy and on-the-go food products [17].

In terms of firm competition, the breakfast cereal market displays common characteristics with fast-moving consumer goods, which compete intensively to introduce new products [18]. Raubitschek described this kind of performance [19] as a model of product proliferation. A group of companies compete to introduce new products into the market, making health benefits and other types of claims, "hoping" that by doing so they will hit the jackpot, i.e., new products introduced into the market will become successful because they are taken up by consumers and remain on retailers' shelves for a long time [19]. However, not all new products survive and between 60% and 80% of these new products eventually disappear from the shelves [20,21].

New breakfast cereals are introduced following consideration of several positional claims trying to attract consumers. For instance, products may contain a high level of fibre and protein content, added vitamins and other nutrients, such as calcium and iron, less sugar and additional multigrain ingredients. Positional claims are strategic in food marketing because, in many cases, food choices are made at the point of purchase [22]. Claims can help consumers identify breakfast cereals with more health benefits and, therefore, assist them in selecting a better diet. It is important to note that nutrition and health labelling is controlled at a European level by Regulation (EC) 1924/2006 and its subsequent amendments. This regulation defines a nutrition claim "as any claim which states, suggests or implies that a food has particular beneficial nutritional properties due to the presence, absence, increased or reduced levels of energy or of a particular nutrient or other substance" while defines a health claim as "any claim that states, suggests or implies that a relationship exists between a food category, a food or one of its constituents and health". This study, following previous research, considers the two different types of claims in a single group.

Previous studies have observed the positive impact of health claims on consumers' evaluations of food [23,24]. Moreover, as reported in previous research, breakfast cereals are one of the food categories that make a higher number of health and nutrition claims [25]. Hieke et al. [26] considered five European countries and reported that almost a third of cereal products in the market make nutritional claims. Later, Sussman et al. [25] revealed that, in the Australian market, 95% of the breakfast cereal products they surveyed made nutrition-, health- or other related claims.

Several studies have audited the prevalence of health claims in different EU and non-EU countries and analysed their impact on healthy diets [26,27]. Other studies have considered the impact of health claims on food choices [28,29]. The purpose of this paper is to focus on the introduction of new products in the breakfast cereals category, as well as examining the role of retailers and manufacturers in encouraging consumers' purchase decisions with the release of new products making particular positional claims. The aim of this is to help the food industry to better understand the potential impact of their innovation strategies and package marketing information on the quality of consumer diets. In particular, the paper aims to answer the following questions:

- To what extent are UK food suppliers working to increase the availability of breakfast cereals with healthy and nutritious attributes?
- What are the companies that are leading the launching of these products? Which is the role of private labels?
- What claims are particularly important for the market success of newly developed breakfast cereals?

The results reveal that manufacturers and retailers are actively working on the introduction of new options with health and nutritional claims. Some of these claims do affect the success of the new product in the market, but no clear pattern can be identified.

The structure of the paper is as follows: it begins with the empirical section, which describes the data and methods. The following section presents a description of the introduction of new products in the UK breakfast cereals sector and the results of the survival analysis, considering the uptake of newly developed breakfast cereals. The final section provides the discussion and conclusions.

2. Materials and Methods

2.1. The Data

The first part of this study employs the Mintel Global New Products Database (GNPD) to present and provide an overview of the introduction of new breakfast cereals launched in the UK between January2000 and September 2019 (understanding it in a broad sense, not just considering new products, but also variations to existing products, and the relaunching and/or repackaging of products). The dataset contains information on 2262 new breakfast cereals launched in different types of retail stores between January 2000 and September 2019 by 282 manufacturing or retailing companies using 700 different brands.

The products are classified into two categories, cold and hot cereals, and into 12 subcategories, as follows: corn cereals, granola, muesli, multigrain cereals, oat-based cereals, porridge, quinoa cereals, rice cereals, rye cereals, spelt cereals, wheat cereals and other. The database also contains information on the type of label (branded, private label), introduction price, and type of packaging. The fact that the dataset also provides information about the positioning claims of each product was particularly important for this study; 96 different claims were found in the dataset for breakfast cereals. This is important because such claims convey information to consumers about the product. For the analysis, these claims were classified into five groups: convenience (e.g., microwaveable), demographic (e.g., if designed for a particular demographic group), health and nutrition (e.g., low in calories), safety (e.g., no additives/preservatives), sustainable (e.g., organic) and others (e.g., limited edition, cobranded). See Figure A1 in Appendix A for more details on the particular claims considered under each claim category.

The second part of the study employs an assembled database combining data from GNPD and the Kantar Worldpanel Dataset (KWDS) for the UK. GNPD data were used to gather information on which new breakfast cereals were launched in the UK market in 2011. KWDS includes weekly records of all foods and beverages that were taken home from supermarkets and similar stores by UK households during the period 2013 to 2015. The breakfast cereals products observed in the GNPD database were identified in KWDS to follow its sales in UK retail and trace their durability in the market.

2.2. Analysis

In order to respond to the first and second research questions—exploring the trend in the introduction of breakfast cereals and identifying the leading suppliers—this research employs descriptive statistics (i.e., frequency distributions and cross-tabulations). To assess the influence of product claims on consumer acceptance, this study uses hazard-based duration models, which provide an explanation on the length of time that launched breakfast cereal products survived in the market. Duration models are based on the survivor function, which in our case is the probability of the product still being available in the market up to a specific time t, as follows:

$$S_{(t)} = pPr\{T \geq t\} = 1 - F(t) = \int_t^{\infty} f(x)dx \tag{1}$$

where T is a continuous random variable, $f(t)$ is its probability density function and $F(t)$ is the cumulative distribution function. The distribution of T can also be expressed as the hazard function, which is the rate of occurrence of the event. In our case, the event means failing or disappearing from the market, and this is expressed as follows:

$$h(t) = \lim_{\Delta t \to \infty} \frac{\Pr(t \leq T < t + \Delta t | T \geq t)}{\Delta t} \quad (2)$$

where the numerator is the conditional probability of occurrence during the period under consideration given that it has not occurred before, and the denominator is the length of the interval under consideration. The hazard function can also be expressed as follows:

$$h(t) = \frac{f(f)}{S(t)} \quad (3)$$

In order to estimate the hazard function, different models can be used depending on the shape of the hazard and the features of the explanatory variables included in the model. A Cox's proportional hazards model [30] was estimated for this research. Using the Cox model, specification of each individual follows its own survival function formulated as follows:

$$h_i(t) = h(t; x_i) = h_0(t) exp\left(x_i' \beta\right) \quad (4)$$

where $h_0(t)$ is a nonparametric function (baseline hazard at time t), $exp\left(x_i' \beta\right)$ is a parametric function, and $x_1, \ldots, x_k$ are a pool of predictor variables. The dependent variable was the period (in years) that a product remained on the market. Table A1 in Appendix A presents the descriptive statistics for the variables considered for the econometric analysis.

3. Results

3.1. *Overview of the Introduction of New Breakfast Cereals in the UK Market*

Figure 1 below presents the evolution of new breakfast cereals introduced into the UK market between 2000 and 2018. A positive trend can be observed from 2000 to 2015, with a particularly large number of products marketed during the period 2011 to 2015. However, in 2016 and 2018, there is a clear drop in the number of new products.

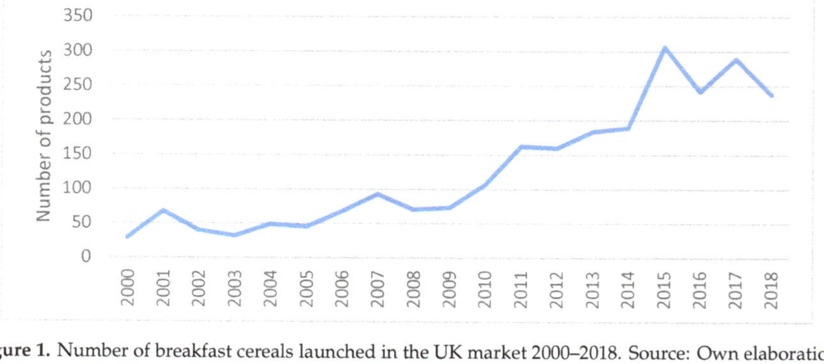

Figure 1. Number of breakfast cereals launched in the UK market 2000–2018. Source: Own elaboration based on Mintel's Global New Products Database (GNPD) database.

Figure 2 disaggregates the information from Figure 1 by subcategories of breakfast cereals. It shows that, overall, multigrain breakfast cereals are the subcategory with the most launched products, comprising 25% of all breakfast cereals introduced during this period. However, since 2012, an increase in the introduction of porridge and granola products can also be observed. Porridge represents only

3% of the breakfast cereals introduced in 2000, while in the period 2012 to 2018, it represents between 20% and 35%.

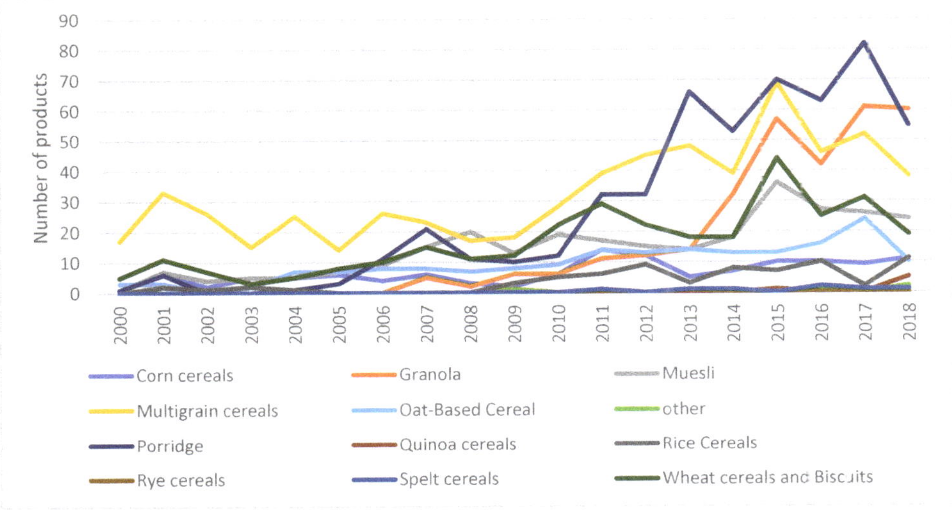

Figure 2. Subcategories of breakfast cereals launched in the UK market 2000–2018. Source: Own elaboration based on Mintel's GNPD database.

Due to the highly concentrated breakfast cereals market, companies invest in research and development in order to differentiate their products and increase their market share. Table 1 indicates that, for the period of January 2000 to September 2019, Kellogg's was the top company for introducing products in the breakfast cereal category. The GNPD data also reveals that ten suppliers were responsible for the introduction of 50% of all the newly developed products during this period. Furthermore, six out of the ten top companies introducing new products are retailers and the new breakfast cereals introduced by retail firms account for 36% of the total. Figure 3 presents the number of products launched over time by the top 12 companies. Although Kellogg's has always been a top developer, the period 2011 to 2016 was when it launched the largest number of products. We can also observe a considerable increase in the participation of retailers in new product development since 2011.

Table 1. Frequency of new products released by companies January 2000–September 2019.

Company	Frequency	Percent
Kellogg's	332	14.68
Weetabix	164	7.25
Tesco	161	7.12
Asda	145	6.41
Sainsbury's	135	5.97
Morrisons	81	3.58
Nestlé	75	3.32
Marks and Spencer	74	3.27
Aldi	73	3.23
Jordans	69	3.05
Rest of Companies	63	51.15

Source: Own elaboration based on Mintel's GNPD database.

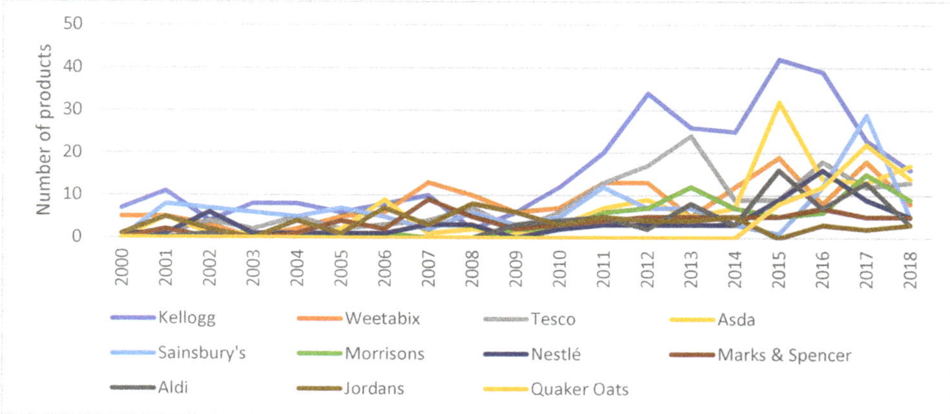

Figure 3. Companies launching breakfast cereals in the UK market 2000–2018. Source: Own elaboration based on Mintel's GNPD database.

It is important to note that 96% of the newly launched products make a positional claim on the packaging. Table 2 lists the claims most frequently made by new breakfast cereals; the top ten claims account for more than 50% of the total claims made on packages during the period January 2000 to September 2019. The most frequent claim carried by newly launched breakfast cereals highlights that the new products are suitable for vegetarians, a demographic claim targeting a sector of consumers. Next, sustainability claims reporting the packaging to be environmentally friendly account for 8% of the total claims associated with new breakfast cereals. The third and fourth most frequent claims are health and nutrition claims, reporting the product to be wholegrain and with high or added fibre. These results are in line with previous reports [25] that the wholegrain claim is the most frequently made health claim on breakfast cereals.

Table 2. Frequency of product claims January 2000–September 2019.

Claims	Frequency	Percent
Vegetarian	1727	10.80
Ethical—Environmentally Friendly Package	1345	8.41
Wholegrain	1043	6.52
High/Added Fibre	970	6.07
No Additives/Preservatives	716	4.48
Ethical—Recycling	712	4.45
Vitamin/Mineral Fortified	549	3.43
Low/No/Reduced Sodium	491	3.07
Social Media	468	2.93
Low/No/Reduced Allergen	446	2.79
Gluten Free	427	2.67
Microwaveable	400	2.50
Ease of Use	361	2.26
Time/Speed	348	2.18
Vegan	304	1.90
Low/No/Reduced Fat	303	1.90
Rest of Claims	5379	33.64

Source: Own elaboration based on Mintel's GNPD database.

Health and nutrition claims represent 27% of all the claims on the newly introduced breakfast cereals for the period January 2000 to September 2019. Figure 4 reveals that health and nutrition claims, together with demographics and sustainability claims, are the most important claims made by breakfast cereals over time. Moreover, in the period 2016 to 2018, health and nutritional claims are more frequent. It is worth noting that health and nutritional claims are often presented in combination with other claims (e.g., sustainability, safety, convenience or demographic claims); just 5% of the products make only health and nutritional claims. For 44% of the products, health and nutritional claims are combined with demographic and sustainability claims on the same package and 16% of the breakfast cereals make all these claims at once—health and nutritional, sustainability, safety, convenience and demographic claims. It is also important to mention that claims classified as safety in this study (i.e., all-natural product, genetically modified-free, hormone-free and no additives/preservatives) can also be considered to be health claims following the EU regulation 1924/2006.

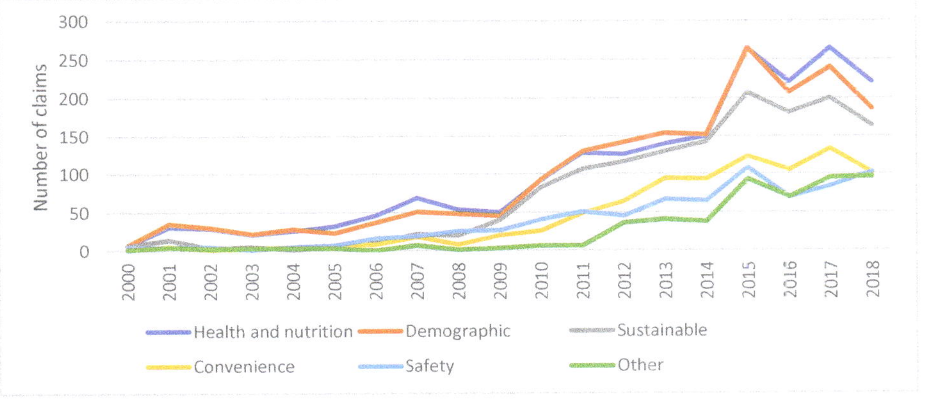

Figure 4. Subcategories of breakfast cereals launched in the UK market 2000–2018. Source: Own elaboration based on Mintel's GNPD database.

The most frequent health claim for the period January 2000 to September 2019 is identifying the product as wholegrain. A total of 40% of the newly introduced products—1043 products—claim to be wholegrain. The claim emphasising a high presence of fibre is the second most frequent, with 37% of the newly launched products carrying this claim—970 new products. Products claiming to be fortified with vitamins and minerals and those declaring they have low or reduced levels of sodium comprise around 20%—549 and 491 products. Claims regarding allergens, such as low/no/reduced allergen or gluten-free are also important; these are made by 17% (446 products) and 16% (427 products) of the new products, respectively. Claims of containing less fat are carried by 12% of the newly developed breakfast cereals (303 products). Finally, the remaining health and nutritional claims are made by less than 10% of the new products. Table 3 shows the frequency of the appearance of health and nutritional claims made on breakfast cereals over time. It is important to note that those related to allergens have been increasing in recent years.

Table 3. Product frequency of health and nutrition claims by year from 2004 to 2018.

	2004	2005	2006	2007	2008	2009	2010	2011	2012	2013	2014	2015	2016	2017	2018
Wholegrain	4	5	23	29	26	25	53	71	62	71	80	151	111	152	96
High/Added Fibre	10	7	21	20	18	2	22	71	66	65	75	132	116	128	105
No Additives/Preservatives	3	6	15	16	25	23	37	46	42	58	54	96	62	77	81
Vitamin/Mineral Fortified	9	14	13	17	13	13	25	33	50	33	30	70	47	58	49
Low/No/Reduced Sodium	3	13	6	23	6	18	33	40	36	33	33	52	54	72	33
Low/No/Reduced Allergen	2	2	2	10	4	12	15	15	18	18	26	51	60	63	89
Gluten Free	1	1	1	9	4	7	13	12	15	15	21	50	63	68	88
Low/No/Reduced Fat	7	11	11	20	11	10	23	23	16	24	21	33	17	29	11
Functional—Cardiovascular	0	0	0	0	0	0	0	0	0	0	0	31	45	58	32
Low/No/Reduced Sugar	1	13	7	16	11	12	19	19	16	17	17	8	0	0	0
Low/No/Reduced Transfat	0	0	0	1	3	4	5	11	14	9	8	33	20	16	16
No Added Sugar	0	0	0	0	0	0	0	0	0	0	0	17	28	27	39
Bone Health	0	0	0	0	0	1	1	4	3	4	13	32	22	20	11
Functional—Energy	0	0	0	0	0	0	0	0	0	0	0	28	31	27	31
Low/No/Reduced Saturated Fat	0	1	5	6	4	5	7	10	7	5	12	9	15	11	9
Cardiovascular (Functional)	0	1	1	4	2	3	14	20	16	15	25	4	0	0	0
Functional—Bone Health	0	0	0	0	0	0	0	0	0	0	0	24	22	20	11
High/Added Protein	0	0	0	0	0	0	0	0	0	0	0	13	21	26	16
Dairy Free	0	0	0	0	0	0	0	0	0	0	0	9	10	28	23
Low/Reduced Sugar	0	0	0	0	0	0	0	0	0	0	0	10	13	16	15
Functional—Other	0	0	0	0	0	0	0	0	0	0	0	14	12	15	13
Rest of Health Claims	11	9	15	17	21	22	31	47	38	49	53	88	87	87	107

3.2. *Effect of Nutrition and Health Claims on the Rate of Success of Newly Developed Breakfast Cereals in the UK*

First, we undertook a descriptive analysis of the success of the new breakfast cereals (see Tables 4–6). Table 4 indicates an overall rate of success of 39.3% even though we can observe differences between the different product categories. The table also shows an index of success where the average rate of success is 100. New products associated with rice cereals, granola, muesli and oat-based cereals were the most successful categories (above the average rate of success).

Table 4. Degree of success of new breakfast cereals by product category.

Categories	Fully Failed	Partial Success	Success	Total	Percentages Failed	Partial	Success	Total	Success Index
Rice Cereals	0	1	5	6	0.0	16.7	83.3	100.0	2.12
Granola	4	1	6	11	36.4	9.1	54.5	100.0	1.39
Muesli	6	2	9	17	35.3	11.8	52.9	100.0	1.35
Oat-Based Cereal	5	5	7	17	29.4	29.4	41.2	100.0	1.05
Porridge	9	9	11	29	31.0	31.0	37.9	100.0	0.97
Corn Cereals	4	5	5	14	28.6	35.7	35.7	100.0	0.91
Wheat Cereals	10	7	9	26	38.5	26.9	34.6	100.0	0.88
Multigrain Cereals	16	11	12	39	41.0	28.2	30.8	100.0	0.78
Spelt Cereals	1	0	0	1	100.0	0.0	0.0	100.0	0.00
Wheat Biscuits	3	0	0	3	100.0	0.0	0.0	100.0	0.00
Total	58	41	64	163	35.6	25.2	39.3	100.0	1.00

Source: Own elaboration based on GNPD and Kantar Worldpanel Dataset (KWDS) database.

Table 5 shows the degree of success by launch type. The greatest success is achieved in the relaunch of products, although there are only five of these. New products and new varieties or range extensions comprise the most launched products, and their degree of success is below the mean. It is also interesting that products that are relaunched with new packaging have a success rate of 50%. This could be related to a strategy of, after launching a new product, if it does not bring in the required sales (or, if it does, launching a new variety or range extensions), trying to apply new formulations, new packaging or relaunching the product. Even then, this does not ensure 100% success, however.

Table 5. Degree of success by launch type.

	Fully Failed	Partial Success	Success	Total	Percentages Failed	Partial	Success	Total	Success Index
New Formulation	5	3	6	14	35.7	21.4	42.9	100.0	1.09
New Packaging	7	11	18	36	19.4	30.6	50.0	100.0	1.27
New Variety/Range Extension	21	16	22	59	35.6	27.1	37.3	100.0	0.95
Relaunch	2	0	3	5	40.0	0.0	60.0	100.0	1.53
New Product	23	11	15	49	46.9	22.4	30.6	100.0	0.78
Total	58	41	64	163	35.6	25.2	39.3	100.0	1.00

Source: Own elaboration based on GNPD and KWDS database.

Table 6 presents the degree of success by claim on branded and private labels. On the branded products, the claims with the higher degree of success have a low glycaemic value (67%) and low transfats (33%). In the case of private-label products, the top claims are reduced allergens (80%) and reduced sugar (50%). In both cases—branded and private label—fibre is also an important claim. It is clear from the table that health claims are an important way to increase the product's degree of success.

Table 6. Degree of success of breakfast cereals by claim on branded and private labels.

Claim	Fully Failed	Partial Success	Success	Total	Percentages Failed	Partial	Success	Total
Branded								
High/Added Fibre	11	7.00	23	41	26.80	17.1	56	100
Low/No/Reduced Allergen	3	3.00	4	10	30.00	30	40	100
Low/No/Reduced Fat	6	1.00	5	12	50.00	8.3	41.7	100
Low/No/Reduced Glycaemic	1	0.00	2	3	33.30	0	66.7	100
Low/No/Reduced Lactose	1	2.00	0	3	33.30	66.7	0	100
Low/No/Reduced Saturated Fat	2	2.00	3	7	28.60	28.6	42.9	100
Low/No/Reduced Sodium	10	7.00	16	33	30.30	21.2	48.5	100
Low/No/Reduced Sugar	6	2.00	9	17	35.30	11.8	52.9	100
Low/No/Reduced Transfat	0	1.00	2	3	0.00	33.3	66.7	100
Private Label								
High/Added Fibre	10	9.00	11	30	33.30	30	36.7	100
Low/No/Reduced Allergen	0	1.00	4	5	0.00	20	80	100
Low/No/Reduced Fat	5	4.00	2	11	45.50	36.4	18.2	100
Low/No/Reduced Saturated Fat	2	0.00	1	3	66.70	0	33	100
Low/No/Reduced Sodium	4	3.00	0	7	57.10	42.9	0	100
Low/No/Reduced Sugar	1	0.00	1	2	50.00	0	50	100
Low/No/Reduced Transfat	3	4.00	1	8	37.50	50	12.5	100
Total								
High/Added Fibre	21	16.00	34	71	29.60	22.5	47.9	100
Low/No/Reduced Allergen	3	4.00	8	15	20.00	26.7	53.3	100
Low/No/Reduced Fat	11	5.00	7	23	47.80	21.7	30.4	100
Low/No/Reduced Glycaemic	1	0.00	2	3	33.30	0	66.7	100
Low/No/Reduced Lactose	1	2.00	0	3	33.30	66.7	0	100
Low/No/Reduced Saturated Fat	4	2.00	4	10	40.00	20	40	100
Low/No/Reduced Sodium	14	10.00	16	40	35.00	25	40	100
Low/No/Reduced Sugar	7	2.00	10	19	36.80	10.5	52.6	100
Low/No/Reduced Transfat	3	5.00	3	11	27.30	45.5	27.3	100.00

Source: Own elaboration based on GNPD and KWDS database.

Second, we examined the factors influencing the success of the newly developed breakfast cereal products in the UK market using a duration model. Table 7 presents the estimated results for the Cox model considering the impact of the variables described in the annex (see Table A1). The goodness-of-fit results show that the model appropriately fits the data. We also implemented the proportional hazards assumption test, which indicated an absence of evidence contradicting the proportionality assumption.

The first variable considered in this analysis was the launch type (i.e., new formulation, new products, new variety or re-launch). The estimation revealed that the only launch strategy with a significant impact on the probability of the success of new breakfast cereals is the introduction of new packaging with a $p < 0.10$. The remaining launch strategies considered were not significant and were removed from the model. We can observe that the rate of failure of products launched with new packaging compared to other launch types (ceteris paribus) is 44% lower. The second and third variables considered were the type of company introducing the product (private company launching branded products or retailer launching private labels) and the introduction price. However, the t values were found to be non-significant for our model and they were, therefore, removed for the final estimation. Next, the origin of the product (UK or not) was considered, and interestingly this variable resulted in significant reduction of the probability of failure. However, its hazard ratio was minimal.

The impact of the type of breakfast cereal (i.e., granola, muesli, multigrain cereals, porridge) on the uptake of the product was also considered. The only category that was found to be significant was muesli, with a 76% lower chance of failure than other categories (ceteris paribus).

Considering the relevance of the ingredients listed on the product packaging (i.e., corn, wheat, oats, rice, rye, spelt and barley), only the wheat and rye results were significant and were, therefore, retained in the final model. We can observe from Table 7 that using rye as an ingredient in a newly developed breakfast cereal has a positive but minimal impact on its market success. On the other hand, including wheat as an ingredient has a negative effect on the uptake of this product The estimated hazard of failure for breakfast cereals listing wheat as an ingredient is 3.4 times the hazard for other breakfast cereals without such a claim.

Table 7. Cox regression time-constant variables label.

	Rate of Failure					
	Coeff.	**St. Err.**	**Haz. Rat.**	**St. Err.**	***t* Ratio**	**Sig.**
New Packaging	−0.579	0.334	0.561	0.187	−1.730	*
No claim	2.260	0.660	9.585	6.324	3.430	***
Muesli	−1.447	0.661	0.235	0.156	−2.190	**
Has wheat as an ingredient	1.211	0.333	3.358	1.119	3.640	***
Has rye as an ingredient	−38.342	0.956	8.20×10^{-18}	7.84×10^{-18}	−40.100	***
Dummy added calcium	−38.693	1.054	5.78×10^{-18}	6.09×10^{-18}	−36.710	***
Dummy all-natural product	−38.736	1.019	5.53×10^{-18}	5.64×10^{-18}	−38.020	***
Dummy brain nervous system	1.773	0.328	5.890	1.933	5.400	***
Dummy ethical animal	−38.457	1.213	7.32×10^{-18}	8.87×10^{-18}	−31.710	***
Dummy environmentally friendly product	1.378	0.419	3.969	1.664	3.290	***
Dummy low/no/reduced glycaemic	−40.508	0.861	9.41×10^{-19}	8.09×10^{-19}	−47.070	***
Dummy low/no/reduced lactose	2.100	0.420	8.166	3.432	5.000	***
Dummy organic	−39.196	0.817	3.49×10^{-18}	2.85×10^{-18}	−47.970	***
Dummy prebiotic	2.352	0.412	10.508	4.334	5.700	***
Dummy instant	1.155	0.653	3.174	2.073	1.770	*
Dummy UK ingredients	−38.114	1.085	1.03×10^{-17}	1.12×10^{-17}	−35.120	***
Log likelihood ratio test	−230.54					
Wald chi2 (15)	12710.55	***				
Number of observations	123					

$p < 0.10$ * $p < 0.05$ ** $p < 0.01$ ***. Coeff. (Coefficient); St. Err. (Standard Error); Haz. Rat. (Hazard Ratio) and Sig. (Significance).

Regarding the utility of including claims, we can observe that the probability of failure for products not including claims is higher than for those making claims. In particular, not listing any claim on the cereal package increases the risk of failure 9.58 times compared to products that do include claims (ceteris paribus). When considering the different claims associated with the newly introduced breakfast cereals, only ten claims had a significant result and were, therefore, included in the final estimation. Half of the significant claims are health- and nutrition-related claims, three are sustainability claims, there is one safety claim and, finally, one convenience claim. The impact that the different claim groups have on the uptake of the products can be either positive or negative, depending on the claim. It is important to highlight that all claims with a positive impact appear to have a very small hazard ratio. This means that its effect is significant but very small; this is the case for products stating to have added calcium and making low/no/reduced glycaemic index health claims. Another claim with a small but positive impact on the success of new breakfast cereals is all-natural products, which we consider as a safety claim, but it can also have some health implications. Sustainability claims, such as ethical animal and organic, also increase the probability of success of new breakfast cereals. Other health and nutritional claims, such as improving the brain and nervous system, with low/no/reduced lactose and prebiotic, increase the probability of failure by 5.9, 8.2 and 10.5 times, respectively, compared with products not labelled with these claims. Finally, being an environmentally friendly product and being instant reduce the probability of success of breakfast cereals 3.174 and 3.969 times, respectively.

Interactions between subcategories and claims were considered in the analysis; the results were non-significant, and these variables were eliminated from the final estimation.

4. Discussion

The present study first explores whether food suppliers (i.e., manufacturers and retailers) are working to increase the range of breakfast cereals they offer and whether these new products are labelled with health and nutritional claims. Second, it identifies which suppliers are leading this process, and third it assesses to what extent health and nutrition claims made by breakfast cereals have an impact on their market success.

An increasing number of new breakfast cereals were introduced into the UK market during the period 2000 to 2015, with a clear reduction in the number of new products from 2016 to 2018. Our results show that 96% of the newly developed breakfast cereals launched from 2000 to 2019 do include positional claims. Therefore, we can state that manufacturers and retailers include health and other information to differentiate their products effectively and, as indicated by previous research [31], potentially help consumers to make informed food purchases.

Among the positional claims, we identified health and nutrition, sustainability, convenience and demographic attributes. The most common are demographic claims, which are included in 77% of all new breakfast cereals. It is essential to note that being suitable for vegetarians comprises 66% of the products with demographic claims. Moreover, if we also include products informing customers about the absence of animal ingredients or reporting to be suitable for vegans, this encompasses 97% of all the new products making demographic claims. Taking into account that a reduction in the use of animal ingredients is an important aspect of achieving a rich and healthy diet, we can state that, by including these claims, suppliers are, in a way, assisting consumers to not only make ethical and sustainable choices but also healthier ones.

The second most numerous claims are health- and nutrition-related, present in 27% of the new breakfast cereals. Although we identified 51 different claims for this group, the most frequent are those reporting that the product is wholegrain, which appears in 40% of the products making health claims. Informing customers about the high level of fibre in the new product is also a very common claim. These two claims are particularly important in improving the nutrition levels of Western diets and they can explain the higher consumption of high-fibre cereals per person compared to other breakfast cereals reported in the introduction section. Health claims reporting the absence of allergens have also become increasingly frequent since 2014. These claims can have a positive impact on the section of the population suffering from food allergies and food intolerances that require them to avoid specific ingredients in their diets. Adverse reactions to foods have been reported to represent a key health-issue in Western societies [32]. It is also important to note that safety claims (i.e., all-natural products, genetically modified-free, hormone free and no additives/preservatives) are made by 32% of the newly introduced breakfast cereals and these claims also have health and nutritional implications.

Our results demonstrate that the breakfast cereal industry in the UK follows the diversification model defined by [19]. A small group of leading suppliers, such as Kellogg's, Weetabix, Nestle and the big retailers (i.e., Sainsbury's, Tesco, Asda, etc.) are directing the research and development in the breakfast cereals sector. The same companies leading research and development are, to a large extent, the ones leading market sales in the breakfast cereals market. This is an important point to consider when developing policies to improve diets. Consumers purchase the options made available to them by retailers and manufacturers [21]. In line with [31,33,34], our results suggest that the research and development departments of manufacturers, and especially, retail companies, can help consumers move towards healthier diets by deciding what to place on the shelves. During the last ten years, companies have been focusing on the development and commercialisation of multigrain breakfast cereals, porridge and granola. The introduction of wheat and corn cereals has been reducing, and it appears to be following a decreasing trend. One of the constraints upon the rich and healthy diet noted as necessary by [1] is the limited consumption of wholegrain and assorted cereals, seeds and fruits. Expansion of the number of multigrain cereals, porridge and granola on offer can help to improve the limitations noted above. Therefore, it can be suggested that companies are working towards an improvement in the availability of healthy and nutritious breakfast options in the UK. However, not all new products are accepted by consumers, and other elements, such as sociodemographic characteristics, education, social pressure, price, marketing and communication of product benefits, will determine the uptake of the new offerings and the consequent impact on consumers' diets [29]. Collaboration between food suppliers, policymakers and health organisations is, therefore, needed to inform citizens regarding healthy diets and to promote healthy food. Consumer knowledge, attitude and purchase intention also shape manufacturers and retailers' research and development decisions.

In line with previous research on the success of newly developed products, our results support the suggestion that a limited proportion of newly launched products succeed in the market in the mid- to long-term. We found that less than 40% of the newly introduced breakfast cereals succeeded in the UK market in the medium–long term. This result highlights the complexity of finding a good new product development strategy and explains why this activity is mainly concentrated among big companies that can afford the economic implications of failing in the food market, and who can also undertake broader market research strategies to capture consumers' needs.

Regarding the impact of health and nutritional claims on the success of newly developed breakfast cereals for the UK market, there is no clear pattern. Our Cox regression results clearly show that placing claims informing consumers about the benefits of the newly developed breakfast cereals on the packaging does positively influence the success of these products in the market. That is, when consumers are informed about what they can get from a newly developed product in comparison to other options, they are more likely to purchase it. This result is in line with [28]. The estimation shows that some health and nutritional claims have a positive effect on the success of new breakfast cereals, while others have a negative impact. The impact of nutritional claims on the success of newly developed breakfast cereals is claim-specific. Moreover, for our sample, all claims with a positive impact appear to have a very small hazard ratio. This may be due to the sample size and the number of years being considered. The health claims found to have a positive effect on the uptake of the new products are added calcium, low/no/reduced glycaemic index, all-natural product, ethical animal and organic. On the other hand, our results show that claims such as improving the brain and nervous system, low/no/reduced lactose, prebiotic, environmentally friendly product and instant increase the probability of failure of newly developed breakfast cereals in the UK market.

We also found that the ingredients listed on the product packaging have an impact on the uptake of the products. Our results show that using wheat as an ingredient reduces the level of success of the newly developed breakfast cereals, while including rye as an ingredient increases its success. This result may be related to the increase in the incidence of celiac disease and the negative information reported on social media about gluten and wheat as a cause of obesity and other health problems [35]. Our results show that newly launched muesli products are more accepted by consumers than other categories of breakfast cereals. Considering that the composition of muesli is grains, fresh or dried fruits, seeds and nuts, it seems that consumers welcome healthy breakfast cereal options. Finally, informing consumers about the UK origin of the cereals was identified as having a positive impact on the purchase of the product.

Interactions between subcategories and claims were considered in the analysis, but the results were non-significant and were, therefore, eliminated from the final estimation. The lack of significance found in these interactions may be due to the sample size. The products introduced over one year might offer limited variability within the different categories and claims.

Taking these results together, we can suggest that manufacturers and retailers are assuming a proactive and important role in research and development to capture consumers' needs and, therefore, help consumers to move towards healthier and better diets. Positional claims carried by food products need to be very clear on the potential benefits and risks of the products in order to be of real help to consumers in their food-purchasing decisions. Future research can consider a longer time period and the nutritional value associated with the newly introduced products.

Author Contributions: Conceptualization, M.C.-F.; C.R.-G.; methodology, M.C.-F.; C.R.-G.; software, M.C.-F.; C.R.-G.; validation, M.C.-F.; C.R.-G.; formal analysis, M.C.-F.; C.R.-G.; investigation, M.C.-F.; C.R.-G.; resources, M.C.-F.; C.R.-G.; data curation, M.C.-F.; C.R.-G.; writing—original draft preparation, M.C.-F.; C.R.-G.; writing—review and editing, M.C.-F.; C.R.-G.; visualization, M.C.-F.; C.R.-G.; supervision, M.C.-F.; C.R.-G.

Funding: This study was funded as part of the Scottish Government Rural Affairs and the Environment Portfolio Strategic Research Programme 2016-21. The work on this paper is part of the strategic research on health and wellbeing (Theme 3), WP 3.1 (new produce and food products with healthy attributes) and WP 3.3 (food security).

Conflicts of Interest: The authors declare no conflict of interest.

Appendix A

Claim group	Positional Claims
Safety	All Natural Product
	GMO-Free
	Hormone Free
	No Additives/Preservatives
Convenience	Convenient Packaging
	Ease of Use
	Economy
	Microwaveable
	On-the-Go
	Portionability
	Seasonal
	Time/Speed
Demographic	Babies & Toddlers (0-4)
	Children (5-12)
	Female
	Halal
	Kosher
	Male
	Maternal
	No Animal Ingredients
	Premium
	Seniors (aged 55+)
	Vegan
	Vegan/No Animal Ingredients
	Vegetarian
Sustainable	Biodegradable
	Carbon Neutral
	Ethical - Animal
	Ethical - Charity
	Ethical - Environmentally Friendly Package
	Ethical - Environmentally Friendly Product
	Ethical - Human
	Ethical - Recycling
	Ethical - Sustainable (Habitat/Resources)
	Ethical - Toxins Free
	Organic
	Refill/Refillable
Other	Cobranded
	Event Merchandising
	Interesting Packaging
	Limited Edition
	Novel
	Social Media

Claim group	Positional Claims
Health and nutrition	Added Calcium
	Anti-Ageing
	Antioxidant
	Beauty Benefits
	Bone Health
	Brain & Nervous System (Functional)
	Cardiovascular (Functional)
	Dairy Free
	Diabetic
	Diet/Light
	Digestive (Functional)
	Functional - Beauty Benefits
	Functional - Bone Health
	Functional - Brain & Nervous System
	Functional - Cardiovascular
	Functional - Digestive
	Functional - Energy
	Functional - Eye Health
	Functional - Immune System
	Functional - Other
	Functional - Skin
	Functional - Slimming
	Functional - Weight & Muscle Gain
	Gluten Free
	High Protein
	High Satiety
	High/Added Fiber
	High/Added Protein
	Immune System (Functional)
	Low/No/Reduced Allergen
	Low/No/Reduced Calorie
	Low/No/Reduced Carb
	Low/No/Reduced Cholesterol
	Low/No/Reduced Fat
	Low/No/Reduced Glycemic
	Low/No/Reduced Lactose
	Low/No/Reduced Saturated Fat
	Low/No/Reduced Sodium
	Low/No/Reduced Sugar
	Low/No/Reduced Transfat
	Low/Reduced Sugar
	Nails & Hair
	No Added Sugar
	Other (Functional)
	Prebiotic
	Probiotic
	Slimming
	Sugar Free
	Vitamin/Mineral Fortified
	Weight & Muscle Gain
	Wholegrain

Figure A1. Claims considered under each claim category.

Table A1. Descriptive statistics.

Variable	Obs.	Mean	St. Dev.	Min	Max
Number of years a product has been sold	152	3.00	2.05	0	5
New Formulation	152	0.28	0.45	1	1
New Packaging	152	0.24	0.43	2	2
New Product	152	0.38	0.49	3	3
New variety	152	0.07	0.26	4	4
Product category (dummies)					
Multigrain cereals	152	0.24	0.43	0	1
Porridge	152	0.06	0.24	0	1
Granola	152	0.18	0.38	0	1
Muesli	152	0.09	0.16	0	1

Table A1. *Cont.*

Variable	Obs.	Mean	St. Dev.	Min	Max
Ingredients (dummies)					
Wheat	152	0.55	0.50	0	1
Corn	152	0.18	0.38	0	1
Oats	152	0.64	0.48	0	1
Rice	152	0.33	0.47	0	1
Rye	152	0.02	0.14	0	1
Spelt	152	0.06	0.24	0	1
Barley	152	0.39	0.49	0	1
Quinoa	152	0.00	0.00	0	1
Dummy branded (0) and private label (1)	152	0.40	0.49	0	1
Dummy 1 UK product or UK ingredients	152	0.01	0.11	0	1
Type of claim (dummies)					
Added Calcium	152	0.01	0.08	0	1
All Natural Product	152	0.03	0.16	0	1
Anti-Ageing	152	0.00	0.00	0	0
Antioxidant	152	0.02	0.14	0	1
Beauty Benefits	152	0.01	0.08	0	1
Bone Health	152	0.03	0.16	0	1
Brain & Nervous System (Functional)	152	0.01	0.11	0	1
Carbon Neutral	152	0.00	0.00	0	0
Cardiovascular (Functional)	152	0.13	0.34	0	1
Children (5–12)	152	0.09	0.29	0	1
Cobranded	152	0.01	0.08	0	1
Convenient Packaging	152	0.04	0.20	0	1
Digestive (Functional)	152	0.06	0.24	0	1
Ease of Use	152	0.07	0.26	0	1
Economy	152	0.07	0.25	0	1
Ethical—Animal	152	0.01	0.11	0	1
Ethical—Charity	152	0.05	0.22	0	1
Ethical—Environmentally Friendly Package	152	0.61	0.49	0	1
Ethical—Environmentally Friendly Product	152	0.07	0.26	0	1
Ethical—Human	152	0.00	0.00	0	0
Event Merchandising	152	0.00	0.00	0	0
Female	152	0.00	0.00	0	0
Gluten-Free	152	0.08	0.27	0	1
GMO-Free	152	0.05	0.22	0	1
Halal	152	0.08	0.27	0	1
High Protein	152	0.00	0.00	0	0
High Satiety	152	0.03	0.16	0	1
High/Added Fiber	152	0.46	0.50	0	1
Immune System (Functional)	152	0.03	0.18	0	1
Interesting Packaging	152	0.00	0.00	0	0
Kosher	152	0.13	0.33	0	1
Limited Edition	152	0.03	0.18	0	1
Low/No/Reduced Allergen	152	0.09	0.29	0	1
Low/No/Reduced Calorie	152	0.00	0.00	0	0
Low/No/Reduced Carb	152	0.00	0.00	0	0
Low/No/Reduced Cholesterol	152	0.00	0.00	0	0
Low/No/Reduced Fat	152	0.14	0.35	0	1
Low/No/Reduced Glycemic	152	0.01	0.11	0	1
Low/No/Reduced Lactose	152	0.01	0.11	0	1
Low/No/Reduced Saturated Fat	152	0.06	0.24	0	1
Low/No/Reduced Sodium	152	0.25	0.43	0	1
Low/No/Reduced Sugar	152	0.11	0.31	0	1
Low/No/Reduced Transfat	152	0.07	0.26	0	1
Male	152	0.00	0.00	0	0
Microwaveable	152	0.20	0.40	0	1
No Additives/Preservatives	152	0.28	0.45	0	1

Table A1. *Cont.*

Variable	Obs.	Mean	St. Dev.	Min	Max
No Animal Ingredients	152	0.07	0.25	0	1
Novel	152	0.00	0.00	0	0
On-the-Go	152	0.01	0.11	0	1
Organic	152	0.06	0.24	0	1
Other (Functional)	152	0.06	0.24	0	1
Portionability	152	0.00	0.00	0	0
Prebiotic	152	0.01	0.11	0	1
Premium	152	0.04	0.20	0	1
Seasonal	152	0.00	0.00	0	0
Seniors (aged 55+)	152	0.00	0.00	0	0
Slimming	152	0.01	0.11	0	1
Social Media	152	0.00	0.00	0	0
Time/Speed	152	0.12	0.32	0	1
Vegan	152	0.07	0.25	0	1
Vegetarian	152	0.76	0.43	0	1
Vitamin/Mineral Fortified	152	0.22	0.41	0	1
Weight & Muscle Gain	152	0.00	0.00	0	0
Wholegrain	152	0.47	0.50	0	1

References

1. GBD 2017 Diet Collaborators. Health effects of dietary risks in 195 countries, 1990–2017: A systematic analysis for the Global Burden of Disease Study 2017. *Lancet* **2019**, *393*, 1958–1972. [CrossRef]
2. Ruxton, C.H.S.; Derbyshire, E. Women's diet quality in the UK. *Nutr. Bull.* **2010**, *35*, 126–137. [CrossRef]
3. Betts, J.; Chowdhury, E.; Gonzalez, J.; Richardson, J.; Tsintzas, K.; Thompson, D. Is breakfast the most important meal of the day? *Proc. Nutr. Soc.* **2016**, *75*, 464–474. [CrossRef] [PubMed]
4. Gibney, M.J.; Barr, S.I.; Bellisle, F.; Drewnowski, A.; Fagt, S.; Livingstone, B.; Masset, G.; Varela Moreiras, G.; Moreno, L.A.; Smith, J.; et al. Breakfast in human nutrition: The international breakfast research initiative. *Nutrients* **2018**, *10*, 559. [CrossRef] [PubMed]
5. Hill, G.M. The impact of breakfast especially ready-to-eat cereals on nutrient intake and health of children. *Nutr. Res.* **1995**, *15*, 595–613. [CrossRef]
6. Liu, Y.; Tooze, J.A.; Zhang, Y.; Leidy, H.J.; Bailey, R.L.; Wright, B.; Ma, M.; Stluka, S.; Remley, D.T.; McCormack, L.A.; et al. Breakfast consumption is positively associated with usual nutrient intakes among food pantry clients living in rural communities. *J. Nutr.* **2019**. [CrossRef]
7. Spence, C. Breakfast: The most important meal of the day? *Int. J. Gastron. Food Sci.* **2017**, *8*, 1–6. [CrossRef]
8. British Dietetic Association. Healthy Breakfast. 2016. Available online: https://www.bda.uk.com/foodfacts/healthy_breakfast (accessed on 24 October 2019).
9. Gaal, S.; Kerr, M.A.; Ward, M.; McNulty, H.; Livingstone, M.B.E. Breakfast consumption in the UK: Patterns, nutrient intake and diet quality. A study from the International Breakfast Research Initiative Group. *Nutrients* **2018**, *10*, 999. [CrossRef]
10. Reeves, C.; Halsey, L.G.; McMeel, Y.; Huber, J.W. Breakfast habits, beliefs and measures of health and wellbeing in a nationally representative UK sample. *Appetite* **2013**, *60*, 51–57. [CrossRef]
11. Gibson, S.A.; Gunn, P. What's for breakfast? Nutritional implications of breakfast habits: Insights from the NDNS dietary records. *Nutr. Bull.* **2011**, *36*, 78–86. [CrossRef]
12. Mintel. *Breakfast Cereals-UK-August 2017*; Mintel: London, UK, 2017.
13. Michels, N.; De Henauw, S.; Beghin, L.; Cuenca-García, M.; Gonzalez-Gross, M.; Hallstrom, L.; Kafatos, A.; Kersting, M.; Manios, Y.; Marcos, A.; et al. Ready-to-eat cereals improve nutrient, milk and fruit intake at breakfast in European adolescents. *Eur. J. Nutr.* **2016**, *55*, 771–779. [CrossRef] [PubMed]
14. Priebe, M.G.; McMonagle, J.R. Effects of ready-to-eat-cereals on key nutritional and health outcomes: A systematic review. *PLoS ONE* **2016**, *11*, 10–35. [CrossRef] [PubMed]
15. Bazzano, L.A.; Song, Y.; Bubes, V.; Good, C.K.; Manson, J.E.; Liu, S. Dietary intake of whole and refined grain breakfast cereals and weight gain in men. *Obes. Res.* **2005**, *13*, 1952–1960. [CrossRef] [PubMed]

16. De La Hunty, A.; Gibson, S.; Ashwell, M. Does regular breakfast cereal consumption help children and adolescents stay slimmer? A systematic review and meta-analysis. *Obes. Facts* **2013**, *6*, 70–85. [CrossRef]
17. Mintel. *Breakfast Cereals-UK-August 2019*; Mintel: London, UK, 2019.
18. Nevo, A. Measuring market power in the ready-to-eat cereal industry. *Econometrica* **2001**, *69*, 307–342. [CrossRef]
19. Raubitschek, R.S. Hitting the jackpot: Product proliferation by multiproduct firms under uncertainty. *Int. J. Gastron. Food Sci.* **1988**, *6*, 469–488. [CrossRef]
20. Dijksterhuis, G. New product failure: Five potential sources discussed. *Trends Food Sci. Technol.* **2016**, *50*, 243–248. [CrossRef]
21. Rudolph, M. The food product development process. *Br. Food J.* **1995**, *97*, 3–11. [CrossRef]
22. Lavriša, Ž.; Pravst, I. Marketing of foods to children through food packaging is almost exclusively linked to unhealthy foods. *Nutrients* **2019**, *11*, 1128. [CrossRef]
23. Saba, A.; Vassallo, M.; Shepherd, R.; Lampila, P.; Arvola, A.; Dean, M.; Lähteenmäki, L. Country-wise differences in perception of health-related messages in cereal-based food products. *Food Qual. Prefer.* **2010**, *21*, 385–393. [CrossRef]
24. Talati, Z.; Pettigrew, S.; Hughes, C.; Dixon, H.; Kelly, B.; Ball, K.; Miller, C. The combined effect of front-of-pack nutrition labels and health claims on consumers' evaluation of food products. *Food Qual. Prefer.* **2016**, *1*, 57–65. [CrossRef]
25. Sussman, R.L.; McMahon, A.T.; Neale, E.P. An audit of the nutrition and health claims on breakfast cereals in supermarkets in the Illawarra region of Australia. *Nutrients* **2019**, *11*, 1604. [CrossRef] [PubMed]
26. Hieke, S.; Kuljanic, N.; Pravst, I.; Miklavec, K.; Kaur, A.; Brown, K.; Egan, B.; Pfeifer, K.; Gracia, A.; Rayner, M. Prevalence of nutrition and health-related claims on pre-packaged foods: A five-country study in Europe. *Nutrients* **2016**, *8*, 137. [CrossRef] [PubMed]
27. Kaur, A.; Scarborough, P.; Matthews, A.; Payne, S. How many foods in the UK carry health and nutrition claims, and are they healthier than those that do not? *Public Health Nutr.* **2015**, *19*, 988–997. [CrossRef]
28. Kaur, A.; Scarborough, P.; Rayner, M. A systematic review, and meta-analyses, of the impact of health-related claims on dietary choices. *Int. J. Behav. Nutr. Phys. Act.* **2017**, *14*, 93. [CrossRef]
29. Steinhauser, J.; Hamm, U. Consumer and product-specific characteristics influencing the effect of nutrition, health and risk reduction claims on preferences and purchase behavior: A systematic review. *Appetite* **2018**, *127*, 303–323. [CrossRef]
30. Cox, D.R. Regression models and life-tables. *J. R. Stat. Soc. Ser. B Stat. Methodol.* **1972**, *34*, 187–220. [CrossRef]
31. Revoredo-Giha, C. Hitting the jackpot and the health agenda: The case of processed potato products. *Procedia Econ. Financ.* **2014**, *14*, 525–533. [CrossRef]
32. De Blok, B.M.; Vlieg-Boerstra, B.J.; Oude Elberink, J.N.G.; Duiverman, E.J.; DunnGalvin, A.; Hourihane, J.O.B.; Dubois, A.E. A framework for measuring the social impact of food allergy across Europe: A EuroPrevall state of the art paper. *Allergy* **2017**, *62*, 733–737. [CrossRef]
33. Dawson, J. Retailer activity in shaping food choice. *Food Qual. Prefer.* **2013**, *28*, 339–347. [CrossRef]
34. Stewart-Knox, B.; Mitchell, P. What separates the winners from the losers in new food product development? *Trends Food Sci. Technol.* **2003**, *14*, 58–64. [CrossRef]
35. Shewry, P.R.; Hey, S.J. Do we need to worry about eating wheat? *Nutr. Bull.* **2016**, *41*, 6–13. [CrossRef] [PubMed]

nutrients

MDPI

Article

The Influence of Health Messages in Nudging Consumption of Whole Grain Pasta

Giovanni Sogari [1,2,*], Jie Li [2], Michele Lefebvre [3], Davide Menozzi [1], Nicoletta Pellegrini [1], Martina Cirelli [1], Miguel I. Gómez [2] and Cristina Mora [1]

1. Department of Food and Drug, University of Parma, 43124 Parma, Italy; davide.menozzi@unipr.it (D.M.); nicoletta.pellegrini@unipr.it (N.P.); martina.cirelli@studenti.unipr.it (M.C.); cristina.mora@unipr.it (C.M.)
2. Charles H. Dyson School of Applied Economics and Management, Cornell University, Ithaca, NY 14850, USA, jl2522@cornell.edu (J.L.); mig7@cornell.edu (M.I.G.)
3. White Lodging School of Hospitality & Tourism Management, Purdue University Northwest, Hammond, IN 46323, USA; mlefebvre@pnw.edu

* Correspondence: giovanni.sogari@unipr.it

Received: 20 November 2019; Accepted: 3 December 2019; Published: 6 December 2019

check for updates

Abstract: Health messages may be an important predictor in the selection of healthier food choices among young adults. The primary objective of our study is to test the impact of labeling whole grain pasta with a health message descriptor displayed at the point-of-purchase (POP) on consumer choice in a campus dining setting. The study was conducted in a large US college dining venue during lunch service; data were collected during a nine-week period, for a total of 18 days of observation. Each day, an information treatment (i.e., no-message condition; vitamin message, fiber message) was alternated assigned to whole grain penne. Over the study period, the selection of four pasta options (white penne, whole grain penne, spinach fettuccine, and tortellini) were recorded and compiled for analysis. Logistic regression and pairwise comparison analyses were performed to estimate the impact of health messages on diners' decisions to choose whole grain penne among the four pasta types. Our results indicate that only the message about vitamin benefits had a significant effect on this choice, with a 7.4% higher probability of selecting this pasta than the no-message condition and 6.0% higher than the fiber message condition. These findings suggest that psychological health claims (e.g., reduction of fatigue) of whole grains seem more attractive than physiological health claims (e.g., maintaining a healthy weight) for university students. In line with the 2015–2020 Dietary Guidelines for Americans, our results suggest that small changes made at the POP have the potential to contribute to significant improvements in diet (e.g., achieving recommended levels of dietary fiber). These findings have important implications for food service practitioners in delivering information with the greatest impact on healthy food choices.

Keywords: choice architecture; dining environment; campus; college students; cereal grains; dietary fiber; vitamins; claim; information

1. Introduction

The move of young adults from living at home with their parents to college life is considered a critical phase for healthy eating behavior and can contribute to altered dietary habits such as high-energy intake and non-nutritious, unbalanced diets [1–3]. This change of life can contribute to the development of noncommunicable diet-related diseases (e.g., obesity, type 2 diabetes, some types of cancer, and cardiovascular diseases) in later stages of life [4,5].

In response, the U.S Department of Agriculture (USDA), has called for a transition toward healthier diets, including foods rich in nutrients (e.g., essential vitamins and minerals as well as dietary fiber

("Nondigestible soluble and insoluble carbohydrates (with three or more monomeric units), and lignin that are intrinsic and intact in plants; isolated or synthetic nondigestible carbohydrates (with three or more monomeric units) determined by FDA to have physiological effects that are beneficial to human health" [6]) that contribute to positive health outcomes [7].

According to health and nutrition experts, diets with a significant intake of whole grains (i.e., "Grains and grain products made from the entire grain seed, usually called the kernel, which consists of the bran, germ, and endosperm. If the kernel has been cracked, crushed, or flaked, it must retain the same relative proportions of bran, germ, and endosperm as the original grain in order to be called whole grain. Many, but not all, whole grains are also sources of dietary fiber." [7] (p. 96)) and low intake of refined grains are associated with reduced risks of chronic disease [8,9] and are associated with increased diet quality [10,11]. Specifically, thanks to the health-promoting components of whole grains, such as dietary fibers, vitamins, minerals, antioxidants, and other bioactive compounds [12–14], the consumption of this type of cereal grain is often associated with a lower risk of metabolic syndrome, type 2 diabetes [15], and cardiovascular diseases [16]. Moreover, whole grain intake has been associated with reduced risks of obesity and weight gain in the general population [8,9], including young adults [17]. One reason is that diets rich in dietary fibers and whole grains tend to have larger food volume, which results in increasing the status of satiety [18], therefore reducing hunger [8]. This increased satiety and reduction in hunger lead to lower calorie intake, resulting in weight loss or weight gain prevention.

In addition, whole grains are also rich sources of vitamins and minerals [7]; generally accepted scientific evidence suggests that these nutrients, especially B vitamins (thiamin, riboflavin, niacin, folic acid), iron, and magnesium contribute to the reduction of tiredness and fatigue [19].

In our study, we use health claims, including messages with better-known whole grain benefits (e.g., fibers have positive effects on weight management) as well as those with less familiar and more specific claims (e.g., the relationship between vitamins and reduced fatigue) [20]. It should be noted that in the US and the European Union, a food-related health claim must be approved by the FDA (the Food and Drug Administration, of the United States Department of Health and Human Services,) and EFSA (the European Food Safety Authority of the European Union), respectively, and it must be supported by a significant body of research showing the relationship between the product and the health claim [21–24].

Past evidence showed that on-site nutrition policies (e.g., information campaigns) can improve dietary intake, resulting in favorable effects on weight-related outcomes [7]. However, awareness and use of nutrition labels among US university students are still not well understood and are highly dependent on the type of message considered [25].

The primary aim of our study is to test the impact of labeling whole grain pasta with a health message displayed at the point-of-purchase (POP) in a US university campus dining setting on diners' choice of whole grain penne among four types of pasta. We hypothesize that a health message increased the probability of selecting the whole grain pasta relative to the no-message condition. A secondary aim is to determine which message intervention (physiological vs. psychological health benefit) would have a larger effect on whole grain pasta choices.

Our paper contributes to the literature by providing insights about the effectiveness of communication interventions about health-promoting food components as a mechanism for encouraging healthy eating behavior, without reducing consumer choice.

2. Background

A wide variety of interventions have been utilized to encourage healthy food choices, including altered choice environments to improve selections of desired options (i.e., nudging). Thaler & Sunstein [26] define a nudge as "any aspect of the choice architecture that alters people's behavior in a predictable way without forbidding any options or significantly changing their economic incentives".

The choice architecture includes all the physical elements in a defined environment that individuals experience when they make choices, including the information provided [27,28].

A systematic review by Vecchio & Cavallo [29] demonstrated that many physical elements have been taken into consideration in experiments targeting individuals, including product placement, the effect of priming, environmental cues, and varying portion sizes. Here, we focus on studies that have used point-of-purchase (POP) information to show the effect of different types of POP messaging on food choices among university students. First, studies have shown that at the POP rational decision behavior is affected by heuristic judgment [28]. For instance, Peterson et al. [30] showed positive changes in students' perceptions and selections of healthy foods by introducing a three-week point-of-selection intervention using "benefit-based messages" in a dining setting. Other studies [31,32] showed how information placement (e.g., 4 × 2.5 in signs placed next to the targeted food), graphics (e.g., eye-catching promotion), and message content (e.g., appeal, relevance and length of the text) are important in influencing healthy food choices. As reported by a systematic review of food environment interventions in university settings [33], useful interventions are mostly focused on nutrition messages and nutrient labeling at the POP.

Only a few studies have investigated potential interventions to specifically promote whole grain products [34,35]. In general, grains, including whole grains, are staple foods in many countries [13] and can be found in single foods (e.g., rice, oatmeal), or as an ingredient in many food products (e.g., bread, cereals, crackers, and pasta) [7]. In our study, we focused on whole grain pasta. The reasons for choosing pasta are its popularity and versatility, which make it one of the most ubiquitous durum wheat-derived products around the world [36,37]. Moreover, pasta is a food item that can be fully produced using whole grains [37]. Finally, previous studies (e.g., Larson et al. [38]) noted that the potential to influence whole-grain intake in a school environment merits investigation.

Few US campus dining venues have promoted whole grains on the menu. However, in recent years, this trend is changing since a campus dining initiative called "Menus of Change: The Business of Healthy, Sustainable, and Delicious Food Choices" was founded in 2012 [39]. The primary focus of this initiative is to achieve healthy and sustainable menus, including the promotion of the use of whole grains across all menus on college campuses. From Menus of Change, the Menus of Change University Research Collaborative (MCURC) was founded jointly by Stanford University and The Culinary Institute of America. MCURC was established with working groups of scholars and campus dining leaders interested in using college and university dining as a platform to establish and accelerate efforts to move campus diners towards healthier and more sustainable menus. One of the working groups' priorities is on "menuing healthy". The primary goal is to use campus dining venues as a platform for education and learning around these principles, and thus educate students and prepare them for a future of healthy and sustainable food decisions. Some of the core principles, in line with the USDA's 2015–2020 Dietary Guidelines for Americans, include making whole, intact grains the new norm, and focusing on whole, minimally processed foods; both of which focus on incorporating more whole grains into the diet. Although the Dietary Guidelines for Americans states that at least half of total grain intake should be provided by whole grains [6], most of the population today still fail to meet the recommended levels of whole grain intake [9,10,19]. In our study, the nudge was implemented in order to draw attention to whole grain penne, which was considered the relatively healthier item among different pasta options in the focal dining venue. The rationale was to inform individuals of the nature and consequences of their choices [40].

3. Materials and Methods

3.1. Data Collection

The study was conducted in a large US college dining venue during lunch (11 am–2 pm), from February to April 2019. The Institutional Review Board (IRB) of the Office of Research Integrity and Assurance of Cornell University approved this study (Protocol ID: 1810008359) and waived informed consent because no identifying information was collected from participants. Data were collected two weekdays per nine weeks (n = 18 days of observation in total) in an 'a la carte' college dining setting where students could choose foods from salad, soup, grill, Mexican, Asian, and pasta stations.

The experiment employed different nudges to promote the purchase of whole grain pasta. The "health claim" nudge informs individuals of the nature and health consequences of their food choices. The field experiment was designed over a nine-week period, as seen in Table 1, such that: (1) Weeks 1, 4, and 7 were a no-message condition weeks (control group) where no nudges were implemented; and (2) Weeks 2, 3, 5, 6, 8, and 9 were the experimental nudge weeks in which the two different treatments were alternately implemented to the whole grain penne dish (i.e., one day the vitamin message and the other day the fiber message, and vice versa).

Table 1. The timeline for the study.

Week	Day 1	Day 2
1	No-message condition	No-message condition
2	Vitamin message	Fiber message
3	Vitamin message	Fiber message
4	No-message condition	No-message condition
5	Fiber message	Vitamin message
6	Fiber message	Vitamin message
7	No-message condition	No-message condition
8	Vitamin message	Fiber message
9	Vitamin message	Fiber message

To achieve the highest level of accuracy, the design was balanced (i.e., each of the treatments occurs the same number of times in each period). Based on previous literature [41], two types of messages were developed: (1) a health claim referring to the function of the body (e.g., increase the sense of satiety) and (2) a health claim referring to psychological functions (e.g., the reduction of tiredness). The content of these messages was based on the higher content in fiber and niacin (vitamin B3) in whole grain wheat flour than in refined grain [42], which is present in whole grain pasta [43]. Specifically, we considered the latest scientific opinions on the substantiation of health claims related to wheat bran fiber and increase in faecal bulk [44], and niacin for reduction of tiredness and fatigue, in situations of inadequate micronutrient status [13,20]. Currently, the claims used in our study are approved by EFSA and not by the FDA.

See Table 2 for the complete information provided.

Table 2. Health messages.

	Physiological Health Benefit	Psychological Health Benefit
Extended message provided on an 11 by 14-in poster intervention located right next to the pasta station (Figure A1 in the Appendix A).	EAT WHOLE GRAIN PASTA Whole grains are rich in fiber, which will make you feel more full. Feeling fuller will help you maintain a healthy weight!	EAT WHOLE GRAIN PASTA Whole grains are rich in B vitamins, which help to reduce fatigue. A reduction in fatigue will help you be more alert!
Short message provided in 3 by 2-in card intervention placed directly in front of the targeted pasta (Figure A2 in the Appendix A).	Whole grains are rich in fiber, which will make you feel more full.	Whole grains are rich in B vitamins, which help to reduce fatigue.

Note: The brand of pasta types served at the college dining venues was Barilla (Ames, IA and Avon, NY). The nutrition composition and fact can be retrieved from its commercial web site (barilla.com/en-us/products/pasta/).

Each day of data collection, two research assistants discretely recorded the number of diners (i.e., students) who selected one of the four pasta options served (white penne, whole grain penne, spinach fettuccine, and tortellini). Customers were given the option to first choose the type of (plain) pasta, then the sauces, and toppings, and finally cheese. Thereby, we assumed that the type pasta (e.g., white vs. whole grain penne) is the main factor influencing the selection because all other choices are made afterwards. As a result, the attribute of taste, which is often more important than health claims in influencing purchasing decisions [45], is taken into consideration only after choosing the type of pasta. It was not possible to control for other determinants of food choices, such as more physiological mechanisms (e.g., hunger and satiety) and habitual patterns (e.g., past choices) [28].

Moreover, no changes were made to how the pasta was prepared or served over the period of data collection. The price for all types of pasta was identical and thus cannot be considered a limiting factor in the choice of pasta. Lastly, no other alterations were made to the pasta menu signage; flavor of sauces, pasta, or toppings was not highlighted in any way to keep the offerings consistent across treatments.

Table 3 shows the conceptual level of our nudging study.

Table 3. Illustration of the whole grain pasta nudging study.

	Elements of Nudging Research
Type of study	Proof of concept (controlled-field experiment at a college dining venue).
Independent variable (nudge)	Placement of two health-promoting component messages about whole grains located at the pasta station. A no-message treatment was included as well.
Experimental design	Nonrandomized study design whereby individuals were exposed to experimental condition or control condition over a nine-week period
Dependent variable	The number of diners who selected whole grain pasta and the other types of pasta. Data was recorded by two researchers who stood next to the pasta station
Mediators explored	No individual data were obtained from the participants except gender (via observation) and the selected type of pasta.
Moderators explored	Gender was explored

3.2. Data Analysis

The data were first recorded in Excel (Microsoft Excel, 2017) using a mobile device at the dining venue. Analyses were carried out using the Statistical Package for Social Sciences (SPSS) version 24 SPSS (SPSS Inc., Chicago, IL, USA) and STATA version 16 (Stata Corporation, College Station, TX, USA).

Data were coded using numbers to represent the different types of pasta (i.e., 'white penne' = 1; 'whole grain penne' = 2; 'tortellini' = 3; and 'fettuccine' = 4), and various types of health message treatments (i.e., "no-health message control group" = 0; "rich in vitamins" = 1; "rich in fiber" = 2). The only moderator explored was gender because it was the only sociodemographic characteristic easily recognized without interacting with subjects in our sample. To investigate the association between the whole grain pasta choice probability and various health message treatments, a logistic regression was used, including treatment and gender as predictors.

Specifically, we employed a generalized linear model with a binomial distribution in a logit link function with post hoc analysis from that model to compare the different probabilities across treatments. The binary dependent variable in the logistic regression model was set to one if whole grain penne was chosen and zero otherwise. The week and days when of data collection were included as fixed effects in the statistical analysis to control of time effects on the pasta choices.

4. Results

First, most diners chose tortellini ($n = 1325$) and spinach fettuccine ($n = 992$), followed by white and whole grain penne. Specifically, 725 and 692 of 3734 total diners selected the white penne (19.4%) and the whole grain penne (18.5%), respectively.

Figure 1 presents the percentage shares of diners choosing different types of pasta across the three treatment groups. The Wilcoxon nonparametric paired tests show that labelling whole grain penne with the vitamin message resulted in a 7.4% higher probability of selecting this pasta than the no-message condition ($p < 0.001$) and 6% higher than the fiber message condition ($p < 0.001$). No significant difference was observed between the fiber message and the selection of whole grain penne. The results also show that when presented with the vitamin message for whole grain penne, the percentage of diners who chose spinach fettuccine decreased by 4% relative to the percentage of diners who chose spinach fettucine when no message was presented ($p < 0.05$). The probability of choosing spinach fettuccine is not significantly different than choosing whole grain penne when presented with the vitamin message ($p = 0.482$). These effects were not observed for tortellini pasta, with its frequency remaining stable across treatments.

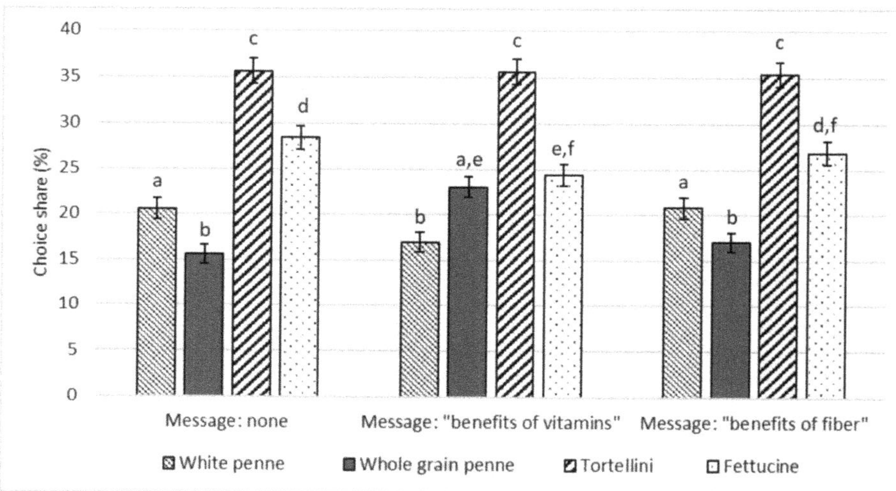

Figure 1. Diners choosing white penne, whole grain penne, tortellini, and fettuccine on the total, by treatment (% values, error bars represent standard errors). Different letters imply values significantly different at $p < 0.05$ probability level between products within the same treatment (Wilcoxon signed rank sum test) and between different treatments (Wilcoxon–Mann–Whitney test).

Table 4 shows the results of the logistic regression, only considering whole grain penne within the total of individuals who chose penne, because white and whole grain penne can be considered close substitutes to each other. The results show that the coefficient for the vitamin message is positive and significant, indicating that the presence of the vitamin message about the whole grain pasta increases the probability that diners chose whole grain penne relative to the control situation (no-message condition). Moreover, the results show that the coefficient of the variable Female is not significant at 5%, indicating that gender has no significant impact on diner's probability in choosing whole grain penne. Finally, the significance of the constant term indicates that ceteris paribus, the choice probability of whole grain penne is lower than for white penne.

Table 4. Regression results of whole grain penne among the total of penne meals.

Independent Variables	Coef.	Std. Err.	z	*p*-Value	95% Conf. Interval
Constant	−0.409	0.180	−2.26	0.024 **	(−0.794 – −0.055)
Vitamin message	0.528	0.249	2.12	0.034 **	(0.039 – 1.017)
Fiber message	0.036	0.249	0.15	0.883	(−0.452 – 0.525)
Female [a]	0.213	0.109	1.95	0.052	(−0.001 – 0.428)

[a] Reference group for gender was male. A total of 1417 observations were used. ** indicates a significant difference at $p < 0.05$.

We also tested the interactions of gender and the different health messages, but did not find any significant interaction effect. This result suggests that males and females respond to information treatments in a similar way.

To further test the impact of information on whole grain penne relative to all other types of pasta, we employed another logit regression considering all pasta types. In this case, the dependent variable was equal to 1 if diners chose whole grain penne, 0 if they chose one of the remaining three types (white penne, spinach fettuccine or tortellini). The results, presented in Table 5, show that the choice probability of whole grain penne is significantly lower than for the other pasta types (constant term significant at $p < 0.001$). However, the choice probability of whole grain penne is higher in the presence of the vitamin message relative to other pasta types. No significant difference was observed with the information treatment about fiber. Considering all pasta types, the gender effect is not significant; in other words, when all pasta types are considered, females and males have the same probability of choosing whole grain penne.

Table 5. Regression of whole grain penne among the total of pasta meals.

Independent Variables	Coef.	Std. Err.	z	*p*-Value	95% Conf. Interval
Constant	−1.771	0.144	−12.25	0.000 ***	(−2.055 – −1.488)
Vitamin message	0.497	0.195	2.55	0.011 **	(0.115 – 0.879)
Fiber message	0.137	0.196	0.70	0.485	(−247 – 0.522)
Female [a]	0.126	0.082	1.53	0.127	(−0.036 – 0.287)

[a] Reference group for gender was male. A total of 3734 observations were used. ** indicates a significant difference at $p < 0.05$; *** indicates a significant difference at $p < 0.001$.

5. Discussion

This paper provides evidence that placing a health message about vitamin benefits (i.e., rich in B vitamins, which help to reduce fatigue) related to whole grain penne in proximity to the targeted pasta significantly increased the number of individuals choosing this option versus white penne. The frequency of the selection of spinach fettuccine decreased when the whole grain penne was labeled with the vitamin message. This suggests that "reduce fatigue" information may change preferences because whole grain penne is perceived to have an added value compared to spinach fettuccine in this context. Nevertheless, no changes in selection frequency were found for the tortellini, which remained constant even when the whole grain penne was labelled with a health message. This result indicates that the type of pasta is the first factor influencing the decision process, especially when it includes other ingredients (e.g., filled pasta such as tortellini). This confirms that palatability of foods is one of the main determinants of food choice [28]. However, our results suggest that connecting the pleasure of eating with healthy food attributes is a good approach [46].

In general, even if whole grain products are generally perceived as healthier [12], the lack of an effective health message campaign promoting these benefits results in low levels of consumer attention and interest towards whole grains [7]. One possible explanation could be that consumers may not understand the health benefits of whole grains and/or may not know how to identify labels and interpret recommendations of such products [20,21].

Therefore, simply making a statement on the label of a cereal grain product such as '100% whole grain' may not be as effective as presenting health-related benefits to influence consumer choice. As suggested from previous studies [47,48], one of the potential barriers to purchase whole grain products is the low knowledge and awareness of their health benefits. Although most people are aware that whole grains are healthy and less processed compared to refined grains [49], there is only a basic understanding of which are the specific health outcomes.

In fact, our results suggest that the presence of only the mention of 'whole grain' is not strong enough to influence student's choice towards whole grain pasta. This could mean that consumers did not associate healthy benefits at the moment of selection, resulting in a lower motivation to choose whole grain penne. Therefore, information about health outcomes from eating whole grains may aid in increasing the likelihood of selecting this pasta over regular types [8].

Even if previous studies (e.g., Marquar et al. [50]) report that individuals usually associated increased fiber intake with the main health benefits of eating whole-grain foods, our results suggest that other health-promoting components (i.e., B vitamins that help to reduce fatigue) have a greater impact in influencing the consumption of whole grains. Our findings also suggest how psychological health claims of whole grains seem more attractive than physiological health claims for university students. Several possible explanations could be drawn. First, the target population might be more interested in "scholarly" benefits, such as the reduction of fatigue, rather than weight management. Nonetheless, it is likely that the information provided will be more effective when it addresses specific needs to the target population [51]. Moreover, when a decision-making process occurs between a long-term benefit (e.g., *'maintain a healthy weight'*) and a short-term benefit (e.g., *'be more alert'*), the latter outcome is likely to be more preferred [27].

Until now, most of the nutrition information provided to consumers focused on macronutrients [52], but our results indicate that providing health benefits of micronutrients (i.e., vitamins) affects consumer food choices. This provides a possible explanation for why the claim referring to vitamins attracted the most interest from consumers [53]. In contrast with previous studies [41,54], which showed how individuals tend to rate psychological health claims as less attractive than physiological health claims, our results indicate the significant impact of psychology/behavior-based benefits (e.g., reduce fatigue). We suggest that attentional account could explain why the latter option is more preferred, while the first appear as less attractive. For instance, the attentional processing is strongly related to different contents of the health benefit [55] as well as the specific topic and audience that the message is addressed to. However, the most common features that may influence persuasion is the perceived argument strength, which refers to individual perceptions of the quality and persuasiveness of the message [56].

Therefore, one of the possible explanations for our findings could be related to how the messages were framed and their ability to evoke recipients' attention and engage them. Perhaps the connection between fiber and satiety ('Whole grains are rich in fiber, which will make you feel more full') was not reported straightforwardly and the content was too vague ('Feeling fuller will help you maintain a healthy weight'), resulting in lower ability to evoke recipients' attention compared to specificity of the vitamin message ('A reduction in fatigue will help you be more alert').

There are, however, other possible explanations. For instance, the content of the vitamin B message is newer (i.e., novelty component) and perhaps this could attract the attention of consumers more.

5.1. Implications

Previous studies [18,57] suggest that whole grain and fiber-rich foods are generally less appealing, but our results show that by keeping the type of pasta constant (i.e., penne), the promotion of whole grain via healthy attribute messaging increases the choice probability of whole grain pasta.

Our findings confirm how the POP nutrition information can help college students make informed and healthy choices [32,58] and could be used to plan interventions in higher education institutions, highlighting the importance of developing messages to influence food choices. For instance, when framing messages (e.g., generic vs. specific health, or benefit vs. risk communication), expert opinion needs to be taken carefully into consideration along with responsiveness from consumers. We showed how more general health benefits (e.g., maintain body weight) is less attractive than a particular outcome (e.g., be more alert).

Consequently, nutrition and dietitian experts, food service management staff, and consumer behavior scientists should work together to develop effective and long-term targeted intervention strategies (e.g., message framing) designed for specific demographics (e.g., university students) within the campus environment (e.g., dining facilities and convenience stores) [1,27,30,31,59,60].

From a marketing perspective, our findings underscore the role of communicating the content and benefits of a food product in an effective and easy-to-read way (e.g., claim related to health benefits) [61]. This approach could be taken into consideration by different stakeholders, including food companies, which can use labels on products to communicate the health benefits of food to consumers [62].

Regulators are also a key stakeholder group which is involved in evidence-based health policy decision-making strategies to improve dietary practices [63]. Nutrition labeling policy could benefit from understanding how specific changes in communication strategies affect consumer behavior.

5.2. Strengths and Limitations

One of the strengths of this field study is that the experiment was carried out in a natural context (a campus dining venue), resulting in the exploration of more natural consumer behavior in response to a nudge [27].

However, this study is not without limitations. First, data collection took place only in a specific setting (college dining venue) with a specific population (university students); thus, the findings should not be generalized to other situations and target groups. Second, we did not collect any sociodemographic information from participants (except gender via observation), resulting in the risk of recurring respondents and a limited possibility to understand potential moderators of the pasta choices. The presence of a specific moderator (i.e., individual permanent trait) could modify the relationship between our dependent variables and the choice probability of whole grain penne [27]. Moreover, our study could be biased by cross-over effects from each message condition. Finally, considering that this is an observational study, we could not be certain that all subjects read the messages; our assumption is that considering the closeness of the information treatment to the pasta station (see Figures A1 and A2 in Appendix A), most diners did notice the health message.

6. Conclusions and Future Studies

In this study, we investigated the role of health-promoting whole grain components in the selection of whole grain penne vs. other types of pasta in a college dining venue.

Our results indicate that targeted information interventions that communicate the health psychology/behavior-based benefits of consuming whole grain (i.e., reduce fatigue) at the POP increased students' whole grain intake.

Moreover, environmental modifications do not involve imposing eating habits and/or limiting food choices, so replacing some refined grain meals with whole grains product consumption should be easier than many other dietary changes [64]. In line with the 2015–2020 Dietary Guidelines For Americans [7], our results suggest that small changes made at the POP can contribute to significant improvements in the diet (e.g., achieving recommended levels of dietary fiber). More generally, our study confirms that it is possible to design a nudge to promote the desired behavior (e.g., eating more whole grains) without conscious effort and still let consumers have the freedom to make a choice [27]. Considering that hard regulation is often not effective to modify food consumption toward a healthy diet [28], the nudge approach could positively change eating habits in specific settings.

Our study sheds light on ways to increase consumption of whole grains, yet more research on this topic is warranted. Future research should investigate individual preferences for whole grain health claims in different eating and purchasing contexts, such as dining venues in worksites and other food service establishments (e.g., restaurants), as well as different target populations (e.g., countries where pasta is a more traditional food). We also recommend that future studies investigate the type of pasta and the healthy attributes separately by testing whole grain vs. nonwhole grain fettuccine as well as whole grain vs. nonwhole grain tortellini.

A natural extension of this work would be to investigate how long this intervention should continue to affect preferences in the long run and create habits among consumers. Considering that the location of the message label is an important factor in the success of the intervention [31], further investigation could also focus on where to place the communication campaigns to be more effective and consider an internal validation to assess whether participants actually read the message. Further work should also take into consideration how participants' subjectively self-reported satiety [65] could influence the choice of whole grain pasta and the impact of different health messages.

Further investigations to address specific nutritional needs could focus on consumers' interest in other types of pasta, such as fortified pasta with protein sources from animal or plant material [66,67]. Finally, future studies should continue to explore how to better frame the content of messages (e.g., argument quality and source credibility), explore attentional account process underlying preferences across different types of messages, and partner flavor messaging with health messaging to determine the combined impact compared to health messaging alone.

Author Contributions: Conceptualization, G.S.; methodology and investigation, G.S., M.L. and M.C.; formal analysis, G.S. and J.L.; writing—original draft preparation, G.S.; writing—review and editing, G.S., M.L., J.L., D.M., N.P., C.M. and M.I.G.; supervision, C.M., N.P. and M.I.G.; project administration, G.S.; funding acquisition, G.S., C.M. and M.I.G.

Funding: This study is part of a wider project called "CONSUMEHealth. Using consumer science to improve healthy eating habits" and has received funding from the European Union's Horizon 2020 Research and Innovation programme under the Marie Sklodowska-Curie grant agreement No 749514.

Acknowledgments: The authors are grateful to all the staff of the Cornell Dining venues for their collaboration and help in carrying out this study. Additionally, we want to express our gratitude to the students who helped in the data collection.

Conflicts of Interest: The authors declare no conflict of interest.

Appendix A

Figure A1. Example label. The left poster depicts the four pasta options available at the station; the right poster shows the long health message about vitamins and the reduction in fatigue.

Figure A2. Example label. This small card depicts the short message about vitamins and the reduction in fatigue placed directly in front of the pasta.

References

1. Sogari, G.; Velez-Argumedo, C.; Gómez, M.; Mora, C. College Students and Eating Habits: A Study Using An Ecological Model for Healthy Behavior. *Nutrients* **2018**, *10*, 1823. [CrossRef] [PubMed]
2. Deliens, T.; Van Crombruggen, R.; Verbruggen, S.; De Bourdeaudhuij, I.; Deforche, B.; Clarys, P. Dietary interventions among university students: A systematic review. *Appetite* **2016**, *105*, 14–26. [CrossRef] [PubMed]
3. Crombie, A.P.; Ilich, J.Z.; Dutton, G.R.; Panton, L.B.; Abood, D.A. The freshman weight gain phenomenon revisited. *Nutr. Rev.* **2009**, *67*, 83–94. [CrossRef] [PubMed]
4. World Health Organization. Obesity and Overweight. Available online: https://www.who.int/en/news-room/fact-sheets/detail/obesity-and-overweight (accessed on 21 October 2019).
5. World Health Organization. *Noncommunicable Diseases: The Slow Motion Disaster*; WHO: Geneva, Switzerland, 2017.
6. Department of Health and Human Service. *Food and Drug Admisistration Federal Register Volume 81, Issue 103 (May 27, 2016)*; Department of Health and Human Service: Washington, DC, USA, 2016
7. U.S. Department of Health and Human Services and U.S. Department of Agriculture. *2015–2020 Dietary Guidelines for Americans*; ODPHP: Washington, DC, USA, 2015.
8. Jones, J.M.; Engleson, J. Whole Grains: Benefits and Challenges. *Annu. Rev. Food Sci. Technol.* **2010**, *1*, 19–40. [CrossRef] [PubMed]
9. Ye, E.Q.; Chacko, S.A.; Chou, E.L.; Kugizaki, M.; Liu, S. Greater whole-grain intake is associated with lower risk of type 2 diabetes, cardiovascular disease, and weight gain. *J. Nutr.* **2012**, *142*, 1304–1313. [CrossRef] [PubMed]
10. O'Neil, C.E.; Nicklas, T.A.; Zanovec, M.; Cho, S.S.; Kleinman, R. Consumption of whole grains is associated with improved diet quality and nutrient intake in children and adolescents: The National Health and Nutrition Examination Survey 1999–2004. *Public Health Nutr.* **2011**, *14*, 347–355. [CrossRef]
11. McGill, C.R.; Fulgoni, V.L., III; Devareddy, L. Ten-Year Trends in Fiber and Whole Grain Intakes and Food Sources for the United States Population: National Health and Nutrition Examination Survey 2001–2010. *Nutrients* **2015**, *7*, 1119–1130. [CrossRef]

12. Shepherd, R.; Dean, M.; Lampila, P.; Arvola, A.; Saba, A.; Vassallo, M.; Claupein, E.; Winkelmann, M.; Lähteenmäki, L. Communicating the benefits of wholegrain and functional grain products to European consumers. *Trends Food Sci. Technol.* **2012**, *25*, 63–69. [CrossRef]
13. European Commission. Whole Grain. Available online: https://ec.europa.eu/jrc/en/health-knowledge-gateway/promotion-prevention/nutrition/whole-grain (accessed on 21 October 2019).
14. Lattimer, J.M.; Haub, M.D. Effects of dietary fiber and its components on metabolic health. *Nutrients* **2010**, *2*, 1266–1289. [CrossRef]
15. De Munter, J.S.L.; Hu, F.B.; Spiegelman, D.; Franz, M.; Van Dam, R.M. Whole grain, bran, and germ intake and risk of type 2 diabetes: A prospective cohort study and systematic review. *PLoS Med.* **2007**, *4*, e261. [CrossRef]
16. Anderson, J.W. Whole grains and coronary heart disease: The whole kernel of truth. *Am. J. Clin. Nutr.* **2004**, *80*, 1459–1460. [CrossRef] [PubMed]
17. Quick, V.; Wall, M.; Larson, N.; Haines, J.; Neumark-Sztainer, D. Personal, behavioral and socio-environmental predictors of overweight incidence in young adults: 10-yr longitudinal findings. *Int. J. Behav. Nutr. Phys. Act.* **2013**, *10*, 37. [CrossRef] [PubMed]
18. Martini, D.; Brusamolino, A.; Del Bo', C.; Laureati, M.; Porrini, M.; Riso, P. Effect of fiber and protein-enriched pasta formulations on satiety-related sensations and afternoon snacking in Italian healthy female subjects. *Physiol. Behav.* **2018**, *185*, 61–69. [CrossRef] [PubMed]
19. Sadler, M. 13-Authorised EU health claims for vitamins and minerals. In *Foods, Nutrients and Food Ingredients with Authorised Eu Health Claims*; Woodhead Publishing: Cambridge, UK, 2014; pp. 274–297. ISBN 9780857098429.
20. EFSA Panel on Dietetic Products, Nutrition and Allergies (NDA) Scientific Opinion on the substantiation of health claims related to niacin and reduction of tiredness and fatigue (ID 47), contribution to normal energy-yielding metabolism (ID 51), contribution to normal psychological functions (ID 55), maintenance of normal blood flow (ID 211), and maintenance of normal skin and mucous membranes (ID 4700) pursuant to Article 13 (1) of Regulation (EC) No 1924/2006. *EFSA J.* **2010**, *8*, 1757.
21. European Commission. *COMMISSION REGULATION (EU) No 432/2012 of 16 May 2012 Establishing a List of Permitted Health Claims Made on Foods, Other Than Those Referring to the Reduction of Disease Risk and to Children's Development and Health*; European Commission: Brussels, Belgium, 2012; Volume 136.
22. European Commission. Regulation (EC) No 1924/2006 of the European Parliament and of the Council of 20 December 2006 on Nutrition and Health Claims Made on Foods. *Off. J. Eur. Union L* **2006**, *404*, 9–25.
23. Food and Drug Administration Guidance for Industry: Evidence-Based Review System for the Scientific Evaluation of Health Claims. Available online: https://www.fda.gov/regulatory-information/search-fda-guidance-documents/guidance-industry-evidence-based-review-system-scientific-evaluation-health-claims (accessed on 21 October 2019).
24. Binns, N. 1-The regulation of health claims in Europe. In *Foods, Nutrients and Food Ingredients with Authorised Eu Health Claims*; Woodhead Publishing: Cambridge, UK, 2014; pp. 3–22. ISBN 9780857098429.
25. Christoph, M.J.; An, R.; Ellison, B. Correlates of nutrition label use among college students and young adults: A review. *Public Health Nutr.* **2016**, *19*, 2135–2148. [CrossRef]
26. Thaler, R.H.; Sunstein, C.R. *Nudge: Improving Decisions about Health, Wealth, and Happiness*; Yale University Press: New Haven, CT, USA, 2008; ISBN 978-0-300-12223-7.
27. Van Kleef, E.; Van Trijp, H.C.M. Methodological Challenges of Research in Nudging. In *Methods in Consumer Research, Volume 1*; Woodhead Publishing: Cambridge, UK, 2018; pp. 329–349.
28. Leng, G.; Adan, R.A.H.; Belot, M.; Brunstrom, J.M.; De Graaf, K.; Dickson, S.L.; Hare, T.; Maier, S.; Menzies, J.; Preissl, H.; et al. The determinants of food choice. *Proc. Nutr. Soc.* **2017**, *76*, 316–327. [CrossRef]
29. Vecchio, R.; Cavallo, C. Increasing healthy food choices through nudges: A systematic review. *Food Qual. Prefer.* **2019**, *78*, 103714. [CrossRef]
30. Peterson, S.; Duncan, D.P.; Null, D.B.; Roth, S.L.; Gill, L. Positive changes in perceptions and selections of healthful foods by college students after a short-term point-of-selection intervention at a dining hall. *J. Am. Coll. Health* **2010**, *58*, 425–431. [CrossRef]
31. Buscher, L.A.; Martin, K.A.; Crocker, S. Point-of-purchase messages framed in terms of cost, convenience, taste, and energy improve healthful snack selection in a college foodservice setting. *J. Am. Diet. Assoc.* **2001**, *101*, 909–913. [CrossRef]

32. Freedman, M.R.; Connors, R. Point-of-purchase nutrition information influences food-purchasing behaviors of college students: A pilot study. *J. Am. Diet. Assoc.* **2010**, *110*, 1222–1226. [CrossRef] [PubMed]
33. Roy, R.; Kelly, B.; Rangan, A.; Allman-Farinelli, M. Food Environment Interventions to Improve the Dietary Behavior of Young Adults in Tertiary Education Settings: A Systematic Literature Review. *J. Acad. Nutr. Diet.* **2015**, *115*, 1647–1681. [CrossRef] [PubMed]
34. De Wijk, R.A.; Maaskant, A.J.; Polet, I.A.; Holthuysen, N.T.E.; Van Kleef, E.; Vingerhoeds, M.H. An In-Store Experiment on the Effect of Accessibility on Sales of Wholegrain and White Bread in Supermarkets. *PLoS ONE* **2016**, *11*, e0151915. [CrossRef] [PubMed]
35. Van Kleef, E.; Seijdell, K.; Vingerhoeds, M.H.; De Wijk, R.A.; Van Trijp, H.C.M. The effect of a default-based nudge on the choice of whole wheat bread. *Appetite* **2018**, *121*, 179–185. [CrossRef] [PubMed]
36. Martini, D.; Ciccoritti, R.; Nicoletti, I.; Nocente, F.; Corradini, D.; D'Egidio, M.G.; Taddei, F. From seed to cooked pasta: Influence of traditional and non-conventional transformation processes on total antioxidant capacity and phenolic acid content. *Int. J. Food Sci. Nutr.* **2018**, *69*, 24–32. [CrossRef] [PubMed]
37. Brennan, C.S. 13-Fibre-enriched and whole wheat pasta. In *Woodhead Publishing Series in Food Science, Technology and Nutrition*; Delcour, J.A., Poutanen, K.B.T.-F.-R., Eds.; Woodhead Publishing: Cambridge, UK, 2013; pp. 273–290. ISBN 978-0-85709-038-6.
38. Larson, N.I.; Neumark-Sztainer, D.; Story, M.; Burgess-Champoux, T. Whole-Grain Intake Correlates among Adolescents and Young Adults: Findings from Project EAT. *J. Acad. Nutr. Diet.* **2010**, *110*, 230–237. [CrossRef] [PubMed]
39. The Culinary Institute of America. *Menus of Change, 2017 Annual Report*; The Culinary Institute of America: New York, NY, USA, 2017.
40. Sunstein, C. Nudging: A Very Short Guide. *J. Consum. Policy* **2014**, *37*, 583–588. [CrossRef]
41. Siegrist, M.; Stampfli, N.; Kastenholz, H. Consumers' willingness to buy functional foods. The influence of carrier, benefit and trust. *Appetite* **2008**, *51*, 526–529. [CrossRef]
42. Truswell, A.S. Cereal grains and coronary heart disease. *Eur. J. Clin. Nutr.* **2002**, *56*, 1–14. [CrossRef]
43. U.S.D.A FoodData. Central Agricultural Research Service. Available online: Fdc.nal.usda.gov (accessed on 21 October 2019).
44. EFSA Panel on Dietetic Products, Nutrition and Allergies (NDA) Scientific Opinion on the substantiation of health claims related to wheat bran fibre and increase in faecal bulk (ID 3066), reduction in intestinal transit time (ID 828, 839, 3067, 4699) and contribution to the maintenance or achievement of a normal body weight (ID 829) pursuant to Article 13 (1) of Regulation (EC) No 1924/2006. *EFSA J.* **2010**, *8*, 1817.
45. Wills, J.M.; Bonsmann, S.; Kolka, M.; Grunert, K.G. Symposium 2: Nutrition and health claims: Help or hindrance: European consumers and health claims: Attitudes, understanding and purchasing behaviour. *Proc. Nutr. Soc.* **2012**, *71*, 229–236. [CrossRef] [PubMed]
46. Köster, E.P. Diversity in the determinants of food choice: A psychological perspective. *Food Qual. Prefer.* **2009**, *20*, 70–82. [CrossRef]
47. Kamar, M.; Evans, C.; Hugh-Jones, S. Factors influencing adolescent whole grain intake: A theory-based qualitative study. *Appetite* **2016**, *101*, 125–133. [CrossRef] [PubMed]
48. McMackin, E.; Dean, M.; Woodside, J.V.; McKinley, M.C. Whole grains and health: Attitudes to whole grains against a prevailing background of increased marketing and promotion. *Public Health Nutr.* **2013**, *16*, 743–751. [CrossRef] [PubMed]
49. Burgess-Champoux, T.; Marquart, L.; Vickers, Z.; Reicks, M. Perceptions of children, parents, and teachers regarding whole-grain foods, and implications for a school-based intervention. *J. Nutr. Educ. Behav.* **2006**, *38*, 230–237. [CrossRef]
50. Marquart, L.; Pham, A.-T.; Lautenschlager, L.; Croy, M.; Sobal, J. Beliefs about whole-grain foods by food and nutrition professionals, health club members, and special supplemental nutrition program for women, infants, and children participants/State fair attendees. *J. Am. Diet. Assoc.* **2006**, *106*, 1856–1860. [CrossRef]
51. Verbeke, W. Agriculture and the food industry in the information age. *Eur. Rev. Agric. Econ.* **2005**, *32*, 347–368. [CrossRef]
52. Grunert, K.G. Do Consumers Care About Micronutrients? A Perspective on the Possible Role of Vitamin E in the Dietary Choices of Consumers. In *Vitamin E in Human Health*; Weber, P., Birringer, M., Blumberg, J.B., Eggersdorfer, M., Frank, J., Eds.; Springer International Publishing: Cham, Switzerland, 2019; pp. 435–444. ISBN 978-3-030-05315-4.

53. Cavaliere, A.; Ricci, E.C.; Banterle, A. Nutrition and health claims: Who is interested? An empirical analysis of consumer preferences in Italy. *Food Qual. Prefer.* **2015**, *41*, 44–51. [CrossRef]
54. Van Kleef, E.; Van Trijp, H.C.M.; Luning, P. Functional foods: Health claim-food product compatibility and the impact of health claim framing on consumer evaluation. *Appetite* **2005**, *44*, 299–308. [CrossRef]
55. Imhof, M.A.; Schmälzle, R.; Renner, B.; Schupp, H.T. How real-life health messages engage our brains: Shared processing of effective anti-alcohol videos. *Soc. Cogn. Affect. Neurosci.* **2017**, *12*, 1188–1196. [CrossRef]
56. Zhao, X.; Strasser, A.; Cappella, J.N.; Lerman, C.; Fishbein, M. A Measure of Perceived Argument Strength: Reliability and Validity. *Commun. Methods Meas.* **2011**, *5*, 48–75. [CrossRef] [PubMed]
57. Jones, J.M.; Reicks, M.; Adams, J.; Fulcher, G.; Weaver, G.; Kanter, M.; Marquart, L. The Importance of Promoting a Whole Grain Foods Message. *J. Am. Coll. Nutr.* **2002**, *21*, 293–297. [CrossRef] [PubMed]
58. Yepes, M.F. Mobile Tablet Menus: Attractiveness and Impact of Nutrition Labeling Formats on Millennials' Food Choices. *Cornell Hosp. Q.* **2014**, *56*, 58–67. [CrossRef]
59. Menozzi, D.; Sogari, G.; Mora, C. Explaining vegetable consumption among young adults: An application of the theory of planned behaviour. *Nutrients* **2015**, *7*, 7633–7650. [CrossRef]
60. Menozzi, D.; Sogari, G.; Mora, C. Understanding and modelling vegetables consumption among young adults. *LWT Food Sci. Technol.* **2017**, *85*, 1–7. [CrossRef]
61. Corallo, A.; Latino, M.E.; Menegoli, M.; Spennato, A. A Survey to Discover Current Food Choice Behaviors. *Sustainability* **2019**, *11*, 5041. [CrossRef]
62. Corallo, A.; Latino, M.E.; Menegoli, M.; De Devitiis, B.; Viscecchia, R. Human Factor in Food Label Design to Support Consumer Healthcare and Safety: A Systematic Literature Review. *Sustainability* **2019**, *11*, 4019. [CrossRef]
63. World Health Organization. Healthy Diet. Available online: https://www.who.int/news-room/fact-sheets/detail/healthy-diet (accessed on 21 October 2019).
64. Marquart, L.; Wiemer, K.L.; Jones, J.M.; Jacob, B. Whole grain health claims in the USA and other efforts to increase whole-grain consumption. *Proc. Nutr. Soc.* **2003**, *62*, 151–160. [CrossRef]
65. Ding, F.; Hamid, N.; Shepherd, D.; Kantono, K. How is Satiety Affected When Consuming Food While Working on A Computer? *Nutrients* **2019**, *11*, 1545. [CrossRef]
66. Fuad, T.; Prabhasankar, P. Role of Ingredients in Pasta Product Quality: A Review on Recent Developments. *Crit. Rev. Food Sci. Nutr.* **2010**, *50*, 787–798. [CrossRef]
67. Liu, T.; Hamid, N.; Kantono, K.; Pereira, L.; Farouk, M.M.; Knowles, S.O. Effects of meat addition on pasta structure, nutrition and in vitro digestibility. *Food Chem.* **2016**, *213*, 108–114. [CrossRef] [PubMed]

nutrients

MDPI

Article

Conflicting Messages on Food and Beverage Packages: Front-of-Package Nutritional Labeling, Health and Nutrition Claims in Brazil

Ana Clara Duran [1,2,*], Camila Zancheta Ricardo [2], Laís Amaral Mais [3], Ana Paula Bortoletto Martins [2,3] and Lindsey Smith Taillie [4]

1 Center for Food Studies (NEPA), University of Campinas, Campinas 13083-852, Brazil
2 Center for Epidemiological Studies in Nutrition and Health (NUPENS), University of Sao Paulo, Sao Paulo 01246-904, Brazil; ca.zancheta@gmail.com (C.Z.R.); anapaula@idec.org.br (A.P.B.M.)
3 Brazilian Institute for Consumers Defense (IDEC), Sao Paulo 05002-000, Brazil; lais.amaral@idec.org.br
4 Department of Nutrition, University of North Carolina in Chapel Hill, Chapel Hill, NC 27516, USA; taillie@unc.edu
* Correspondence: anaduran@unicamp.br; Tel.: +55-19-3521-2177

Received: 1 October 2019; Accepted: 30 November 2019; Published: 5 December 2019

Abstract: We assessed the prevalence of front-of-package (FoP) claims in the Brazilian packaged food supply and examined whether foods with claims were more likely to be high in critical nutrients. Using data from a random 30% subsample of 11,434 foods and beverages collected in the five largest food retailers in Brazil in 2017 (n = 3491), we classified claims into nutrition, health, and environment. We examined whether foods with claims were more likely to be high in critical nutrients using 95% confidence intervals. Claims were found in 41.2% of the products. Nutrition claims were the most prevalent (28.5%), followed by health (22.1%), and environment-related claims (5.2%). More than 85% of the breakfast cereals, granola bars, and nectars contained claims, which were found in 51% of the dairy beverages. Foods with nutrition claims were more likely to be high in critical nutrients (65.3%; 95% CI 62.3, 68.2% vs. 54.1% 95% CI 52.1, 56.0). Products with health (52.9%; 95% CI 49.3, 56.4% vs. 58.5%; 95% CI 56.6, 60.3) and environment-related claims (33.5%; 95% CI 27.0, 40.8% vs. 58.6%; 95% CI 56.9, 60.2%) were less likely to be high in critical nutrients. FoP claims are prevalent in Brazil, and nutrition claims are more likely to have a poorer nutritional profile. Policymakers should consider restricting the presence of claims on unhealthy products.

Keywords: claims; labeling; nutrition; consumer behavior; policy

1. Introduction

In recent decades, consumption of ultra-processed foods (UPF) in Brazil has rapidly increased [1], with current intake of these foods estimated to contribute 20–30% of a Brazilian's daily caloric intake [2,3]. UPF, which are industrially formulated products usually containing additives, tend to have more sugar, saturated fat, and sodium; less fiber, and higher energy density [1]. Higher intake of UPF has also been associated with higher body mass index (BMI), increased odds of obesity, and excess weight in the Brazilian population [2,4]. In addition, a recent tightly controlled randomized trial found that UPF diets increased weight gain [5].

As part of a larger response to this increase in UPF in the Brazilian food supply, the Brazilian Health Regulatory Agency (Anvisa) is currently discussing the implementation of front-of-package (FoP) nutritional labeling in the country to help inform consumers about the nutritional content of packaged foods and beverages. Similar to the Chilean FoP warning label system, as well as systems adopted in Uruguay, Peru, and currently under final discussions in Mexico and Canada, the proposed labels in Brazil will be able to identify unhealthy foods only and would be applied to products containing excessive levels of added sugar, saturated fat, and sodium with the words "high in (sugar, saturated fat, or sodium)" [6]. Warning labels have been shown to be easier to understand than competing systems such as the traffic light among various groups, including Brazilians [7,8]. In the same proposal, Anvisa lays out options for restricting the use of nutrition claims in foods high in critical nutrients [9].

These proposed regulations have the potential to substantially alter the Brazilian food supply as well as consumers' understanding, perceptions, purchase intentions, and intake of UPF. A recent randomized controlled trial in Brazil found that the exposure to FoP warning labels that only identify unhealthy foods helped consumers understand the unhealthiness of UPF and reduced their intentions to purchase this food [8], consistent with a growing body of literature that shows that warning labels improve consumers' ability to assess the healthfulness of products [10,11] and help them make healthier choices [12–15]. Conversely, nutrition and health claims have been shown to make consumers think a product is healthier than it actually is, working as a marketing strategy rather than an information tool (i.e., the "health halo" effect) [16,17]. This body of research suggests that limiting the use of health and nutrition claims on products with FoP nutritional labeling could further improve consumers' ability to assess the healthfulness of foods and beverages.

Thus, one important question is the prevalence of health and nutrition claims in the Brazilian food supply, and the proportion of these products that would carry FoP nutritional labeling under the proposed regulation. A related question is whether the nutritional profile of products with claims is more favorable compared with products without claims. Understanding the prevalence and nutritional profile of products which would be affected by the proposed regulation is important for developing the most effective policy to improve consumers' ability to make healthier food decisions.

Considering much of the existing evidence is available for high-income countries [18–22], our paper contributes to the literature by showing the prevalence of nutrition, health, and environment-related claims in the food supply of a large middle-income country; and whether foods that carry those claims can be considered healthy.

The objectives of this study are to: (1) quantify the prevalence and type of front-of-package health, nutrition, and environment-related claims in the Brazilian food supply; and (2) examine the prevalence of foods high in critical nutrients and therefore eligible to receive front-of-package nutritional labeling on products with and without claims.

2. Methods

2.1. Database of Food Products Available in Brazilian Market

This was a cross-sectional study that took advantage of information from packaged foods and beverages sold in supermarket chains with the largest market share in Brazil in 2017. Supermarkets account for 59% of the calories purchased in the country [23], and the five supermarkets chains with the largest market share and that control 70% of the Brazilian market were identified using annual food retail sales organized by Euromonitor International in 2016 [24,25]. São Paulo, located in the Southeast region of the country, was chosen as the primary study area as it is the largest city in Brazil. Because one of the five top supermarket chains in Brazil only had stores in the Northeast region of the country, for that specific chain, data collection took place in Salvador, their largest market.

Data on the location of every store of the selected supermarkets chains in the cities of São Paulo and Salvador were gathered from each company's website or customer service, and the addresses were geocoded. The neighborhood of each store was defined as a one-kilometer buffer (using Euclidean distance) around each store location. We used information on income from the household top earner from the latest Brazilian Census conducted in 2010 [26] to determine the mean neighborhood income around each store location. Stores were distributed in tertiles of neighborhood income, and the largest store in square meters of each chain in both the first and the third tertile was selected, except for one chain that only allowed us to collect data in its distribution center, where all the products sold in the stores are found. Formal permission to collect data was obtained from all the supermarkets chains included in this study.

Data were collected between April and July 2017 by trained fieldworkers, according to methods proposed by Kanter et al. (2017) [27]. All packaged foods and beverages were included, and around 13,000 different items had all sides of their package photographed. Data on composition information were entered between July and November 2017 by trained nutritionists in the online platform RedCap, using a template developed by the University of North Carolina at Chapel Hill (UNC) from the United States of America (USA) and by the *Instituto de Nutrición y Tecnología de los Alimentos* (INTA) from Chile, and adapted to be used in Brazil. In this stage, information collected included product information, package size, nutrition facts panel information, list of ingredients, and information about reconstitution (when applied). After the exclusion of items available in more than one package size, products without nutrition information, multipack with different items, products without a list of ingredients, and products with missing values for portion size and/or calories, 11,434 records were maintained in our database.

2.2. Claims Information and Classification

The taxonomy developed by The International Network for Food and Obesity/Non-communicable Diseases (NCDs) Research, Monitoring and Action Support (INFORMAS) was used to classify different types of claims featuring on food and beverages packages [28]. The INFORMAS taxonomy divides claims into three major categories: nutrition claims, health claims, and other claims, which includes other health-related claims and environment-related claims. In our study, we decided to combine other health-related claims with health claims. In addition, we used sub-categories for nutrition and health claims (Table 1).

Table 1. Categories of claims.

Categories of Claims	Subcategories of Claims	Content of Claims
Health Claims	General health claims	Claims related to general beneficial health effects (e.g., healthy, fit, nutritious).
	Claims for special diets [1]	Claims related with allergies/intolerance (e.g., gluten free, dairy free), vegetarian/vegan.
	Claims for 'Natural' [1]	Claims related to natural/pure products, and absence of additives, pesticides, and hormones.
	Others	Nutrient and function claims, reduction of disease risk claims.
Nutrition Claims	Nutrients of concern [2]	Nutrient comparative or nutrient content claims related to saturated fat, trans fat, sugar, and sodium.
	Nutrients not of concern	Nutrient comparative or nutrient content claims related to fiber, vitamins and minerals, unsaturated fatty acids, and protein.
	Others	Health-related ingredients (e.g., quantity of fruits, nuts, wholegrains), claims related to nutrients not included in the other sub-categories created (e.g., carbohydrates).
Environment-Related Claims		Organic, biodiversity, genetically modified organism free.

[1] Although special diets and natural-related claims are classified as "Other Claims" in the INFORMAS taxonomy, we combined them with the other health-related claims; [2] We considered nutrients of concern those for which excess intake is associated with increased non-communicable disease risk [29].

All visible text was allowable to be coded as a claim, including brand names and slogans. We did not consider as a claim the mention of substances in the list of ingredients, and the mention of nutrients as a mandatory part of nutrition labeling. In the case of claims that could be classified as more than one type of claim, the following hierarchy was applied for classification: health claim, nutrition claim, and other claim. For this study, we only considered claims in the front of the package.

We gathered information available on the food labels on health, nutrition, and environmental claims in a random subsample of 30% of all the products food in the main supermarket chains in

Brazil. A 30% random sample was drawn from each of the 128 categories of food primarily used in data entering, yielding 3491 products. We did not find any statistical differences in food composition when we compared this random sample with the universe of photographed food packages.

2.3. Classification of Products According to Their Nutritional Profile

The Pan American Health Organization (PAHO) nutrient profile model (NPM) was used to classify foods according to their nutritional profile because it was developed to be used in various food and nutrition policies in Latin America, including labeling, and identifies unhealthy foods being aligned with the proposed FoP regulations under discussion in Brazil [30]. This NPM considers the level and degree of industrial processing, according to the NOVA classification, as an eligibility criterion. It classifies food products as containing or not excessive amounts of five nutrients: free sugar, total fats, saturated fats, trans fat, and sodium. In addition, the presence of nonnutritive sweeteners in the list of ingredients is also considered in the model. The thresholds determined in the model are applied on the ratio between the content of critical nutrients and the content of energy, and is based on the World Health Organization (WHO) recommendations to prevent obesity and chronic diseases [30]. In our study, we included a modification to the original PAHO NPM, in which a food or beverage was eligible to be regulated and therefore receive FOP warning signs based on the Chilean nutritional labeling law eligibility criteria (Law 20.606/2015). In the Chilean law, only foods and beverages with added salt, sugar, or saturated fat were eligible to receive FOP warning signs for "high in" critical nutrients [31]. Culinary ingredients, as sugar, salt, oils, butter, and milk creams were only included if the product had the addition of another critical nutrient in excessive amounts (for instance, butter made with milk cream and salt is eligible to be regulated and receives a warning sign for high content of sodium if this nutrient is in excess—however, it does not receive a warning sign related to the high content of fats). For this modified PAHO NPM, we used the same targeted five nutrients (free sugar, total fats, saturated fats, trans fat, and sodium) as well as nonnutritive sweeteners and applied the same threshold levels as the PAHO NPM, and the model behaved similarly to the originally proposed PAHO NPM in terms of identifying foods high in critical nutrients [32].

The PAHO NPM considers free sugars, information that is not available on food labels of products sold in Brazil. We thus estimated the amount of free sugars using the method proposed by PAHO that considers the information on the amount of total sugars declared on food labels [30]. In this method, foods are classified by the information available on the nutrition facts panel (total sugars or added sugars) and by food category. For instance, if information on total sugars is available and the product has no or a minimal amount of naturally occurring sugars, such as soda and sports drinks, then the total amount of added sugars is considered free sugars. For milk or yogurt with any type of sugar in the list of ingredients, 50% of the declared added sugars were considered free sugars, so lactose, galactose, and other types of naturally occurring sugars were less likely to be considered free sugars. In Brazil, the content of total sugar is not required to be present on nutrition facts panel in the country, and analyses that considered free sugars were conducted for 10% of the sample that provided this information.

In order to apply the PAHO nutrient profile model, products were classified as containing added sugar, sodium, fat, and nonnutritive sweeteners on the basis of keyword searches in the list of ingredients. Briefly, ingredients used as a proxy for added sugars included sugar, honey, syrups, molasses, maltodextrin, glucose, fructose, and concentrated fruit and vegetables juices, as well as chocolate and milk fondant. Ingredients for the addition of salt included salt, sodium chloride, cheeses, and processed meats. For fat, we considered oils, olives, butter, creams, and animal and vegetal fats. Nonnutritive sweeteners included aspartame, saccharin, sucralose, cyclamate, acesulfame k, stevia, polydextrose, maltitol, mannitol, isomaltose, neotame, xylitol, thaumatin, and advantame. All searches were made in Portuguese.

2.4. Reliability Analysis

To assess inter-rater and test–retest reliability of food composition data, 10% of the foods were double-entered. We used the intraclass coefficient to assess inter and test–retest reliability and found strong inter-rater and test-retest reliability (ICC ≥ 0.90) for all assessed nutrients [33]

Information on claims was entered twice for all 3491 products. We used Cohen's kappa to assess inter-rater and test–rest reliability of the claims' entered information. Considering that coefficients above 0.80 are considered to show a strong to an almost perfect agreement [34,35], we found strong reliability for all assessed claims (Cohen's kappa ≥ 0.82), except for claims related to the reduction of disease risk, which had coefficients ranging from 0.55 to 0.76.

2.5. Analysis

We first assessed the prevalence of health, nutrition, and environment-related claims in the front-of-package overall, and by food categories. Then, considering the current discussions on the Brazilian nutritional labeling regulatory process [9], that, among other things, includes requirements for FoP nutritional labeling, we first described how many foods with FoP claims were high in critical nutrients and therefore eligible to receive FoP nutritional labeling. Thirdly, we examined whether foods and beverages with different types of claims were more likely to be high in critical nutrients using 95% confidence intervals. We checked whether different types of nutrition claims (for nutrients of concern or not of concern) and health claims (general health claims, claims for special diets, and for natural foods) were more likely to be high in critical nutrients using 95% confidence intervals. Finally, we present stratified analysis of the proportion of different types of claims by foods high in free sugars, saturated fat, and sodium.

3. Results

Figure 1 shows the presence of claims in the front of the package of foods and beverages by food categories. Overall, 41.2% of the assessed products presented claims, however, in some categories, claims were present in more than 80% of the products, such as breakfast cereals and granola bars (93.7%), fruit juices and nectars (92.5%), and fruit-flavored drinks (84.1%). Nutrition claims were more frequently found in Brazilian packaged foods and beverages (28.5%), followed by health claims (22.1%). Although environment-related claims were found to be used less frequently in packaged foods and beverages (5.2%), within fruit juices and packaged fruits and vegetables, around 20% of the products presented such claims. We found nutrition and health claims in more than a third of sweetened and unsweetened dairy products and in over 50% of juices, nectars, and other sweetened beverages such as those plant-based beverages. Nutrition claims were found in 40.0% of carbonated beverages. Among cookies, 40.4% and 23.1% of them had nutrition and health claims, respectively; and 87.4% of the breakfast cereals and granola bars had nutrition claims, while 44.2% had health claims.

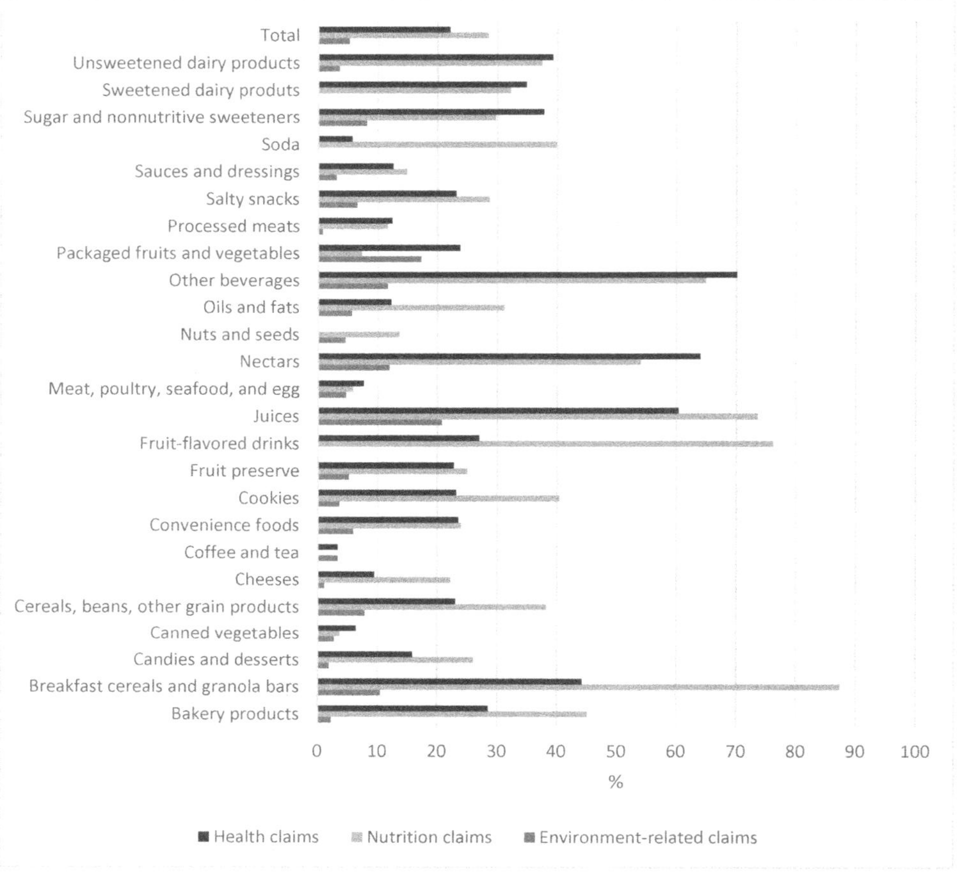

Figure 1. Presence of claims (%) in the front of the package of Brazilian packaged foods and beverages, 2017.

We found claims in a quarter (23.5%) of the assessed products that were high in any of the critical nutrients according to the PAHO NPM. The percentages of foods and beverages high in any of the assessed critical nutrients (free sugar, total fats, saturated fats, trans fat, and sodium) among those with or without claims in the front of the package are shown in Figure 2. More products with nutrition claims (65.3%; 95% CI 62.3, 68.2% vs. 54.1% 95% CI 52.1, 56.0) were high in critical nutrients, and therefore eligible to receive FoP warning signs, as compared with those not eligible to receive FoP warning signs. On the other hand, fewer products with health claims (52.9%; 95% CI 49.3–56.4% vs. 58.5%; 95% CI 56.6, 60.3) and environment-related claims (33.5%; 95% CI 27.0, 40.8% vs. 58.6%; 95% CI 56.9, 60.2%) were high in critical nutrients.

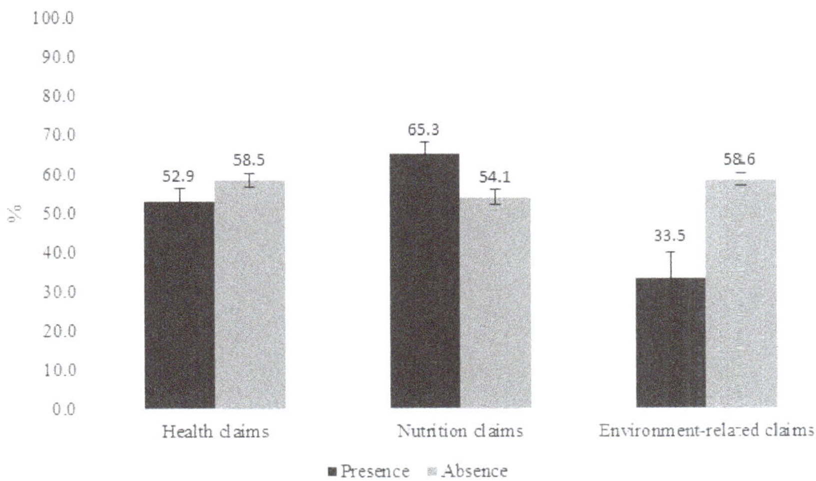

Figure 2. Presence of high level of critical nutrients according to front-of-package claims in Brazilian packaged foods and beverages, 2017. Note: Bars denote 95% confidence intervals.

Table 2 shows the foods and beverages high in critical nutrients among those with or without claims in the front of the package by food categories. Among the products presenting health claims, salty snacks, candies and desserts, and fruit-flavored drinks had fewer products high in critical nutrients. More sweetened dairy products high in critical nutrients were found among those with nutrition claims. For fruit-flavored drinks and soda, we found similar results: of those with nutrition claims, more than 80% were high in critical nutrients as compared with 46.7% and 33.3% of those without claims, respectively. Of the foods with information for the content of free sugars and claims, 38.6% had health claims and 44.4% had nutrition claims. Similarly, of the foods high in sodium, a quarter (24.5%) had health claims and a third (29.2%) had nutrition claims. Claims were slightly less present in foods high in saturated fat (18–21%) (Table S1 in Supplementary Materials).

We show in Tables 3 and 4 findings for the subcategories of health and nutrition claims, respectively. General health claims were found in 8.5% (95% CI 7.6, 9.5) of the assessed products. Claims for special diets and "natural" were found in 7.0% (95% CI 6.2, 7.9), and 9.0% (95% CI 8.1, 10.0) of the products, respectively. Among products with claims for "natural", fewer products (43.5%; 95% CI 38.3, 48.9 vs. 58.7%; 95% CI 57.0, 60.4) were high in critical nutrients. The proportion of foods with general health claims and claims for special diets was similar across products with and without high levels of critical nutrients.

Nutrition claims were divided in claims for nutrients of concern—present in 12.7% of the assessed products and that have been associated with increased cardiovascular disease risk—and not of concern (present in 13.1% of the assessed products) (Table 4). Of foods and beverages with claims for sodium, sugar, saturated fat, and trans-fat (of concern), three quarters were high in critical nutrients—a larger proportion when compared with foods that were not high in critical nutrients (74.0%; 95% CI 69.7, 77.9 vs. 54.8%; 95% CI 53.0, 56.6). Such differences were not found for claims for nutrients that were considered not of concern.

Table 2. Foods and beverages high in critical nutrient by the presence of different types of claims in the front of the package, Brazil, 2017.

Food Category	Health Claims						Nutrition Claims						Environment-Related Claims					
	Yes			No			Yes			No			Yes			No		
	%	95% CI		%	95% CI		%	95% CI		%	95% CI		%	95% CI		%	95% CI	
Breakfast Cereals and Granola Bars	57.1	41.8	71.2	62.3	48.5	74.3	61.4	50.5	71.3	50.0	23.5	76.5	50.0	21.2	78.8	61.2	50.4	71.0
Bakery Products	92.0	80.4	97.0	87.2	80.1	92.0	91.1	82.5	95.7	86.5	78.0	92.0	50.0	9.3	90.7	89.5	83.9	93.3
Convenience Foods	83.9	71.8	91.5	90.7	85.47	94.12	87.7	76.3	94.1	89.5	84.1	93.2	85.7	55.7	96.6	89.3	84.5	92.7
Unsweetened Dairy Products	4.5	0.6	27.1	0.0	.	.	4.8	0.6	28.2	0.0	.	.	0.0	.	.	1.9	0.3	12.2
Sweetened Dairy Products	48.1	34.8	61.6	43.3	33.8	53.4	77.1	63.0	86.9	29.7	21.6	39.4	0.0	.	.	45.0	37.1	53.0
Salty Snacks	76.0	55.3	89.0	96.4	89.3	98.8	83.9	66.3	93.2	94.8	86.9	98.1	100.0	.	.	91.1	83.7	95.3
Cookies	92.3	81.1	97.1	92.5	87.5	95.6	92.3	84.7	96.3	92.5	86.7	95.9	100.0	.	.	92.2	87.7	95.1
Canned Vegetables	71.4	29.8	93.6	86.4	78.3	91.8	75.0	18.0	97.6	85.8	77.8	91.3	100.0	.	.	85.0	76.9	90.7
Oils and Fats	30.8	11.5	60.3	12.9	7.5	21.4	36.4	21.7	54.0	5.5	2.1	13.8	0.0	.	.	16.0	10.0	24.6
Sauces and Dressings	75.8	58.2	87.5	85.6	80.4	89.6	92.3	78.4	97.5	83.0	77.4	87.4	62.5	26.3	88.6	85.0	80.1	88.9
Coffee and Tea	0.0	.	.	0.0	.	.	0.0	.	.	0.0	.	.	0.0	.	.	0.0	.	.
Candies and Desserts	0.0	0.0	0.0	71.7	66.4	76.4	78.9	69.5	86.0	69.3	63.5	74.5	57.1	20.7	87.2	72.1	67.2	76.5
Cereals, Beans, Other Grain Products	3.8	0.9	14.1	87.0	81.18	91.22	0.0	.	.	11.3	7.0	17.6	5.6	0.7	32.2	11.3	7.7	16.4
Packaged Fruits and Vegetables	5.5	1.7	15.8	0.0	.	.	0.0	.	.	0.0	.	.	0.0	.	.	0.0	.	.
Meat, Poultry, Seafood, and Egg	0.0	.	.	0.0	.	.	0.0	.	.	0.0	.	.	0.0	.	.	0.0	.	.
Sugar and Nonnutritive Sweeteners	0.0	.	.	21.7	9.2	43.4	90.9	53.6	98.9	15.4	5.8	35.0	0.0	.	.	41.2	25.9	58.3
Processed Meats	64.3	36.6	84.9	92.4	88.2	95.3	93.3	76.5	98.4	93.0	88.8	95.6	100.0	.	.	92.9	89.1	95.5
Juices	96.9	80.3	99.6	23.8	10.0	46.7	7.7	2.5	21.6	28.6	10.7	57.1	27.3	8.5	60.4	9.5	3.6	23.0
Nectars	6.3	1.5	22.2	27.8	11.7	52.7	29.6	15.3	49.4	21.7	9.2	43.4	16.7	1.8	68.1	27.3	16.1	42.3
Fruit-Flavored Drinks	25.0	12.9	42.9	87.0	73.7	94.1	87.5	74.7	94.3	46.7	23.4	71.4	0.0	.	.	77.8	65.8	86.4
Soda	52.9	29.6	75.0	54.5	37.4	70.7	92.9	61.1	99.1	33.3	16.5	55.9	0.0	.	.	57.1	40.3	72.5
Other Beverages	100.0	.	.	78.3	56.6	90.8	80.0	66.5	89.0	44.4	27.0	63.4	66.7	31.3	89.8	67.6	55.6	77.7
Nuts and Seeds	63.0	49.3	74.8	31.8	15.7	53.9	33.3	2.6	90.5	31.6	14.6	55.5	0.0	.	.	33.3	16.5	55.9
Cheese	82.4	56.3	94.4	91.4	86.0	94.9	92.5	78.9	97.6	90.0	83.8	94.0	100.0	.	.	90.4	85.2	94.0
Fruit Preserve	32.3	18.1	50.6	7.6	3.8	14.5	32.4	18.7	49.8	6.9	3.3	13.8	0.0	.	.	14.0	9.0	21.1

Abbreviation: CI, confidence interval.

Table 3. Foods and beverages high in critical nutrients by the presence of subcategories of health claims in the front of the package, Brazil, 2017.

Food Category	General Health Claims						Claims for Special Diets						Claims for "Natural"					
	Yes			No			Yes			No			Yes			No		
	%	95% CI		%	95% CI		%	95% CI		%	95% CI		%	95% CI		%	95% CI	
Breakfast Cereals and Granola Bars	50.0	30.6	69.4	63.4	51.6	73.8	66.7	31.3	89.8	59.3	48.6	69.2	53.8	27.2	78.5	61.0	50.0	70.9
Bakery Products	88.9	63.7	97.3	88.5	82.5	92.7	90.0	66.6	97.6	88.4	82.3	92.6	88.9	63.7	97.3	88.5	82.5	92.7
Convenience Foods	93.3	63.0	99.1	88.8	83.9	92.3	88.5	69.1	96.3	89.2	84.2	92.7	86.1	70.3	94.2	89.6	84.6	93.1
Unsweetened Dairy Products	0.0	.	.	1.9	0.3	12.2	0.0	.	.	2.5	0.3	16.1	12.5	1.5	57.5	0.0	.	.
Sweetened Dairy Products	46.7	29.6	64.5	44.5	35.8	53.6	64.3	36.5	84.9	43.0	34.9	51.5	37.5	17.4	63.1	45.9	37.6	54.4
Salty Snacks	40.0	8.2	83.2	94.2	87.6	97.4	75.0	48.1	90.7	94.6	87.5	97.7	80.0	43.7	95.4	92.9	85.7	96.6
Cookies	92.0	72.3	98.1	92.5	87.9	95.4	92.6	74.1	98.2	92.4	87.8	95.4	75.0	35.0	94.4	93.1	88.8	95.8
Canned Vegetables	50.0	1.9	98.1	86.1	78.2	91.5	100.0	.	.	85.2	77.1	90.7	75.0	17.9	97.6	85.8	77.8	91.3
Oils and Fats	28.6	6.3	70.3	14.1	8.5	22.5	0.0	.	.	15.5	9.7	23.9	33.3	2.5	90.5	14.6	9.0	22.8
Sauces and Dressings	40.0	8.2	83.2	85.2	80.3	89.1	85.7	55.7	96.6	84.3	79.2	88.3	66.7	42.1	84.6	85.7	80.7	89.5
Coffee and Tea	0.0	0.0	0.0	0.0	.	.	0.0	.	.	0.0	.	.	0.0	.	.	0.0	.	.
Candies and Desserts	73.3	45.6	90.0	71.7	66.8	76.2	73.3	54.6	86.3	71.6	66.6	76.2	60.0	33.9	81.4	72.3	67.4	76.7
Cereals, Beans, Other Grain Products	0.0	.	.	12.4	8.5	17.8	4.3	0.6	26.3	11.6	7.9	16.7	6.3	0.8	35.2	11.2	7.6	16.2
Packaged Fruits and Vegetables	0.0	.	.	0.0	.	.	0.0	.	.	0.0	.	.	0.0	.	.	0.0	.	.
Meat, Poultry, Seafood, and Egg	0.0	.	.	0.0	.	.	0.0	.	.	0.0	.	.	0.0	.	.	0.0	.	.
Sugar and Nonnutritive Sweeteners	40.0	8.2	83.2	37.5	22.5	55.4	0.0	.	.	38.9	24.4	55.7	50.0	1.9	98.1	37.1	22.8	54.2
Processed Meats	100.0	.	.	92.6	88.5	95.3	100.0	.	.	92.9	89.1	95.5	93.8	64.8	99.2	92.9	88.9	95.6
Juices	0.0	.	.	14.9	7.2	28.3	0.0	.	.	13.5	6.5	25.8	6.7	1.6	23.6	21.7	9.2	43.4
Nectars	40.0	8.2	83.2	24.4	14.0	39.2	0.0	.	.	26.0	15.6	40.0	26.7	13.7	45.4	25.0	10.6	48.5
Fruit-Flavored Drinks	81.8	47.3	95.8	76.9	63.5	86.5	0.0	.	.	77.8	65.8	86.4	12.5	1.5	57.5	87.3	75.5	93.9
Soda	0.0	.	.	57.1	40.3	72.5	0.0	.	.	57.1	40.3	72.5	100.0	.	.	54.5	37.4	70.6
Other Beverages	61.8	44.4	76.6	72.1	56.8	83.5	53.8	27.2	78.5	70.3	58.0	80.3	50.0	27.8	72.2	72.9	60.1	82.7
Nuts and Seeds	0.0	.	.	31.8	15.7	53.9	0.0	.	.	31.8	15.7	53.9	0.0	0.0	0.0	31.8	15.7	53.9
Cheese	100.0	.	.	90.4	85.1	94.0	83.3	50.4	96.1	91.1	85.7	94.6	75.0	17.9	97.6	90.9	85.7	94.4
Fruit Preserve	20.0	2.1	74.5	13.0	8.2	19.9	54.5	25.6	80.7	9.6	5.5	16.2	29.4	12.4	55.1	10.9	6.4	17.9
Total	50.7	45.0	56.4	57.9	56.1	59.6	63.0	56.8	68.9	56.8	55.1	58.5	43.5	38.3	48.9	58.7	57.0	60.4

Abbreviation: CI, confidence interval.

Table 4. Foods and beverages high in critical nutrients by the presence of the subcategories of nutrition claims in the front of the package.

Food Category	Claims for Nutrients of Concern [1]						Claims for Nutrients not of Concern [2]					
	Yes			No			Yes			No		
	%	95% CI		%	95% CI		%	95% CI		%	95% CI	
Breakfast Cereals and Granola Bars	64.7	47.3	79.0	57.4	44.7	69.2	48.0	34.5	61.8	73.3	58.5	84.3
Bakery Products	96.0	85.1	99.0	85.6	78.3	90.8	95.0	81.8	98.8	86.7	79.8	91.5
Convenience Foods	87.1	69.8	95.2	89.4	84.4	92.9	84.0	63.7	94.0	89.7	84.8	93.1
Unsweetened Dairy Products	33.3	2.6	90.5	0.0	0.0	0.0	7.7	1.0	41.2	0.0	.	.
Sweetened Dairy Products	88.9	46.7	98.6	42.1	34.2	50.5	81.8	59.7	93.2	38.6	30.5	47.4
Salty Snacks	85.0	61.6	95.2	93.2	85.6	96.9	70.0	35.8	90.7	93.9	87.0	97.2
Cookies	91.7	79.6	96.9	92.7	87.7	95.7	100.0	.	.	91.0	86.0	94.3
Canned Vegetables	50.0	1.9	98.1	86.1	78.2	91.5	100.0	.	.	85.3	77.3	90.8
Oils and Fats	61.1	37.1	80.7	5.7	2.4	13.0	10.0	1.2	49.7	15.6	9.6	24.4
Sauces and Dressings	85.7	63.1	95.5	84.2	79.1	88.3	100.0	.	.	84.3	79.3	88.2
Coffee and Tea	0.0	.	.	0.0	.	.	0.0	.	.	0.0	.	.
Candies and Desserts	92.0	80.3	97.0	68.6	63.2	73.5	79.5	63.8	89.5	70.9	65.7	75.5
Cereals, Beans, Other Grain Products	35.3	16.3	60.4	8.9	5.8	13.6	0.0	.	.	14.6	10.1	20.8
Packaged Fruits and Vegetables	0.0	.	.	0.0	.	.	0.0	.	.	0.0	.	.
Meat, Poultry, Seafood, and Egg	0.0	.	.	0.0	.	.	0.0	.	.	0.0	.	.
Sugar and Nonnutritive Sweeteners	75.0	17.9	97.6	33.3	19.3	51.1	0.0	.	.	37.8	23.7	54.4
Processed Meats	91.7	56.3	98.9	93.1	89.1	95.6	94.4	68.0	99.3	92.9	88.8	95.5
Juices	7.7	1.9	26.7	18.5	7.8	37.9	0.0	.	.	16.7	8.1	31.3
Nectars	58.3	29.6	82.3	15.8	7.2	31.2	27.3	12.5	49.6	25.0	12.2	44.3
Fruit-Flavored Drinks	100.0	.	.	70.2	55.6	81.6	86.1	70.4	94.2	66.7	46.9	81.9
Soda	100.0	.	.	34.8	18.1	56.2	66.7	9.5	97.4	56.3	38.7	72.3
Other Beverages	73.9	52.2	88.0	64.8	51.2	76.4	80.0	51.8	93.7	64.5	51.8	75.4
Nuts and Seeds	0.0	.	.	31.8	15.7	53.9	0.0	.	.	35.0	17.3	58.0
Cheese	0.0	.	.	91.6	86.5	94.9	100.0	.	.	90.0	84.5	93.7
Fruit Preserve	44.0	25.9	63.8	6.3	3.0	12.7	40.0	18.6	66.1	9.9	5.7	16.7
Total	74.0	69.7	77.9	54.8	53.0	56.6	57.4	52.8	61.9	57.2	55.5	59.0

Abbreviation: CI, confidence interval; [1] Includes nutrition claims for saturated fats, trans fats, sodium, and sugar; [2] Includes nutrition claims for vitamins, minerals, protein, and unsaturated fats.

4. Discussion

Our study provides an overview of the extent and nature of nutrition and health claims on the FoP in packaged foods and beverages sold in the Brazilian largest retailers. Overall, 41% of the products featured claims on the FoP, indicating extensive use of nutrition, health, and to a lesser extent environment-related claims on packaged foods sold in Brazil. We found claims in a quarter (23.5%) of the assessed products that were high in any of the critical nutrients according to the PAHO NPM; and that more foods and beverages with nutrition claims were high in critical nutrients, as compared with those low in critical nutrients.

In other words, should FoP nutritional labeling be implemented on foods high in critical nutrients in Brazil, more products with such labeling would have more nutrition claim than products without FoP nutritional labeling if restrictions for the presence of nutrition claims in foods high in critical nutrients are not implemented concomitantly. These products would thus have conflicting messages in the front-of-package. On the other hand, we found that fewer products with health claims and environment-related claims were high in critical nutrients.

Nutrition claims were indeed the most prevalent type of claim found in the Brazilian packaged food supply, covering almost a third of the assessed products as found in other countries [19–21,36,37]. Health claims were also quite prevalent and were found in 22% of the assessed products. The proportion of claims found in some of the food categories, particularly those with a high consumption among children and adolescents in the country, such as sweetened dairy and non-dairy beverages [38,39], were even higher. Nutrition claims were found in 32% of the sweetened dairy products, whereas health claims were found in 34% of the same products. In a study conducted in five European countries, nutrition and/or health claims were found in 46% of the dairy products—the largest proportion among the studied products [19].

Nutrition and health claims were also found in more than half of most types of sweetened beverages, such as juices, nectars, carbonated beverages, and other beverages (i.e., plant-based beverages). Considering sweetened beverages contribute with 49% of the added sugars consumption in Brazil [38], allowing such foods to carry health and nutrition claims can jeopardize the ability of consumers to make informed healthier choices and contribute to increasing their consumption of added sugars. Such issue could particularly impact the consumption of added sugars among Brazilian children and adolescents. For example, a third of Brazilian children under the age of two consume sweetened beverages [40], and adolescents report sweetened dairy beverages among the most consumed food items [41]. In fact, the proportional contribution of sweetened dairy beverages to added sugars intake among adolescents is higher than that found for adults and elders [38].

Forty percent of the assessed cookies and 29% of the salty snacks, two other highly consumed items among adolescents in Brazil [2], had nutrition claims, comparable to previously found in a smaller sample from Brazil [37]. Interestingly, of the 16 assessed countries in the same survey, the authors found the second largest proportion of nutrition claims among cookies and chips sold in Brazil (50%) [37]. The prevalence of marketing geared towards children in these same foods sold in Brazil was also among the highest found in the same survey [37]. Another study conducted in one single Brazilian supermarket found that half of the products marketed towards children bore nutrient claims and 95% of them were classified as ultra-processed foods [42].

Additionally, we found that more foods with nutrition claims—more specifically, those foods with claims for nutrients considered of concern—that have been associated with increased cardiovascular disease risk, such as saturated fat, trans fats, sugar, and sodium [29,30], were high in critical nutrients when compared with foods without such claims. On the other hand, we did not find any difference in foods with presence of nutrition claims for other nutrients (e.g., vitamins, minerals, unsaturated fats). Future analyses could consider looking more closely at nutritional claim sub-types (e.g., presence of beneficial nutrients such as Vitamin C vs. absence of nutrients of concern, such as low sugar) to better understand the link between claims and the nutritional profile of products.

Although by a small margin, more products with health claims had a better nutritional profile that those without such claims, as found elsewhere [19–21]. When we looked at subcategories of health claims, only products with health claims for "natural" were less likely to be high in critical nutrients, which may have driven the overall small difference in foods with health claims by the presence of high levels of critical nutrients.

Taken together, our findings from a large sample of Brazilian packaged foods and beverages, sold in supermarkets in the country that control 70% of the retail market share contribute to the literature by showing the pervasive presence of nutrition and health claims in unhealthy products, even more so among those highly consumed by children and adolescents [2,25,40]. Although FoP labeling in the format of warning signs seems to be effective in nudging consumers to choose healthier products [8,11,14,43], when these signs are accompanied by nutrition claims for the same nutrient for which a warning sign is present in the package, this can undermine the efficacy of FoP labeling [17]. Health and nutrition claims have been associated with purchase behavior [44]. Nonetheless, the misleading nature of nutrition claims has been depicted in previous studies [45]. Groups with special dietary needs as well as those with illnesses and parents refer to be more likely to benefit from health and nutrition claims and consider that overall nutrition and health claims made them more interested in a product which they considered to be healthy [46]. Restricting, thus, the presence of claims—particularly nutrition claims in foods high in critical nutrients—should be part of regulatory processes that aim to help consumers make healthier choices at the point of purchase.

Less information in the literature is found for environment-related claims. In our study, claims such as organic, biodiversity, and genetically modified, organism free were classified as being environmentally related. Other studies may have included them as part of health claims. We found these specific claims in a smaller proportion of products (5%) as compared with health and nutrition claims, but a fifth of fruit juices and packaged fruits and vegetables depicted them. Fewer foods and beverages with environment-related claims were high in any of the assessed critical nutrients (free sugar, total fats, saturated fats, trans fat, and sodium).

Claims for "organic" are associated with the food production method, but this claim seems to represent a cluster of attributes that goes beyond production-specific characteristics (e.g., pest management, fertilizer usage, and soil treatment). Organic products or other "eco-friendly" foods are associated with ethical, health, and environmental concerns, as well as nutrition and food safety aspects. They have been shown to be higher in antioxidants and reduce consumers' exposure to pesticides more than their conventional counterparts [47]. Nonetheless, an "eco-friendly" or "organic" claim can lure people into believing it is healthier than an alternative, when this may not be true [48]. Consumers can infer proprieties that are unrelated to the production method, perceiving organic foods as healthier, tastier, and less caloric than those produced conventionally [49]. Perceived healthfulness of a food product and use of food labeling, in turn, influences food intake [50].

Our study has a few limitations. First, we used a random subsample to evaluate claims, which may have introduced bias in our results. Nonetheless, we were cautious in selecting our subsample, which did not statistically differ from the total packaged food supply sample in terms of the proportion of foods in each food category or on the average content of nutrients. Second, we did not weigh the products by market share and did not specifically consider the most consumed foods. However, we included a considerable sample (over 10,000 items) of foods sold by the five top grocery retailers in Brazil. Third, we only included packaged foods that have information on the ingredients list and a nutrition facts panel in our sample, however, in Brazil, all packaged foods are required to depict the ingredients list and the nutrition facts panel in their label. Forth, information on the content of total sugar is not required to be present in the nutrition facts panel in Brazil. Analyses that considered total or free sugars were conducted in 10% of the sample that provided information for the content of total sugars in the nutrition facts panel, which may have underestimated the number of products that would be required to receive FoP labeling for high content of sugar. The method we used to estimate free sugars may have also introduced bias in the proportion of foods classified as high in sugar. The method we used classifies all foods and beverages in only four categories and defines the proportion of total sugar that should be considered as free sugars (zero, 50%, 75%, or 100%) based on food groups. This method is accurate for products without any free sugar and for those in which all the sugar comes from a source of free sugar. However, for dairy and fruit-based products such as fruit preserves and juices, for example, that have added and intrinsic sugars, the estimation could be biased. Although a few other methods to estimate free sugars are available, no method is considered standard, free of limitations, and applicable to all contexts [51,52].

Strengths of our paper include our sample that draws from all available packaged foods and beverages found in the five largest food retailers in the country, and the taxonomy we used to classify claims proposed by INFORMAS for which we found strong inter-rater reliability. This standardized classification of health-related labelling components has been used in different countries and therefore allows for international comparison [28].

In conclusion, we found a pervasive presence of health and nutrition claims in the Brazilian food supply and a larger proportion of nutrition claims than other types of claims. Moreover, a quarter of the Brazilian food supply that carries a claim was found to be high in critical nutrients and would therefore receive FoP nutritional labeling should a FoP nutritional labeling system be implemented in Brazil. Additionally, we found that more foods with nutrition claims—in particular, those foods with claims for nutrients that have been associated with increased cardiovascular disease risk—were high in critical nutrients when compared with foods without such claims. Restricting the presence of nutrition claims in foods high in critical nutrients should be part of regulatory processes that aim to help consumers make healthier choices at the point of purchase and an essential part of discussions to implement clearer FoP nutritional labeling.

Supplementary Materials: The following are available online at http://www.mdpi.com/2072-6643/11/12/2967/s1, Table S1: Presence of claims in the front of the package of Brazilian packaged foods and beverages high in free sugars, saturated fat, and sodium. Brazil, 2017.

Author Contributions: Conceptualization, A.C.D., L.S.T.; Methodology, C.Z.R., A.C.D., L.S.T.; Analysis, C.Z.R., A.C.D.; Data Curation, A.C.D., C.Z.R., L.S.T.; Writing-Original Draft, A.C.D.; Final draft review, A.C.D., C.Z.R., L.S.T., L.A.M., A.P.B.M.

Funding: This study was funded by a grant awarded by Bloomberg Philanthropies through a sub award agreement (5104695) between the University of North Carolina at Chapel Hill and Fundação Faculdade de Medicina, Brazil and with the aid of a grant from the International Development Research Center (108166-001). The funders had no role in the study design, data collection and analysis, decision to publish, or preparation of the manuscript.

Acknowledgments: We thank North Carolina TraCS Institute, University of North Carolina, Chapel Hill UL1RR025747 for support for the REDCap electronic data capture tool.

Conflicts of Interest: The authors declare no conflict of interest.

References

1. Monteiro, C.A.; Levy, R.B.; Claro, R.M.; de Castro, I.R.; Cannon, G. Increasing consumption of ultra-processed foods and likely impact on human health: Evidence from Brazil. *Public Health Nutr.* **2011**, *14*, 5–13. [CrossRef]
2. da Costa Louzada, M.L.; Baraldi, L.G.; Steele, E.M.; Martins, A.P.B.; Canella, D.S.; Moubarac, J.-C.; Levy, R.B.; Cannon, G.; Afshin, A.; Imamura, F.; et al. Consumption of ultra-processed foods and obesity in Brazilian adolescents and adults. *Prev. Med.* **2015**, *81*, 9–15. [CrossRef] [PubMed]
3. da Costa Louzada, M.L.; Ricardo, C.Z.; Steele, E.M.; Levy, R.B.; Cannon, G.; Monteiro, C.A. The share of ultra-processed foods determines the overall nutritional quality of diets in Brazil. *Public Health Nutr* **2018**, *21*, 94–102. [CrossRef] [PubMed]
4. Canella, D.S.; Levy, R.B.; Martins, A.P.; Claro, R.M.; Moubarac, J.C.; Baraldi, L.G.; Cannon, G.; Monteiro, C.A. Ultra-processed food products and obesity in brazilian households (2008–2009). *PLoS ONE* **2014**, *9*, e92752. [CrossRef] [PubMed]
5. Hall, K.D.; Ayuketah, A.; Brychta, R.; Cai, H.; Cassimatis, T.; Chen, K.Y.; Chung, S.T.; Costa, E.; Courville, A.; Darcey, V.; et al. Ultra-Processed Diets Cause Excess Calorie Intake and Weight Gain: An Inpatient Randomized Controlled Trial of Ad Libitum Food Intake. *Cell Metab.* **2019**, *30*, 226. [CrossRef] [PubMed]

6. Reyes, M.; Garmendia, M.L.; Olivares, S.; Aqueveque, C.; Zacarias, I.; Corvalan, C. Development of the Chilean front-of-package food warning label. *BMC Public Health* **2019**, *19*, 906. [CrossRef] [PubMed]
7. Nieto, C.; Jáuregui, A.; Contreras-Manzano, A.; Arillo-Santillan, E.; Barquera, S.; White, C.M.; Hammond, D.; Thrasher, J.F. Understanding and use of food labeling systems among Whites and Latinos in the United States and among Mexicans: Results from the International Food Policy Study, 2017. *Int. J. Behav. Nutr. Phys. Act.* **2019**, *16*, 87. [CrossRef]
8. Khandpur, N.; de Morais Sato, P.; Mais, L.A.; Bortoletto Martins, A.P.; Spinillo, C.G.; Garcia, M.T.; Urquizar Rojas, C.F.; Jaime, P.C. Are Front-of-Package Warning Labels More Effective at Communicating Nutrition Information than Traffic-Light Labels? A Randomized Controlled Experiment in a Brazilian Sample. *Nutrients* **2018**, *10*, 688. [CrossRef]
9. Agência Nacional de Vigilância Sanitária (ANVISA). Gerência-Geral de Alimentos. Relatório de Análise de Impacto Regulatório sobre Rotulagem Nutricional. ANVISA: Brasília. 2019. Available online: http://portal.anvisa.gov.br/documents/10181/3882585/Relat%C3%B3rio+de+An%C3%A1lise+de+Impacto+Regulat%C3%B3rio+sobre+Rotulagem+Nutricional/1c57c446-fcff-4adf-8465-021e69a91296 (accessed on 1 November 2019).
10. Machín, L.; Aschemann-Witzel, J.; Curutchet, M.R.; Giménez, A.; Ares, G. Traffic Light System Can Increase Healthfulness Perception: Implications for Policy Making. *J. Nutr. Educ. Behav.* **2018**, *50*, 668–674. [CrossRef]
11. Arrúa, A.; Machín, L.; Curutchet, M.R.; Martínez, J.; Antúnez, L.; Alcaire, F.; Giménez, A.; Ares, G. Warnings as a directive front-of-pack nutrition labelling scheme: Comparison with the Guideline Daily Amount and traffic-light systems. *Public Health Nutr.* **2017**, *20*, 2308–2317. [CrossRef]
12. Ares, G.; Aschemann-Witzel, J.; Curutchet, M.R.; Antúnez, L.; Machín, L.; Vidal, L.; Martínez, J.; Giménez, A. Nutritional warnings and product substitution or abandonment: Policy implications derived from a repeated purchase simulation. *Food Qual. Prefer.* **2018**, *65*, 40–48. [CrossRef]
13. Bollard, T.; Maubach, N.; Walker, N.; Mhurchu, C.N. Effects of plain packaging, warning labels, and taxes on young people's predicted sugar-sweetened beverage preferences: An experimental study. *Int. J. Behav. Med.* **2016**, *13*, 95. [CrossRef] [PubMed]
14. Machín, L.; Aschemann-Witzel, J.; Curutchet, M.R.; Giménez, A.; Ares, G. Does front-of-pack nutrition information improve consumer ability to make healthful choices? Performance of warnings and the traffic light system in a simulated shopping experiment. *Appetite* **2018**, *121*, 55–62. [CrossRef]
15. Arrúa, A.; Curutchet, M.R.; Rey, N.; Barreto, P.; Golovchenko, N.; Sellanes, A.; Velazco, G.; Winokur, M.; Giménez, A.; Ares, G. Impact of front-of-pack nutrition information and label design on children's choice of two snack foods: Comparison of warnings and the traffic-light system. *Appetite* **2017**, *116*, 139–146. [CrossRef]
16. Chandon, P.; Wansink, B. Does food marketing need to make us fat? A review and solutions. *Nutr. Rev.* **2012**, *70*, 571–593. [CrossRef]
17. Acton, R.B.; Hammond, D. Do manufacturer 'nutrient claims' influence the efficacy of mandated front-of-package labels? *Public Health Nutr.* **2018**, *21*, 3354–3359. [CrossRef]
18. Hieke, S.; Kuljanic, N.; Pravst, I.; Miklavec, K.; Kaur, A.; Brown, K.A.; Egan, B.M.; Pfeifer, K.; Gracia, A.; Rayner, M. Prevalence of Nutrition and Health-Related Claims on Pre-Packaged Foods: A Five-Country Study in Europe. *Nutrients* **2016**, *8*, 137. [CrossRef]
19. Kaur, A.; Scarborough, P.; Hieke, S.; Kusar, A.; Pravst, I.; Raats, M.; Rayner, M. The nutritional quality of foods carrying health-related claims in Germany, The Netherlands, Spain, Slovenia and the United Kingdom. *Eur. J. Clin. Nutr.* **2016**, *70*, 1388–1395. [CrossRef]
20. Al-Ani, H.H.; Devi, A.; Eyles, H.; Swinburn, B.; Vandevijvere, S. Nutrition and health claims on healthy and less-healthy packaged food products in New Zealand. *Br. J. Nutr.* **2016**, *116*, 1087–1094. [CrossRef]
21. Franco-Arellano, B.; Labonte, M.E.; Bernstein, J.T.; L'Abbe, M.R. Examining the Nutritional Quality of Canadian Packaged Foods and Beverages with and without Nutrition Claims. *Nutrients* **2018**, *10*, 832. [CrossRef]
22. Lopéz-Galán, B.; De Magistris, T. Prevalence of relevant nutritional claims related to prevention of obesity in Spanish market. *Nutr. Hosp.* **2017**, *34*, 154–164. [CrossRef] [PubMed]
23. Machado, P.P.; Claro, R.M.; Canella, D.S.; Sarti, F.M.; Levy, R.B. Price and convenience: The influence of supermarkets on consumption of ultra-processed foods and beverages in Brazil. *Appetite* **2017**, *116*, 381–388. [CrossRef] [PubMed]

24. Euromonitors International. *Grocery Retailers in Brazil*; Euromonitors International: Vilnius, Lithuania, 2016.
25. Planet Retail. *Brazil: Retail Sales 2013–2022*; Planet Retail: Frankfurt, Germany, 2018.
26. Instituto Brasileiro de Geografia e Estatística. *Censo Demográfico 2010*; IBGE: Rio de Janeiro, Brazil, 2010.
27. Kanter, R.; Reyes, M.; Corvalán, C. Photographic methods for measuring packaged food and beverage products in supermarkets. *Curr. Dev. Nutr.* **2017**, *1*, e001016. [CrossRef] [PubMed]
28. Rayner, M.; Vandevijvere, S. INFORMAS Protocol: Food Labelling Module. *J. Contrib.* **2017**. [CrossRef]
29. World Health Organization. *Diet, nutrition and the prevention of chronic diseases: Report of a Joint WHO/FAO Expert Consultation*; World Health Organization: Geneva, Switzerland, 2003.
30. Pan American Health Organization. *Pan American Health Organization Nutrient Profile Model*; PAHO: Washington, DC, USA, 2016.
31. Ministerio de la Salud. *Manual de Etiquetado Nutricional de Alimentos*, 1st ed.; Ministerio de Salud: Santiago, Chile, 2017.
32. Duran, A.C.; Ricardo, C.Z.; Mais, L.A.; Martins, A.P.B. Role of Different Nutrient Profiling Models in Identifying Targeted Foods for Front-Of-Package Food Labeling in Brazil. *Public Health Nutr.* In Press.
33. Hallgren, K.A. Computing Inter-Rater Reliability for Observational Data: An Overview and Tutorial. *Tutor. Quant. Methods Psychol.* **2012**, *8*, 23–34. [CrossRef]
34. McHugh, M.L. Interrater reliability: The kappa statistic. *Biochem. Med.* **2012**, *22*, 276–282. [CrossRef]
35. Gisev, N.; Bell, J.S.; Chen, T.F. Interrater agreement and interrater reliability: Key concepts, approaches, and applications. *Res. Soc. Adm. Pharm.* **2013**, *9*, 330–338. [CrossRef]
36. Pulker, C.E.; Scott, J.A.; Pollard, C.M. Ultra-processed family foods in Australia: Nutrition claims, health claims and marketing techniques. *Public Health Nutr.* **2018**, *21*, 38–48. [CrossRef]
37. Mayhew, A.J.; Lock, K.; Kelishadi, R.; Swaminathan, S.; Marcilio, C.S.; Iqbal, R.; Dehghan, M.; Yusuf, S.; Chow, C.K. Nutrition labelling, marketing techniques, nutrition claims and health claims on chip and biscuit packages from sixteen countries. *Public Health Nutr.* **2016**, *19*, 998–1007. [CrossRef]
38. Pereira, R.A.; Duffey, K.J.; Sichieri, R.; Popkin, B.M. Sources of excessive saturated fat, trans fat and sugar consumption in Brazil: An analysis of the first Brazilian nationwide individual dietary survey. *Public Health Nutr.* **2014**, *17*, 113–121. [CrossRef] [PubMed]
39. Pereira, R.A.; Souza, A.M.; Duffey, K.J.; Sichieri, R.; Popkin, B.M. Beverage consumption in Brazil: Results from the first National Dietary Survey. *Public Health Nutr.* **2015**, *18*, 1164–1172. [CrossRef]
40. Jaime, P.C.; Prado, R.R.D.; Malta, D.C. Family influence on the consumption of sugary drinks by children under two years old. *Rev. Saude Publica* **2017**, *51*, 13s. [CrossRef] [PubMed]
41. Souza Ade, M.; Pereira, R.A.; Yokoo, E.M.; Levy, R.B.; Sichieri, R. Most consumed foods in Brazil: National Dietary Survey 2008–2009. *Rev. Saude Publica* **2013**, *47*, 190s–199s.
42. Rodrigues, V.M.; Rayner, M.; Fernandes, A.C.; de Oliveira, R.C.; Proenca, R.P.; Fiates, G.M. Nutritional quality of packaged foods targeted at children in Brazil: Which ones should be eligible to bear nutrient claims? *Int. J. Obes.* **2017**, *41*, 71–75. [CrossRef]
43. Tortora, G.; Machin, L.; Ares, G. Influence of nutritional warnings and other label features on consumers' choice: Results from an eye-tracking study. *Food Res. Int.* **2019**, *119*, 605–611. [CrossRef]
44. Steinhauser, J.; Hamm, U. Consumer and product-specific characteristics influencing the effect of nutrition, health and risk reduction claims on preferences and purchase behavior—A systematic review. *Appetite* **2018**, *127*, 303–323. [CrossRef]
45. Talati, Z.; Pettigrew, S.; Neal, B.; Dixon, H.; Hughes, C.; Kelly, B.; Miller, C. Consumers' responses to health claims in the context of other on-pack nutrition information: A systematic review. *Nutr. Rev.* **2017**, *75*, 260–273. [CrossRef]
46. Benson, T.; Lavelle, F.; McCloat, A.; Mooney, E.; Bucher, T.; Egan, B.; Dean, M. Are the Claims to Blame? A Qualitative Study to Understand the Effects of Nutrition and Health Claims on Perceptions and Consumption of Food. *Nutrients* **2019**, *11*, 2058. [CrossRef]
47. Baranski, M.; Srednicka-Tober, D.; Volakakis, N.; Seal, C.; Sanderson, R.; Stewart, G.B.; Benbrook, C.; Biavati, B.; Markellou, E.; Giotis, C.; et al. Higher antioxidant and lower cadmium concentrations and lower incidence of pesticide residues in organically grown crops: A systematic literature review and meta-analyses. *Br. J. Nutr.* **2014**, *112*, 794–811. [CrossRef]
48. Sörqvist, P.; Haga, A.; Langeborg, L.; Holmgren, M.; Wallinder, M.; Nöstl, A.B.; Seager, P.; Marsh, J.E. The green halo: Mechanisms and limits of the eco-label effect. *Food Qual. Prefer.* **2015**, *43*, 1–9. [CrossRef]

49. Prada, M.; Garrido, M.V.; Rodrigues, D. Lost in processing? Perceived healthfulness, taste and caloric content of whole and processed organic food. *Appetite* **2017**, *114*, 175–186. [CrossRef] [PubMed]
50. Shangguan, S.; Afshin, A.; Shulkin, M.; Ma, W.; Marsden, D.; Smith, J.; Saheb-Kashaf, M.; Shi, P.; Micha, R.; Imamura, F.; et al. A Meta-Analysis of Food Labeling Effects on Consumer Diet Behaviors and Industry Practices. *Am. J. Prev. Med.* **2019**, *56*, 300–314. [CrossRef]
51. Louie, J.C.Y.; Moshtaghian, H.; Boylan, S.; Flood, V.M.; Rangan, A.M.; Barclay, A.W.; Brand-Miller, J.C.; Gill, T.P. A systematic methodology to estimate added sugar content of foods. *Eur. J. Clin. Nutr.* **2015**, *69*, 154–161. [CrossRef]
52. Wanselius, J.; Axelsson, C.; Moraeus, L.; Berg, C.; Mattisson, I.; Larsson, C. Procedure to Estimate Added and Free Sugars in Food Items from the Swedish Food Composition Database Used in the National Dietary Survey Riksmaten Adolescents 2016–2017. *Nutrients* **2019**, *11*, 1342. [CrossRef]

nutrients

MDPI

Article

Organic Consumer Choices for Nutrient Labels on Dried Strawberries among Different Health Attitude Segments in Norway, Romania, and Turkey

Valérie L. Almli [1], Daniele Asioli [2,*] and Celia Rocha [3,4]

1 Department of Innovation, Sensory and Consumer Sciences, Nofima AS, NO-1431 Ås, Norway; valerie.almli@nofima.no

2 Department of Applied Economics and Marketing, School of Agriculture, Policy and Development, University of Reading, Reading RG6 6AR, UK

3 GreenUPorto—Sustainable Agrifood Production Research Centre & LAQV-REQUIMTE/DGAOT, Faculty of Sciences, University of Porto, Edifício das Ciências Agrárias (FCV2), 4485-646 Vila do Conde, Portugal; crocha@fc.up.pt

4 Sense Test, Lda., 4400-345 Vila Nova de Gaia, Portugal

* Correspondence: d.asioli@reading.ac.uk; Tel.: +44-0-118-378-5426

Received: 12 November 2019; Accepted: 25 November 2019; Published: 4 December 2019

Abstract: Consumer interest towards healthy food is driving the growth of the organic food market because consumers perceive organic food products to improve their personal health. Berries have well-known health benefits and show increasing market shares in European markets. This manuscript investigates for the first time how health attitudes relate to organic consumers' choices for nutrient labels of organic dried strawberry products. We conducted an online survey with 614 consumers from Norway, Romania, and Turkey. All participants consumed and liked strawberries and purchased organic food at least once a month. Participants filled out attitudinal questionnaires and conducted an experimental choice task featuring paired images of packaged organic dried strawberries varying in nutrients content label and other factors. The pooled sample was split into three groups of varying health attitudes for profiling and choice analysis. The results show that broad variations exist in health attitudes among Norwegian, Romanian, and Turkish organic consumers. A non-linear effect of health attitude is revealed, where a moderate health attitude is more strongly associated with the selection of products with increased nutrients content than either a low or a high health attitude. The results highlight the complexity in targeting nutrition labels to organic consumers. Finally, implications and suggestions for organic food operators are discussed along with future research avenues.

Keywords: organic consumer choices; organic dried strawberries; nutrient labels; health attitudes; consumer heterogeneity; cross-cultural comparison

1. Introduction

Diet-related health problems evidenced by the increasing number of food-linked diseases as well as the consumer concerns towards the growing industrialized food production have dramatically increased over the last decades [1–5]. This has raised the importance of nutritional information as a relevant criterion affecting consumers' food choices [1]. Indeed, one of the most important trends that is affecting food consumption is the increasing consumer interest towards food products that provide health benefits [6–8]. This growing interest is also signaled by the increasing number of food products on the market that feature health claims or other labels that consumers infer as beneficial to health (e.g., functional foods, organic products, clean labels, or local food) [9]. A typical way to convey information about the healthiness of food products is through the back-of-pack nutrition label [4,10,11].

Such nutrition labels are an important tool that food marketers can use to communicate information about the nutritional value and composition of the products and at the same time they are a valuable tool to help consumers make informed decisions [11]. Yet studies have shown that few consumers read nutrition labels [12], and marketers as well as governments are prone to use front-of-pack labels to highlight specific nutritional features [13,14].

Berries have a well-known balanced nutritional composition and healthiness properties due to their richness in vitamins, fibers, antioxidants, polyphenols, and minerals that consumers appreciate as part of a healthy diet [15]. Thus, berries are suggested to potentially reduce the risk of cardiovascular diseases, inflammation, obesity, diabetes, cancer, and other chronic diseases [16–18]. Strawberries in particular are among the most popular and widespread berries on the food market [19] as their sensory [20,21] and nutritional properties [22] are highly appreciated by consumers [23,24]. However fresh berries are highly perishable and strawberries typically have a shelf life of approximately one week only, assuming chilled storage conditions (typically 6 °C or less depending on the variety) [25]. Several strategies can be used in order to increase shelf life of strawberries, including antimicrobial edible coating, gamma-irradiation, modified atmosphere packaging and antimicrobial active packaging of fresh berries [26–30], transformation into puree or juice possibly combined with high pressure and thermal processing [31], or drying [32–34]. A conventional drying process such as convective air-drying allows to extend shelf life considerably, but not without loss of nutritional and sensory qualities of the berry [35]. Recent mild processing technologies (e.g., microwave-vacuum) are able to dry the berries while preserving their nutritional quality by retaining more of the initial ascorbic acid, anthocyanins and phenolic compounds, and antioxidant capacity [36–38].

Organic foods are a product category that consumers perceive to be healthy [39–41]. Indeed, a large number of studies show that organic consumers' choices are mainly driven by the perceived naturalness and healthiness of these products compared to conventional food products [41–48]. The European organic food market is currently one of the most rapidly expanding food markets (+114% between 2006 and 2017) [49], and Europe represents the second largest organic agricultural area in the world (13.5 million hectares) after Oceania (27.3 million hectares) [50]. However, large discrepancies exist within Europe and it is therefore important to consider multicultural studies when it comes to the European organic consumer. In particular, in 2016 the organic food market retail sales weighed 394 million Euros in Norway (population: 5.2 million inhabitants), with 4.8% of total agricultural land dedicated to organic production [50]. In contrast, the organic market in Romania (population: 19.6 millions) represented 80 million Euros (Data from 2011) with 1.7% of agricultural land dedicated to organic production, and 4 million Euros in Turkey (Data from 2009) (population: 82 millions) with 1.4% of organic land [50]. These three countries exemplify therefore different maturity levels of the organic market in Europe.

This manuscript investigates organic consumers' choices for nutrient labels of dried strawberries based on their health attitude. While a large number of studies have investigated consumers' preferences and willingness to pay (WTP) for organic food products [51–53] and explored drivers of consumption [47,54–57], personal determinants [58,59], subjective norms and attitudes [58], sensory attributes [44,60–63], and organic labeling [64], no prior study has investigated their preferences for nutrient labels in relation to their health attitudes. Eagly and Chaiken [65] define attitude as "a psychological tendency that is expressed by evaluating a particular entity with some degree of favor or disfavor". In practice, attitudes are normative, valuative, subjective and may expand to opinion, habits, tendencies to act, impulses to act, inhibitive impulses, feelings, wishes, and values [66]. Attitudes are learned and not inherited [67]. Besides forming from experience, they also derive from social influence, where social attitudes define a group of similar status in a specific domain [67]. Further, Hong [68] presents health consciousness as an "important psychographic variable (...) [that] predicts a variety of health attitudes and behaviors" and "is closely related to how [an individual] seeks and responds to health information". According to this author, health consciousness is a five-dimensional concept related to integration of health behavior, attention to one's health, health information seeking and usage, personal health responsibility, and health motivation.

The relationship between organic consumers and health has been investigated by several authors. While health consciousness has been found to predict attitudes and intention of purchasing organic foods [69,70], the relationships between health consciousness, attitudes, and consumers' preferences are ambiguous. Indeed, Michaelidou and Hassan [71] found that health consciousness is the least important motive in predicting attitudes towards organic produce while Magnusson et al. [70] indicated that it is a motive for shaping attitude towards organic produce. In addition, Hansen, Sørensen, and Eriksen [41] reported that health consciousness has a high positive influence on organic food identity. It is therefore not clear if and how different health attitudes may drive organic consumers' preferences towards nutrition labels. This information may help organic-food marketers to better target nutrition labeling strategies at consumer segments of different health attitudes.

Our study aims to highlight the heterogeneity in health attitudes among multicultural organic consumers, and establish possible relationships between health attitudes and consumers' attractiveness for nutrition labels. New mild processing technologies may better retain the original nutrients content in organic products such as dried strawberries—our underlying question is: will organic consumers be attracted to more nutrients, and is that attractiveness related to their health attitude? We utilize data from Norway, Romania, and Turkey, where consumers' choices for organic dried strawberry products of varying nutrients labels as well attitudinal, behavioral, and socio-demographic characteristics were collected. We profile three consumer groups of low, medium, and high health attitudes in terms of attitudinal, behavioral, and socio-demographic characteristics. We then model consumer choices for organic dried strawberry products labeled with "more natural nutrients" versus "natural nutrients" labels with regard to health attitude groups. The results show that broad differences exist in terms of health attitude among European organic consumers, and that health attitude has a non-linear relationship to the selection of products with increased nutrients content. We discuss the results in light of previous literature and conclude in terms of implications for the promotion of nutritional foods to organic consumers.

2. Materials and Methods

An online consumer survey was conducted in the Winter of 2017 in the three target countries (i.e., Norway, Romania, and Turkey). Respondents completed a choice task for dried strawberry properties and responded to attitudinal, behavioral, and socio-demographic questions.

2.1. Choice Task

Consumers participated in an online choice experiment where several mock-up images of dried strawberry packages were used. All packages included an organic logo. The images varied in nutrients labels, as nutrients is one of the key attributes that consumers consider when purchasing organic foods [72] and berry products [21,73]. Two levels of nutrient content were specified: "Natural nutrients (Antioxidants, Vitamin C and Fibres)" or "More natural nutrients (Antioxidants, Vitamin C and Fibres)". Except for the word 'more', the two labels were kept identical with regards to size, fonts, colors, and remaining text in order to elicit consumer responses to nutrients content and nothing else (Figure 1). In addition to nutrients, the dried strawberry packages varied in origin (Europe or own country), technology (air drying or microwave drying) and price (three levels varying from 10% to +10% of the average market prices in each target country) following a D-optimal choice design of 11 choice sets. Each choice set was composed of two product alternatives (options A and B) and an 'opt-out' option (option C) (see an example in Supplementary Materials Figure S1). To make sure that respondents would notice and understand the different product attributes, an introduction screen presented one of the mock-ups and the four attributes.

In the present paper, focus is brought on consumer choices for products labeled with "more natural nutrients" as opposed to "natural nutrients". For more details on the choice design and analysis of consumer preferences with regard to all four attributes, see Asioli et al. [74].

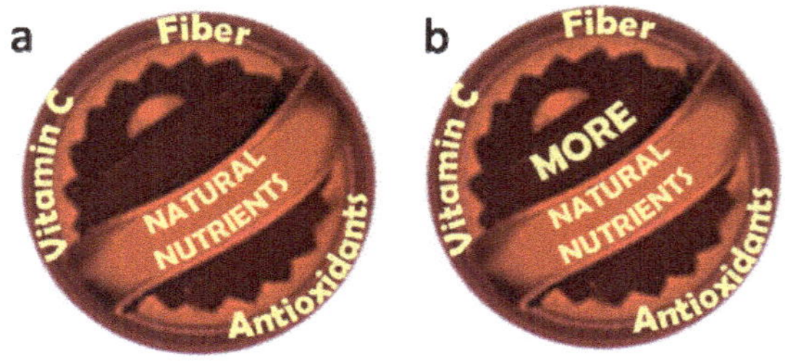

Figure 1. "Natural nutrients" (**a**) and "More natural nutrients" (**b**) labels utilized on organic dried strawberry packages in the experimental choice task (English translation).

2.2. Attitudinal and Behavioral Questionnaire

After the choice task, we included in the survey selected attitudinal question items related to consumers' health (HAQ) and natural (NAQ) attitudes selected from the Health and Taste Attitude Questionnaire [75], consumers' ecological attitudes using the Food-Related Lifestyle (FRL) scale [76] and consumers' attitude towards new food technology using the Food Technology Neophobia Scale (FTNS) [77] (Table 1). All attitudinal questions were collected using scales anchored from 1 (strongly disagree) to 7 (strongly agree). Furthermore, the questionnaire included a set of behavioral items with regards to purchase and consumption habits of strawberry products (i.e., purchase frequency, point of sale, search criteria) (Table 1). The survey development was executed in English, then translated in Norwegian, Romanian, and Turkish by the local research teams, then carefully checked against the original English questionnaire. The global survey was pretested by 6 to 19 respondents (colleagues of the local teams) in each country during Fall 2016. After adjustments, the survey was pilot-tested (n = 20 per country), before the full data collection took place in February 2017. The full questionnaire is available upon request.

2.3. Data Collection

The data were collected during Winter 2017 involving a total of 614 consumers in Norway (N = 204), Romania (N = 206), and Turkey (N = 204) using the online platform Userneeds (Copenhagen, Denmark). Consumers where randomly recruited by Userneeds using sampling quotas in terms of age and gender. Participants were informed about the opportunity to participate in a survey on consumers' valuation of berry products. An overall sample of 13,070 consumers were invited by e-mail to participate to the web-survey, wherein 614 consumers (4.7% of the invited consumers) fulfilled the eligibility criteria and completed the survey. We recruited only consumers aged between 18 and 65 years, who consume and like strawberries, purchase organic foods once a month or more often ("I purchase organic food products … At least once a week/At least once a month/A few times a year/Rarely or never"), and had purchased and consumed dried nuts, fruits and/or berries at least once during the last three months prior to the survey. Dried strawberries is not a common product as such in the three surveyed countries to date, but may be encountered in, e.g., breakfast cereals. We recruited only consumers who eat and like strawberries and purchased dried fruits, dried berries and/or nuts in the last three months, as these may represent potential customers of dried strawberries if the product expands on the market. Respondent age groups were equally balanced (18–29 years old: 25.9%, 30–41 y.o.: 25.9%, 42–53 y.o.: 23.1%, 54–65 y.o.: 25.1%), 50.2% of the respondents were males, and 79.1% of the pooled sample reported a university education or equivalent. These sociodemographic characteristics were similarly distributed across the three countries. For further details about the sociodemographic characteristics of the sample investigated, see [74]. During the test, the participants first received the recruitment questionnaire. Eligible participants proceeded to the description of the attributes and levels in an introductory screen, then to the choice task. Upon completion of the series of choice tasks, they filled out the attitudinal, behavioral, and socio-demographic questionnaire. The survey took approximately 8–10 min to answer. The data were collected with compliance to the ethical principles of the declaration of Helsinki.

Table 1. Attitudinal and behavioral items from the consumer questionnaire.

Variable	Attitudinal Items [1]
HAQ	
	1. The healthiness of food has little impact on my food choices (R)
	2. I am very particular about the healthiness of food I eat
	3. I eat what I like and I do not worry much about the healthiness of food (R)
	4. It is important for me that my diet is low in fat
	5. I always follow a healthy and balanced diet
	6. It is important for me that my daily diet contains a lot of vitamins and minerals
	7. The healthiness of snacks makes no difference to me (R)
	8. I do not avoid foods, even if they may raise my cholesterol (R)
	9. I do not avoid foods, even if they may be high in sugar content (R)
	10. I pay attention to the salt intake in my diet
NAQ	
	1. I try to eat foods that do not contain additives
	2. I do not care about additives in my daily diet (R)
	3. I do not eat processed foods, because I do not know what they contain
	4. I would like to eat only organically grown vegetables
	5. In my opinion, artificially flavored foods are not harmful for my health (R)
	6. In my opinion, organically grown foods are no better for my health than those grown conventionally (R)
FRL	
	1. I always buy organically grown food products if I have the opportunity
	2. I make a point of using natural or ecological food products
	3. I do not mind paying a premium for ecological products
FTNS	
	1. New food technologies are something I am uncertain about
	2. New foods are not healthier than traditional foods
	3. The benefits of new food technologies are often grossly overstated
	4. There are plenty of tasty foods around, so we do not need to use new food technologies to produce more
	5. New food technologies decrease the natural quality of food
	6. New food technologies are unlikely to have long term negative health effects (R)
	7. New food technologies give people more control over their food choices (R)
	8. New products using new food technologies can help people have a balanced diet (R)
	9. New food technologies may have long term negative environmental effects
	10. It can be risky to switch to new food technologies too quickly (R)
	11. Society should not depend heavily on technologies to solve its food problems (R)
	12. There is no sense trying out high-tech food products because the ones I eat are already good enough
	13. The media usually provides a balanced and unbiased view of new food technologies (R)
	Behavioral Items [2]
PurFreBer	Purchase of berry products: _Fresh berries in the warm seasons
PurBerJam	Purchase of berry products: _Berry jams
PurFroBer	Purchase of berry products: _Frozen berries
PurDrinkBer	Purchase of berry products: _Soft drink or smoothie with berries
FreBerOnl	Where do you usually buy [fresh] berries?_Online grocery store
FroBerGre	Where do you usually buy [frozen] berries?_Greengrocer's
FroBerSup	Where do you usually buy [frozen] berries?_Supermarket/hypermarket
FroBerOnl	Where do you usually buy [frozen] berries?_Online grocery store
Imp_NutClaim	When you buy dried fruits or dried berries, how important are the following characteristics for you?_Nutrition Claim
ImpNutrients	When you buy food products based on dried fruits or on dried berries, how important are the following characteristics for you?_Nutrients
NoHarmSub	When you buy food products based on dried fruits or on dried berries, how important are the following characteristics for you?_No health-harming substances
ConBerSna	In which context do you consume dried berries?_Snacking
ConBerDes	In which context do you consume dried berries?_Dessert

HAQ: health attitudes from the Health and Taste Attitude Questionnaire [75]; NAQ: natural attitudes from the Health and Taste Attitude Questionnaire [75]; FRL: ecological attitudes from the Food-Related Lifestyle (FRL) scale [76]; FTNS: attitude towards new food technology from the Food Technology Neophobia Scale (FTNS) [77]. [1] (R) indicates reversed items for the computation of attitudinal scores. [2] Only the attitudinal and behavioral items relevant for the results section are shown.

2.4. Data Modeling

2.4.1. Attitudinal Groups

To start with, individual mean attitudinal scores were derived from the HAQ, NAQ, FRL, and FTNS scales by summing up the respondents' answer scores across the different items within each scale. Negatively-phrased question items were reversed before summing (Table 2). Furthermore, in order to develop models for different levels of health attitude, three consumer groups (HAQ1, HAQ2, and HAQ3) were defined cross-nationally based on question items for HAQ by using the 25% and 75% quartiles as cut-off thresholds. Also attitudinal groups related to NAQ (two groups split at median), FRL (two groups split at median), and FTNS (two groups split at median) were defined in order to profile our health attitude groups in relation to natural, organic and technology attitudes. It should be noted that all attitudinal variables presented nearly normal distributions (skewness HAQ: 0.15, NAQ: −0.38, FRL: −0.50, FTNS: 0.08). The group splits are therefore arbitrary and only defined to highlight different effects between lower/higher attitudinal scores in the respondent sample in further analyses.

Table 2. Attitudinal characteristics of the consumers in Norway, Romania, and Turkey and pooled sample, and attitudinal group thresholds for further analysis

Attitudes	Norway (n = 204)	Romania (n = 206)	Turkey (n = 204)	Pooled (n = 614)	Attitudinal Groups (n = 614)
	Mean (SD)	Mean (SD)	Mean (SD)	Mean (SD)	[Mean Score Range], Group Size
Health attitude (HAQ)	4.64 [a] (0.79)	4.94 [b] (0.93)	5.07 [b] (0.93)	4.89 (1.00)	HAQ1: [1.6–4.2], 174 HAQ2: [4.3–5.4], 265 HAQ3: [5.5–7.0], 173
Natural attitude (NAQ)	4.44 [a] (1.16)	5.26 [b] (0.96)	5.31 [b] (1.03)	5.00 (1.12)	Low: [1.0–5.0], 314 High: [5.1–7.0], 298
Ecological attitude (FRL)	4.15 [a] (1.40)	5.64 [b] (1.11)	5.35 [c] (1.25)	5.05 (1.42)	Low: [1.0–5.0], 318 High: [5.1–7.0], 294
Food technology neophobia scale (FTNS)	4.18 [a] (0.91)	4.60 [b] (0.78)	4.43 [c] (0.91)	4.40 (0.89)	Low: [1.2–4.3], 324 High: [4.4–7.0], 288

[a,b,c] Significant differences based on chi-squared and Pearson chi-squared tests. Same letter indicates that there is no statistical significant difference at a 5% level.

2.4.2. Characterization of Health Attitude Groups with Partial Least Squares Regression (PLS-R)

We related the respondents' variations in health attitude levels to socio-demographic characteristics, other attitudinal dimensions (i.e., NAQ, FRL, and FTNS scores) and prioritized factors in the choice task (i.e., individual coefficients from the ML model). A partial least squares regression (PLS-R) was used for the purpose as it is able to handle highly correlated variables [78]. The base model included all socio-demographics and questionnaire variables as independent variables (X) and HAQ as the dependent variable (Y). Both models utilizing HAQ as a continuous variable as well as models utilizing HAQ1, HAQ2, and HAQ3 as binary variables coding for group belonging were attempted. In all models, cross-validation with 20 random segments and uncertainty testing based on jack-knifing (95% confidence interval) were used for model validation and variable selection [79]. Only the refined models for HAQ1 and HAQ3 are presented in the results section as these best identified respondent characteristics associated to variations in health attitude levels. The models were run in The Unscrambler X 10.4.1 (Camo Software AS, Oslo, Norway).

2.4.3. Individual Choices for "More Nutrients"

The data collected in the choice tasks were analyzed using the mixed logit (ML) model [80], estimating the utility of each product alternative for each individual in each choice occasion. This model is based on all 11 choice sets, and includes main effects and two-way interactions of the attributes origin, technology, nutrition content, and price. Full details on this model and its results are given in [74] and will not be discussed here. For the purpose of this study, individual parameter estimates (beta coefficients) for attribute nutrients were then extracted from the ML model and utilized as a measure of consumers' nutrient label preferences in further analyses. This approach allows focusing on attribute nutrients alone in the present report while accounting for the impact of the other attributes during the choice experiment. The ML model was estimated using the Stata module *mixlogit* while the individual parameter estimates were extracted using the Stata module *mixlbeta* run in STATA 15.0 software (StataCorp LP, College Station, TX, USA).

2.4.4. Analysis of Variance (ANOVA)

We utilized analysis of variance (ANOVA) to investigate socio-demographic and attitudinal effects related to respondents' selection of organic dried strawberry products with either standard or increased nutrients contents in the choice task. A first ANOVA was conducted in a general linear model with factors country (three levels), gender (two levels), age (four levels: 18–29; 30–41; 42–53; and 54–65 years), organic consumption frequency (two levels), HAQ (three levels), NAQ (two levels), FRL (two levels), FTNS (two levels), income (four levels), urban area (two levels), and university (two levels) as well as their two-way interactions. Individual ML model estimates for design factor "nutrients" were used as a dependent variable. The primary ANOVA model was refined in a second model by withdrawing non-significant main effects that were not involved in interactions and non-significant interactions (p-value > 0.05). Only the final model is presented in the results section, which includes the factors country, gender, age, HAQ, FTNS, income, urban area, university and selected two-way interactions. ANOVA models were run in Minitab 18.1 (Minitab, Inc., State College, PA, USA).

3. Results

3.1. Attitudinal Characteristics in Norway, Romania, and Turkey

Table 2 reports descriptive statistics (i.e., mean and standard deviations) of the attitudinal questions computed across the three countries. The results show that Turkish and Romanian organic consumers have significantly larger pro-health and pro-natural product interest compared to Norwegian consumers, and in turn Romanian consumers have significantly larger pro-ecological attitudes compared to Turkish and Norwegian consumers. Romanian organic consumers also show a significantly higher food technology neophobia compared to Turkish and Norwegian consumers, and in turn Turkish consumers have significantly higher food technology neophobia compared to Norwegians. Table 2 also reports the group thresholds used for further data analysis of attitudinal variables (see Section 2).

3.2. Characteristics of the Health Attitude Segments

Discriminative characteristics of health attitude segments HAQ1 and HAQ3 were extracted by means of PLS-R modeling. The models showed 19% and 21% explained Y-variance on the first PLS component, respectively (validation variance 18% and 19%, resp.). Only variables that showed statistically significant relationships to HAQ in the jack-knifing test are reported; for an elaboration of the variables' abbreviations, see Table 1. Figure 2 shows that respondents in the low health attitude group (HAQ1) were typically males and households with children (Hous-18), and not in the eldest age group (54–65 years old). Their berry purchases consisted in particular of convenient, long shelf life products (frozen berries, berry-based drinks, and berry jams) in addition to fresh berries, while they seldom consumed berries as a snack or as a dessert. These consumers were more often online shoppers than the other groups. Furthermore, these respondents expressed a low importance of nutrients, nutrition claims, or the absence of health-harming substances as factors in their choice of berry products. Finally, respondents in the HAQ1 group showed low natural attitudes (NAQ) and low organic and ecological attitudes (FRL) compared to the rest of the respondent sample, while they did not stand out in terms of food technology neophobia (FTNS). We may also note that none of the three countries was systematically related to low health attitude.

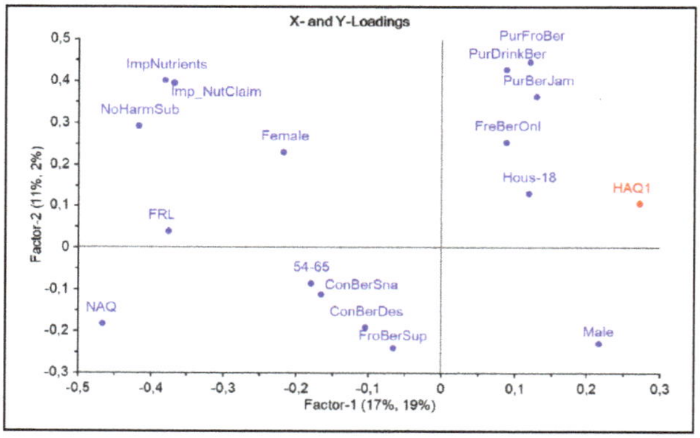

Figure 2. Significant attitudinal, behavioral and socio-demographic factors characterizing organic consumers in the low health attitude segment (HAQ1) in the PLS-R model. Variables projected to the right positively correlate to belonging to the (HAQ1) group; variables projected to the left negatively correlate to this group.

On the opposite side of the health attitude scale, respondents in the high interest group (HAQ3) were typically females, highly educated, seldom private-sector workers, and in the 54–65 age group but not in the 18–29 age group (Figure 3). Their berry purchases consisted of fresh berries, but neither of frozen berries (neither online nor in grocery stores) nor of berry jam. These respondents expressed a high importance of nutrients content and nutrition claims as factors in their choices of berry products. Finally, respondents in the HAQ3 group showed high Natural attitudes (NAQ) and high Organic and Ecological attitudes (FRL) scores compared to the rest of the respondent sample, and rejected dried strawberries from microwave-drying technology in the choice task. These respondents were more typically from Romania and less typically from Norway.

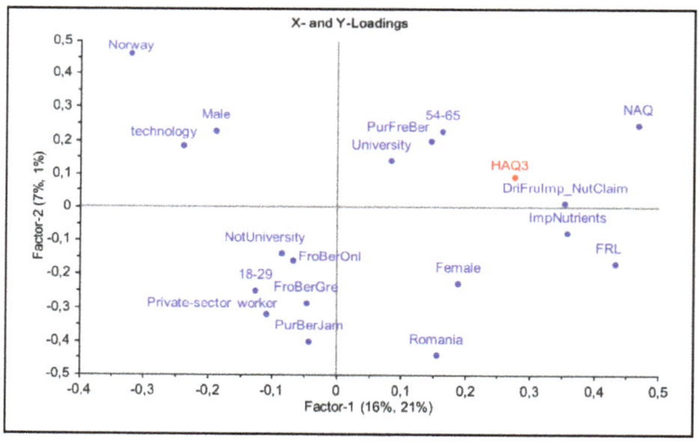

Figure 3. Significant attitudinal, behavioral and socio-demographic factors characterizing organic consumers in the high health attitude segment (HAQ3) in the PLSR model. Variables projected to the right positively correlate to belonging to the (HAQ3) group; variables projected to the left negatively correlate to this group.

The HAQ2 group with middle health attitude was poorly modelled (Y-calibration variance: 3%; Y-validation: 2%), indicating low systematic structure in this group's responses. No salient socio-demographic, attitudinal, or behavioral characteristics emerged to characterize this segment (results not shown).

3.3. Factors Connected to Nutrient Labels Choices in Organic Consumers

The refined ANOVA model reveals eight factors significantly involved in main effects or interactions connected to organic consumers' choices of nutrient labels in the choice task (Table 3). Figures 4 and 5 display the size and direction of selected effects. In these plots, the y-axis corresponds to the variable of individual ML model estimates for design factor 'nutrients', which is used as the dependent variable in the ANOVA (see section Materials and Methods). Firstly, Romanian and Turkish consumers more often selected products with a "more natural nutrients" label than Norwegian consumers (Figure 4). Country also interacts with income, as in Norway high-income respondents make fewer 'more natural nutrients' choices than low-income respondents while no income effect is observed in Romania and Turkey (results not shown). Secondly, a two-way interaction between sex and age reveals that in younger age groups (18–41 years) females preferred packs with the 'more natural nutrients' label compared to males, middle-age respondents (42–53 years) responded equally across genders, while in the older age group (54–65 years) males more often chose packs with 'more natural nutrients' than females (results not shown). Thirdly, a significant effect of health attitude segments occurs where the HAQ2 group (middle health attitude) more frequently chose the 'more natural nutrients' label than HAQ1 and HAQ3 (Figure 4). Additionally, factor HAQ interacts with urban area: respondents of moderate health attitude (HAQ2 group) who live in an urban area made more choices of "more natural nutrients"-labeled products than respondents from rural areas across all three health attitude groups (Figure 5a). Fourthly, food technology neophobia attitudes interact with income (high-income respondents make fewer choices of more nutrients products than low-income respondents; results not shown) and education groups, where a university degree reverses the skepticism observed for more nutrients in respondents with technology neophobia (Figure 5b). Finally, one may note that consumption frequency of organic products was not significantly related to consumers' nutrition label choices, neither as a main effect nor in interactions. This variable was therefore not retained in the refined model.

Table 3. Analysis of variance (ANOVA) model for nutrient label choices

Source	DF	Adj SS	Adj MS	F-Value	p-Value
Country	2	1.9831	0.991554	15.22	0.000
Sex	1	0.0001	0.000095	0.00	0.970
Age	3	0.1052	0.035055	0.54	0.656
HAQ	2	0.6907	0.345329	5.30	0.005
FTNS	1	0.0275	0.027495	0.42	0.516
Income	3	0.3056	0.101861	1.56	0.197
Urban area	1	0.0063	0.006307	0.10	0.756
University	1	0.1661	0.166091	2.55	0.111
Country × Income	6	1.0368	0.172796	2.65	0.015
Sex × Age	3	0.5851	0.195020	2.99	0.030
HAQ × Urban area	2	0.7095	0.354760	5.45	0.005
FTNS × Income	3	0.6797	0.226572	3.48	0.016
FTNS × University	1	0.3869	0.386895	5.94	0.015
Error	535	34.8531	0.065146		
Lack-of-Fit	367	28.0640	0.076469	1.89	0.000
Pure Error	168	6.7891	0.040411		
Total	564	42.2876			

HAQ: Health attitude score, from the Health and Taste Attitude Questionnaire [74]. FTNS: Food technology neophobia score, from the Food Technology Neophobia Scale (FTNS) [76]. DF: degrees of freedom; Adj SS: adjusted sum of squares; Adj MS: adjusted mean squares; F-value: the test statistic used to determine whether the term is associated with the response; p-value: a probability that measures the evidence against the null hypothesis (source: https://support.minitab.com/).

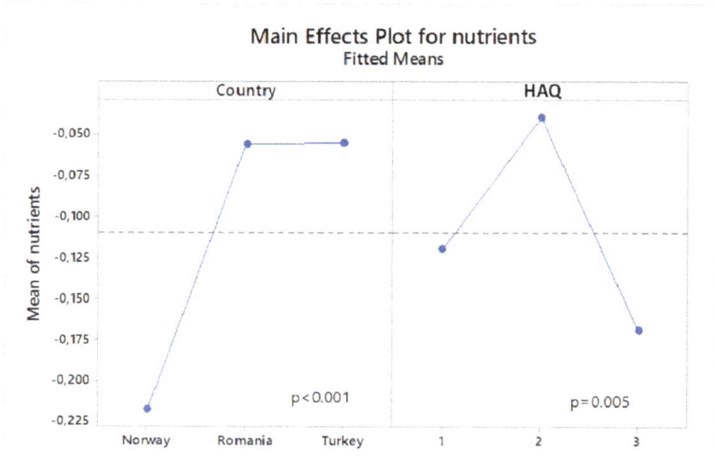

Figure 4. Significant main effects of (Country left) and Health attitude (HAQ score, right) on organic consumers' choices for the "more natural nutrients (antioxidants, vitamin C, and fibers)" options in the choice task.

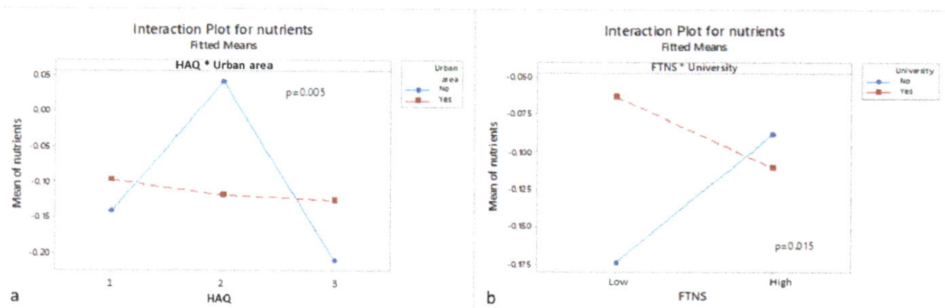

Figure 5. Significant interaction effects of health attitude (HAQ) * Urban area (**a**) and food technology neophobia (FTNS) * University (**b**) on organic consumers' choices for the 'more Natural nutrients (antioxidants, vitamin C, and fibers)' options in the choice task.

4. Discussion

Much evidence has been put forward in the literature that establishes an association between health interest and organic consumption, and shows that consumers infer healthiness properties to organic products in an "organic = healthy" assumption (see e.g., [54,70]). However, previous studies have reported somewhat conflicting results on the role of health interest as a primary driver for organic food consumption [41,69,71]. This manuscript investigated the heterogeneity in health attitudes among multicultural organic consumers, and the relationships between health attitudes and consumers' choices for varying nutrition labels on organic dried strawberry products.

4.1. Organic Consumers and Health Attitudes

A simplified view of organic consumers assumes they are a fairly homogeneous segment with strong concerns to health and sustainability. Our analysis conducted on a cross-national sample of 614 consumers reveals a different reality, where large individual variations in health attitudes were observed among our respondents. On a cross-cultural level, Romanian and Turkish organic consumers showed larger health, natural and ecological interests and more often selected "more natural nutrients" labels than Norwegian consumers. One possible explanation is that the higher maturity and extent of the organic market in Norway, where organic products are available in any typical supermarket, may result in a broader range of organic consumers who are not primarily driven by health, natural, or ecological values. Specifically, previous research indicates a large variation in consumer motives for choosing organic products, including the conviction of a better taste, food safety, naturalness, quality assurance, and status signaling [63,81–84]. Interestingly, recent research has also highlighted how internal versus external personal health locus of control (i.e., the degree to which people believe that they have control over their health, as opposed to external forces beyond their control (source: Wikipedia)) moderates organic food intake [85]. While consumers with an external locus of control are misled by the label and consume more unhealthy food if marked as organic, consumers with an internal locus of control consume less [85].

The section of the pooled consumer sample into health attitude groups led to two distinct consumer profiles corresponding to the low and high health attitude groups. Corroborating previous literature, respondents in the low health attitude group (HAQ1) were often young and males [75,86,87]. They reported low natural and ecological attitudes, expressed low interest in nutrients and nutrition claims in the habitudinal questionnaire, and preferred 'natural nutrients' labels in the choice task. This group therefore presents high internal validity throughout the survey. Also in line with the literature, respondents in the high attitude group (HAQ3) were typically females, highly educated, in the elder age group [75,86,87]. As could be expected, these respondents reported high natural and ecological attitudes, and expressed high interest in nutrients and nutrition claims in the habitudinal questionnaire. Paradoxically, this group also preferred 'natural nutrients' labels in the choice task as is discussed below.

4.2. Nutrition Label Choices

The diverging health attitudes that we observed were related to different selections of nutrients label alternatives in the choice tasks. More specifically, a non-linear effect of health attitude was revealed, where both the low and high health attitude groups rejected the healthier label offering more nutrients, while the moderate health attitude group more often preferred this alternative. It is not surprising that HAQ1 did not consider nutrient claim as important, in line with their attitudes. According to the literature, these consumers may lack knowledge: having low nutritional knowledge and/or being unaware that claims are regulated may lead to a disbelief of the label that is seen as a pure marketing tool [88]. Corroborating the importance of knowledge, Dominick and colleagues [89] conducted a national representative survey in the US with 1000 respondents to investigate the effect of an "all-natural" label on a set of nine product categories. These authors report than being male and having too little information both decreased the likelihood of purchase.

Furthermore, we expected a clear preference for more nutrients in the HAQ3 group, which was disproved. Several explanations may be brought forward from the literature. The first one is that claims have been found to decrease tastiness, attractiveness, and naturalness perceptions [88]. More specifically, respondents in this group may have interpreted the 'more natural nutrients' label as a sign for ultra-processing and less naturality, which are highly in contradiction with their values. This may all the more be the case as processing technology was one of the product attributes under evaluation in the choice task, which may have primed these consumers' skepticism for technology. Another possibility is that these consumers may not have felt in the target group for this product alternative. In a recent qualitative study, Benson and colleagues [88] identified that consumers perceive nutrition claims as mostly relevant and beneficial to specific populations presenting health issues (e.g., diabetes) or in a life phase characterized by the need for nutritional adjustments, such as dieting for weight loss, pregnancy or old age. In line with these results, Berhaupt-Glickstein and colleagues [90] found that health-risk reduction claims increased purchase intentions for green tea particularly in consumers who had modified their diet in the past year due to health concerns.

Respondents in the HAQ2 group did not express strong attitudes neither with regards to nutrients claims, nor with regards to naturality. Consequently, in the choice task, these respondents may have simply seen the indication of more nutrients as a valuable feature of the product. More research is needed to investigate whether moderate health attitudes consistently lead to an increased attractivity towards nutrient labels in other settings and product types.

To our knowledge, very few research studies have previously investigated organic consumers' choices for organic products with nutrition labels. Aschemann-Witzel and colleagues [91] investigated the impact of combining functional food claims with organic labels on organic consumers' product choices, and concluded that nutrition claims can be beneficial in the marketing of organic products. More specifically, these authors report that organic consumers choose organic foods with claims over those without claims, and especially so when it comes to occasional organic buyers. Their study did not investigate the role of organic consumer's health attitude on their choices. In contrast, our data presents no evidence of the role of organic consumption frequency on consumers' nutrient label choices, when choosing between two nutrient labels on a healthy product.

4.3. Limitations

In this study, the choice task investigating nutrition labels used organic dried strawberries as a case product. Effects of nutrition claims on consumer purchase decisions have been reported to be product dependent in terms of product category [85,88]. Another limitation of the study relates to the packaging design in general and to the label design in particular. Previous research has highlighted the impact of forms, colors, font types, etc. on the perception of food packaging information [92,93] including on the communication of nutrient or nutritional content [94,95]. Finally, the organic food market differs significantly across nations as a result of different geographies, agricultural policies, and public attention [50,96]. Consequently, conclusions from the current experiment may not be directly transferable to other product types and in particular to non-healthy products, other nutrient labels, and other cultures.

5. Conclusions

This research highlighted the heterogeneity in health attitudes among multicultural organic consumers, and revealed a non-linear relationship between health attitudes and consumers' choices for nutrient content in organic dried strawberries. These findings underline the complexity in targeting nutrition labels to organic consumers. Three organic consumer groups of different health attitudes levels were profiled and characterized in terms of salient habitudinal, attitudinal, behavioral, and socio-demographic characteristics as well as in terms of stated nutrient content choices. This information may help organic-food marketers to better target nutrition labeling strategies at consumer segments of different health attitudes. Complementary research is recommended to test the robustness of our findings across several product categories, label designs, as well as across several cultures. Moreover, further research is needed to uncover what attitudes, motives, and knowledge drive organic consumers of different health attitudes towards different nutrition label choices.

Supplementary Materials: The following are available online at http://www.mdpi.com/2072-6643/11/12/2951/s1, Figure S1: An example of choice set (English translation).

Author Contributions: Conceptualization, V.L.A. and D.A.; Methodology, V.L.A., D.A., and C.R.; Data formatting and analysis: V.L.A., D.A., and C.R.; Writing—Original draft preparation, V.L.A., D.A., and C.R.; Writing—review and editing, V.L.A. and D.A.; Supervision, V.L.A.; Project administration, V.L.A.; Funding acquisition, V.L.A.

Funding: This research was funded by funding bodies within the FP7 ERA-Net CORE Organic Plus, and with co-funds from the European Commission.

Acknowledgments: Author C.R. acknowledges Industry Doctoral grant no. SFRH/BDE/100483/2014, funded by FCT, Portuguese Foundation for Science and Technology (FCT). Author D.A. acknowledges the European Commission through the Marie Curie Actions Intra European Fellowship (IEF), call FP/-PEOPLE-I2012-IEF – project title "Innovative Methodologies for New Food Product Development: combining Sensory Science and Experimental Economics—NEFOMET" for the support. Author V.L.A. acknowledges the Norwegian Agriculture Agency for support through the Fund for Research Fees for Agricultural Products (FoodSMaCK, project nr. 262308).

We thank Rungsaran Wongprawmas for the support in the experimental choice design as well as Mona Elena Popa and Gogus Fahrettin for their support in preparing the surveys in Romania and Turkey.

Conflicts of Interest: The authors declare no conflict of interest.

References

1. Drichoutis, A.; Lazaridis, P.; Nayga, R., Jr. Consumers' Use of Nutritional Labels: A Review of Research Studies and Issues. *Acad. Mark. Sci. Rev.* **2006**, *9*, 93–118.
2. Weis, T. *The Global Food Economy: The Battle for the Future of Farming*; Fernwood Publishing: Black Point, NS, Canada, 2007; ISBN 978-1-55266-228-1.
3. Kearney, J. Food Consumption Trends and Drivers. *Philos. Trans. R. Soc. B Biol. Sci.* **2010**, *365*, 2793–2807.
4. Ikonen, I.; Sotgiu, F.; Aydinli, A.; Verlegh, P.W.J. Consumer Effects of Front-of-Package Nutrition Labeling: An Interdisciplinary Meta-Analysis. *J. Acad. Mark. Sci.* **2019**. [CrossRef]
5. WHO. Obesity and Overweight Fact Sheet. Available online: http://www.who.int/en/news-room/fact-sheets/detail/obesity-and-overweight (accessed on 17 April 2019).
6. Grunert, K.G. Trends in Food Choice and Nutrition. In *Consumer Attitudes to Food Quality Products: Emphasis on Southern Europe*; Klopčič, M., Kuipers, A., Hocquette, J.-F., Eds.; Wageningen Academic Publishers: Wageningen, The Netherlands, 2013; pp. 23–30. [CrossRef]
7. Grunert, K.G.; Wills, J.M. A Review of European Research on Consumer Response to Nutrition Information on Food Labels. *J. Public Health* **2007**, *15*, 385–399. [CrossRef]
8. Annunziata, A.; Vecchio, R. Consumer Perception of Functional Foods: A Conjoint Analysis with Probiotics. *Food Qual. Prefer.* **2013**, *28*, 348–355. [CrossRef]
9. Asioli, D.; Aschemann-Witzel, J.; Caputo, V.; Vecchio, R.; Annunziata, A.; Næs, T.; Varela, P. Making Sense of the "Clean Label" Trends: A Review of Consumer Food Choice Behavior and Discussion of Industry Implications. *Food Res. Int.* **2017**, *99*, 58–71. [CrossRef]
10. Caswell, J.A.; Mojduszka, E.M. Using Informational Labeling to Influence the Market for Quality in Food Products. *Am. J. Agric. Econ.* **1996**, *78*, 1248–1253. [CrossRef]
11. Wills, J.M.; Schmidt, D.B.; Pillo-Blocka, F.; Cairns, G. Exploring Global Consumer Attitudes toward Nutrition Information on Food Labels. *Nutr. Rev.* **2009**, *67*, S102–S106. [CrossRef]
12. Varela, P.; Antúnez, L.; Cadena, R.S.; Giménez, A.; Ares, G. Attentional Capture and Importance of Package Attributes for Consumers' Perceived Similarities and Differences among Products: A Case Study with Breakfast Cereal Packages. *Food Res. Int.* **2014**, *64*, 701–710. [CrossRef]
13. Carrillo, E.; Varela, P.; Fiszman, S. Effects of Food Package Information and Sensory Characteristics on the Perception of Healthiness and the Acceptability of Enriched Biscuits. *Food Res. Int.* **2012**, *48*, 209–216. [CrossRef]
14. Ares, G.; Varela, F.; Machin, L.; Antúnez, L.; Giménez, A.; Curutchet, M.R.; Aschemann-Witzel, J. Comparative Performance of Three Interpretative Front-of-Pack Nutrition Labelling Schemes: Insights for Policy Making. *Food Qual. Prefer.* **2018**, *68*, 215–225. [CrossRef]
15. Schnettler, B.; Miranda, H.; Sepúlveda, J.; Denegri, M.; Mora, M.; Lobos, G. Preferences for Berries Among Consumers in Southern Chile: Blueberries Are Produced but Are They Consumed? *J. Food Sci.* **2011**, *76*, S458–S464. [CrossRef]
16. Jia, N.; Kong, B.; Liu, Q.; Diao, X.; Xia, X. Antioxidant Activity of Black Currant (*Ribes nigrum* L.) Extract and Its Inhibitory Effect on Lipid and Protein Oxidation of Pork Patties during Chilled Storage. *Meat Sci.* **2012**, *91*, 533–539. [CrossRef]
17. Wu, R.; Frei, B.; Kennedy, J.A.; Zhao, Y. Effects of Refrigerated Storage and Processing Technologies on the Bioactive Compounds and Antioxidant Capacities of 'Marion' and 'Evergreen' Blackberries. *LWT—Food Sci. Technol.* **2010**, *43*, 1253–1264. [CrossRef]
18. Szajdek, A.; Borowska, E.J. Bioactive Compounds and Health-Promoting Properties of Berry Fruits: A Review. *Plant Foods Hum. Nutr.* **2008**, *63*, 147–156. [CrossRef]
19. Invenire Market Intelligence. *Berries in the WorldIntroduction to the International Markets of Berries*; Sitra: Helsinki, Finland, 2008; ISBN 978-951-563-649-2.
20. Aday, M.S.; Caner, C. Individual and Combined Effects of Ultrasound, Ozone and Chlorine Dioxide on Strawberry Storage Life. *LWT Food Sci. Technol.* **2014**, *57*, 344–351. [CrossRef]

21. Tanase, E.E.; Popa, V.I.; Popa, M.E.; Geicu-Cristea, M.; Popescu, P.; Drăghici, M.; Miteluț, A.C. Identification of the Most Relevant Quality Parameters for Berries—A Review. *Sci. Bull. Ser. F Biotechnol.* **2016**, *20*, 222–236.
22. Basu, A.; Nguyen, A.; Betts, N.M.; Lyons, T.J. Strawberry As a Functional Food: An Evidence-Based Review. *Crit. Rev. Food Sci. Nutr.* **2014**, *54*, 790–806. [CrossRef]
23. Häkkinen, S.H.; Törrönen, A.R. Content of Flavonols and Selected Phenolic Acids in Strawberries and Vaccinium Species: Influence of Cultivar, Cultivation Site and Technique. *Food Res. Int.* **2000**, *33*, 517–524. [CrossRef]
24. Meyers, K.J.; Watkins, C.B.; Pritts, M.P.; Liu, R.H. Antioxidant and Antiproliferative Activities of Strawberries. *J. Agric. Food Chem.* **2003**, *51*, 6887–6892. [CrossRef]
25. Baicu, A.; Popa, M. Trends in Prolonging the Post-Harvest Life of Strawberries—A Review. In *Annals of the University Dunarea de Jos of Galati, Fascicle VI: Food Technology*; Galati University Press: Galati, Romania, 2018; Volume 42, pp. 9–16.
26. Khan, I.; Tango, C.N.; Chelliah, R.; Oh, D.-H. Development of Antimicrobial Edible Coating Based on Modified Chitosan for the Improvement of Strawberries Shelf Life. *Food Sci. Biotechnol.* **2019**, *28*, 1257–1264. [CrossRef]
27. Silva, I.; Prado, N.; Melo, P.; Arantes, D.; Zeni, M.; Otaguro, H.; Pasquini, D. Edible Coatings Based on Apple Pectin, Cellulose Nanocrystals, and Essential Oil of Lemongrass: Improving the Quality and Shelf Life of Strawberries (Fragaria Ananassa). *J. Renew. Mater.* **2019**, *7*, 73–87. [CrossRef]
28. Panou, A.A.; Karabagias, I.K.; Riganakos, K.A. Effect of Gamma-Irradiation on Sensory Characteristics, Physicochemical Parameters, and Shelf Life of Strawberries Stored under Refrigeration. *Int. J. Fruit Sci.* **2019**. [CrossRef]
29. Matar, C.; Gaucel, S.; Gontard, N.; Guilbert, S.; Guillard, V. Predicting Shelf Life Gain of Fresh Strawberries 'Charlotte Cv' in Modified Atmosphere Packaging. *Postharvest Biol. Technol.* **2018**, *142*, 28–38.
30. Duran, M.; Aday, M.S.; Zorba, N.N.D.; Temizkan, R.; Büyükcan, M.B.; Caner, C. Potential of Antimicrobial Active Packaging 'Containing Natamycin, Nisin, Pomegranate and Grape Seed Extract in Chitosan Coating' to Extend Shelf Life of Fresh Strawberry. *Food Bioprod. Process.* **2016**, *98*, 354–363. [CrossRef]
31. Aaby, K.; Grimsbo, I.H.; Hovda, M.B.; Rode, T.M. Effect of High Pressure and Thermal Processing on Shelf Life and Quality of Strawberry Purée and Juice. *Food Chem.* **2018**, *260*, 115–123. [CrossRef]
32. Szadzińska, J.; Kowalski, S.J.; Stasiak, M. Microwave and Ultrasound Enhancement of Convective Drying of Strawberries: Experimental and Modeling Efficiency. *Int. J. Heat Mass Transf.* **2016**, *103*, 1065–1074. [CrossRef]
33. Pellicer, J.A.; Fortea, M.I.; Trabal, J.; Rodríguez-López, M.I.; Gabaldón, J.A.; Núñez-Delicado, E. Stability of Microencapsulated Strawberry Flavour by Spray Drying, Freeze Drying and Fluid Bed. *Powder Technol.* **2019**, *347*, 179–185. [CrossRef]
34. Adak, N.; Heybeli, N.; Ertekin, C. Infrared Drying of Strawberry. *Food Chem.* **2017**, *219*, 109–116. [CrossRef]
35. Sablani, S.S. Drying of Fruits and Vegetables: Retention of Nutritional/Functional Quality. *Dry. Technol.* **2006**, *24*, 123–135. [CrossRef]
36. Guo, Q.; Sun, D.-W.; Cheng, J.-H.; Han, Z. Microwave Processing Techniques and Their Recent Applications in the Food Industry. *Trends Food Sci. Technol.* **2017**, *67*, 236–247. [CrossRef]
37. Wojdyło, A.; Figiel, A.; Oszmiański, J. Effect of Drying Methods with the Application of Vacuum Microwaves on the Bioactive Compounds, Color, and Antioxidant Activity of Strawberry Fruits. *J. Agric. Food Chem.* **2009**, *57*, 1337–1343. [CrossRef]
38. Böhm, V.; Kühnert, S.; Rohm, H.; Scholze, G. Improving the Nutritional Quality of Microwave-Vacuum Dried Strawberries: A Preliminary Study. *Food Sci. Technol. Int.* **2006**, *12*, 67–75. [CrossRef]
39. Monier-Dilhan, S.; Bergès, F. Consumers' Motivations Driving Organic Demand: Between Self-Interest and Sustainability. *Agric. Resour. Econ. Rev.* **2016**, *45*, 522–538. [CrossRef]
40. Apaolaza, V.; Hartmann, P.; D'Souza, C.; López, C.M. Eat Organic—Feel Good? The Relationship between Organic Food Consumption, Health Concern and Subjective Wellbeing. *Food Qual. Prefer.* **2018**, *63*, 51–62. [CrossRef]
41. Hansen, T.; Sørensen, M.I.; Eriksen, M.-L.R. How the Interplay between Consumer Motivations and Values Influences Organic Food Identity and Behavior. *Food Policy* **2018**, *74*, 39–52. [CrossRef]

42. Kretzschmar, U.; Schmid, O. Quality and Safety Aspects of Organic and Low-Input Food Processing: Results of a Delphi Survey from an Expert Consultation in 13 European Countries. *NJAS Wagening. J. Life Sci.* **2011**, *58*, 111–116. [CrossRef]
43. Bravo, C.P.; Cordts, A.; Schulze, B.; Spiller, A. Assessing Determinants of Organic Food Consumption Using Data from the German National Nutrition Survey {II}. *Food Qual. Prefer.* **2013**, *28*, 60–70. [CrossRef]
44. Asioli, D.; Canavari, M.; Pignatti, E.; Obermowe, T.; Sidali, K.L.; Vogt, C.; Spiller, A. Sensory Experiences and Expectations of Italian and German Organic Consumers. *J. Int. Food Agribus. Mark.* **2014**, *26*, 13–27. [CrossRef]
45. Asioli, D.; Pignatti, E.; Canavari, M. Sensory Attributes and Marketing: An Exploration of Italian Organic Producers' Perspective. *Bodenkultur* **2012**, *63*, 7–16.
46. Oates, L.; Cohen, M.; Braun, L. Characteristics and Consumption Patterns of Australian Organic Consumers. *J. Sci. Food Agric.* **2012**, *92*, 2782–2787. [CrossRef]
47. De Magistris, T.; Gracia, A. The Decision to Buy Organic Food Products in Southern Italy. *Br. Food J.* **2008**, *110*, 929–947.
48. Ditlevsen, K.; Sandøe, P.; Lassen, J. Healthy Food Is Nutritious, but Organic Food Is Healthy Because It Is Pure: The Negotiation of Healthy Food Choices by Danish Consumers of Organic Food. *Food Qual. Prefer.* **2019**, *71*, 46–53. [CrossRef]
49. IFOAM. *Into the Future. Consolidated Annual Report*; IFOAM Organics International: Bonn, Germany, 2015.
50. Willer, H.; Lernoud, J. *The World of Organic Agriculture. Statistics and Emerging Trends 2018*; Research Institute of Organic Agriculture (FiBL), Frick, Switzerland and IFOAM—Organics International: Bonn, Germany, 2018; ISBN 978-3-03736-068-2.
51. Van Loo, E.J.; Caputo, V.; Nayga, R.M.; Meullenet, J.-F.; Ricke, S.C. Consumers' Willingness to Pay for Organic Chicken Breast: Evidence from Choice Experiment. *Food Qual. Prefer.* **2011**, *22*, 603–613. [CrossRef]
52. De Magistris, T.; Gracia, A. Consumers' Willingness-to-Pay for Sustainable Food Products: The Case of Organically and Locally Grown Almonds in Spain. *J. Clean. Prod.* **2016**, *118*, 97–104. [CrossRef]
53. Denver, S.; Jensen, J.D. Consumer Preferences for Organically and Locally Produced Apples. *Food Qual. Prefer.* **2014**, *31*, 129–134. [CrossRef]
54. Zanoli, R.; Naspetti, S. Consumer Motivations in the Purchase of Organic Food: A Means-End Approach. *Br. Food J.* **2002**, *104*, 643–653.
55. Shaw Hugner, R.; McDonagh, P.; Prothero, A.; Shultz, C.J.; Stanton, J. Who Are Organic Food Consumers? A Compilation and Review of Why People Purchase Organic Food. *J. Consum. Behav.* **2007**, *6*, 94–110.
56. Baudry, J.; Péneau, S.; Allès, B.; Touvier, M.; Hercberg, S.; Galan, P.; Amiot, M.-J.; Lairon, D.; Méjean, C.; Kesse-Guyot, E. Food Choice Motives When Purchasing in Organic and Conventional Consumer Clusters: Focus on Sustainable Concerns (The NutriNet-Santé Cohort Study). *Nutrients* **2017**, *9*, 88. [CrossRef]
57. Zagata, L. Consumers' Beliefs and Behavioural Intentions towards Organic Food. Evidence from the Czech Republic. *Appetite* **2012**, *59*, 81–89.
58. Aertsens, J.; Verbeke, W.; Mondelaers, K.; van Huylenbroeck, G. Personal Determinants of Organic Food Consumption: A Review. *Br. Food J.* **2009**, *111*, 1140–1167.
59. Stolz, H.; Stolze, M.; Janssen, M.; Hamm, U. Preferences and Determinants for Organic, Conventional and Conventional-plus Products—The Case of Occasional Organic Consumers. *Food Qual. Prefer.* **2011**, *22*, 772–779.
60. Ekezie, F.-G.C.; Sun, D.-W.; Han, Z.; Cheng, J.-H. Microwave-Assisted Food Processing Technologies for Enhancing Product Quality and Process Efficiency: A Review of Recent Developments. *Trends Food Sci. Technol.* **2017**, *67*, 58–69. [CrossRef]
61. Asioli, D.; Wongprawmas, R.; Pignatti, E.; Canavari, M. Can Information Affect Sensory Perceptions? Evidence from a Survey on Italian Organic Food Consumers. *AIMS Agric. Food* **2018**, *3*, 327–344. [CrossRef]
62. Fillion, L.; Arazi, S. Does Organic Food Taste Better? A Claim Substantiation Approach. *Nutr. Food Sci.* **2002**, *32*, 153–157.
63. Hemmerling, S.; Asioli, D.; Spiller, A. Core Organic Taste: Preferences for Naturalness-Related Sensory Attributes of Organic Food Among European Consumers. *J. Food Prod. Mark.* **2016**, *22*, 824–850. [CrossRef]
64. Janssen, M.; Hamm, U. Product Labelling in the Market for Organic Food: Consumer Preferences and Willingness-to-Pay for Different Organic Certification Logos. *Food Qual. Prefer.* **2012**, *25*, 9–22. [CrossRef]

65. Eagly, A.H.; Chaiken, S. *The Psychology of Attitudes*; Harcourt Brace Jovanovich College Publishers: Orlando, FL, USA, 1993.
66. Bain, R. An Attitude on Attitude Research. *Am. J. Sociol.* **1928**, *33*, 940–957.
67. Shrigley, R.L.; Koballa, T.R., Jr.; Simpson, R.D. Defining Attitude for Science Educators. *J. Res. Sci. Teach.* **1988**, *25*, 659–678. [CrossRef]
68. Hong, H. Scale Development for Measuring Health Consciousness: Re-Conceptualization. In Proceedings of the 12th Annual International Public Relations Research Conference, Coral Gables, FL, USA, 11–14 March 2009; pp. 212–233.
69. Magnusson, M.K.; Arvola, A.; Hursti, U.; Aberg, L.; Sjödén, P.-O. Attitudes towards Organic Foods among Swedish Consumers. *Br. Food J.* **2001**, *103*, 209–227.
70. Magnusson, M.K.; Arvola, A.; Hursti, U.-K.K.; Åberg, L.; Sjödén, P.-O. Choice of Organic Foods Is Related to Perceived Consequences for Human Health and to Environmentally Friendly Behaviour. *Appetite* **2003**, *40*, 109–117. [CrossRef]
71. Michaelidou, N.; Hassan, L.M. The Role of Health Consciousness, Food Safety Concern and Ethical Identity on Attitudes and Intentions towards Organic Food. *Int. J. Consum. Stud.* **2008**, *32*, 163–170.
72. Bourn, D.; Prescott, J. A Comparison of the Nutritional Value, Sensory Qualities, and Food Safety of Organically and Conventionally Produced Foods. *Crit. Rev. Food Sci. Nutr.* **2002**, *42*, 1–34.
73. Farruggia, D.; Crescimanno, M.; Galati, A.; Tinervia, S. The Quality Perception of Fresh Berries: An Empirical Survey in the German Market. *Agric. Agric. Sci. Procedia* **2016**, *8*, 566–575. [CrossRef]
74. Asioli, D.; Rocha, C.; Wongprawmas, R.; Popa, M.; Gogus, F.; Almli, V.L. Microwave-Dried or Air-Dried? Consumers' Stated Preferences and Attitudes for Organic Dried Strawberries. A Multi-Country Investigation in Europe. *Food Res. Int.* **2019**, *120*, 763–775. [CrossRef]
75. Roininen, K.; Lähteenmäki, L.; Tuorila, H. Quantification of Consumer Attitudes to Health and Hedonic Characteristics of Foods. *Appetite* **1999**, *33*, 71–88. [CrossRef]
76. Bredahl, L.; Grunert, K.G. *Food-Related Lifestyle Trends in Germany 1993–1996*; Handelshøjskolen i Århus, MAPP Centeret: Århus, Denmark, 1997; Available online: https://pure.au.dk/portal/files/32299533/wp50.pdf (accessed on 27 November 2019).
77. Cox, D.N.; Evans, G. Construction and Validation of a Psychometric Scale to Measure Consumers' Fears of Novel Food Technologies: The Food Technology Neophobia Scale. *Food Qual. Prefer.* **2008**, *19*, 704–710. [CrossRef]
78. Martens, H.; Naes, T. *Multivariate Calibration*; John Wiley & Sons: Chichester, UK, 1992.
79. Martens, H.; Martens, M. Modified Jack-Knife Estimation of Parameter Uncertainty in Bilinear Modelling by Partial Least Squares Regression (PLSR). *Food Qual. Prefer.* **2000**, *11*, 5–16. [CrossRef]
80. Train, K. *Discrete Choice Methods with Simulation*; Press, C.U., Ed.; Discrete Choice Methods with Simulation; Cambridge University Press: New York, NY, USA, 2009; Volume 47. [CrossRef]
81. Puska, P.; Kurki, S.; Lähdesmäki, M.; Siltaoja, M.; Luomala, H. Sweet Taste of Prosocial Status Signaling: When Eating Organic Foods Makes You Happy and Hopeful. *Appetite* **2018**, *121*, 348–359. [CrossRef]
82. Bernard, J.C.; Liu, Y. Are Beliefs Stronger than Taste? A Field Experiment on Organic and Local Apples. *Food Qual. Prefer.* **2017**, *61*, 55–62. [CrossRef]
83. Bryła, P. Organic Food Consumption in Poland: Motives and Barriers. *Appetite* **2016**, *105*, 737–746. [CrossRef]
84. Hemmerling, S.; Obermowe, T.; Canavari, M.; Sidali, K.L.; Stolz, H.; Spiller, A. Organic Food Labels as a Signal of Sensory Quality—Insights from a Cross-Cultural Consumer Survey. *Org. Agric.* **2013**, *3*, 57–69. [CrossRef]
85. Lee, H.-C.; Chang, C.-T.; Cheng, Z.-H.; Chen, Y.-T. Will an Organic Label Always Increase Food Consumption? It Depends on Food Type and Consumer Differences in Health Locus of Control. *Food Qual. Prefer.* **2018**, *63*, 88–96. [CrossRef]
86. Roininen, K.; Tuorila, H.; Zandstra, E.H.; de Graaf, C.; Vehkalahti, K.; Stubenitsky, K.; Mela, D.J.C. Differences in health and taste attitudes and reported behaviour among Finnish, Dutch and British consumers: a cross-national validation of the Health and Taste Attitude Scales (HTAS). *Appetite* **2001**, *37*, 33–45. [CrossRef]
87. Robinson, R.; Smith, C. Associations between Self-Reported Health Conscious Consumerism, Body-Mass Index, and Attitudes about Sustainably Produced Foods. *Agric. Hum. Values* **2003**, *20*, 177–187. [CrossRef]

88. Benson, T.; Lavelle, F.; McCloat, A.; Mooney, E.; Bucher, T.; Egan, B.; Dean, M. Are the Claims to Blame? A Qualitative Study to Understand the Effects of Nutrition and Health Claims on Perceptions and Consumption of Food. *Nutrients* **2019**, *11*, 2058. [CrossRef]
89. Dominick, S.R.; Fullerton, C.; Widmar, N.J.O.; Wang, H. Consumer Associations with the "All Natural" Food Label. *J. Food Prod. Mark.* **2017**. [CrossRef]
90. Berhaupt-Glickstein, A.; Hooker, H.N.; Hallman, K.W. Qualified Health Claim Language Affects Purchase Intentions for Green Tea Products in the United States. *Nutrients* **2019**, *11*, 921. [CrossRef]
91. Aschemann-Witzel, J.; Maroscheck, N.; Hamm, U. Are Organic Consumers Preferring or Avoiding Foods with Nutrition and Health Claims? *Food Qual. Prefer.* **2013**, *30*, 68–76. [CrossRef]
92. Piqueras-Fiszman, B.; Spence, C. Sensory Expectations Based on Product-Extrinsic Food Cues: An Interdisciplinary Review of the Empirical Evidence and Theoretical Accounts. *Food Qual. Prefer.* **2015**, *40*, 165–179. [CrossRef]
93. Ares, G.; Piqueras-Fiszman, B.; Varela, P.; Marco, R.M.; López, A.M.; Fiszman, S. Food Labels: Do Consumers Perceive What Semiotics Want to Convey? *Food Qual. Prefer.* **2011**, *22*, 689–698. [CrossRef]
94. Tait, P.; Saunders, C.; Guenther, M.; Rutherford, P.; Miller, S. Exploring the Impacts of Food Label Format on Consumer Willingness to Pay for Environmental Sustainability: A Choice Experiment Approach in the United Kingdom and Japan. *Int. Food Res. J.* **2016**, *23*, 1787–1796.
95. Ducrot, P.; Julia, C.; Méjean, C.; Kesse-Guyot, E.; Touvier, M.; Fezeu, L.K.; Hercberg, S.; Péneau, S. Impact of Different Front-of-Pack Nutrition Labels on Consumer Purchasing Intentions: A Randomized Controlled Trial. *Am. J. Prev. Med.* **2016**, *50*, 627–636. [CrossRef]
96. Rossi, R. *Facts and Figures on Organic Agriculture in the European Union*; European Commission, DG Agriculture and Rural Development, Unit Economic Analysis of EU Agriculture: Brussels, Belgium, 2013.

nutrients

MDPI

Article

Consumers' Trade-Off between Nutrition and Health Claims under Regulation 1924/2006: Insights from a Choice Experiment Analysis

Rosaria Viscecchia [1,*], Giuseppe Nocella [2], Biagia De Devitiis [1], Francesco Bimbo [1], Domenico Carlucci [3], Antonio Seccia [4] and Gianluca Nardone [1]

1 Department of Science of Agriculture, Food and Environment University of Foggia, 71121 Foggia, Italy; biagia.dedevitiis@unifg.it (B.D.D.); francesco.bimbo@unifg.it (F.B.); gianluca.nardone@unifg.it (G.N.)
2 School of Agriculture, Policy and Development, University of Reading, Reading RG6 6BZ, UK; g.nocella@reading.ac.uk
3 Department of Agricultural and Environmental Sciences, University of Bari "Aldo Moro", 70126 Bari, Italy; domenico.carlucci@uniba.it
4 Department of Humanities University of Foggia, 71121 Foggia, Italy; antonio.seccia@unifg.it
* Correspondence: rosaria.viscecchia@unifg.it; Tel.: +39-088-158-9313

Received: 1 October 2019; Accepted: 22 November 2019; Published: 26 November 2019

check for updates

Abstract: The aim of this paper is to investigate consumers' preferences and willingness to pay (WTP) for functional mozzarella cheese whose health benefits (reduced fat and enrichment in omega-3) are communicated by using nutrition claims (article 8) and health claims (articles 13 and 14) of the EU Regulation 1924/2006. In order to achieve the stated objective a choice survey was developed and administered to a sample of Italian respondents. The product attributes and attribute levels included in the choice experiment were obtained from in-depth interviews conducted with stakeholders working on the development of this new product in the Italian region of Puglia. Results show that many participants were not aware of functional food. Marketing segmentation performed via latent class analysis indicates that the development of this hypothetical product should be based on the addition of naturally enriched omega-3. In terms of health communication under Regulation 1924/2006, heterogeneity of preferences of the nine identified segments reveals that respondents have a clear preference for products from the Puglia region, for the combined nutrition claim over single nutrition claims and for the reduction of disease health claim (article 14) over the health claim (article 13). In monetary terms, willingness to pay for health claims is higher than nutrition claims.

Keywords: nutrition claim; health claim; functional food; WTP; segmentation

1. Introduction

The "EU Regulation 1924/2006 on nutrition and health claims made on foods" makes a clear distinction between nutrition claims (article 8), health claims (article 13), and reduction of disease health claims (article 14). Nutrition claims inform consumers that a food has particular beneficial nutritional properties (e.g., either contains or does not contain a certain nutrient); permitted nutritional claims are listed in the Annex of Regulation 1924/2006. Health claims inform consumers that a relationship exists between a food or one of its constituents and health, while the reduction of disease health claims inform consumers that the consumption of a food or one of its constituents reduces a risk factor in the development of a human disease. To be included in the community list of permitted health claims, claims must be based on generally accepted scientific evidence and well understood by the average consumer [1–3].

The scientific substantiation of health claims (articles 13 and 14) is a challenging and long process because it requires a substantial effort in terms of financial and human resources on behalf of the food industry to get these claims approved by the European Food Safety Authority (EFSA). It has been estimated that in the European Union (EU), the costs of the scientific substantiation process of health claims range from €4.51 to €7.65 million, without taking into account clinical trials and other costs necessary to provide proprietary data to support applications to EFSA [4]. In order to use these claims manufacturers also face challenges related to changing lists of ingredients, wording of claims, and limited financial and research and development resources [5].

The list of scientific authorised health claims is continuously updated with regulation amendments (Table 1) based on assessments provided by the EFSA panel on dietary product, nutrition, and allergies. These evaluations have created uncertainty among food companies investing in research and development of new functional food products because of the high rate of rejections of submitted health claims [6,7]. Between July 2008 and March 2010 the European Commission (EC), after examining over 44,000 claims supplied by the Member States, submitted a list of 4.637 "general function" health claims to EFSA for scientific evaluation. However, at the end of June 2012 only 222 health claims (article 13) were authorised by EFSA [8]. As Table 1 shows, so far 243 health claims and 22 reduction of disease risk claims have been authorised under articles 13 and 14 of Regulation 1924/2006, respectively.

Manufacturers, using authorised food health claims, transfer information to consumers about positive effects that may be obtained from consuming their healthy products [9–11]. However, since the implementation of EU Regulation 1924/2006, it is not yet clear how information and beneficial effects communicated by nutrition and health claims can be perceived and understood by the average consumer [12].

A large body of literature explores how socio-demographic and economic characteristics of consumers influence their preferences towards nutrition and health claims. Even if some general patterns can be identified, the influence of socio-demographic and economic factors (gender, education, age, and income) remains undetermined [1,13–23].

Despite the fact that EU Regulation 1924/2006 was introduced in 2007, it appears that so far no studies have simultaneously compared consumers' preferences and willingness to pay (WTP) for these three types of claims in the context of this regulation. Some studies explored WTP for different nutrition claims or compared consumers' WTP for claims under article 8 with claims under articles 13 or 14. For example, Barreiro et al. (2010) [24] found a higher WTP for nutrition and health claims used on less healthy products (sausages) and that nutrition claims were not significant in the choice of the healthy product (yoghurt). Instead, Van Wezemael et al. (2014) [25] observed higher WTP for steaks conveying nutrition and health claims but also cross-cultural differences and heterogeneity of preferences both for types of nutrient (iron, fat, or protein) and claims (nutrition or health claim).

Table 1. Amendments of EU Regulation 1924/2006 from 2007 to 2017.

Year of Amendment	Amendments of EU Regulation 1924/2006			
	Article 8 Nutrition Claims	**Article 13 General Function (Health Claims)**	**Article 14.1a Reduction of Disease Risk (Health Claims)**	**Article14 1b Health Claims on Children's Development and Health (Health Claims)**
2007	Annex of nutrition claims			
2008	Regulation EC No 107/2008 and Regulation EC No 109/2008			
2009			Regulation 983/2009 (2 claims) Regulation 1024/2009 (1 claim)	Regulation 983/2009 (5 claims) Regulation 1024/2009 (1 claim)
2010	Regulation EC No 116/2010		Regulation 384/2010 (1 claim)	Regulation 957/2010 (2 claims)
2011	Regulation EC No 1169/2011			
2011			Regulation 665/2011 (2 claims) Regulation 1160/2011 (1 claim)	Regulation 440/2011 (1 claim)
2012	Regulation EC No (1047/2012)	Regulation EC No 432/2012 222 authorised health claims	Regulation 1048/2012 (1 claim)	
2013		3 Amendments Regulation EC No 432/2012 10 authorised health claims		
2014		1 Amendment Regulation EC No 432/2012 1 authorised health claims	Regulation 1135/2014 (1 claim) Regulation 1226/2014 (1 claim) Regulation 1228/2014 (3 claims)	
2015		3 Amendments Regulation EC No 432/2012 3 authorised health claims		
2016		3 Amendments Regulation EC No 432/2012 5 authorised health claims		
2017		2 Amendments Regulation EC No 432/2012 2 authorised health claims		
Total of approved health claims		243	13	9

Source: Our Elaboration on EFSA data (www.efsa.europa.eu) [26]. EFSA (European Food Safety Authority).

De Magistris and Lopez-Galan (2016) [27] found that consumers showed a positive willingness to pay for cheese with a reduced-fat claim, and a negative willingness to pay for cheese with a low salt claim. If both the claims "reduced-fat" and "low salt" were presented on the same package consumers were willing to pay a premium price. Lemken (2017) [28] also observed differences in consumers' WTP for diverse types of nutrition claims conveyed on legumes pasta. He found that legume pasta advertised with nutrition claims showed a higher WTP than conventional pasta and in particular the fibre claim appeared to be superior to that of the protein claim. In a recent study, Lopez-Galan and de Magistris (2019) [29] found that emotional eating can have a negative impact in purchasing behaviour related to nutritional claims.

In light of this overview of past studies, this paper aims at evaluating how consumers' preferences and WTP are influenced by the simultaneous presence of nutrition and health claims regulated by EU legislation 1924/2006. To achieve this objective, we explore the following research questions. To what extent, do nutrition and health claims under Regulation 1924/2006 influence consumer purchasing behaviour? Do consumers prefer nutrition, health, or reduction risk claims? Are consumers willing to pay more for claims regulated by article 8, 13, or 14? Do different nutrition and health claims generate heterogeneity of preferences? Is WTP of consumers who pay attention to claims the same as of those who do not attend information reported on food products? What socio-economic factors influence heterogeneity of preferences?

2. Materials and Methods

In order to achieve the stated objective a survey was developed on the basis of four in-depth interviews conducted with a food scientist, an animal husbandry expert, a dairy producer, and a marketing expert working on product development of this functional product. Experts stated that ongoing experiments were developing a functional mozzarella cheese whose consumption could help consumers to maintain normal levels of cholesterol because of an enhanced lipid composition obtained via reduced saturated fatty acids and/or enrichment of omega-3 (polyunsaturated fatty acids and conjugated linoleic acid). The enrichment of omega-3 was experimented either by adding this nutrient directly to milk or naturally feeding cows with flax seeds. Functional mozzarella cheese obtained in these two ways did not show any difference in terms of organoleptic characteristics with conventional mozzarella. Both methods did not impact very much on costs even if the natural enrichment of omega-3 was more expensive than when it was added directly to the milk. The only variation observed was linked to the production process that in comparison to conventional produce required a reduced temperature and spinning time of functional mozzarella cheese. Insights from in-depth interviews were fundamental to develop the questionnaire not only in relation to understanding what attributes and attribute levels this hypothetical functional mozzarella cheese might contain, but also to cherry pick the most appropriate nutritional and health claims for this new functional product from those approved under EU Regulation 1924/2006 (see Section 2.1).

The final questionnaire contained four sections. The first section collected information about respondents' family history (heart attacks, strokes, and other pathologies) and concern for risk factors of cardiovascular diseases (e.g., high blood pressure, obesity, etc.), and knowledge of functional food products. The second section elicited information about participants' consumption habits of mozzarella cheese. The third section presented the hypothetical market scenario for this product and the choice experiment. The last section collected information about socio-demographic and economic characteristics of respondents (gender, age, education, occupational status, income, etc.). The questionnaire was administered in 13 Italian cities (Bari, Bergamo, Bitritto, Cisterna di Latina, Fiumicino, Latina, Milano, Roma, Sanremo, Taranto, Treviso, Verona, Villorba) distributed in the four Italian Nielsen Areas. Data were collected between July and September 2013 by face-to-face interviews conducted by trained interviewers of an Italian marketing research company using the computer aided personal interview method. The sample was representative of the Italian population, stratified by age and sex according to demographic statistics provided by the national statistical institute.

2.1. The Choice Experiment and Econometric Analysis

The choice experiment was developed on the basis of information collected via in-depth interviews. As shown in Table 2, the attributes and relative levels used to develop the choice experiment design were: Origin (Not specified and made in Puglia), nutritional claims (No information, high in omega-3, reduced saturated fatty acids, high in omega-3, and reduced saturated fatty acids), health claims (No information, contributes to the maintenance of normal blood cholesterol levels, reduces cardiovascular disease risk), source of nutrient (No information, omega-3 added to the milk, and omega-3 present naturally in the milk), and price (+5%, and +10%, +20%, +40%).

Table 2. Attributes and attribute levels used in the choice experiment.

Attributes	Levels
Origin	Not specified; produced in Puglia
Nutritional claim	No information; high omega-3; reduced saturated fatty acids; high in omega-3, and reduced saturated fatty acids
Health claim	No information; contributes to the maintenance of normal blood cholesterol levels; helps to reduce risk of cardiovascular disease.
Information source of omega-3	No information, omega-3 added to the milk; omega-3 naturally present in the milk.
Price	+5%; +10%; +20%; +40%

The choice experiment design was obtained by performing a *D-efficient* design using the NGene software [30]. The final design contained 36 choice sets blocked into groups of six, so we had six final versions of the survey instrument. Table 3 shows an example of the unlabelled choice set presented to respondents.

Table 3. Example of a choice set.

Please evaluate the attributes of these products and indicate your choice carefully. If none of these products satisfies your preferences indicate 'none of these products'.				
	Product 1	**Product 2**	**Product 3**	
Origin	Produced in Puglia	Not specified	Produced in Puglia	None of these products
Source of omega-3	No information	Omega-3 added to the milk	Omega-3 naturally present in the milk	
Nutritional claim	No information	High in omega-3 and reduced in saturated fatty acids	High in omega-3	
Health claim (Articles 13 and 14)	No information	Helps to reduce risk of cardiovascular diseases (Article 14).	Contributes to the maintenance of normal blood cholesterol levels (Article 13).	
Price	+ 5%	+ 40%	+ 20%	
Your choice	□	□	□	□

Consumers' preferences for this hypothetical product were analysed by means of a discrete choice model [31,32]. The estimation of discrete choice models is based on Lancaster's theory which postulates that consumers do not derive satisfaction from goods themselves but from their attributes and attribute levels. The Lancastrian approach assumes that each consumer chooses a single option yielding the greatest utility [33,34]. This class of econometric models has its basis in the random utility models, which represent the fundamental econometric approach to the analysis of consumer preferences within a discrete choice multi-attribute context [35]. Random utility models are based on the hypothesis that individuals make choices according to attributes of alternatives characterized by a degree of randomness [36]. According to Greene (2003) [37], random utility models for multiple choices indicate that for the i_{th} consumer faced with J alternatives, the utility of alternative j (U_{ij}) is

$$U_{ij} = \mathbf{Z}_{ij}^{Att}\boldsymbol{\beta}_{ij} + \varepsilon_{ij} \tag{1}$$

where

- $\mathbf{Z}_{ij}^{Att}\boldsymbol{\beta}_{ij}$ is the deterministic component, i.e., a function of the sets of observable functional mozzarella attributes ($\mathbf{Z}_{ij}^{Att}$) and a set of associated population parameters ($\boldsymbol{\beta}_{ij}$) to be estimated;
- ε_{ij} is the random component that captures variations in choices determined by omitted variables, measurement errors, and within- and between-individual variances.

The deterministic component was modelled as a function of functional mozzarella cheese products' attributes assuming the Lancastrian approach that respondents can assign a subjective utility value to each product attribute, sum these values for each product to obtain a monotonic utility index, and select the product that gives them the highest utility. Thus, if a respondent makes her/his choice for a hypothetical functional mozzarella cheese product j, then, it is assumed that it is the maximum among the J utilities. The specification of this model depends on various conditions and assumptions about the distribution of the random component and on the way in which choice sets are developed and presented to respondents [37].

In this study, heterogeneity of preferences was estimated performing a latent class analysis using Latent Gold Choice 5.1 (Statistical Innovation Inc.). The number of classes was determined on the basis of statistical information criteria (Bayesian information criterion and the Consistent Akaike information criterion) [32]. Latent class analysis allows researchers to identify segments of individuals with identical preferences estimating the probability of membership to each class along with their respective class-specific preference weights. Thus, this segmentation technique is based on a likelihood model that permits statistical inference and is considered superior to traditional segmentation multivariate statistical techniques [34,38]. The following conditional logit model was estimated

$$P\left(y_{ij}\middle|\mathbf{Z}_{ij}\right) = \sum\nolimits_{x=1}^{k} P(x|\mathbf{Z}_{i}^{Cov}) \prod\nolimits_{j=1}^{Ji} P(y_{it}|x, \mathbf{Z}_{ij}^{Att}) \tag{2}$$

where $P(y_{ij}|\mathbf{Z}_{ij})$ is the probability of choosing alternative j conditional on the vectors of all covariate values of respondent i, x is the latent class that can take on values $1 \leq x \leq$ K, $\mathbf{Z}_{i}^{Cov}$ refers to all characteristics of individual i and $\mathbf{Z}_{ij}^{Att}$ is the vector of attribute values of respondent i.

Finally, a growing number of studies recognises that respondents do not consider all attributes presented in the choice sets and that this kind of behaviour (non-attendance attribute) can lead to biased estimates of consumers' preferences and willingness to pay [39–43]. In order to take account of this issue, we asked respondents to state what attributes they attended after having completed their choice tasks. Since the objective of this study is not to contrast modelling approaches that can account for stated or inferred attribute non-attendance, we only compare a latent class analysis where we include participants who stated to attend all attributes with a latent class analysis where participants who stated that they did not attend at least one attribute. Thus, for individuals who attended all attributes, the $\mathbf{Z}_{i}^{Cov}$ of Equation (2) will only include the socio-demographic characteristics of respondents.

3. Results and Discussion

The final sample size for this study is 601 with an average age of participants of 46 (s = 14.8). Table 4 shows that the majority of participants were females (51.1%), with high school education (60.9%), more than two family members (56.2%), without children below the age of 14 (74.9%), from Northern Italy (50.0%), white collar (30.4%) and with a monthly net income less than €2000 (26.6%). During the survey many participants refused to disclose their income and thus for this variable 41.1% missing values were observed. For this reason, occupational status will be used as a proxy of income when estimating consumers' preferences for functional mozzarella cheese.

Table 4. Socio-demographic characteristics of respondents (N = 601).

Socio-Demographic and Economic Characteristics	Frequency	Percent
Gender (Female):	307	51.1
Education (High school or degree):		
Less than high school	108	19.6
High school	366	60.9
Degree	107	19.5
Family size (More than two members)	338	56.2
Presence of Children younger than 14 (Yes)	151	25.1
Occupational status:		
Unemployed	62	10.3
Blue collar	131	21.8
White collar	183	30.4
Manager	60	10.0
Pensioner	86	14.3
Housewife	79	13.1
Income:		
Under €2000	160	26.6
Between €2000 and €2999	116	19.3
Between €3000–€3999	38	6.3
€4000 or above	39	6.5
Missing values	248	41.3
Geographic area:		
North	301	50.0
Middle	150	25.0
South	150	25.0

3.1. *Concern about Cardiovascular Disease and Knowledge of Functional Foods*

About 38% of participants stated that in their families there were cases of heart attack, stroke, high blood pressure, or other illnesses such as angina pectoris, aneurism, etc In addition, Table 5 shows to what extent respondents were worried about selected biological and lifestyle factors affecting cardiovascular diseases. The mean values of stress (4.52), cholesterol (4.21), blood pressure (4.12), and physical inactivity (4.08) indicate that participants are rather concerned about these risk factors. On the other hand, obesity, smoking, and diabetes with about one-quarter of respondents 'not concerned at all' and a mean of just above the mid-point of the proposed scale appear to be of less of a concern than previous risk factors. A test of reliability was performed (Cronbach's Alpha = 0.929) to create an index of participants' concern for cardiovascular diseases. The cardiovascular diseases concern index, obtained summing and averaging scores presented in Table 4, shows that on the average respondents ($\overline{X}$= 4.02; s = 1.64) were somewhat concerned about these risk factors.

Table 5. Concern for risk factors affecting cardiovascular diseases (N = 601).

Concern / Risk Factors	Not at All Concerned 1	2	3	4	5	6	Extremely Concerned 7	Mean(s)
Obesity	26.5%	11.8%	12.6%	12.1%	13.5%	9.5%	14.1%	3.59 (2.14)
Smoking	26.3%	8.8%	9.0%	11.3%	15.1%	12.1%	17.3%	3.86 (2.23)
Blood pressure	18.0%	8.2%	9.0%	18.3%	18.0%	12.6%	16.0%	4.12 (2.04)
Diabetes	23.0%	9.3%	12.5%	15.0%	15.0%	10.6%	14.6%	3.80 (2.10)
Cholesterol	11.6%	10.8%	12.1%	18.8%	17.5%	15.0%	14.1%	4.21 (1.90)
Triglycerides	14.6%	11.5%	11.6%	19.6%	18.0%	12.3%	12.3%	4.01 (1.92)
Stress	8.2%	5.0%	15.0%	20.1%	19.0%	16.3%	16.5%	4.52 (1.78)
Physical inactivity	13.8%	9.8%	11.1%	22.0%	18.1%	13.1%	12.0%	4.08 (1.88)

Furthermore, it was interesting to find out that only 66% of the participants had never heard the term functional foods before and that 53.7% had never consumed these products. The relationship between knowledge and consumption of functional foods was significant to the chi-square test (χ^2 = 31.80; p = 0.0001). Cross tabulating the yes/no categories of these two variables in the following segments of consumers were identified: No knowledge/no consumption (unknowing), knowledge/no consumption (apathetic), buyers without knowledge, and informed purchasers. These four segments are illustrated in Figure 1 where it can be observed that the ignorant represent the largest segment (40.9%) of the sample, followed by buyers without knowledge (25.1%), apathetic (12.8%), and informed purchasers (21.1%). The identified segments are more or less consistent across Italian geographic areas with the ignorant category slightly bigger in the Middle in comparison to Northern and Southern Italy. There are more males than females in the unknowing and apathetic segments, while the opposite pattern was observed both for buyers without knowledge and informed purchasers. Age also affects these four segments with apathetic and informed buyers being on the average younger than ignorant and buyers without knowledge. Finally, the apathetic participants are more concerned about cardiovascular diseases than other segments with differences significant to the ANOVA one-way (F = 6.21; df = 3; $p < 0.001$).

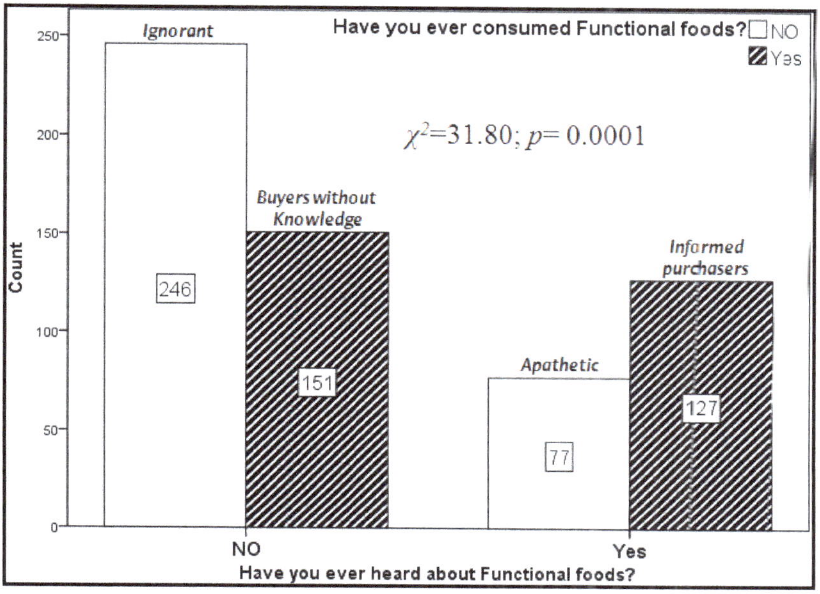

Figure 1. Relationship between knowledge and consumption of functional foods.

3.2. Consumption Habits and Heterogeneity of Preferences for Nutritional and Health Claims

Mozzarella cheese or 'Fiordilatte' is a very popular Italian cheese which is consumed regularly by many consumers as confirmed by the results of this study where about 2/3 of respondents consume this product at least once a month. The average weekly quantity of mozzarella cheese purchased by respondents is 438 g (s = 406) with 43.4% of them spending weekly less than €5.00, 40.1% between €5.00 and €5.99, 12.5% between €10.00 and €14.99, and the remaining 4% more than €14.99. Mozzarella cheese is purchased exclusively in big retailers by about 57% of respondents, but 20% of them only buy this product in traditional shops and 4% only in local markets and specialized shops, while 19% of respondents buy mozzarella cheese both in traditional shops and big retailers It is not surprising to observe that many respondents buy "Fiordilatte" in traditional and specialized shops because in these shops they can find unpackaged and very fresh mozzarella cheese that when consumed the day in which it is produced is a delicacy. In fact, 20.2% of respondents only buy the loose product, 41.4% only the packaged product, and 38.4% both loose and packaged mozzarella cheese.

The analysis of choice data show that 197 respondents stated to attend all attributes and 404 participants did not attend at least one attribute. Thus, in our sample a significant number of individuals did not process in full information provided in the choice sets. Taking into account the estimate of the 95% confidence interval for the proportion ($p < 0.05$), there should be between 63.5% and 71% individuals in the Italian population who ignore information for at least one attribute. This finding is confirmed also in other studies where the percentage of individuals who do not pay attention to at least one attribute seem to be more or less of the same magnitude 50% [39–43].

The development of functional mozzarella cheese designs and its marketing decisions depend on the selection of the right number of consumers' segments that might buy this product. However, retaining the correct number of segments is a difficult task because experimental studies have shown that information criteria are influenced by sample sizes, parameter estimates, and model complexity [44,45]. Furthermore, the decision about the number of segments most appropriate for this data must also be informed by other factors, such as the pattern of significant parameters and relative signs can be informative [46]. Considering all these aspects and information criteria (LL, BIC, and CAIC) presented in Table 6, the structure of the data suggests the presence of up to four groups with different preferences for participants who stated attribute full attendance (SAFA) and five classes for those who stated non-attendance (SANA) for at least one of these attributes. In Appendix A, we also report the estimates of the full model in order to allow readers to make a comparison with the estimates of the SAFA and SANA models.

Table 6. Information criteria to select the number of classes for SAFA and SANA LC models *.

	SAFA LC Models				SANA LC Models			
Number of classes	**Npar**	**LL**	**BIC**	**CAIC**	**Npar**	**LL**	**BIC**	**CAIC**
1	10	−1289.44	2631.70	2641.70	10	−3190.41	6440.84	6450.84
2	23	−1190.75	2503.02	2526.02	28	−2848.49	5865.01	5893.01
3	36	−1153.23	2496.65	2532.65	46	−2671.54	5619.14	5665.14
4	49	−1116.25	2491.38	2540.38	64	−2560.59	5505.26	5569.26
5	62	−1086.41	2500.38	2562.38	82	−2490.60	5473.32	5555.32
6	75	−1067.68	2531.60	2606.60	100	−2439.32	5478.78	5578.78

* Npar = Number of estimated parameters.; LC = Latent class; LL = Log-likelihood; BIC (LL) = Bayesian information criterion based on the log-likelihood; CAIC (LL) = Consistent Akaike information criterion, based on the log-likelihood; SAFA = stated attribute full attendance; SANA = stated non-attendance.

Table 7 reports the parameter estimates for the attribute levels of the SAFA and SANA models, where 'No information' is the baseline for all discrete attributes of the choice experiment. The SAFA model shows four classes whose size ranges from 29% to 20% and according to the order of extraction have been named A (29%), B (26.9%), C (24.1%), and D (20%). The five classes of the SANA model show more variability in terms of size ranging from 29.8% to 14.4% and have been named E (29.8%), F (24.5%), G (15.9%), H (15.4%), I (14.4%). For both models, socio-demographic and economic characteristics were not significant other than the geographic area. The covariates of the SANA model also include the stated attribute non-attendance for origin, nutrition claims, health claims, source of omega-3, and price.

With regards to parameters estimates of the SAFA model, the origin Puglia is significant in all four segments probably because SAFA respondents know that this region is renowned for mozzarella cheese production. The nutrition claim 'Reduced fat' is not significant in all classes, while price is negative and significant only in classes B and D. Moreover, the beta parameters of the other attributes of class A, are not significant or negative and significant (Omega-3 naturally present in the milk) and thus these participants were named 'pro-tradition consumers' because they are not attracted by functional mozzarella cheese. In class B, beta parameters are significant for the article 8 'Rich in omega-3' and 'Rich in Omega3 plus Reduced saturated fatty acids', the article 14 'Helps to reduce cardio-vascular disease risk' and for the attribute 'Omega-3 added directly to milk'. These participants were named 'pro-industry articles 8 and 14' consumers because in addition to being interested in nutrition and health claims they show preferences for a production process where omega-3 is added directly to the milk. In class C, beta parameters are significant for the combined nutrition claim and for both health claims, but they are not interested in the way in which omega-3 is added to functional mozzarella. Participants belonging to this class were named 'pro-EU Regulation 1924/20016' consumers and they are likely to be from Southern Italy. In class D, beta parameters are significant for the combined articles 8 'Rich in Omega3 and Reduced saturated fatty acids', the article 14 and for the attribute 'Omega-3 already contained in the milk produced by cows'. These participants were named 'pro-nature articles 8 and 14' consumers because differently from those belonging to class B they are interested in a production process where omega-3 is naturally contained in milk because cows are fed with flax seeds. They are likely to be from Northern Italy.

Table 7. Comparison of maximum likelihood estimates between SAFA and SANA for consumer preferences towards functional mozzarella cheese.[1.]

Attributes \ Classes	SAFA (N = 197)				SANA (N = 404)				
	$\beta_{A\,(29\%)}$	$\beta_{B\,(26.9\%)}$	$\beta_{C\,(24.1\%)}$	$\beta_{D\,(20\%)}$	$\beta_{E\,(29.8\%)}$	$\beta_{F\,(24.5\%)}$	$\beta_{G\,(15.9\%)}$	$\beta_{H\,(15.4\%)}$	$\beta_{I\,(14.4\%)}$
Origin: Puglia	1.66 (6.17)***	1.07 (7.45)***	0.36 (3.27)**	0.47 (2.92)**	0.90 (8.46)***	1.72 (7.71)***	−0.34 (3.01)**	0.67 (3.94)***	−0.05 (0.47)
Nutrition claims (Article 8):									
Reduced fat	0.03 (0.12)	−0.041 (1.19)	0.32 (1.06)	−0.26 (0.59)	0.16 (0.79.)	−0.18 (0.74)	0.11 (0.47)	−0.68 (1.57)	−0.69 (1.17)
Rich in omega-3	0.21 (1.27)	0.49 (2.31)*	−0.65 (2.74)**	0.42 (1.68)	0.32 (2.68)	0.11 (0.62)	−0.48 (2.48)	−0.44 (1.36)	0.33 (1.07)
Rich in omega-3 and reduced fat	0.44 (1.82)	0.48 (2.39)**	1.27 (6.40)***	0.52 (2.39)**	0.45 (3.68)***	0.60 (2.89)**	−0.17 (1.11)	0.39 (1.82)	0.96 (3.92)***
Health claims (Articles 13 and 14):									
Contributes to the maintenance of normal blood cholesterol levels (Article 13).	0.18 (0.96)	0.18 (0.87)	0.75 (4.07)***	−0.09 (0.47)	0.26 (2.38)**	0.25. (1.45)	−0.28 (1.33)	−0.16 (0.77)	0.39 (2.11)**
Helps to reduce cardiovascular diseases risk (Article 14).	−0.20 (1.02)	0.43 (2.15)**	0.47 (3.11)**	0.65 (3.49)***	0.44 (4.09)***	0.47 (2.36)**	−0.08 (0.51)	−0.03 (0.13)	0.24 (1.38)
Information Source of omega-3:									
Omega-3 added to the milk	0.27 (1.75)	0.35 (2.22)**	0.22 (1.37)	−0.10 (0.64)	−0.03 (0.35)	0.18 (1.24)	0.11 (0.57)	0.32 (1.64)	0.23 (1.54)
Omega-3 naturally present in the milk	−0.60 (2.99)**	0.35 (1.47)	0.12 (0.62)	0.65 (2.98)**	0.58 (4.30)***	−0.15 (0.71)	−0.31 (0.87)	−0.04 (0.15)	0.46 (2.07)**
Price	0.36 (0.71)	−10.35 (7.22)***	0.10 (0.23)	−2.74 (4.98)***	−2.05 (5.31)***	−13.49 (8.77)***	0.51 (1.04)	−24.19 (5.71)***	−9.93 (6.46)***
No buy	−1.08 (2.61)**	−1.11 (5.70)***	−4.17 (0.97)	−1.27 (5.46)***	−1.23 (8.57)	−1.57 (7.27)***	−0.16 (0.98)	−0.51 (2.12)**	−1.65 (7.02)***
Intercept	1.00 (2.21)**	0.53 (1.10)	−0.60 (0.65)	−0.92 (0.96)	1.08 (2.22)**	−1.24 (0.81)	0.39 (0.61)	0.34 (0.72)	−0.56 (0.98)
Covariates									

Table 7. *Cont.*

Attributes \ Classes	SAFA (N = 197)				SANA (N = 404)				
	$\beta_{A\ (29\%)}$	$\beta_{B\ (26.9\%)}$	$\beta_{C\ (24.1\%)}$	$\beta_{D\ (20\%)}$	$\beta_{E\ (29.8\%)}$	$\beta_{F\ (24.5\%)}$	$\beta_{G\ (15.9\%)}$	$\beta_{H\ (15.4\%)}$	$\beta_{I\ (14.4\%)}$
Italian geographic areas:									
Northern Italy	−0.06 (0.09)	0.83 (1.19)	−3.30 (1.87)	2.54 (2.35)**	−0.02 (0.10)	−0.62 (2.47)	0.75 (1.72)	0.14 (0.64)	−0.25 (0.85)
Central Italy	−0.33 (0.70)	0.14 (0.28)	0.51 (0.54)	−0.32 (0.31)	−0.18 (0.61)	−0.09 (0.41)	0.27 (0.57)	0.04 (0.14)	−0.04 (0.09)
Southern Italy	0.39 (0.57)	−0.97 (1.25)	2.79 (2.64)**	−2.22 (1.25)	0.20 (0.61)	0.72 (2.28)**	−1.03 (1.23)	−0.18 (0.54)	0.29 (0.75)
Stated attribute non-attendance:									
Origin	-	-	-	-	0.39 (0.88)	2.24 (1.51)	−0.63 (1.53)	−0.55 (1.33)	−1.44 (3.36)***
Nutrition claims	-	-	-	-	0.48 (2.58)**	0.15 (0.95)	−0.20 (1.04)	−0.67 (3.20)***	0.23 (1.00)
Health claims	-	-	-	-	−0.17 (1.00)	−0.11 (0.65)	−0.34 (1.90)	−0.07 (0.43)	0.70 (2.59)**
Source of omega-3	-	-	-	-	0.29 (1.76)	−0.09 (0.54)	−0.50 (2.41)**	−0.29 (1.78)	0.59 (2.44)**
Price	-	-	-	-	−0.97 (4.97)***	0.30 (1.09)	−0.44 (1.85)	-0.02 (0.08)	1.12 (2.82)**

[1] In brackets z values: *** Significant at 0.001; ** Significant at 0.01; * Significant at 0.05.

As regards parameters estimates of the SANA model, origin is positive and significant in classes E, F, and H, negative and significant in class G, and not significant in class I. The nutrition claim 'Reduced fat' and the attribute 'Omega-3 added to the milk' are not significant in all classes, while price is negative and significant in all classes other than classes G. Class G appears to be a very distinct class of the SANA model because beta parameters are either not significant or negative and significant for the attributes 'Puglia' and the nutrition claim 'Rich in omega-3'. Thus, participants belonging to this class were named 'disinterested consumers' as they preferred to have no information of this hypothetical product. The other classes of the SANA model appear to be more or less similar to classes of the SAFA model in relation to their preferences for nutrition and health claims. For example, respondents belonging to class E showed positive preferences for 'omega-3 naturally present in the milk' and for all types of claims and thus they were named 'pro-nature EU regulation 1924/2006'. Participants in class F were only influenced positively by the combined nutrition claim and article 14 (pro-wellbeing article 8 and 14), and respondents in class I only by the combined nutrition claim and article 13 (pro-nature articles 8 and 13). Preferences of participants belonging to class H were only influenced by the Puglia provenance and thus they were named 'pro-tradition' as respondents of class A of the SAFA model.

The covariates of attribute non-attendance are significant and negative for 'price' in Class E, for 'source of Omega-3' in class G, for 'nutrition claims' in class H, and for 'origin in class I. This indicates that there is a high probability that respondents of these classes did not attend these attributes. On the other hand, 'nutrition claims' are positive and significant in class E and so 'health claims', 'source of omega-3', and 'price' in class I. In this case, the likelihood that participants attended these attributes is very high. Despite the fact that covariates results are difficult to interpret because several parameters are not significant, parameters that are significant can help us to evaluate the willingness to pay as presented in Table 8.

Table 8. Comparison of WTP (willingness to pay) for mozzarella functional cheese attributes between SAFA and SANA models*.

Classes / Attribute Levels	SAFA WTP (βattribute/βprice)		SANA WTP (βattribute/βprice)			
	Class B	Class D	Class E	Class F	Class H	Class I
Puglia	0.10	0.17	0.44	0.13	0.03	NS
No information	−0.05	−0.25	−0.46	−0.04	0.03	−0.02
Reduced fat	NS	NS	NS	NS	NS	NS
Rich in omega-3	0.05	NS	0.16	NS	NS	NS
Reduced fat+Rich in Omega-3	0.05	0.19	0.22	0.04	NS	0.04
No information	NS	NS	−0.34	−0.05	NS	−0.03
Article 13	0.04	0.24	0.12	NS	NS	0.02
Article 14	0.04	0.24	0.22	0.03	NS	NS
No information	NS	NS	−0.27	NS	NS	−0.03
Added directly to milk	0.03	NS	NS	NS	NS	NS
Naturally enriched	NS	0.24	0.28	NS	NS	0.02

* Values in EURO; NS = Not significant.

In monetary terms, Table 8 shows that, in both models, participants are willing to pay less when information is not provided, and that Puglia appears to be the most rewarding attribute. In regard to the SAFA model, intensity of preferences of respondents in class D is higher than those in class B for all attributes. For 100 g of functional mozzarella cheese, respondents in class D are willing to pay €0.07 more for Puglia, €0.14 for the combined nutrition claim, €0.20 for health claims, and €0.24 more omega-3 naturally enriched than participants in class B. Thus, combining the WTP of the various attributes, respondents in class B and class D might be willing to pay for this hypothetical product, respectively 2.20 and €8.40/Kg more than the product that they generally buy. In the SANA model, WTP estimates of class E are the largest than all classes identified in both models. Combining WTP of the various attributes, respondents of this class are willing to pay €11.60/Kg more than the product that they generally buy. Since respondents of this class did not attend the price it is likely that WTP estimates are biased. For class F and I WTP estimates of nutrition and health claims are a bit smaller than those observed in class B. Finally, the parameters estimates of the full model presented in the Appendix A show that for the seven classes identified results are similar to those of the SAFA and SANA models. The first three classes account for 67.2% of the make share. Origin appears to be the most attractive attribute and participants respond negatively to price increases other than for class E, which is price insensitive. Moreover, for the full model we observe a preference for the combination of nutrition claims and the article 14 claim. Instead, WTP estimates of the full model results are higher than those observed for the SAFA model.

4. Conclusions

Even though large retailers, supermarkets, and shops market an enormous variety of functional food products, our results show only 21% of respondents were informed about functional food products. Most participants are uninformed about functional food products or buy these products without appropriate knowledge. Since functional food delivers health benefits that must be communicated with nutrition and health claims, manufactures should invest more in educational initiatives and marketing campaigns that aim at increasing an awareness of these products. By increasing consumers' knowledge and awareness of functional foods and their health benefits, it is likely that consumers can pay more attention to nutrition and health claims. In regard to consumer preferences for functional mozzarella cheese, our study indicates that participants have a clear preference for Puglia, a well-known Italian region for the quality of this product. In relation to the investment on how to produce this product, only the 'pro-industry articles 8 and 14' participants showed a preference for omega-3 added directly to milk. On the other hand, omega-3 added naturally to milk is an attribute appealing both to 'pro-nature articles 8 and 14' participants of the SAFA model and to 'pro-nature EU regulation 1924/2006' and 'pro-nature articles 8 and 13' respondents of the SAFA model. Thus, stakeholders should invest more in developing breeding systems where cows are fed directly with flax seeds. In monetary terms this seems the most rewarding attribute because participants of these three classes are willing on the average to pay €1.80/Kg more than the product they usually buy.

In terms of choices about how to communicate health benefits of functional mozzarella cheese, participants' preferences offer a clear indication in terms of article 8 of Regulation 1924/206. The single nutrition claim 'reduced fat' is not significant in all classes, while 'Rich in omega-3' is positive and significant in classes B and E, and negative and significant in classes C and G. Thus, the communication of health benefits obtained by adding nutrients (rich in omega-3) seems to attract consumers more than those gained by reducing nutrients (Reduced saturated fatty acids) for this specific product. Furthermore, the combined nutrition claim is positive and significant in six out of the nine classes identified in both models and on average participants are willing to pay about €1.00/Kg for communication delivered by the proposed combined claim. This result also corroborates a previous study [27] where consumers are WTP more for combined nutrition claims than a single nutrition claim reduced fat. In regard to health claims, results show that articles 13 and 14 were never negative and significant and thus it seems that in general, health claims influence consumers' preferences more positively than nutrition claims. Comparing preferences between article 13 and article 14, the reduced risk health claim seems to be the most preferred being significant in five out of nine classes, while the former only in three classes. However, in monetary terms the average WTP for both health claims is about €1.30/Kg and thus a bit higher than the combined article 18 (€1.00/Kg). Moreover, in this case, our results confirm previous findings [24,47] in terms of preferences between article 13 and 14 of Regulation 1924/2006.

Finally, even if our WTP estimates appear to be acceptable in monetary terms, the high WTP observed for respondents belonging to class E of the SANA model confirms that attribute non-attendance can lead to biased results especially when participants do not attend price. In terms of health communication, marketers should pay more attention both to the Annex of EU Regulation 1924/2006 to see whether they can take advantage of combined nutrition claims and to the list of validated health claims because they seem to be more rewarding than nutrition claims.

Author Contributions: Conceptualization, B.D.D., G.N. (Giuseppe Nocella), and R.V.; Literary review, F.B.; Methodology, G.N. (Giuseppe Nocella), and R.V.; Investigation, B.D.D., G.N. (Giuseppe Nocella), and R.V.; Data curation, B.D.D., G.N. (Giuseppe Nocella), and R.V.; Software and formal analysis, G.N. (Giuseppe Nocella); Writing—original draft preparation, G.N. (Giuseppe Nocella) and R.V.; Writing—review and editing, G.N. (Giuseppe Nocella), A.S., and R.V.; Visualization, F.B., G.N. (Giuseppe Nocella), and R.V.; Supervision, F.B., D.C., and B.D.D.; Project administration and funding acquisition, G.N. (Gianluca Nardone).

Funding: This research received funds from project PON 01_00851 "Bioinnovation for high healthy value dairy production" of the Italian Ministry of Education, University and Research (MIUR).

Acknowledgments: Authors are grateful to Marcella Viscecchia for the support of data collection and to the research activity execution.

Conflicts of Interest: The authors declare no conflict of interest. The founding sponsors had no role in the design of the study; in the collection, analysis, or interpretation of data; in the writing of the manuscript, and in the decision to publish the results.

Appendix A

Table A1. Information criteria to select the number of classes for SAFA and SANA LC models *.

		LC Models of the Full Sample		
Number of Classes	**Npar**	**LL**	**BIC**	**CAIC**
1	10	−4569.83	9203.64	9213.64
2	28	−4126.16	8431.49	8459.49
3	46	−3917.62	8129.58	8175.58
4	64	−3780.12	7969.75	8033.75
5	82	−3692.52	7909.72	7991.72
6	100	−3619.36	7878.59	7978.59
7	118	−3560.7246	7876.48	7994.48
8	136	−3514.1645	7898.54	8034.54

* Npar = Number of estimated parameters.; LL = Log-likelihood; BIC (LL) = Bayesian information criterion based on the log-likelihood; CAIC (LL) = Consistent Akaike information criterion, based on the log-likelihood.

Table A2. Maximum likelihood estimates of the full sample for consumer preferences towards functional mozzarella cheese.[1].

Classes	Full sample						
Attributes	$\beta_{A\ (27.8\%)}$	$\beta_{B\ (24.3\ \%)}$	$\beta_{C\ (15.1\ \%)}$	$\beta_{D\ (12\%)}$	$\beta_{E\ (9.2\%)}$	$\beta_{F\ (7.5\%)}$	$\beta_{G\ (4.1\%)}$
Origin: Puglia	1.30 (10.83)***	1.35 (13.91)***	−0.15 (1.77)	0.47 (4.53)**	−0.21 (1.77)	1.25 (5.22)**	0.22 (0.96)
Nutrition claims (Article 8):							
Reduced fat	0.02 (0.13)	−0.17 (0.88)	−0.14 (0.55)	0.30 (1.14)	0.18 (1.75)	−0.74 (1.31)	−2.69 (0.39)
Rich in omega-3	0.24 (2.38) **	0.23 (1.73.)	0.20 (1.20)	−0.35. (1.52)	−0.59 (2.62)**	−0.82. (1.71)	1.82 (0.77)
Rich in omega-3 and reduced fat	0.36 (2.96) **	0.54 (4.42) ***	0.23 (1.74)	1.20 (6.23)***	0.10 (0.50)	0.28 (0.92)	1.95 (0.84)
Health claims (Articles 13 & 14):							
Contributes to the maintenance of normal blood cholesterol levels (Article 13).	0.09 (1.01)	0.28 (2.10)	0.05 (0.46)	0.69 (3.72)***	−0.37 (2.40)**	0.04 (0.16)	−0.53 (1.30)
Helps to reduce cardiovascular diseases risk (Article 14).	0.15 (1.41)	0.40 (3.01)***	0.32 (2.75 **	0.52 (3.88)***	−0.04 (0.22)	0.15 (0.46)	0.73 (1.74.)
Information Source of omega-3:							
Omega-3 added to the milk	0.06 (0.69)	0.26 (2.58)**	0.03 (0.33)	0.18 (1.37)	0.28 (1.67)	0.59 (1.91)	0.18 (0.53)
Omega-3 naturally present in the milk	0.04 (0.34)	−0.18 (1.24)	0.54 (3.68)***	0.27 (1.51)	−0.47 (2.33)**	−0.05 (0.11)	0.66 (1.48)
Price	−0.44 (3.14) ***	−13.28. (12.48)***	−2.22 (5.20)***	−0.10 (0.23)	0.94 (2.54)**	−51.97 (6.29)***	−40.21 (5.27)***
No buy	−0.85 (5.88) ***	−1.79 (11.29)***	−1.66 (7.95)***	−4.28 (1.15)	0.48 (4.16)***	−1.78 (4.62)***	−2.19 (1.75)
Intercept	1.67 (2.16)**	1.21 (2.60)**	2.03 (2.97)**	5.14 (1.67)	0.66 (0.86)	1.15. (1.60)	−1.57 (0.78)
Covariates							

15. Ann-Marie, L.; McKevitt, A.; Gibney, M.J. Irish consumers' use and perception of nutrition and health claims. *Public Health Nutr.* **2011**, *14*, 2213–2219.
16. Dean, M.; Lampila, P.; Shepherd, R.; Arvola, A.; Saba, A.; Vassallo, M.; Claupein, E.; Winkelmann, M.; Lahteenmaki, L. Perceived relevance and foods with health-related claims. *Food Qual. Prefer.* **2012**, *24*, 129–135. [CrossRef]
17. Sabbe, S.; Verbeke, W.; Deliza, R.; Matta, V.; Van Damme, P. Effect of a health claim and personal characteristics on consumer acceptance of fruit juices with different concentrations of açaí (Euterpe oleracea Mart.). *Appetite* **2009**, *53*, 84–92. [CrossRef]
18. Siegrist, M.; Stampfli, N.; Kastenholz, H. Consumers' willingness to buy functional foods. The influence of carrier, benefit and trust. *Appetite* **2008**, *51*, 526–529. [CrossRef]
19. Lalor, F.; Madden, C.; McKenzie, K.; Wall, P.G. Health claims on foodstuffs: A focus group study of consumer attitudes. *J. Funct. Foods* **2011**, *3*, 56–59. [CrossRef]
20. Cavaliere, A.; Ricci, E.C.; Banterle, A. Nutrition and health claims: Who is interested? An empirical analysis of consumer preferences in Italy. *Food Qual. Prefer.* **2015**, *41*, 44–51. [CrossRef]
21. Vecchio, R.E.; Van Loo, J.; Annunziata, A. Consumers' willingness to pay for conventional, organic and functional yogurt: Evidence from experimental auctions. *Int. J. Consum. Stud.* **2016**, *40*, 368–378. [CrossRef]
22. Bimbo, F.; Bonanno, A.; Nocella, G.; Viscecchia, R.; Nardone, G.; De Devitiis, B.; Carlucci, D. Consumers' acceptance and preferences for nutrition-modified and functional dairy products: A systematic review. *Appetite* **2017**, *113*, 141–154. [CrossRef]
23. Steinhauser, J.; Hamm, U. Consumer and product-specific characteristics influencing the effect of nutrition, health and risk reduction claims on preferences and purchase behavior–A systematic review. *Appetite* **2018**, *127*, 303–323. [CrossRef] [PubMed]
24. Barreiro-Hurlé, J.; Gracia, A.; De-Magistris, T. Does nutrition information on food products lead to healthier food choices? *Food Policy* **2010**, *35*, 221–229. [CrossRef]
25. Van Wezemael, L.; Caputo, V.; Nayga, R.M., Jr.; Chryssochoidis, G.; Verbeke, W. European consumer preferences for beef with nutrition and health claims: A multi-country investigation using discrete choice experiments. *Food Policy* **2014**, *44*, 167–176. [CrossRef]
26. EFSA Data. Available online: www.efsa.europa.eu (accessed on 12 June 2019).
27. De-Magistris, T.; Lopéz-Galán, B. Consumers' willingness to pay for nutritional claims fighting the obesity epidemic: The case of reduced-fat and low salt cheese in Spain. *Public Health* **2016**, *135*, 83–90. [CrossRef]
28. Lemken, D.; Knigge, M.; Meyerding, S.; Spiller, A. The value of environmental and health claims on new legume products: A non-hypothetical online auction. *Sustainability* **2017**, *9*, 1340. [CrossRef]
29. López-Galán, B.; de-Magistris, T. Testing Emotional Eating Style in Relation to Willingness to Pay for Nutritional Claims. *Nutrients* **2019**, *11*, 1773. [CrossRef]
30. ChoiceMetrics. *NGENE 1.1.2. User Manual and Reference Guide*; ChoiceMetrics: Sidney, Australia, 2014.
31. McFadden, D. *Conditional Logit Analysis of Qualitative Choice Behaviour*; Academic Press New York: New York, NY, USA, 1973.
32. Greene, W.H.; Hensher, D.A. A latent class model for discrete choice analysis: Contrasts with mixed logit. *Transp. Res. Part B Methodol.* **2003**, *37*, 681–698. [CrossRef]
33. McFadden, D. Economic choices. *Am. Econ. Rev.* **2001**, *91*, 351–378. [CrossRef]
34. Train, K. *Discrete Choice Methods with Simulation*; Cambridge University Press: New York, NY, USA, 2003.
35. McFadden, D. Conditional logit analysis of qualitative choice behavior. In *Frontiers in Econometrics*; Zarembka, P., Ed.; Academic Press: New York, NY, USA, 1974; pp. 105–142.
36. McFadden, D. The choice theory approach to market research. *Mark. Sci.* **1986**, *5*, 275–297. [CrossRef]
37. Greene, W.H. *Econometric Analysis*; Prentice Hall: Englewood Cliffs, NJ, USA, 2003.
38. Magidson, J.; Vermunt, J. Latent class models for clustering: A comparison with K-means. *Can. J. Mark. Res.* **2002**, *20*, 36–43.
39. Balcombe, K.; Burton, M.; Rigby, D. Skew and attribute non-attendance within the Bayes-ian mixed logit model. *J. Environ. Econ. Manag.* **2011**, *62*, 446–461. [CrossRef]
40. Hensher, D.; Rose, J.; Greene, W. Inferring attribute non-attendance from stated choicedata: Implications for willingness to pay estimates and a warning for stated choice experimentdesign. *Transportation* **2012**, *39*, 235–245. [CrossRef]

41. Scarpa, R.; Zanoli, R.; Bruschi, V.; Naspetti, S. Inferred and stated attribute non-attendance in food choice experiments. *Am. J. Agric. Econ.* **2013**, *95*, 165–180. [CrossRef]
42. Caputo, V.; Van Loo, E.J.; Scarpa, R.; Nayga, R.M., Jr.; Verbeke, W. Comparing serial, and choice task Stated and inferred attribute non-attendance methods in food choice experiments. *J. Agric. Econ.* **2018**, *69*, 35–37. [CrossRef]
43. Kragt, M.E. Stated and Inferred Attribute Attendance Models: A Comparison with Environmental Choice Experiments. *J. Agric. Econ.* **2013**, *64*, 719–736. [CrossRef]
44. Andrews, R.L.; Imran, S.C. Retention of latent segments in regression-based marketing models. *Int. J. Res. Mark.* **2003**, *20*, 315–321. [CrossRef]
45. Yang, Z. On the log-convexity of two-parameter homogeneous functions. *Math. Inequal. Appl.* **2007**, *10*, 499. [CrossRef]
46. Scarpa, R.; Thiene, M. Destination choice models for rock climbing in the Northeastern Alps: A latent-class approach based on intensity of preferences. *Land Econ.* **2005**, *81*, 426–444. [CrossRef]
47. Verbeke, W.; Scholderer, J.; Lähteenmäki, J. Consumer appeal of nutrition and health claims in three existing product concepts. *Appetite* **2009**, *52*, 684–692. [CrossRef]

nutrients

MDPI

Article

Making Sense of Information Overload: Consumer Ranking of Nutritional Claims in Cereal Based Products

Azucena Gracia [1,2,*] and Jesús Barreiro-Hurlé [3]

[1] Unidad de Economía Agroalimentaria y de los Recursos Naturales, Centro de Investigación y Tecnología Agroalimentaria de Aragón (CITA), 50059 Zaragoza, Spain
[2] Instituto Agroalimentario de Aragón—IA2, CITA-Universidad de Zaragoza, 50059 Zaragoza, Spain
[3] European Commission, Joint Research Centre (JRC), 41092 Seville, Spain; jesus.barreirohurle@gmail.com
* Correspondence: agracia@aragon.es; Tel.: +34-976-71-6350

Received: 6 November 2019; Accepted: 14 November 2019; Published: 21 November 2019

Abstract: As a result of increased consumer awareness, demand for healthier food products is increasing day by day. Consumers seek healthier versions of food products which they relate to reduced presence of unhealthy components or increased presence of healthy ones. As a result, the food industry has not only increased the variety of products available but also uses nutritional claims to signal the presence of more substances. As an average consumer at the supermarket devotes just a few seconds to selecting each product, they are only able or willing to process that information that immediately attracts their attention or that is felt to be more important to them. This paper analyses how consumers rank different nutritional claims for two processed cereal products. Five claims were chosen to reflect the current market landscape of availability, and that relates to both "healthy" (i.e., fiber) and "unhealthy" (i.e., fat) substances. We use a direct ranking preference method with data from a survey conducted with consumers in a Spanish region in 2017. Results show that the ranking of claims differs between the two products (biscuits and pastries) and across consumers. However, consumers prefer those that show reduced presence of unhealthy substances above those that highlight the presence of healthy ones. Therefore, policy to maximize the impact of nutritional labelling should be product-specific.

Keywords: consumer preferences; nutritional labels; biscuits; pastries; Aragón

---

1. Introduction

Consumers' awareness of the healthiness of foods is increasing and as a result they are starting to demand healthier food alternatives. Industry responds to this demand by offering variants that reduce or eliminate unhealthy components (such as fat) and/or add beneficial ingredients (such as fibre) thereby increasing the presence of food products that carry one or more nutritional claims on supermarket shelves. Results from a survey of the presence of claims on food products stated that food products normally provide more than one claim, either by repeating the claim several times on the package or including claims on two or more substances [1]. The latter happens because food products have been produced with added or removed substances (e.g., 'high in fibre', 'low fat'). In the review reported it was also noted that 21% of food products in the market carried at least one nutritional claim and they often carried multiple claims, in particular, the average number of nutritional claims carried by the product was two in the European countries and 2.1 in Spain. This increased diversity in supply contrasts with the limited time consumers spend in the store to select a food product from the aisle, on average, between 23 and 29 s [2,3]. The presence of multiple claims can lead to a situation of information overload which complicates the selection of healthy food in a supermarket and, as a response, the average consumer generally reverts to a heuristic which involves routinely picking out the same product in few seconds [3]. This routine selection of products implies that consumers are only able or willing to process the information that attracts their attention or that is perceived to be more important for them [4]. Against this landscape of multiple claims per product, it is clear that both from a food industry and a public health perspective it is necessary to understand which nutritional claims drive consumer choices. This is the main objective of this paper.

In particular, the importance attached by consumers to several nutritional claims on processed cereal products was assessed. The selection of the processed cereal products to be analysed and the nutrients to be claimed is based on the results of existing literature reviews on the prevalence of nutritional claims in the market and previous empirical papers on consumers and nutritional claims. Most of these reviews study only a single country being respectively the UK, Ireland, Slovenia, and Spain [5–8]. The only multi-country review [1] covers Germany, the Netherlands, Slovenia, Spain, and the United Kingdom. This review concludes that for the five countries surveyed the food category with a higher prevalence of nutritional claims is that of cereals and cereal products (31%) closely followed by dairy products (28%). Our application focuses on two products from the cereal category. This review also stated that apart from minerals and vitamins, the most frequent nutritional claims were related to fat content (24%), sugar (12%), and fiber (9%) [1]. In addition, previous research on consumers identify that the most important nutritional claims for them are those related to the content of fat, fiber, and sugar while the level of interest is lower for salt in Italy [9]. However, in Spain, consumers attach the highest importance to the fat-free and source of calcium claims and the least to the low sugar in the case of yogurts [10]. Similarly, a more recent research stated that apart from the health claims, consumers gain the highest utility from the high in fiber and fat-free nutritional claims [11]. Considering the high availability of nutritional claims, we selected several types of claims related to fat, fiber, sugar, and salt.

The importance given by consumers to the selected claims when shopping two based cereal products was assessed using a direct ranking preference method with primary data from a survey administrated to consumers in a region of Spain in 2017. As prior research has identified that preferences for nutritional claims are heterogeneous across consumers [12–16] and products [17], the specification of the model accommodates differences in preferences across both domains.

2. Materials and Methods

2.1. Product Selection

A field survey of cereal products (biscuits, pastries, breakfast cereals) available in different supermarkets in the region was undertaken to compile the available nutritional claims at the time this study was carried out. The information on cereal products was gathered by direct observation of the aisles of main supermarkets and hypermarkets in the region. Based on the results of this survey, two specific cereal products were chosen. First, biscuits were selected as the product category with the highest prevalence of nutritional claims. A second product category, pastries, was chosen as it is a close substitute of the biscuit category but with a lower perceived level of healthiness by consumers. Last, we needed to select the type of claims related to the substances previously specified (fat, fiber, sugar, and salt). We selected three prevalent nutritional claims for the cereal products (source of fiber, reduced fat, and with no added sugar), and one less prevalent (reduced saturated fat). In addition, although salt content was of less value by consumers, the "low salt" claim was also included based on the recommendation for low salt intake in vulnerable consumers groups (children and elderly).

2.2. Data Collection and Survey

Information on consumers was collected using an online survey administrated in the Spanish region of Aragon (a NUTS2 region in North-eastern Spain) in 2017. The target population was established as people living in the region older than 18 years mainly responsible for food purchase.

This region was selected as its socio-demographics are similar to those of the Spanish Census of Population (Table A1 in the Appendix A), then, considered to be representative of the Spanish population. Consumers were stratified by gender, age, and province of residence (Aragon is a NUTS2 region which is composed of 3 NUTS3 provinces (Huesca, Teruel and Zaragoza).). Sample size for an error of ± 5%, (95.5% (k = 2), and $p = q = 0.5$ was established at 400. The questionnaire design builds on the authors' prior experience with nutrition and food consumption questionnaires and was structured in four parts. First, consumers where asked about their food consumption and purchase habits (i.e., How often do you eat each of the following food product? How many meals per day do you eat?). Second, direct rank questions were asked. In the third part, questions on nutritional information behavior and knowledge, and on food diet and health were included (i.e., Do you pay attention to the nutritional information on food products when shopping? Do you read nutritional information on food products? How strong do you consider your nutritional knowledge to be? What do you believe your health status to be? How much impact do you believe your food intake has on your health?). Finally, a set of questions on socio-demographic characteristics were asked (relating to sex, age, education level, income level, province of residence). Before the final questionnaire was distributed additional validity and reliability tests were undertaken. First, the questionnaire was discussed in a focus group of five consumers and, second, a pilot survey was conducted with a sample of respondents (n = 15) to test for understanding and response time. The feedback gathered in the focus group and the analysis of the pilot survey concluded that consumers had a clear understanding of what was being asked of them and that they faced no major difficulties when answering. The final survey was conducted by a market research company hired to provide a representative sample of food consumers stratified by gender, age, and province of residence from their on-line panel. The company sought informed consent from participants before they responded to the questionnaire.

2.3. Utility Theory Framework

A direct ranking approach was used where respondents have to rank the analyzed nutritional claims from most to least preferred. In order to do so, they were presented a list of nutritional claims and give the preference ranking (from 1; most preferred; to 5; least preferred) for both product categories.

With the ranking information, two aggregate indicators of importance can be calculated, probability of ith rank (for i ranging from 1 to 5), and ranking means. These indicators provide the ordering of preferences, but they are not able to deal with heterogeneity across respondents because of their aggregate nature. To overcome this, rank ordering can be broken into several choice sets taking some ad hoc assumptions. The new data obtained is transformed into a sequence of choice behavior that can be used to estimate any choice model that incorporates the possibility of addressing heterogeneity in preferences (mixed logit). Therefore, consumer preference heterogeneity can be investigated.

As one of our objectives was to understand individual heterogeneity in claim ranking, besides estimating probabilities and rank means which would reflect average consumer preferences, ranking data were re-coded by considering each rank as a sequential process where respondents take a discrete choice between options to estimate individual preferences. Consumers' ranking were broken down into sequences of choices and a rank-ordered mixed logit was estimated [18]. The estimated parameters of the importance of the analyzed nutritional claims for each of the respondents were utilized to segment consumers into homogenous groups. These consumer segments were characterized using consumption habits, nutrition information behavior and knowledge, food diet and health and personal consumer characteristics.

When faced with a direct ranking question, respondents rank several alternatives from the most to the least preferred. This ranking is assumed to be based on the utility that respondents get from each option. This gathered information can be used within the utility theory framework defined by the random utility model (RUM) assuming that each individual n faces a choice among J options, and he/she gains utility (Unj) from choosing option j over other options. To transform the direct ranking into specific choices the original ranking information of the different options must be transformed into "pseudochoices" or "pseudo-observations". Thus, for the first pseudo-observation, the choice set includes J options (in our case 5, one for each of the claims), and the dependent variable corresponds with the option ranked as the most important; for the second pseudo-observation, the option ranked first is discarded, leading to a choice set composed of $J-1$ options, and the alternative ranked second becomes the chosen alternative. The process continues until the choice set consisted of only two options. Then, the ranking of J options can be represented as $J-1$ independent choices, and the new dataset includes $J-1$ choices for each respondent [18].

Utility (U_{nj}) under the random utility model (RUM) has two parts: One, observed by the researcher (V_{nj}) and the other unobserved and random (ε_{nj}) distributed i.i.d. extreme value (as for a logit model). According to Lancaster's model, consumer utility from a product can be split into attribute specific partial utilities, in our case the characteristics from which consumer derives utility are the nutritional claims included in the direct ranking question [19]. To take this into account the utility function is represented as:

$$U_{nj}= \beta'_n X_{nj} + \varepsilon_{nj}, \quad (1)$$

where β'_n is a vector of parameters of the exogenous variables X_{nj}, and ε_{nj} is an independent identically distributed (i.i.d.) error term over individual and options. The β'_n coefficients for each individual take into account the heterogeneity in preferences. In other words, the vector of parameters β'_n is random, with a density $g(\beta/\theta)$ where θ represents the parameters of the distribution (i.e., mean and standard deviation).

2.4. Rank-Ordered Mixed Logit

Within this framework, a mixed rank-ordered logit (MRL) is estimated with the new data from the ranking observations taking into account heterogeneous preferences.

Under the assumptions of a standard logit, the probability of individual n ranking J options from most to least important as $j_1; \ldots ; j_m; \ldots ; j_J$, where jm represents the option chosen at the ranking order m, can be expressed as the product of logit choice probabilities:

$$\mathrm{Prob}(ranking \quad j_1, \ldots, j_m, \ldots, j_J) = \mathrm{Prob}(U_{j_1} > \ldots > U_{j_m} > \ldots > U_{j_J}) = \prod_{m=1}^{J-1} \frac{e^{V_{njm}}}{\sum_{k=m}^{J} e^{V_{njm}}}. \quad (2)$$

This expression represents the probability for an individual n of choosing a specific ranking conditional on β. The unconditional probability is the integral of that product of probabilities over the density of β:

$$\mathrm{Prob}(ranking \quad j_1, \ldots, j_m, \ldots, j_J) = \int \prod_{m=1}^{J-1} \frac{e^{V_{njm}}}{\sum_{k=m}^{J} e^{V_{njk}}} \times g(\beta|\theta) d\theta. \quad (3)$$

Equation (3) is estimated using the new transformed data described above where the $J-1$ pseudo-observations for each ranking are considered as $J-1$ choices in a panel. The mixed logit takes into account that each respondent has his own coefficients that affect his ranking in the way that the pseudo-observations are correlated [18]. To estimate the rank-ordered mixed logit the Nlogit software was used.

2.5. Preference Heterogeneity

Estimated parameters for the rank-ordered mixed logit for each of the respondents (β'_n) are utilized to segment them using a cluster k-means approach. These segments are characterized by consumption habits, nutritional information behavior and knowledge, food diet and health and consumer socio-demographics characteristics defined in Tables 1 and 2. This characterization was done using chi-square or Bonferroni test [20], for discrete and continuous variables respectively, to test whether statistically significant differences across segments exist for the different characterization variables. Descriptive and bivariate statistics were calculated using STATA software.

3. Results

The final sample consisted of 400 consumers and the summary statistics for its characteristics together with some of the general population's characteristics are shown for comparison in Table 1.

Table 1. Sample demographic characteristics (%, unless stated).

Characteristics	Sample (n = 400)	Population
Gender [a]		
Male	50.0	49.4
Female	50.0	50.6
Age [a] (average, standard deviation)	48.0 (14.0)	N/A
18–34	21.0	21.1
35–44	21.0	19.1
45–54	19.9	18.6
≥ 55	38.5	41.3
Studies level b		
Primary	27.5	24.6
Secondary	32.5	50.0
Higher	40.0	25.4
Income range		
≤ 1000 €/month	10.0	N/A
1001–2500 €/month	42.2	N/A
2501–4500 €/month	17.5	N/A
> 4500 €/month	3.3	N/A
Do not know/refuse to answer	27.0	N/A
Household size (average, standard deviation)	2.7 (1.1)	N/A
Province of residence [a]		
Huesca	14.2	17.0
Teruel	6.8	11.0
Zaragoza	79.0	72.0
Body Mass Index (BMI) [c]	25.8 (4.6)	N/A
Less than 25	48.2	47.5
25–30	35.7	38.8
More than 30	16.1	15.7
Practice exercise or walk more than 30 min at least five times a week	66.5	N/A

[a] Instituto Nacional de Estadística, INE (2017a) [21]; [b] Instituto Aragonés de Estadística, IAEST (2018) [22], [c] INE (2017b) [23]. N/A: not available.

The sample is representative of the population in terms of age, sex, and province of residence. Our sample's average age is 48 years and half of respondents were female (50%). Table 2 shows that 15% of respondents consumed pastries, cookies, and cakes either daily or never, respectively. In addition, half of the respondents consume those products less than once a week (31.0%) or once week (22.2%). Only a minority of respondents admitted to snacking often (9%) (This figure has to be taken with caution as snacking is considered a non-desirable behaviour and when asked directly for these under-reporting is common [24]) and 17% declared that they never snack. Half of the respondents stated that they pay attention to nutritional information on food packages but only 20% of them always read this information.

Table 2. Food consumption and health habits (%, unless stated).

Consumption Habits	
Number of meals (average)	3.7 ± 0.94
Frequency of consumption pastries, cookies, and cakes	
Never	14.8
Less than one a week	31.0
Once a week	22.2
Several times a week	16.2
Daily	15.8
Snacking	
Never	17.2
Sometimes	73.8
Often	9.0
Nutritional information	
Pay attention (yes)	49.0
Read always (% of those paying attention)	20.4
Nutritional knowledge (average)[a]	2.8 ± 0.93
Diet and health	
Perceived impact of food diet on health (average) [a]	4.3 ± 0.74
Follow a healthy diet	
Very unhealthy diet	0.5
Unhealthy diet	6.2
Neutral	41.0
Healthy diet	48.8
Very healthy diet	3.5
Self-reported health status	
Very unhealthy	0.5
Unhealthy	1.5
Neutral	21.3
Healthy	66.7
Very healthy	10.0

[a] In a 5 point increasing scale where 1 indicates the lowest level and 5 the highest.

Respondents' self-reported nutritional knowledge is not very high (2.8 in a scale of 5) and they reported a high awareness of the impact their diet had on their health (4.3 in a scale of 5). Around half of the respondents declared that they follow a healthy or very healthy diet (48.8% and 3.5%, respectively). Finally, most of respondents believe that they are healthy or very healthy with only 2% indicating that their health is bad or very bad.

3.1. Attribute Importance Ranking

Tables 3 and 4 show the percentage of respondents that ranked the different nutritional claims in the different levels, together with their mean of the ranks for biscuits and pastries, respectively. It is worth recalling that due to the scale used (order from most important to least important) a lower value of the mean indicates the highest importance for the claim.

Table 3. Probability of ranks and ranking means: Biscuits.

	Rank #1 (%)	Rank #2 (%)	Rank #3 (%)	Rank #4 (%)	Rank #5 (%)	Mean ± SD
Source of fibre [a]	30.3	15.5	16.5	16.2	21.5	2.83 ± 1.54
Reduced saturated fat [b]	30.8	21.0	21.2	14.2	12.8	2.57 ± 1.38
With no added sugar [b]	27.2	23.8	21.5	19.2	8.3	2.57 ± 1.29
Reduced fat [c]	6.5	26.8	24.5	23.2	19.0	3.21± 1.21
Low salt [d]	5.2	13.0	16.2	27.0	38.5	3.80 ± 1.23

[a,b,c,d] Superscript letters mean that importance means are statistically different among nutritional claims using the *t*-test. Note: Rank #1 indicates ranked as the most important (in the first position).SD, Standard Deviation.

Table 4. Probability of ranks and ranking means: Pastries.

	Rank #1 (%)	Rank #2 (%)	Rank #3 (%)	Rank #4 (%)	Rank #5 (%)	Mean ± SD
Source of fibre [a]	21.2	11.8	19.2	23.5	24.3	3.17 ± 1.46
Reduced saturated fat [b]	41.0	24.2	12.3	13.2	9.3	2.25 ± 1.35
With no added sugar [b]	25.2	24.3	25.2	16.5	8.8	2.59 ± 1.27
Reduced fat [c]	8.0	31.0	26.2	21.5	13.3	3.01 ± 1.17
Low salt [d]	4.5	8.8	17.0	25.2	44.5	3.96 ± 1.17

[a,b,c,d] Superscript letters mean that importance means are statistically different among nutritional claims using the *t*-test. Note: Rank #1 indicates ranked as the most important (in the first position).

Looking at the probability of ranks it can be concluded that the most and least important claim is the same for both product categories "reduced saturated fat" and "low salt", respectively. Heterogeneity is found for the claim ranked second as for biscuits, that is "source of fiber" while for pastries it is "with no added sugar". It seems that for the less healthy product (pastries) consumers place more importance on avoiding unhealthy ingredients while for the reference product (biscuits) the importance is placed on healthy components. However, if we focus on the rank means ordering homogeneity is found for the two most important claims ("reduced saturated fat" and "with no added sugar") which cannot be considered different based on a paired sample *t*-test.

To identify heterogeneity across products, we test whether the mean importance for the different claims differ between the two tested products and we found that only for the claim "with no added sugar" the rank means were statistically the same between biscuits and pastries (last column in Table 5). For the remaining claims, the rank means were statistically different. In particular, the rank means for "source of fiber" and "low salt" were higher for pastries, indicating that the preference for these claims was lower for pastries compared to the biscuits. On the other hand, the rank means for "reduced saturated fat" and "reduced fat" were higher for breakfast biscuits than those for pastries, therefore, these claims are preferred in the case of pastries. This pattern shows that for unhealthy products claims signaling a lower presence of unhealthy components play a greater role in the purchase decision, while for the reference product the presence of healthy substances are more valued.

Table 5. Test of differences for the nutritional claims between biscuits and pastries.

	Breakfast Biscuit	Pastries	Paired *t*-Test (*p*-Value) for Differences between Columns
Source of fibre	2.83 ***	3.17 ***	−5.56 (0.00)
Reduced saturated fat	2.57 ***	2.25 ***	5.23 (0.00)
With no added sugar	2.57 ***	2.59 ***	−0.26 (0.39)
Reduced fat	3.21 ***	3.01 ***	3.31 (0.00)
Low salt	3.80 ***	3.96 ***	−2.75 (0.00)

Note: *** denotes statistical significance at 1% significance level within-columns.

3.2. Consumers Heterogeneity

As mentioned above, we expected that heterogeneity in the ranking of importance of different claims would exist not only between products but also between individuals. Looking at the results reported in Tables 3 and 4 this heterogeneity becomes evident as the standard deviations of the means rankings are quite high. To gain additional insights on the drivers of this heterogeneity, a ranked-order mixed logit was estimated using NLOGIT 5.0 (Econometric Software INC, Plainview, USA) assuming random parameters following a normal distribution. To avoid the issue of multicollinearity the least preferred nutritional claim ("low salt") was considered the reference level and excluded in the final specification.

Table 6 shows the mean and the standard deviation of the estimated parameters for the ranked-order mixed logit. The standard deviations of the estimated coefficients were statistically significant at the 1% significance level, meaning that the consumer importance attached to the different claims were heterogeneous. The mean of the estimated parameters were positive and statistically significant indicating that the importance attached to the included nutritional claims were statistically different from the importance given to the reference claim ("low salt"). These estimations corroborate findings based on aggregate rankings reported in Tables 3 and 4 as the most preferred nutritional claims remained "reduced saturated fat" and "with no added sugar" for both products. The least preferred claims were "source of fiber" and "reduced fat" which occupies the third and fourth position, respectively in the case of biscuits and the fourth and third in the case of pastries. However, the ranked-order mixed logit model provides statistically significant differences across participants' estimated coefficients.

Table 6. Estimation results of the rank-ordered mixed logit.

	Biscuits		Pastries	
Parameters Estimates	**Coefficient**	**Z-Ratio**	**Coefficient**	**Z-Ratio**
Source of fibre	1.7542	4.07 ***	1.1954	3.46 ***
Reduced saturated fat	2.2311	4.31 ***	2.8022	4.11 ***
With no added sugar	2.1990	4.31 ***	2.1017	4.06 ***
Reduced fat	1.1985	3.90 ***	1.6510	4.04 ***
	Standard deviation of parameters			
Source of fibre	3.4485	3.40 ***	2.3215	2.59 **
Reduced saturated fat	3.6425	4.26 ***	3.1126	3.10 **
With no added sugar	2.3897	3.58 ***	1.8733	2.16 **
Reduced fat	2.8731	7.62 ***	2.3041	2.54 **
Number of observations	1600		1600	
Log likelihood (at convergence)	−1749.8		−1694.5	
McFadden Pseudo R-square	0.32		0.34	

Note: ***, ** denotes statistical significance at 1% and 5%, significance levels respectively.

The estimated parameters for each of the participants (β'_n) were used to segment them into homogeneous consumers groups using a k-means cluster analysis [20]. From the cluster analysis, we obtained four segments for both products (Table 7). The size of the clusters for the two products can be seen in the first two rows in Table 7 and the mean values of the estimated parameters for each cluster in the subsequent rows.

The latter mean values were used to name the different clusters using the profile of the claim importance. For biscuits, we observed that consumers in the first two clusters attached the least importance to the "low salt" claim while the other two clusters have different least important claims: Reduction of fat ("reduced saturated fat" and "reduced fat") for cluster 3 and "source of fiber" for cluster 4. Based on the least preferred claim we named clusters 3 and 4 "fat careless" and "fiber careless". Then, because cluster 1 showed the highest preference for the "source of fiber" claim it was named "fiber lovers", despite the lower preference for "low salt". Similarly, as cluster 2 showed the highest preference for the "reduced saturated fat" closely followed by the "reduced fat" claim this cluster was named "fat avoiders". For pastries, we also observed that two clusters (cluster 2 and cluster 3) attached the least importance to the "low salt" claim while the other two clusters attached less importance to reduction in fat content (cluster 4 "reduced saturated fat" and "reduced fat") and "source of fiber" (cluster 1). These last two clusters were named "fat careless" and "fiber careless". On the other hand, as cluster 2 attached the highest preference for the "reduced saturated fat" followed by the "reduced fat" claim this cluster was named "fat avoiders". For cluster 3, the only distinguishing preference pattern is showing the lowest preference for the "low salt" claim while the estimated means for the rest of the claims were not much different, therefore we named "salt careless".

Table 7. Segmentation of consumer according to nutritional claim importance.

	Cluster 1	Cluster 2	Cluster 3	Cluster 4
	Cluster size (%)			
Breakfast biscuits	27	27	27	19
Pastries	18	27	30	25
	Source of fibre			
Breakfast biscuits	4.65 [a]	1.43 [b]	2.11 [c]	−1.88 [d]
Pastries	−0.94 [a]	1.17 [b]	2.49 [c]	1.18 [d]
	Reduced saturated fat			
Breakfast biscuits	2.87 [a]	5.61 [b]	−1.38 [c]	1.60 [d]
Pastries	3.22 [a]	5.74 [b]	2.84 [c]	−0.56 [d]
	With no added sugar			
Breakfast biscuits	2.62 [a]	2.92 [b]	1.41 [c]	1.70 [d]
Pastries	1.98 [a]	2.35 [b]	2.61 [c]	1.37 [d]
	Reduced fat			
Breakfast biscuits	1.62 [a]	3.83 [b]	−1.51 [c]	0.72 [d]
Pastries	1.81 [a]	3.76 [b]	1.79 [c]	−0.77 [d]

[a,b,c,d] Superscript letters mean that importance means are statistically different among clusters.

3.3. Segments Profiling

Tables 8 and 9 shows the chi-square and ANOVA (Bonferroni) test results among the four segments and the consumers' characteristics that will allow profiling of the clusters. In particular, the consumers' characteristics displayed in Tables 1 and 2 that were found statistically different across clusters at least at a 10% significance level are included in the tables.

Table 8. Biscuits: Profiling consumer segments (%, unless stated).

Characteristics	Fiber Lovers	Fat Avoiders	Fat Careless	Fiber Careless	Total Sample
Gender					
Female **	54.2	57.8	46.8	40.2	50.0
Age (average) ***	47.0	44.6	49.1	51.6	48.0
Frequency of consumption pastries, cookies, and cakes (%)					
Never **	14.0	10.1	11.7	22.4	14.8
Daily ***	14.0	25.7	13.0	9.4	15.8
Nutritional knowledge (average) **	2.7	2.6	3.0	2.7	2.8
Follow a healthy diet (%)					
Very unhealthy and unhealthy diet	7.2	11.2	1.9	6.8	6.7
Neutral	28.9	32.4	49.5	51.0	41.0
Healthy diet	61.5	50.9	43.9	41.2	48.8
Very healthy diet	2.4	5.5	4.7	1.0	3.5

Note: ***, ** means statistical significance at 1% and 5%, significance levels, respectively.

Table 9. Pastries: Profiling consumer segments (%, unless stated).

Characteristics	Fiber Careless	Fat Avoiders	Salt Careless	Fat Careless	Total Sample
Gender Female ***	37.5	61.3	59.8	35.0	50.0
Age (average) ***	47.8	43.6	47.4	53.5	48.0
Household size (average) *	2.6	2.9	2.5	2.7	2.7
Practice exercise or walk more than 30 min at least five times a week **	62.5	57.5	69.7	75	66.5
Frequency of consumption pastries, cookies, and cakes (%)					
Never ***	19.4	5.6	15.6	20.0	14.8
Perceived impact of food diet on health (average) **	4.2	4.4	4.3	4.2	4.3

Note: ***, **, * means statistical significance at 1%, 5%, 10% significance levels, respectively.

For breakfast cereals, Table 8 shows that the "fiber lovers" and the "fat avoiders" include a higher proportion of young and female individuals than the other two segments. In addition, a larger proportion of consumers in these segments declared following a healthy or very healthy diet (63.9% for "fiber lovers" and 56.4% for "fat avoiders") in comparison to the other two segments (48.6 and 42.2, respectively). Thus, it seems that there is some correlation between the importance given to claims signaling healthier product versions and healthy dietary habits.

The main difference between these clusters is that a higher proportion of consumers in the "fat avoiders" segment claimed eating pastries, cookies, and cakes on a daily basis. In contrast, consumers in the segments "fat careless" and "fiber careless" were characterized by being older males with an average perception of following a healthy diet. The main difference between these two clusters was that a higher proportion of consumers in the "fiber careless" cluster stated that they never eat pastries, cookies, and cakes (22.4%).

For pastries, Table 9 shows that "fat careless" consisted of a lower proportion of older females. In addition, a higher proportion of consumers in this segment state that they exercise (75%) and never eat pastries cookies and cakes (20%) in comparison with the other three clusters. We observed similar personal characteristics of consumers in this cluster and the segment "fiber carless" namely that of the proportion of females, the frequency of consumption of pastries, cookies and cakes, and the perceived impact of the food diet on health. On contrary, the "fat avoiders" segment consisted of a higher proportion of females, the youngest among clusters, and living in bigger households. Consumers in this cluster also presented with the highest awareness that food diet has an impact on health. In addition, the "fat avoiders" segment consisted of the lowest proportion of people that exercise and the lowest proportion of consumers that never eat pastries, cookies, and cakes. Finally, the "salt careless" segment consisted of a higher proportion of females living in households of the smallest size. Consumers in this cluster also presented with a high awareness that food diet has an impact on health and consisted of a high proportion of people that exercise.

4. Discussion

This study aimed to assess the importance consumers attached to several nutritional claims related to the most prevalent claimed nutrients, some of them beneficial (fiber) and others harmful to health (saturated fat, sugar, fat, and salt). The selected food carriers for the claims were biscuits and pastries because of their different perceived healthiness. Finally, a highly prevalent claim for a beneficial nutrient ("source of fiber"), highly prevalent claims for harmful nutrients ("reduced fat", "with no added sugar"), and less prevalent ones for harmful nutrients ("reduced saturated fat" and "low salt") were selected.

The results indicated that for the average consumer the most important nutritional claims for the two cereal products were "reduced saturated fat" and "with no added sugar". On the other hand, the least important claim was "low salt". The importance given to the other two claims "source of fiber" and "reduced fat" differs between cereals products. While "source of fiber" is more important than "reduced fat" for biscuits, the opposite was found to be true for pastries. This finding is consistent with previous results for cheese where the "reduced fat" claim is positively valued while the "low salt" negatively [14]. In addition, on breakfast biscuits consumers positively value the claims "high in fiber" and "reduced saturated fat" [15], likewise on cheese the valuation for the low saturated fat claim was higher than the valuation for the low fat claim [25]. On the contrary, a study using a hedonic prices approach found no market valuation for the nutritional claims related to fat, sugar and fiber in the case of yogurts [26]. In particular, they stated that market prices for food products with the nutritional claims "fat free", "no sugar", and "source of fiber" are not statistically different from food products without these claims. In addition, we find that preferences differ across consumers, and four segments of consumers were identified based on the estimated importance for the different nutritional claims. Except for one, segments did not differ between biscuits and pastries. These segments were named "fat careless", "fiber careless", and "fat avoiders". In the case of biscuits, a further segment was identified as "fiber lovers" and in the case of pastries, "salt careless". Heterogeneity and different consumers' segments according to the consumers' valuation for nutritional claims were also found for breakfast biscuits and yogurt, respectively [13,15]. In particular, for biscuits, two segments of consumers named "reduced saturated fat lovers" and "fiber lovers" were found because consumers' valuation for the reduced saturated fat claim was higher than that of the valuation of high in fiber in the "reduced saturated fat lovers" segment and the contrary for the "fiber lovers" [15]. In the case of yogurt also two clusters were detected where consumers preferred yogurt with low or medium sugar content but they differed in the preference for fat content [13]. While one preferred yoghurt with the lowest fat content, the other preferred yoghurt with high fat content. The third segment also preferred yoghurt with the lowest fat content but high in sugar. Our segments also follow this pattern, with some consumers preferring claims that signal negative substances and other that prefer the signal of positive substances. In addition, we find some groups of consumers do not care about the different substances evaluated (fiber careless, salt careless).

As for the characterization of the consumers in the different clusters, the "fat avoiders" segment includes younger females and a more intense consumption of pastries, cookies, and cakes. Whereas, the "fiber careless" segment is characterized by being older males with a higher proportion of households that never consume pastries, cookies, and cakes. In addition, the "fat careless" segment is also characterized by younger females but with a higher proportion of households that never consume pastries, cookies, and cakes.

An interesting result is there are no differences between clusters for education, weight, and health status of individuals, eating habits (snacking), or use of nutritional information. Similar results were also found for Spain where no differences between obese, overweight, and normal weight consumers were detected in the valuation of "high in fiber" and "reduced saturated fat" claims for breakfast biscuits [15]. This result indicates that the weight status itself did not affect the preferences for nutritional claims in the case of cereal products. However, for hard cheese, the valuation of low salt and reduced fat nutritional claims was different for differences in education, income, and body mass index levels [14]. This corroborates that the importance of different claims are product-specific.

Our findings are relevant for food producers and from a public health perspective. Food producers that want to increase the attractiveness of their products for consumers should focus on highlighting the absence, or low levels, of unhealthy substances (fats) for cereal products. This is particularly relevant as consumers with a high consumption of these products attach more importance to these claims. However, this will not stimulate consumption in non-consuming segments, as they tend to place little importance on these claims. From a public health perspective, these findings raise some concerns with regards to moral licensing and the consumption of the less healthy cereal product (pastries) as those consumers that attach more importance to avoidance claims (low fat) tend to consume more products whilst also following a less healthy lifestyle with less exercise. If public authorities and consumer organizations want to avoid that claims are used as an excuse to follow less healthy diets and lifestyles, promotion activities emphasizing the importance of limiting the consumption of these products, even in their healthier versions, are needed. This seems to be a general trend for less healthy products (sausages) [14]. Based on the profile of the "fat avoider" cluster, these promotional initiatives should target only male and younger consumers, communicating them that, although they do not have yet health problems related to their diet, following a healthy diet is the best way of preventing.

Finally, we would like to highlight the strengths and limitations of our study. Regarding the former, our sample is an exact representation of the target population for all age ranges, however it underrepresents respondents with secondary education with a bias towards the more educated strata. This bias is common in the majority of studies using primary data as this strata is more disposed to respond to questionnaires [27]. Moreover, the sample is also representative of the population in terms of the Body Mass Index (BMI) as the percentage of our sample of consumers belonging to the three BMI groups was very similar to the percentage in the general population. In addition, we have shown how claim valuation changes within a specific product category between healthy and unhealthy products. As for limitations, while our results hold for the specific population and products selected, the generalization of the results are not straightforward. While some of the socio-economic and behavioral drivers seem to be constant with regards to the valuation of nutritional claims, the implications regarding consumption can be population specific. More studies are needed to confirm that nutritional claims in less healthy products can be driving moral licensing for their consumption. Moreover, the study of consumers' importance for the nutritional claims was applied only to two food products within a specific food product category (cereals). As preferences for nutritional claims were found to be product-specific, conclusions and recommendations might vary when applied to other food categories.

5. Conclusions

The ranking of nutritional claims varies across consumers and differs between the two products (biscuits and pastries). Consumers prefer the nutritional claims that show reduced presence of unhealthy substances above those that highlight the presence of healthy ones. For the average consumer, the most important nutritional claims for the two cereal products were "reduced saturated fat" and "with no added sugar" while the least important claim was "low salt". The importance given to the other two claims "source of fiber" and "reduced fat" differs between cereals products. While "source of fiber" is more important than "reduced fat" for biscuits, the opposite was found to be true for pastries. Four segments of consumers were identified based on the estimated importance for the different nutritional claims. Except for one, segments did not differ between biscuits and pastries. These segments were named "fat careless", "fiber careless", and "fat avoiders". In the case of biscuits, a further segment was identified as "fiber lovers" and in the case of pastries, "salt careless". The characterization of the segments indicate that the "fat avoiders" includes younger females and a more intense consumption of pastries, cookies, and cakes. Whereas, the "fiber careless" segment is characterized by being older males with a higher proportion of households that never consume pastries, cookies, and cakes. In addition, the "fat careless" segment is also characterized by younger females but with a higher proportion of households that never consume pastries, cookies, and cakes.

Author Contributions: A.G., development of methods, data gathering, data analysis, and drafting the paper; J.B.-H., development of methods, data analysis, and drafting the paper.

Funding: Funding for the research was provided by the Gobierno de Aragón via the Programa de Desarrollo Rural 2014–2020. The views expressed are purely those of the authors and may not in any circumstances be regarded as stating an official position of the European Commission.

Conflicts of Interest: The authors declare no conflict of interest.

Appendix A

Table A1. Population by sex and age in Spain and in the region (%).

	Total	Sex		Age				
		Female	Male	18–34	35–44	45–54	55–64	More Than 64
Spain	46,572,132	51.0	49.0	22.9	20.2	19.0	15.2	22.9
Region	1,308,750	50.6	49.4	21.1	19.1	18.6	15.5	25.8

Source: Spanish Census of Population, 2017. Instituto Nacional de Estadística (www.ine.es), Spain.

References

1. Hieke, S.; Kuljanic, N.; Pravst, I.; Miklavec, K.; Kaur, A.; Brown, K.A.; Rayner, M. Prevalence of nutrition and health-related claims on pre-packaged foods: A five-country study in Europe. *Nutrients* **2016**, *8*, 137. [CrossRef] [PubMed]
2. Grunert, K.G.; Wills, J.M.; Fernandez-Celemin, L. Nutrition knowledge, and use and understanding of nutrition information on food labels among consumers in the UK. *Appetite* **2010**, *55*, 177–189. [CrossRef] [PubMed]
3. Saarela, A.M. Change of behaviour when selecting food products in a supermarket environment after reminding consumers about weight management. *Public Health Nutr.* **2013**, *17*, 1147–1155. [CrossRef] [PubMed]
4. Grunert, K.G.; Wills, J.M. A review of European research on consumer response to nutrition information on food labels. *J. Public Health* **2007**, *15*, 385–399. [CrossRef]
5. Kaur, A.; Scarborough, P.; Matthews, A.; Payne, S.; Mizdrak, A.; Rayner, M. How many foods in the UK carry health and nutrition claims, and are they healthier than those that do not? *Public Health Nutr.* **2015**, *19*, 1–10. [CrossRef] [PubMed]
6. Lalor, F.; Kennedy, J.; Flynn, M.A.; Wall, P.G. A study of nutrition and health claims—A snapshot of what's on the irish market. *Public Health Nutr.* **2010**, *13*, 704–711. [CrossRef] [PubMed]
7. Pravst, I.; Kušar, A. Consumers' exposure to nutrition and health claims on pre-packed foods: Use of sales weighting for assessing the food supply in Slovenia. *Nutrients* **2015**, *7*, 9353–9368. [CrossRef]
8. Lopéz-Galán, B.; Magistris, T.D. Prevalencia de las declaraciones nutricionales en la prevención de la obesidad en el mercado español. *Nutr. Hosp.* **2017**, *34*, 154–164. [CrossRef]
9. Cavaliere, A.; Ricci, E.C.; Banterle, A. Nutrition and health claims: Who is interested? An empirical analysis of consumer preferences in Italy. *Food Qual. Prefer.* **2015**, *41*, 44–51. [CrossRef]
10. Ballco, P.; de-Magistris, T.; Caputo, V. Consumer preferences for nutritional claims: An exploration of attention and choice based on an eye-tracking choice experiment. *Food Res. Int.* **2019**, *116*, 37–48. [CrossRef]
11. Ballco, P.; Caputo, V.; de-Magistris, T. Consumer valuation of European nutritional and health claims: Do taste and attention matter? *Food Qual. Prefer.* **2020**, *79*, 103793. [CrossRef]
12. Almli, V.L.; Øvrum, A.; Hersleth, M.; Almøy, T.; Næs, T. Investigating individual preferences in rating and ranking conjoint experiments. A case study on semi-hard cheese. *Food Qual. Prefer.* **2015**, *39*, 28–39. [CrossRef]
13. Miklavec, K.; Pravst, I.; Grunert, K.G.; Klopčič, M.; Pohar, J. The influence of health claims and nutritional composition on consumers' yoghurt preferences. *Food Qual. Prefer.* **2015**, *43*, 26–33. [CrossRef]
14. De-Magistris, T.; Lopéz-Galán, B. Consumers' willingness to pay for nutritional claims fighting the obesity epidemic: The case of reduced-fat and low salt cheese in Spain. *Public Health* **2016**, *135*, 83–90. [CrossRef]
15. Jurado, F.; Gracia, A. Does the valuation of nutritional claims differ among consumers? Insights from Spain. *Nutrients* **2017**, *9*, 132. [CrossRef]
16. Żakowska-Biemans, S.; Tekień, A. Free range, organic? Polish consumers preferences regarding information on farming system and nutritional enhancement of eggs: A discrete choice based experiment. *Sustainability* **2017**, *9*, 1999. [CrossRef]
17. Barreiro-Hurle, J.; Gracia, A.; De-Magistris, T. The effects of multiple health and nutrition labels on consumer food choices. *J. Agric. Econ.* **2010**, *61*, 426–443. [CrossRef]
18. Train, K. *Discrete Choice Methods with Simulation*; Cambridge University Press: Cambridge, UK, 2003.
19. Lancaster, K.J. A new approach to consumer theory. *J. Political Econ.* **1966**, *74*, 132–157. [CrossRef]

20. Hair, J.F.; Anderson, R.E.; Tathan, R.L.; Black, W.C. *Multivariate Data Analysis*; Prentice-Hall International: Bergen County, NJ, USA, 1998.
21. Instituto Nacional de Estadística. Padrón Continuo a 1 de enero de. 2017. Available online: https://www.ine.es/dyngs/NEbase/es/operacion.htm?c=Estadistica_C&cid=1254736177012&menu=resultados&secc=1254736195461&idp=1254734710990 (accessed on 27 November 2018).
22. Instituto Aragonés de Estadística. Estadística Local de Aragón. Ficha Territorial Aragón. Available online: http://bonansa.aragon.es:81/iaest/fic_mun/pdf/P50.pdf (accessed on 28 November 2018).
23. Instituto Nacional de Estadística. Encuesta Nacional de Salud. 2017. Available online: https://servicios3.aragon.es/iaeaxi/tabla.do (accessed on 11 September 2019).
24. Junger-Tas, J.; Marshall, I. The self-report methodology in crime research. *Crime Justice* **1999**, *25*, 291–367. [CrossRef]
25. Øvrum, A.; Alfnes, F.; Almli, V.L.; Rickertsen, K. Health information and diet choices: Results from a cheese experiment. *Food Policy* **2012**, *37*, 520–529. [CrossRef]
26. Ballco, P.; de-Magistris, T. Valuation of nutritional and health claims for yoghurts in Spain: A hedonic price approach. *Span. J. Agric. Res.* **2018**, *16*, e0108. [CrossRef]
27. Verhoef, P.C. Explaining purchases of organic meat by Dutch consumers. *Eur. Rev. Agric. Econ.* **2005**, *32*, 245–267. [CrossRef]

nutrients

MDPI

Article

Relationships Between Health and Environmental Information on the Willingness to Pay for Functional Foods: The Case of a New Aloe Vera Based Product

Elena Castellari [1], Elena Claire Ricci [2], Stefanella Stranieri [3,*], Stéphan Marette [4], Martina Sarnataro [1] and Claudio Soregaroli [1]

1 Department of Agri-Food Economics, Università Cattolica del Sacro Cuore, Piacenza, Via Emilia Parmense, 84, 29122 Piacenza, Italy; elena.castellari@unicatt.it (E.C.); m.sarnataro123@gmail.com (M.S.); claudio.soregaroli@unicatt.it (C.S.)

2 Department of Business Administration, Università degli Studi di Verona, Via Cantarane 24, 37129 Verona, Italy; elenaclaire.ricci@univr.it

3 Department of Environmental Science and Policy, Università degli Studi di Milano, Via Celoria 2, 20133 Milano, Italy

4 UMR Économie Publique, INRA, Université Paris-Saclay, F-78850 Grignon, France; stephan.marette@agroparistech.fr

* Correspondence: stefanella.stranieri@unimi.it; Tel.: +39-0250316460

Received: 29 September 2019; Accepted: 11 November 2019; Published: 15 November 2019

Abstract: There is an increasing interest in healthy and sustainable product characteristics. Consumers determine their dietary intake and frame production systems with their choices. However, little is known about the relationships between health and environmental information in influencing these choices, especially when considering functional foods. This study assessed the influence of health-related and environmental-friendliness-related product information on the willingness to pay (WTP) for functional foods. To this end, a WTP elicitation experiment was set up using a jam-like fruit compote enriched with *Aloe vera* gel. Participants were provided with different messages related to the health and environmental benefits of *Aloe vera* products, and were also asked to taste the product. Results indicated that providing new information significantly increased the WTP for the enriched compote. This increase was significant for both health and environmentally based benefits, with the health message leading to a higher WTP. Combining health and environmental messages produced an additive effect on WTP which was independent of the sequential order in which the two messages were given. Results contrasted the view that health messages are the main drivers of WTP, and open a broader range of communication in terms of marketing strategies and sustainable policy objectives.

Keywords: food choices; functional food; health information; environmental information; choice experiment; sustainability

1. Introduction

Consumers are increasingly interested in healthy and sustainable product characteristics when buying food [1]. Most consumers have concerns about environmental issues and look for sustainable products with environmentally friendly characteristics which can also positively affect their health [2,3].

In recent decades, the prevalence of food with nutritionally enriched characteristics and functional ingredients aimed at improving human wellbeing has increased considerably [4]. Functional food captures this trend of food demand by offering products that impact positively on human health.

The market for functional food is steadily increasing, even if making an estimate of the market dimension for these products is complicated, mainly because of the lack of a common, internationally recognized definition of such products [5]. In Europe, functional food is defined as "natural or processed foods that contain known or unknown biologically active compounds which, in defined, effective non-toxic amounts, provide a clinically proven and documented health benefit for the prevention, management, or treatment of chronic disease" [6]. Diplock et al. [7] considered a broader definition by considering products to be functional when it has been satisfactorily proven that they positively affect human health and wellbeing beyond nutritional effects. Moreover, Poulsen [8] introduced an even broader definition, identifying four characteristics that could cause a product to be considered functional: (i) the enrichment of food with a substance which is already part of the product; (ii) the substitution of a nutrient with another one; (iii) the adding of a new substance in the product; or (iv) the elimination of a component of the product.

Plant food supplements like aloe-based food products can be considered products with functional characteristics according to the above definitions. Plant food supplements are foodstuffs with a high concentration of botanical preparations that have nutritional or physiological effects, alone or in combination with vitamins, minerals, and other substances which are not plant-based.

Literature to date has explored the market response of functional food mainly by investigating the factors influencing consumer acceptance of such products [9–11]. This includes socioeconomic characteristics such as gender [12–14], age [15,16], knowledge [17–19], and lifestyles [5], as well as cognitive and attitudinal drivers [5,17,20–24] including cultural factors [25], and the role of beliefs [26]. In addition, some quality product characteristics, such as price, convenience, and taste [27] have been found to influence consumer acceptance for functional foods.

A relevant body of literature has also explored the effect of labeled health information on consumer preferences and willingness to use novel foods with functional characteristics. Most of the empirical results have highlighted a positive influence of health claims on the evaluation and choice of functional foods [28,29]. Other studies have concentrated their attention on the information most able to enhance consumer response to health claims for functional foods, highlighting how physiology-related health benefits [30], or health benefits more generally [24,31], are positively impacting motivation to purchase or are evaluated better than nutrition claims or disease-risk claims [17]. At the same time, the negative trade-off between healthy characteristics and taste seems to reduce the acceptance of functional foods over time [27].

To the best of our knowledge, there is currently a gap in the evaluation of other types of information, apart from health-based, on consumer intention to use functional foods. For example, the effect of information about the environmentally sustainable attributes of such products has received only limited attention, even if consumers have been shown to have interest in such characteristics in food products [32–34]. Moreover, there is also little evidence on the possible synergic or additive effects of using different types of information on the willingness to pay for products with functional characteristics. Only a recent study conducted by Goetzke et al. [35] has studied the effect of healthy consumer lifestyle on the consumption of organic and functional food. More generally, it appears that the impact of environmental information on food consumption has been much less researched compared to health-related information and the combination of these two types of information has garnered very little attention [36].

Study on the effects of health-related and environmental sustainability information is particularly important as it contributes to the knowledge that could favor a transition toward healthier diets and sustainable food systems [37]. This is a widely recognized goal among scientists, institutions, and the public. Information influences consumer demand, and this demand decides the healthiness of diets and frames production systems [38,39]. There is a growing understanding of the interrelationships between diet, health, and the environment [40]. Actions in this direction are already visible. For example, as highlighted by Hoek et al. [41], although dietary guidelines worldwide are mostly focused on health, some governmental bodies are starting to introduce indications from both environmental and nutritional science to frame new guidelines.

On the basis of the above research gaps, the aims of the present paper were: (i) to assess the influence of health-related and environmental friendliness product information on the willingness to pay for functional foods; (ii) to empirically estimate the presence of possible synergic or additive effects of health and environmental-friendliness-related information on the overall evaluation of functional foods.

In our empirical analysis we referred to a unique, jam-like, aloe-based fruit compote that has not been commercialized. The health characteristics of this product relate to the absence of added sugar and to a high concentration of *Aloe vera*. The gel has a wide array of pharmacological attributes including: anti-viral, anti-bacterial, laxative, protection against radiation, anti-oxidant, anti-inflammation, anticancer, anti-diabetic, anti-allergenic, and immuno-stimulation activities. As far as the food industry is concerned, the potential use of *Aloe vera* gel has mainly focused on the development of functional foods due to its beneficial properties in treating constipation, coughs, diabetes, headaches, arthritis and immune system deficiencies, and digestive effects [42]. The environmentally friendly properties of such products relate to the sustainable cultivation practices of *Aloe vera*, as the plant can grow in arid and marginal areas without the need for chemicals and with a very limited water consumption.

The paper is organized as follows: Section 2 describes the experimental design and the data analysis procedure, Section 3 presents the results, and Section 4 discusses the results in the context of the available literature, and highlights suggestions for future research and study limitations.

2. Materials and Methods

2.1. Sample and Products

The experiment was conducted across several sessions in June–July 2017 in Italy. We interviewed 115 respondents in groups of about 15 each, but given the incomplete replies of three participants, the analysis was carried out on a sample of 112. Participants were all aware of the experiment and agreed to be included in the study before participating. Data were treated anonymously and used for research purposes only, with no way of connecting responses to specific individuals.

The experiment focused on two products: (1) a 250 g jar of fruit compote made with *Aloe vera* gel (40%), Sultanina grape (30%), plums (27.5%), and orange peel (2.5%); (2) a 250 g jar of fruit compote with the same composition as the previous one, but with the *Aloe vera* gel replaced with pectin. Pectin was used as it conferred a similar texture to the final product and minimally altered the taste of the product with respect to the other ingredients.

The aloe-based fruit compote was developed by university students taking part in the Ecotrophelia project as an innovative and eco-friendly product called "AloeSpoon", completely new and original in its category, as at the time the experiment took place, no jam or jam-like products with at least 40% *Aloe vera* gel concentration were available on the market. AloeSpoon is a jam-like fruit compote product that could possibly be assigned to the functional food category thanks to the health benefits conferred by the presence of aloe. The product, as law requires (art. 2, para. 4, Legislative Decree 20 February 2004, no. 50), belongs to the category of "fruit compote" because of its fruit content, which is equal to or higher than 65%. More information about AloeSpoon is provided in Appendices A and B.

For the purpose of the experiment, what was relevant was the ability to control ceteris paribus for the presence of *Aloe vera* gel. Both products were made in the university laboratory of the Department for Sustainable Food Process of the Università Cattolica del Sacro Cuore, Italy, and were provided to the authors for this experiment. The two fruit compotes were presented using the same type of glass jar, no brand was indicated, and they were visually indistinguishable from each other. Therefore, the only differences perceived by the participants were in the information provided in the experimental design and in the product tasting.

2.2. Research Aims and Experimental Design

This study evaluated the willingness to pay (WTP) for attributes provided by a new (aloe-based) product, AloeSpoon, and the relevance of the information provided. More precisely, the aim of this study was to measure consumers' WTP for the health and environmental attributes provided by the fruit compote, and to evaluate whether the WTP changed when consumers were informed of the health benefits and of the environmentally friendly characteristics of the product. Moreover, since the product had not been commercialized at the time of the experiment, consumers were not familiar with its taste. This allowed assessment of the possible effect of the *Aloe vera* gel in changing the WTP as a result of product tasting. As explained in Figure 1, product tasting was randomly assigned to two groups (G3 and G4) with different information schemes. The tasting took place at the beginning of the experiment, during the first elicitation round (where no information was provided).

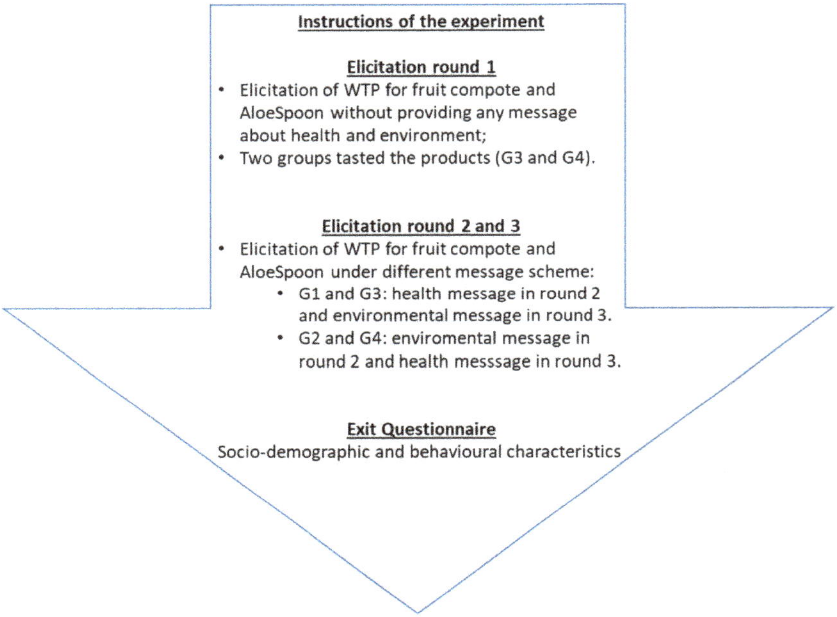

Figure 1. Timeline of the experiment.

The hypotheses we aimed to test were as follows:

Hypothesis 1. *Consumers assign a higher value to the product enriched with Aloe vera, i.e., they show a higher WTP for such a product compared to its fruit-only counterpart;*

Hypothesis 2. *Providing additional product information on the health-related properties of Aloe vera and/or on the environmental impacts of its production affects the WTP;*

Hypothesis 3. *A health message directed at highlighting the private benefits for consumers is more effective than one about environmental sustainability, which is associated with public benefits, i.e., the health message induces a greater increase in WTP;*

Hypothesis 4. *The two types of information interact and impact the size of their effects;*

Hypothesis 5. *The ordering of the information provided, i.e., whether the health-related or the environmental message is provided first, impacts the evaluation of the product (manifested by different WTP values).*

To this aim, the questionnaire consisted of different rounds of WTP elicitations using a multiple price list (MPL) approach. The MPL approach relies on eliciting WTP asking to the participants their willingness to buy a specific product in an array of ordered prices ranging from a maximum to a minimum. For each price the subject is asked to indicate his willingness to buy indicating "yes", "no" or "maybe". The WTP results on the value where the subject switch from "yes" to "no" or "maybe". In case the subject answer to any price of the array either always "yes" or always "no" or "maybe", the WTP corresponds either to the maximum or to the minimum of the listed prices. For a further explanation of the MPL and its applications please refer to Andersen et al. [43].

The study was conducted in accordance with the principles of the Declaration of Helsinki. At the beginning of the experiment, initial explanations were read, and participants decided whether they wanted to provide their consent and thus agree to take part in the study. They were informed that all of their replies were anonymous, since they were identified only by a number. Participants were asked to indicate choices as if they were in a supermarket. It was clarified that there were no "good" or "bad" replies, so they were strongly suggested to freely indicate choices reflecting their preferences.

The framing of the experiment in its timeline is illustrated in Figure 1. Following Castellari et al. [44], after providing instructions for the experiment, the first section of WTP elicitation was carried out without any message related to the health or environmental benefits of *Aloe vera*. For this first round, instructions about the experiment were given, with only a few indications describing each product which were basically focused on the composition of the fruit compote.

The health and environmental messages were written after studying articles from the nutrition, agronomic, and environmental fields. The messages were relatively short, because previous works have underlined the benefit of providing a concise message when conveying complex information [45]. The messages provided were the following:

(a) Health-related message about the benefits of Aloe vera *consumption*

"Aloe gel has been used for centuries for its healing and therapeutic properties. Frequent use of aloe-based foods brings benefits to health. The gel has anti-viral, anti-bacterial, anti-oxidant, anti-inflammatory, anti-diabetic, anti-allergic, immune- stimulatory, and wetting and wound healing properties, and burn action. It is also used to support the health of the digestive tract. The use of aloe gel in the food industry is mainly focused on the development of functional foods. Thanks to its beneficial properties, it is mainly used in the treatment of constipation, cough, diabetes, migraine, arthritis, and immune system deficiency. Aloe-based products are suitable for consumers of all ages and offer a diverse range of health-related properties".

(b) Environmental message about the environmentally friendly characteristics of Aloe vera

"Aloe vera is an evergreen, xerophytic, greasy plant that has a tissue in the leaves allowing it to store a high water content and to survive in dry regions with reduced rainfall. Thus, it grows and can be cultivated even in arid and/or marginal areas without the need for chemicals, and hence with a highly sustainable agricultural process. It follows, therefore, that the cultivation of aloe has no negative effects on the environment".

The layout of these messages was precisely controlled by equally varying their order across four different groups of participants (Group1 (G1), Group 2 (G2), Group 3 (G3), and Group 4 (G4). Two groups started with the health-related message preceding the environmental message (G1 and G3), and two other groups started with the environmental message followed by the health-related message (G2 and G4). Furthermore, two of these groups tasted the products during the first round (G3 and G4). Participants were randomly assigned to one of the four groups (G1, G2, G3, G4). (For a clearer picture of the survey structure, refer to Table A2 in Appendix A).

At the end of the third round, participants were asked to fill in a questionnaire aimed at identifying the consumer in terms of sociodemographic characteristics. They were asked to answer general questions concerning their age, gender, job type, family composition, monthly income, purchasing habits, and their previous knowledge about *Aloe vera* and *Aloe-vera*-based products.

2.3. Elicitation Mechanism

A multiple price list (payment card) was used to elicit WTP for each product. At the beginning of each round, participants were asked to indicate whether they would be willing to buy the product displayed for prices varying from €2.00 to €18.00 per unit. Research for reference prices was mostly carried out online, since aloe-based products are not usually sold in hypermarkets or large retail chain shops. Price research was also carried out in specialty stores and pharmacies. The multiple price list was characterized by increments of 50 cents between successive prices. For each price, participants were asked to select either "yes", "no", or "maybe" regarding their purchase willingness. For each product and each round of choice R, with $R = \{1 \ldots 3\}$, the WTP was determined by taking the highest price linked to a "yes" choice (with the next highest price on the paper sheet implying a reply "no" or "maybe"). If a participant replied only "no" or "maybe" to each line, the selected WTP was taken as equal to €2.00. If a participant only replied "yes" to each line, the selected WTP was taken as equal to €18.00.

The multiple price list method is a direct survey category which allows consumers to be provided with a product's description as well as information about its use and benefits. The use of this experimental method has several advantages: First, participants are able to focus on the product attributes rather than just on price. Second, they perceive their role less heavily, since they have some information to rely on. Moreover, participants can better judge the product's price based on the data provided. Finally, it is relatively easy for subjects to see that truthful revelation is in their best interests: If the subject believes that their responses have no effect on which row is chosen, then the task collapses to a binary choice in which the subject gets what they want if they answer truthfully [46].

Furthermore, the MPL has several attractive characteristics as an elicitation procedure since it is relatively easy to explain to subjects and to implement.

However, some possible disadvantages can be associated with the use of MPL. Firstly, MPL only elicits WTP valuations at intervals instead of "point" estimates [43]. In this experiment, we adopted a €0.5 interval in order to provide an adequate degree of precision for the WTP elicitation.

Secondly, subjects can switch back and forth from row to row, implying potentially inconsistent valuations [43]. This problem was not consistent with the way the consumers were surveyed in this study. Indeed, papers containing different kinds of information were given to consumers separately, in sequential steps.

Lastly, MPL could be susceptible to framing effects, as subjects are drawn to the middle of the ordered table irrespectively of their true values [43]. This can be controlled by changing the price list boundaries; however, we did not do this in the current study [43]. Further analyses by Anderson et al. [46] indicate that multiple price lists perform relatively well, obtaining precise product valuations that are also robust to framing effects.

2.4. Sample Description

The final sample analyzed in the experiment was made up of 112 respondents (Table 1). Women represented 60% of the sample, while 40% were men. Of the respondents, 22% of respondents were younger than 30 years of age, 30% were between 30 and 40 years of age, and 47% of the respondents were older than 40 years. A total 36% of the sample had an education up to high school, 32% had obtained a bachelor's-level degree, and 46% had reached a higher level of education.

Table 1. Sample description.

Variable Name	Variable Definition	Frequency	Percentage
male	1 = if respondent = male; 0 = otherwise	45	40%
		67	60%
younger	1 = if repondent age < 30 years old; 0 = otherwise	25	22%
adult	1 = if respondent was between 30–40 years old; 0 = otherwise	34	30%
older	1 = if respondent was more than 40 years old; 0 = otherwise	53	47%
middle school	1 = if the respondent's maximum level of education was middle school; 0 = otherwise	3	3%
diploma	1 = if the respondent's maximum level of education was high school; 0 = otherwise	26	23%
bachelor	1 = if the respondent's maximum level of education was a bachelor's degree; 0 = otherwise	32	29%
university	1 = if respondent had a university degree higher than bachelor; 0 = otherwise	51	46%
l_income	1 = if household monthly income was less than €2000, 0 = otherwise	31	28%
m_income	1 = if household monthly income was between €2000–5000; 0 = otherwise	65	58%
h_income	1 = if household monthly income was more than €5000, 0 = otherwise	16	14%
member	1 = if size of family was up to three members; 0 = otherwise	69	62%
		43	38%
unemployed	1 = if respondent was unemployed; 0 = otherwise	15	5%
		321	95%
knowledge	1 = if respondent knew about *Aloe vera*; 0 = otherwise	246	73%
		90	27%

These distributions indicate that globally, the sample might have slightly favored consumer segments more interested in the nutritional and environmental aspects of food (for example, for gender and education levels). However, having included such variables in the analyses controlled for possible effects related to sample representitaveness.

2.5. Data Analysis

Data analysis included three steps. A first, exploratory step in which descriptive statistics were analyzed was done to provide an initial indication of how consumer preferences changed in the different treatments and rounds.

We then adopted inferential tools to investigate the role of information provision and other variables on consumer willingness to pay. More specifically, we used two random effects Tobit regression models to evaluate whether there was an increased willingness to pay for the compote enriched with *Aloe vera* and whether, and to what extent, this willingness to pay was affected by information provision regarding the health benefits of *Aloe vera*, information provision about the low environmental impacts of *Aloe vera*, or individual socio-demographic characteristics. The choice of the Tobit model was related to the fact that our data was left-censored at €2 and right-censored at €18 because of the experimental design. The random effect was associated to individuals and it was introduced to model possible positive dependencies among answers by the same individuals induced by individual specific preferences toward fruit compotes. The dataset indeed included six answers by each individual, given that answers were pooled for the elicitation round (Round 1 to 3). In detail, we estimated the model as follows:

$$\begin{gathered} WTP^*_{Ri} = \beta_0 + \beta_1 X_R + \beta_2 X_i + \nu_i + \epsilon_{Ri} \\ WTP_{Ri} = WTP^*_{Ri} \text{ if } 2 < WTP^*_{Ri} < 18 \\ WTP_{Ri} = 2 \text{ if } WTP^*_{Ri} \leq 2 \\ WTP_{Ri} = 18 \text{ if } WTP^*_{Ri} \geq 18 \end{gathered} \quad (1)$$

where β_0 is the intercept, β_1 is a vector of a parameter associated with the experimental design variables (related to the presence of *Aloe vera*, the information provided, and tasting of the product), X_R is the vector of dummy variables coding the experimental design, β_2 is a vector of a parameter associated with individual characteristics (age, gender, education, income, and occupation), *Xi* is the vector of regressors associated with the individual sociodemographic characteristics, v_i is the individual specific random effect, and ϵ_{Ri} is the error term.

More specifically, the two estimated models differed in how we modeled the experimental design. In the first model, X_R was a vector of dummy variables indicating the presence of *Aloe vera* (aloe), the provision of a health-related message for the fruit-only/conventional compote (h_conv) or for the *Aloe vera* compote (h_aloe), and the provision of an environmental-friendliness-related message for the fruit-only/conventional compote (en_conv) or for the *Aloe vera* compote (en_aloe). For all of the above dummies, 1 indicated the presence of *Aloe vera* or of the information provision. The same was true for the variable tasting (in both models) which was meant to control for the difference in taste of the two compotes.

Model 2 deepened the investigation into the impact of information provision and, in particular, possible saturation and/or order effects in relation to the health and environmental aspects. In this case, X_R was a vector of two variables: treatment and tasting. Tasting (taste variable) was codified as a dummy variable that was assigned a value of 1 when respondents tasted the compote (in groups G3 and G4), and 0 otherwise (in groups G1 and G2). The "treatment" was codified through five dummy variables describing the six treatments: (i) when the WTP was elicited for the fruit-only compote, (ii) for the *Aloe vera* fruit compote with no additional information (Round 1), (iii) for the *Aloe vera* fruit compote with only the health-related information (Round 2 for G1 and G3), (iv) for the *Aloe vera* fruit compote with only the environment-related information (Round 2 for G2 and G4), (v) for the *Aloe vera* fruit compote with the health-related information plus the environmental information—in this order (Round 3 for G1 and G3), and (vi) for the *Aloe vera* fruit compote having received both types of information, but in the opposite order (Round 3 for G2 and G4). This new set of variables allowed us to evaluate ceteris paribus the willingness to pay across the different treatments. As before, we coded 1 when the condition was verified and 0 when it was not.

Finally, we used the delta method to verify via hypothesis tests whether the coefficients that emerged and their differences were statistically different from one other. All analyses were performed using Stata 15.

3. Results

Data were first analyzed via descriptive statistics. In particular, we have reported mean, standard deviation, and extreme values for the elicited willingness to pay for the two types of compote in the different experimental conditions (Table 2). What emerged is that the WTP for the aloe compote was always greater that for the conventional fruit-only compote. Moreover, the WTP for the enriched compote increased as more information on the positive properties of *Aloe vera* was given, i.e., from Round 1 to Round 2, and from Round 2 to Round 3 in all cases. However, there seemed to be a greater increase in average WTP when the information regarded health, at least when the first information was provided.

Table 2. Descriptive statistics of elicited willingness to pay for the two compotes in the different experimental conditions.

Variable	Experimental Condition	Round	Obs	Mean	Std. Dev.	Min	Max
WTP_conv	No information	1	112	6.0	1.3	2	9
WTP_aloe	No information	1	112	8.1	1.8	4	13
WTP_conv	Health information	2	60	6.2	1.3	3.5	9
WTP_conv	Environmental information	2	52	5.9	1.1	3	8
WTP_aloe	Health information	2	60	9.9	2.1	6	15.5
WTP_aloe	Environmental information	2	52	9.2	1.9	6	15
WTP_conv	Environmental + health information	3	52	6.0	1.3	3	8.5
WTP_conv	Health + environmental information	3	60	6.2	1.4	3.5	9.5
WTP_aloe	Environmental + health information	3	52	10.1	2.2	6	15.5
WTP_aloe	Health + environmental information	3	60	11.0	2.2	7	15.5

Notes: In this table we report the descriptive statistics of the willingness to pay (WTP) elicited for the two compotes (conventional fruit-only compote and aloe-enriched compote) in the different treatment conditions. The treatments differed in information provision. The information the subjects were given is reported in the second column. WTPs are measured in euros for a 250 g jar.

To further investigate the impacts of information and other variables on the willingness to pay for the fruit compotes, we applied a set of regression models. Table 3 reports the results of the first model, while Table 4 presents the results of the second model.

Table 3. Tobit model with random effect for the WTP for the fruit compote (Model 1).

WTP	*Coef.*		*Std. Err.*	*p-Value*
aloe	1.82	***	0.37	0.000
h_aloe	1.94	***	0.29	0.000
en_aloe	1.80	***	0.41	0.000
h_conv	−0.22		0.29	0.422
en_conv	−0.18		0.41	0.666
taste	−0.27		0.32	0.402
younger	−0.81	**	0.33	0.013
older	−0.06		0.29	0.835
male	−0.63	***	0.18	0.001
l_income	−0.10		0.23	0.639
h_income	1.00	***	0.18	0.000
diploma	−0.04		0.32	0.894
bachelor	0.19		0.35	0.582
university	0.32		0.37	0.379
unemployed	−0.69	**	0.30	0.024
knowledge	0.57		0.35	0.108
_cons	6.37		0.42	0.000
Obs.	672			
Prob > chi2	0.000			
LR test of sigma_u = 0	0.000			

Significance levels: *** $p \leq 0.01$; ** $0.01 < p \leq 0.05$. Notes: Reference categories: age between 30 and 40; income level between €2000–5000; lowest education level (middle school). This table reports the estimated coefficients, standard errors, and related *p*-values associated with all the independent variables included in Model 1. The model applied was a random effect Tobit model censored between €2 and €18 investigating the willingness to pay (WTP) for the fruit compote depending on its attributes (e.g., with/without *Aloe vera*; with/without the health and/or environmental information).

Table 4. Tobit model with random effect for the WTP for the fruit compote (Model 2).

WTP	*Coef.*		*Std. Err.*	*p-Value*
Treatment				
aloe_no info	1.78	***	0.36	0.000
aloe_health info	3.40	***	0.37	0.000
aloe_env info	2.87	***	0.37	0.000
aloe_health+env	4.47	***	0.37	0.000
aloe_env+health	4.54	***	0.37	0.000
taste	−0.25		0.31	0.427
Control variables				
younger	0.03		0.23	0.876
older	0.06		0.19	0.761
male	−0.15		0.14	0.288
l_income	−0.05		0.18	0.775
h_income	0.08		0.15	0.576
diploma	−0.05		0.25	0.852
bachelor	−0.14		0.27	0.592
university	0.06		0.28	0.816
unemployed	−0.05		0.23	0.826
knowledge	0.60	*	0.35	0.084
_cons	6.10		0.31	0.000
Obs.	672			
Prob > chi2	0.000			
LR test of sigma_u = 0	0.000			

Significance levels: *** $p \leq 0.01$; * $0.05 < p < 0.1$. Notes: This table reports the estimated coefficients, standard errors, and related p-values associated with all the independent variables included in Model 2. The model applied was a random effect Tobit model censored between €2 and €18 investigating the willingness to pay (WTP) for the fruit compote compared to the *Aloe-vera*-enriched one in the different treatment conditions.

What emerged from the results of Model 1m reported in Table 3, was that the presence of aloe induced a higher willingness to pay for the fruit compote (confirming H.1). Moreover, providing information on the health benefits of *Aloe vera* did not significantly affect the willingness to pay of the conventional fruit compote, but it did significantly increase the willingness to pay for the *Aloe-vera*-enriched compote. Similarly, the provision of the environmental message also did not impact the valuation for the conventional compote, but it increased that for the enriched one (confirming H.2). Tasting the compote with *Aloe vera* did not seem to have a significant effect. Among the individual characteristics, younger respondents seem to value the product less than older ones. Table 3 also indicates that the predicted value of willingness to pay for male respondents was lower compared to that of women. Similar results were found for unemployed consumers. Respondents with higher levels of income were found to have a higher estimated WTP.

The results reported in Table 4 confirmed the strong role of the presence of *Aloe vera* and information provision on the willingness to pay. Indeed, compared to the conventional fruit-only compote, adding *Aloe vera* significantly increased the WTP (confirming H.1). Providing one set of information on either the health benefits or the low environmental impact also significantly increased the WTP for the *Aloe-vera*-enriched compote (confirming H.2). The health message showed a higher coefficient and a greater increase compared to the no information treatment (Table 4). The difference was confirmed statistically by the hypothesis test reported in Table 5 (confirming H.3).

Table 5. Hypothesis testing on the statistical difference among Model 2 regression coefficients β.

Test	*Coef.*		*Std. Err.*	*p-Value*
βAloe_Health Info - βAloe_No Info (Test H.2)	1.63	***	0.12	0.000
βAloe_env info - βAloe_no info (test H.2)	1.10	***	0.13	0.000
(βAloe_health info - βAloe_no info) - (βAloe_env info - βAloe_no info) (test H.3)	0.53	***	0.16	0.001
(βAloe_env + health - βAloe_env info) - (βAloe_health + env - βAloe_health info)	0.50	***	0.18	0.006
(βAloe_env info - βAloe_no info) - (βAloe_health + env - βAloe_health info) (test H.4)	−0.06		0.17	0.701
(βAloe_health info - βAloe_no info) - (βAloe_env + health - βAloe_env info) (test H.4)	−0.04		0.17	0.804
(βAloe_health + env - βAloe_env + health) (test H.5)	0.03		0.15	0.859

Significance levels: *** $p \leq 0.01$. *Notes:* Tests were performed using the delta method on the regression coefficients β derived from Model 2. The model is reported in Equation (1) (Section 2.5), while the estimated regression coefficients are shown in Table 4.

Moving to Round 3 of the experiment, i.e., when an additional informational message was provided, we found that, again, both types of additional information increased the WTP for the *Aloe-vera*-enriched compote (Table 4). Again, the increase was higher when the message regarded the health benefits compared to the low environmental message (confirming H.3).

Moreover, when we tested the differences between the two health-information-related increases in WTP, we found that they were not significantly different from each other (Table 5). A similar result was found for the environmental messages. Thus, there seemed not to be a diminishing return for the additional information provided nor an amplification of the informational effect; instead, the two effects seem to be additive (confuting H.4).

Furthermore, the final WTP was the same whether the health information was given before or after the environmental information. Indeed, the two coefficients associated with "full" information provision (associated with the variables "aloe_health + env" and "aloe_env + health" in Table 4) were not statistically different from each other (as emerging from Table 5). Thus, the order in which the information was provided did not seem to influence the consumer evaluation of the product elicited as the willingness to pay for it (confuting H.5).

Furthermore, the results reported in Table 4 also indicated that previous knowledge about *Aloe vera* may have had a positive significant effect on the willingness to pay for the *Aloe-vera*-enriched compote.

Table 5 reports the results of the seven hypothesis tests that used the delta method to test whether the regression coefficients β (or some combinations of these) emerging from Model 2 (Table 4) were statistically different from each other.

4. Discussion

The presented results lead to several interpretations and implications for practice. First of all, without considering the effect of information, adding a food ingredient such as *Aloe vera* increased the WTP for the fruit compote. This result suggests that the new ingredient changed the reference price thinking of the potential consumer [47]. The absolute value of the price change, which was about €2, also suggested that the product had the potential to move away from the fruit compote category to a "functional" food category. Therefore, it seems that the attribute *Aloe vera* was already perceived as a quality-enhancing attribute, independently of the level of knowledge of the product itself. This introduces a first implication for management: It appears that enhancing an existing product with the *Aloe vera* ingredient might result in a concrete repositioning of the product within its category or even move the product to a different category.

Although *Aloe vera* was perceived to be a quality attribute, the results suggest that there was no full awareness and understanding of its properties. Providing new information significantly increased the WTP for the enriched compote. This increase was significant for both health and environmental-friendliness-based benefits, with the health message leading to a higher WTP as compared to the environmental one. These results confirmed findings highlighting the higher impact of healthiness attributes on motives or WTP of functional food choice [5,24,27] and, more generally, of food categories [48,49]. Low price, taste, and health food attributes are usually seen as better driving food choices compared to environmental ones, because they rely on egoistic motives rather than altruistic ones [40,50,51]. Similarly, in the organic food category, a large body of literature shows that motives for the purchase of organic products are linked mainly to taste and health attributes, and less to environmental aspects [50,52,53]. The importance of egoistic vs. altruistic factors also depends on consumer segments, with the more "conscious" cluster giving additional weight to environmental dimensions [38].

Literature assessing the combined effect of health and environmental messages is scarce. Hoek et al. [36] used a hypothetical choice experiment to investigate the effect of prices and health/environmental logos and information at the point of purchase. Their results emphasized that the effect was very dependent on the consumer segment and on the similarity between standard and alternative products. However, when products were similar, results showed that the combination of health and the environment had a larger effect than when considered independently. The remaining literature has focused more on the potential synergies between healthiness and sustainability attributes, where environmental friendliness attributes could be health-driven and vice versa [40]. For example, Verain et al. [38] highlighted how the perception of healthiness could be increased by a sustainability attribute. They explained it by referring to the so-called halo effect, which is a positive effect on a quality aspect originating from a positive perception of an unrelated attribute of the same product [54–56]. In terms of implications for practice and sustainability-enhancing policies, synergies generally lead to the suggestion of preferring egoistic attributes, such as health, to leverage sustainability [38]. Hoek at al. [41] suggested what they called a stealth intervention, where the communication focus remains on health as the main benefit and driver of behavioral change, and environmental friendliness is a side effect.

The presented results are in line with those of Hoek et al. [36]. They showed that a combined effect of health and environmental messages had a larger effect than messages given in insolation. Moreover, findings highlighted how health and environmental messages add up. Both effects were large in magnitude even if, as stated before, the health message had a larger impact. However, the final WTP was independent of the sequential order in which the two messages were given. It could be that both concepts supported each other in a symmetrical way, or that no synergy effect occurred at all. In any case, this result contrasted with the view that health messages should be the main drivers and opened a broader range of communications in terms of marketing strategies and policy objectives. In fact, both concepts could be leveraged and lead to the same outcome, at least in terms of WTP. This result suggests that further investigation is needed on this aspect of functional foods in order to understand the reasons behind this observation and its external validity.

Finally, it has to be noted that tasting was never significant in the presented results. This suggested that the tangible attributes of the augmented product did not significantly affect WTP decisions. The two proposed products were considered very similar, which gave a substantial advantage in identifying the effect of the information component. Controlling for taste was also important as sensorial factors appear to discriminate consumers more than preferences toward health and environment. Consumers need to make a trade-off between decisive factors such as taste and additional attributes such as health [27] or environment [40]. If taste has a neutral effect, it implies that consumers perceive the two products as homogeneous, making it is easier to create a switch from the standard product to the augmented one. Therefore, for a product like the one tested, actions in terms of product pricing could favor a replacement of the standard products with healthier and more environmentally friendly alternatives. Of course, this should consider the cost of the food ingredients, but from a policy perspective this is in line with the results of Hoek et al. [36] showing that a price subsidy for healthy and sustainable food products is effective in shifting consumer choices when products are perceived to be similar.

The presented research also carried some limitations and areas in need of further investigation. First of all, there is a difference between stated WTP and actual behavior. Participants reported their answers in a controlled environment, which might be substantially different from the contextual situation during a purchase. Information might not be read or noticed in the same way. For this reason, natural field experiments could help to improve the external validity of this type of study. Secondly, functional foods are not a homogeneous category in the eyes of consumers [24], and results should not be generalized to the category. Moreover, the environmental and health-related information was very specific to the combination of the product and *Aloe vera* attribute. Again, external validity could be gained from extending the analysis to other products in the functional domain. Finally, the analysis would benefit from a larger sample that could allow the definition of different clusters of consumers and take cultural effects into account [25,26]. This would allow a better tailoring of marketing and policy communications as, as suggested by Verain et al. [38], health and environmental concepts do not always go hand in hand across consumer segments.

Author Contributions: Conceptualization, E.C., E.C.R., S.S., S.M. and C.S.; Data curation, M.S.; Formal analysis, E.C. and E.C.R.; Investigation, C.S.; Methodology, E.C., S.M. and C.S.; Supervision, C.S.; Writing—original draft, E.C., E.C.R., S.S. and C.S.; Writing—review & editing, S.S.

Funding: This research received no external funding.

Acknowledgments: Authors thank Luigi Lucini and the Department for Sustainable Food Process of Università Cattolica del Sacro Cuore, Italy, for providing the fruit compotes used in this study and the information concerning, product composition, nutritional contents, and production process.

Conflicts of Interest: The authors declare no conflict of interest.

Appendix A. Additional Information on the Experimental Settings

Figure A1. Fruit Compote Visual Appearance. Note: (1) visual appearance did not change between the fruit compotes (Aloe vera gel vs. pectin); (2) no specific labels were used and the jars were the same for the two products.

Table A1. Average nutritional contents of the fruit compotes used in the experiments.

Average Nutritional Values	Per 100 g (% RI)	Per serving (20 g), (% RI)
Calories	795 kJ/190 kcal (9.5%)	159 kJ/38 kcal (1.9%)
Total fat	0.83 g (1%)	0.16 g (<1%)
Saturated fat	0 g (0%)	0 g (0%)
Total carbohydrate	39.7 g (15%)	7.94 g (3%)
Sugars	36.97 g (41%)	7.4 g (8%)
Dietary fibre	4 g	0.8 g
Protein	3.87 g (8%)	0.77 g (1%)
Sodium	1.01 mg (<1%)	0.202 mg (<1%)

Source: elaborations of the nutritional lab of Department for Sustainable Food Process, Università Cattolica del Sacro Cuore, Italy. Note: (1) this information was available during the experiment but not provided to participants unless explicitly requested; (2) if asked, to better control for the effect of information, the nutritional contents of the two fruit compotes were generically stated to be "similar"; (3) only a few participants asked this information and only at the end of the experiment.

Table A2. Experimental Conditions.

		Treatment Name			
		G1	G2	G3	G4
Information provided	Round 1	No info	No info	No info	No info
	Round 2	Health	Environment	Health	Environment
	Round 3	Environment	Health	Environment	Health
Taste		No	No	Yes	Yes
Participants (n)		30	30	30	22

Appendix B. Fruit Compote Composition and Production

AloeSpoon is made of *Aloe vera* gel (40%), Sultanina grape (30%), plums (27.5%), and orange peel (2.5%). The product only contains the sugars that are naturally present in the fruits used in the production process. Aloe gel contains between 97.5% and 98.5% water, followed by beta-polysaccharides; therefore, the caloric value of *Aloe vera* gel is very low and the consumption of one serving (about 200 mL) of gel contributes less than 5 kcal. The biggest intake of simple sugars is provided by Sultanina grape: it is a dehydrated food, so it has a low water content and is a real nutritional and caloric concentrate, in particular because of its high fructose content. Finally, the presence of orange peels allowed several goals to be achieved including a better taste, given that *Aloe vera* tastes naturally bitter.

The production process includes fruit mixing and concentration under heating until the desired level of Brix, measured by refractometer, is reached. *Aloe barbadensis* Mill. gel is then added. The final compote is dosed in the jar, pasteurized, and cooled to ensure the creation of vacuum. The product can be stocked at room temperature.

References

1. Rousseau, S. The role of organic and fair trade labels when choosing chocolate. *Food Qual. Prefer.* **2015**, *44*, 92–100. [CrossRef]
2. Grunert, K.G.; Hieke, S.; Wills, J. Sustainability labels on food products: Consumer motivation, understanding and use. *Food Policy* **2014**, *44*, 177–189. [CrossRef]
3. Del Giudice, T.; Stranieri, S.; Caracciolo, F.; Ricci, E.C.; Cembalo, L.; Banterle, A.; Cicia, G. Corporate Social Responsibility certifications influence consumer preferences and seafood market price. *J. Clean. Prod.* **2018**, *178*, 526–533. [CrossRef]
4. Annunziata, A.; Vecchio, R. Consumer perception of functional foods: A conjoint analysis with probiotics. *Food Qual. Prefer.* **2013**, *28*, 348–355. [CrossRef]
5. Szakály, Z.; Kovács, S.; Peto, K.; Huszka, P.; Kiss, M. A modified model of the willingness to pay for functional foods. *Appetite* **2019**, *138*, 94–101. [CrossRef]
6. Martirosyan, D.M.; Singh, J. A new definition of functional food by FFC: What makes a new definition unique? *Funct. Foods Health Dis.* **2015**, *5*, 209–223.
7. Diplock, A.; Aggett, P.; Ashwell, M.; Bornet, F.; Fern, E.; Robertfroid, M. Scientific concepts of functional foods in Europe: Consensus document. *Br. J. Nutr.* **1999**, *81*, 1–27.
8. Poulsen, J. *Danish Consumers' Attitudes Towards Functional Foods*; University of Aarhus, Aarhus School of Business, The MAPP Centre: Aarhus, Denmark, 1999.
9. Siegrist, M.; Stampfli, N.; Kastenholz, H. Consumers' willingness to buy functional foods. The influence of carrier, benefit and trust. *Appetite* **2008**, *51*, 526–529. [CrossRef]
10. Frewer, L.; Scholderer, J.; Lambert, N. Consumer acceptance of functional foods: Issues for the future. *Br. Food J.* **2003**, *105*, 714–731. [CrossRef]
11. Labrecque, J.; Doyon, M.; Bellavance, F.; Kolodinsky, J. Acceptance of functional foods: A comparison of French, American, and French Canadian consumers. *Can. J. Agric. Econ. Rev. Can. D'agroecon.* **2006**, *54*, 647–661. [CrossRef]
12. Urala, N.; Lähteenmäki, L. Consumers' changing attitudes towards functional foods. *Food Qual. Prefer.* **2007**, *18*, 1–12. [CrossRef]

13. Niva, M. 'All foods affect health': Understandings of functional foods and healthy eating among health-oriented Finns. *Appetite* **2007**, *48*, 384–393. [CrossRef] [PubMed]
14. Ares, G.; Gámbaro, A. Influence of gender, age and motives underlying food choice on perceived healthiness and willingness to try functional foods. *Appetite* **2007**, *49*, 148–158. [CrossRef] [PubMed]
15. Annunziata, A.; Vecchio, R. Functional foods development in the European market: A consumer perspective. *J. Funct. Foods* **2011**, *3*, 223–228. [CrossRef]
16. Bhaskaran, S.; Hardley, F. Buyer beliefs, attitudes and behaviour: Foods with therapeutic claims. *J. Consum. Mark.* **2002**, *19*, 591–606. [CrossRef]
17. Verbeke, W. Consumer acceptance of functional foods: Socio-demographic, cognitive and attitudinal determinants. *Food Qual. Prefer.* **2005**, *16*, 45–57. [CrossRef]
18. Wansink, B.; Westgren, R.E.; Cheney, M.M. Hierarchy of nutritional knowledge that relates to the consumption of a functional food. *Nutrition* **2005**, *21*, 264–268. [CrossRef]
19. Ricci, E.C.; Stranieri, S.; Casetta, C.; Soregaroli, C. Consumer preferences for Made in Italy food products: The role of ethnocentrism and product knowledge. *AIMS Agric. Food* **2019**, *4*, 88–110. [CrossRef]
20. Cox, D.N.; Koster, A.; Russell, C.G. Predicting intentions to consume functional foods and supplements to offset memory loss using an adaptation of protection motivation theory. *Appetite* **2004**, *43*, 55–64. [CrossRef]
21. Huotilainen, A.; Seppälä, T.; Pirttilä-Backman, A.M.; Tuorila, H. Derived attributes as mediators between categorization and acceptance of a new functional drink. *Food Qual. Prefer.* **2006**, *17*, 328–336. [CrossRef]
22. Korzen-Bohr, S.; Jensen, K.O. Heart disease among post-menopausal women: Acceptability of functional foods as a preventive measure. *Appetite* **2006**, *46*, 152–163. [CrossRef] [PubMed]
23. Urala, N.; Lähteenmäki, L. Attitudes behind consumers' willingness to use functional foods. *Food Qual. Prefer.* **2004**, *15*, 793–803. [CrossRef]
24. Krystallis, A.; Maglaras, G.; Mamalis, S. Motivations and cognitive structures of consumers in their purchasing of functional foods. *Food Qual. Prefer.* **2008**, *19*, 525–538. [CrossRef]
25. Siegrist, M.; Shi, J.; Giusto, A.; Hartmann, C. Worlds apart. Consumer acceptance of functional foods and beverages in Germany and China. *Appetite* **2015**, *92*, 87–93. [CrossRef] [PubMed]
26. Pappalardo, G.; Lusk, J.L. The role of beliefs in purchasing process of functional foods. *Food Qual. Prefer.* **2016**, *53*, 151–158. [CrossRef]
27. Verbeke, W. Functional foods: Consumer willingness to compromise on taste for health? *Food Qual. Prefer.* **2006**, *17*, 126–131. [CrossRef]
28. Ares, G.; Giménez, A.; Gámbaro, A. Consumer perceived healthiness and willingness to try functional milk desserts. Influence of ingredient, ingredient name and health claim. *Food Qual. Prefer.* **2009**, *20*, 50–56. [CrossRef]
29. Williams, P.; Ridges, L.; Batterham, M.; Ripper, B.; Hung, M.C. Australian consumer attitudes to health claim–food product compatibility for functional foods. *Food Policy* **2008**, *33*, 640–643. [CrossRef]
30. Van Kleef, E.; van Trijp, H.C.; Luning, P. Functional foods: Health claim-food product compatibility and the impact of health claim framing on consumer evaluation. *Appetite* **2005**, *44*, 299–308. [CrossRef]
31. Marette, S.; Roosen, J.; Blanchemanche, S.; Feinblatt-Mélèze, E. Functional food, uncertainty and consumers' choices: A lab experiment with enriched yoghurts for lowering cholesterol. *Food Policy* **2010**, *35*, 419–428. [CrossRef]
32. Horne, R.E. Limits to labels: The role of eco-labels in the assessment of product sustainability and routes to sustainable consumption. *Int. J. Consum. Stud.* **2009**, *33*, 175–182. [CrossRef]
33. Nash, H.A. The European Commission's sustainable consumption and production and sustainable industrial policy action plan. *J. Clean. Prod.* **2009**, *17*, 496–498. [CrossRef]
34. Tobler, C.; Visschers, V.H.M.; Siegrist, M. Eating green. Consumers' willingness to adopt ecological food consumption behaviors. *Appetite* **2011**, *57*, 674–682. [CrossRef] [PubMed]
35. Goetzke, B.; Nitzko, S.; Spiller, A. Consumption of organic and functional food. A matter of well-being and health? *Appetite* **2014**, *77*, 96–105. [CrossRef]
36. Hoek, A.C.; Pearson, D.; James, S.W.; Lawrence, M.A.; Friel, S. Healthy and environmentally sustainable food choices: Consumer responses to point-of-purchase actions. *Food Qual. Prefer.* **2017**, *58*, 94–106. [CrossRef]
37. Banterle, A.; Ricci, E.C.; Cavaliere, A. Environmental sustainability and the food system. In *Regulating and Managing Food Safety in the EU—A Legal-Economic Perspective*; Bremmers, H.J., Purnhagen, K., Eds.; Springer International Publishing: Cham, Switzerland, 2018.

38. Verain, M.C.; Sijtsema, S.J.; Antonides, G. Consumer segmentation based on food-category attribute importance: The relation with healthiness and sustainability perceptions. *Food Qual. Prefer.* **2016**, *48*, 99–106. [CrossRef]
39. Banterle, A.; Stranieri, S. Sustainability standards and the reorganization of private label supply chains: A transaction cost perspective. *Sustainability* **2013**, *5*, 5272–5288. [CrossRef]
40. Aschemann-Witzel, J. Consumer perception and trends about health and sustainability: Trade-offs and synergies of two pivotal issues. *Curr. Opin. Food Sci.* **2015**, *3*, 6–10. [CrossRef]
41. Hoek, A.C.; Pearson, D.; James, S.W.; Lawrence, M.A.; Friel, S. Shrinking the food-print: A qualitative study into consumer perceptions, experiences and attitudes towards healthy and environmentally friendly food behaviours. *Appetite* **2017**, *108*, 117–131. [CrossRef]
42. Sahu, P.K.; Giri, D.D.; Singh, R.; Pandey, P.; Gupta, S.; Shrivastava, A.K.; Kumar, A.; Pandey, K.D. Therapeutic and medicinal uses of Aloe vera: A review. *Pharmacol. Pharm.* **2013**, *4*, 599. [CrossRef]
43. Andersen, S.; Harrison, G.W.; Lau, M.I.; Rutström, E.E. Elicitation using multiple price list formats. *Exp. Econ.* **2006**, *9*, 383–405. [CrossRef]
44. Castellari, E.; Marette, S.; Moro, D.; Sckokai, P. The Impact of Information on Willingness to Pay and Quantity Choices for Meat and Meat Substitute. *J. Agric. Food Ind. Organ.* **2018**, *17*. [CrossRef]
45. Wansink, B.; Sonka, S.T.; Hasler, C.M. Front-label health claims: When less is more. *Food Policy* **2004**, *29*, 659–667. [CrossRef]
46. Anderson, S.; Harrison, G.W.; Lau, M.I.; Elisabet, R.E. Valuation using multiple price list formats. *Appl. Econ.* **2007**, *39*, 675–682. [CrossRef]
47. Mazumdar, T.; Raj, S.P.; Sinha, I. Reference price research: Review and propositions. *J. Mark.* **2005**, *69*, 84–102. [CrossRef]
48. Markovina, J.; Stewart-Knox, B.J.; Rankin, A.; Gibney, M.; de Almeida, M.D.V.; Fischer, A.; Kuznesof, S.A.; Poínhos, R.; Panzone, L.; Frewer, L.J. Food4Me study: Validity and reliability of Food Choice Questionnaire in 9 European countries. *Food Qual. Prefer.* **2015**, *45*, 26–32. [CrossRef]
49. Sautron, V.; Péneau, S.; Camilleri, G.M.; Muller, L.; Ruffieux, B.; Hercberg, S.; Méjean, C. Validity of a questionnaire measuring motives for choosing foods including sustainable concerns. *Appetite* **2015**, *87*, 90–97. [CrossRef]
50. Magnusson, M.K.; Arvola, A.; Hursti, U.K.K.; AAberg, L.; Sjödén, P.O. Choice of organic foods is related to perceived consequences for human health and to environmentally friendly behaviour. *Appetite* **2003**, *40*, 109–117. [CrossRef]
51. Stranieri, S.; Baldi, L. Shelf Life Date Extension of Fresh-Cut Salad: A Consumer Perspective. *J. Food Prod. Mark.* **2017**, *23*, 939–954. [CrossRef]
52. Gracia, A.; De Magistris, T. The demand for organic foods in the South of Italy: A discrete choice model. *Food Policy* **2008**, *33*, 386–396. [CrossRef]
53. Chryssohoidis, G.M.; Krystallis, A. Organic consumers' personal values research: Testing and validating the list of values (LOV) scale and implementing a value-based segmentation task. *Food Qual. Prefer.* **2005**, *16*, 585–599. [CrossRef]
54. Richetin, J.; Demartini, E.; Gaviglio, A.; Ricci, E.C.; Stranieri, S.; Banterle, A.; Perugini, M. The biasing effect of evocative attributes at the implicit and explicit level: The tradition halo and the industrial horn in food products evaluations. *J. Retail. Consum. Serv.* **2019**, 101890. [CrossRef]
55. Nisbett, R.E.; Wilson, T.D. The halo effect: Evidence for unconscious alteration of judgments. *J. Pers. Soc. Psychol.* **1977**, *35*, 250. [CrossRef]
56. Demartini, E.; Ricci, E.C.; Mattavelli, S.; Stranieri, S.; Gaviglio, A.; Banterle, A.; Richetin, J.; Perugini, M. Exploring Consumer Biased Evaluations: Halos Effects of Local Food and of Related Attributes. *Int. J. Food Syst. Dyn.* **2018**, *9*, 375–389.

nutrients

MDPI

Article

Do Consumers Care about Nutrition and Health Claims? Some Evidence from Italy

Azzurra Annunziata * and Angela Mariani

Department of Economic and Legal Studies, University of Naples Parthenope, 80133 Naples, Italy; mariani@uniparthenope.it

* Correspondence: azzurra.annunziata@uniparthenope.it

Received: 22 October 2019; Accepted: 7 November 2019; Published: 11 November 2019

Abstract: This study investigates Italian consumer knowledge and use of nutrition and health claims (NHCs). Six specific claims are examined on the basis of a web survey carried out on a sample of 504 consumers. Our results show that there is little attention to NHCs and their use is not widespread; objective knowledge of the selected claims is fairly scant, generating misinterpretation and confusion about their real meaning. K-means cluster analysis allowed us to identify three segments of consumers, characterized by different levels in attention and use frequency of NHCs, with a specific profile in terms of motivation and nutritional knowledge. Our results suggest the advisability of policy interventions and communication efforts which target the three segments with a view to achieving greater attention to NHCs. In conclusion, to boost knowledge concerning the actual meaning of the claims and their relation with a healthy diet, especially to reach non-users, information should be provided both simply and clearly, avoiding the use of complex scientific terminology.

Keywords: nutritional claims; health claims; cluster analysis

1. Introduction

Non-communicable diseases constitute a major cause of death, disease, and disability, and one in five deaths globally can be attributed to an unhealthy diet [1]. Among other measures to supply consumers with appropriate information, provision of nutrition information on packaged foods is an important instrument in promoting healthier eating habits [2]. Worldwide, three main formats for providing information can be identified: Nutrition facts panels, front-of-pack labels, and nutrition and health claims (NHCs). NHCs include any form of short text describing the specific nutritional contents or health benefits of a food product. NHCs have the potential to make consumers more aware in making choices, increasing understanding of specific nutrient–disease relationships, and are used by food manufacturers as marketing devices. In Europe, EC Regulation 1924/2006 reports the specific definitions for nutrition and health claims in order to help consumers make healthy and informed food choices, stimulate and protect innovation in the food market, and facilitate the circulation of foods bearing claims across EU member states [3].

For nutrition claims the Regulation stipulates that such claims are only permitted if they are listed in its annex and fulfil the conditions laid down in the Regulation itself. As regards health claims, the Regulation defines two main different types of claims: Those specified in Article 13.1 that pertain to 'general function' relating to the role of a nutrient or other substance in growth, development, and the functions of the body (13.1a), or psychological and behavioral functions (13.1b), and those specified in Article 14.1 that refer to disease risk reduction (14.1.a), or children's development (14.1.b). The Regulation also specifies that the use of a nutrition and health claim is only permitted if the average consumer can be expected to understand the beneficial effects expressed in the claim. At the end of 2018, there were 261 authorized health claims in the EC (2018).

According to previous research, consumer demand for food products carrying NHCs has grown rapidly [4,5], but at the same time a high market failure rate of new products with this kind of claim is registered. In this regard, Lähteenmäki (2013) [6] suggests that claims can only lend added value if consumers understand the benefit(s) concerned and whether the benefit is relevant to themselves, suggesting the need for in-depth understanding of the main motivations underlying consumer preferences and the heterogeneity in their demand.

Recently the EU-funded FP project CLYMBOL examined a large number of HCs with reference to 10 EU countries (not including Italy), reaching interesting conclusions on the need for more targeted information and greater communication efforts to make claims easier to understand and effective in supporting informed choices by consumers [7–9]. Other research proposes a cross-country approach including Italy [10–12] while specific studies on Italian consumers are few and far between [13–15].

In this regard, this paper enriches the existing literature, providing results from an Italian representative sample testing the familiarity, credibility, and objective knowledge of specific NHCs. Italy represents an interesting case study which is worth investigating: Previous research showed that Italian consumers differed from their European counterparts in their attitudes towards products with NHCs, showing less familiarity with, and attraction towards, such products and little positive influence of NHCs on their healthiness perception [10–12].

The paper is structured as follows: Section 2 presents the research background; Section 3 illustrates the materials used in the study and the methodology; Section 4 reports the results; Section 5 contains a discussion of the main findings and implications, and the paper ends with some concluding remarks.

2. Research Background

The impact of NHCs on consumer preferences and purchase behavior has been extensively analyzed. As highlighted in a recent review, such studies have yielded somewhat divergent findings on the influence of both consumer characteristics and product-specific characteristics [16].

From a theoretical perspective, in accordance with the Motivation–Ability–Opportunity framework [17] consumers' responses to NHCs are affected by the opportunity, i.e., the availability of NHCs on the market, the motivation to engage in processing the information and their ability to process the information, which is related to their nutritional knowledge and level of understanding of claims [7,18].

As for motivation, research has shown the following positive correlation: The more consumers feel the need for information about food, diet and health and, in particular, are interested in a healthy diet, the more likely they are to spend time searching for information [18,19]. Interest in healthy eating has been found to be a prominent driver in explaining consumer use of NHCs [7,12]. In accordance, the same correlation was found with reference to personal health or that of family and friends, as experiencing illnesses (included overweight problems) is a major motivation for seeking health-related information which has increased interest in NHCs [14,15,20,21].

As for the ability to process, previous research has shown that nutrition knowledge strongly influences food label use [22] and an understanding of nutrition information [23], while specifically for knowledge and NHCs the results are contradictory. Consumers with in-depth objective nutritional knowledge according to Ares and colleagues (2008) [24] evaluated food products with an NHC as healthier, and had a higher purchase intention; furthermore, lack of nutrition knowledge limits consumers' abilities to understand or evaluate a health claim, thus leading to lower perceived credibility of such claims. Others found that higher levels of nutrition knowledge led to less trust in health claims [25]. Consumers with a low nutritional knowledge have been found in Italy to have greater interest in nutritional claims [14]; other studies found for health claims no correlation at all [26].

With reference to opportunity, in accordance with the existing literature, other factors closely linked to consumer responses to NHCs are familiarity, in the sense of previous exposure, and credibility [6,27], which in turn affects the importance and reported use of claims [12].

Other research has stressed the role of consumer knowledge and understanding of nutrition and health claims already in use. With regard to subjective understanding, Williams (2005) and Trijp and van der Lans (2007) [28,29], focusing on four EU countries, show that health claims are perceived by consumers to be somewhat difficult to understand. By contrast, focusing on objective understanding, i.e., whether consumer understanding is in accordance with the scientific dossier on the claim, some studies have provided evidence that people do not always understand health and nutrition claims as they are intended [28,30–33]. Recent research [34] confirms such evidence, finding that consumers tend to interpret health and nutrition claims differently from scientific experts and regulators.

Finally, as regards socio-demographic characteristics, it is widely agreed that older consumers and women are more interested in, have a greater preference for, or a greater intention to buy products labelled with NHCs [16]. The structure of the household and the presence of young children have also been found to be significant with respect to interest in health [25] and nutrition claims [15], while with reference to education and income, research provides contradicting indications. Some studies have found that less educated consumers and those with lower incomes are more interested in products with NHCs [15,35]; on the other hand, other studies refute any correlation with education [36].

Against this background our detailed research objectives were: (a) To analyze the degree of attention and use of NHCs by Italian consumers; (b) to analyze consumers' evaluation and objective knowledge of specific NHCs; (c) to identify and evaluate interpersonal diversity, through the identification and profiling of consumer segments with different levels of nutrition and health claims interest, knowledge and use.

3. Materials and Methods

3.1. Questionnaire Content

In order to achieve the research objectives described above, a quantitative online survey was performed with a representative sample of the Italian population by age, gender and region. Based on the existing literature and in accordance with the Motivation–Ability–Opportunity framework, a structured questionnaire was designed and organized in different sections to measure the following aspects: (a) Motivation to pay attention to NHCs; (b) nutritional knowledge; (c) general attention and use of NHCs; (d) evaluation and knowledge of specific NHCs; (e) socio-demographics and health status (Table 1).

Motivations investigated in the first section concerned consumers' general health interest in food choices and the need for health-related information. The former was assessed using four items adapted from the General Health Interest scale [37]. The latter was measured by means of three items proposed in previous studies on NHCs [7,8]. In both cases we used a five-point agreement Likert scale (1 = not at all–5 = completely agree). The reliability of the scale was verified with Cronbach's α test. The items used and Cronbach's α scores are reported in Table 2.

Respondents' nutritional knowledge was assessed through the use of five multiple choice questions extracted from the nutrition knowledge questionnaire [38] in line with previous studies carried out on the same topic [14,15,35]. In particular, in accordance with [14,15], two questions aimed to assess consumers' knowledge on general nutritional recommendations (i.e., the optimal number of fruit and vegetable portions to be consumed daily and the type of fats that must be reduced) and three concerned specific knowledge on carbohydrate, protein and fat content of food. The decision to select these five questions was due to their relevance to, and consistency with, the Mediterranean Diet model which is widely followed in Italy. In order to obtain a synthetic measure on nutritional knowledge a normalized index was constructed using the scores obtained by each question (0 if wrong, 1 if correct) ranging from 0 to 5.

Respondents' attention and use of NHCs when shopping were detected by asking consumers how often they paid attention to NHCs on labels and their purchasing frequency of such products using a 5-point Likert scale (1 = never–5 = always) as in previous research [14,15]. Respondents were then asked to express their perceived ability to process NHCs using three items selected from previous research [7] (see Table 2). Also in this case the scale reliability was verified with Cronbach's α test.

In the fourth section respondents had to evaluate six specific claims, respectively three NCs and three HCs selected from the European Commission register of authorized claims (reported in Table 3). The selection of claims was based on their presence on the Italian market [39] but also considered the diversity in nutrients and the difference in wording. In the case of NCs we deliberately chose claims related to so-called "less is more" products whose sales, according to a recent Nielsen survey, have increased considerably on the Italian market, with particular reference to no sugar added products, low in sodium, and energy-reduced products [40].

For HCs we chose both general function claims (Art. 13.1.a) and disease risk reduction (Art. 14.1.a). Claims related to children's development and health (14.1b) were deliberately excluded in order to avoid the answers being affected by the different household composition of the sample. For each claim respondents expressed their degree of familiarity and credibility using a 5-point Likert Scale (1 = not at all–5 = very much), as proposed in previous research [7,8]. Subsequently, for each of the selected claims objective individual knowledge was assessed with multiple-choice questions. As for nutritional knowledge, two indices were computed, counting the number of correct answers for both health and nutrition claims. In addition, an index of objective claim knowledge was created (knowledge score ranging from 0 to 6).

The last section of the questionnaire included socio-demographic and health status-related variables that are considered important predictors of consumers' use of nutritional and health information on labels [21,31]. Among socio-demographic data, gender, age, education level, occupation, family size, presence of children in the household and income were included. As for health status, in accordance with previous research, personal diet-related needs and the presence of health problems in households were investigated [8,12].

Questionnaire understanding and length were pre-tested with a pilot sample of 20 consumers before proceeding with the main survey.

3.2. Data Collection

Data were collected in June 2019 by a national market research company (Astra Ricerche) through computer-assisted web interviewing. A quota sampling method was applied. Respondents were selected based on age (limited to the range 18–70 years) and place of residence (based on the four Nielsen areas: Northwest, Northeast, Central, and South of Italy), in accordance with statistics reported by the Italian National Institute of Statistics (ISTAT) for the 2017 resident population [41]. The inclusion criterion used was that respondents should be responsible for their household food shopping.

The procedures for contacting participants and administering questionnaires were managed by the same market research company. Over 950 individuals were invited to participate in a survey by email. About 47% of individuals contacted did not accept the invitation or did not complete the questionnaire. The final sample consists of 504 respondents.

Data collection procedures were performed in accordance with the ethical standards protocol of the data collection company in full compliance with the 1964 Helsinki declaration and its later amendments. Participants gave their informed consent to participate in the study to the data collection company. All data were collected and processed anonymously, and each participant was associated with a specific temporary identifier code.

3.3. Data Analysis

Data analysis included descriptive statistics (frequency distributions), bivariate (i.e., chi-square test, t-test, one-way analysis of variance) and multivariate analysis. The internal consistency and reliability of the scales used were measured with Cronbach's alpha coefficients. Normal distribution of the data was checked using a graphical test (histogram with normality curve), skewness and kurtosis indices.

K-means cluster analysis, performed using variables related to general attention to and use of NHCs, allowed the statistical units to be classified into a set of 'exclusive and exhaustive' clusters so as to maximize their internally homogeneous nature and externally heterogeneous nature and identify different consumer profiles.

Subsequently, in order to profile each cluster in terms of socio-demographic and attitudinal variables, cross-tabulation with Chi-square statistics and one-way ANOVA comparison of means with post-hoc Tukey tests were performed (Table 4; Table 5). All analyses were conducted with IBM SPSS Statistics 24.

4. Results

4.1. Socio-Demographic Characteristics and Health Status

Table 1 shows the socio-demographic characteristics of the survey respondents: 50.2% are female, the mean age is 45.38 years, and mean household size is three individuals. With reference to the level of education, 58% had a high school diploma while 15% had a bachelor's degree. Almost 45% of respondents live in northern Italy; as regards employment, those employed are the largest category (21.3%), followed by pensioners (14%) and housewives (12.7%); 28% of respondents had children under 12 living at home and in 63.3% of cases respondents stated that they had an average annual income in line with the national average.

With regard to the variables related to special dietary needs, 33% of respondents stated that they were influenced in their food choices by health reasons. As regards personal and/or household members' health status, 43% of the sample stated that they did not suffer from any of the pathologies, while high cholesterol (27%) and high blood pressure (26%) were the main personal or household pathologies.

Table 1. Sample characteristics.

		Sample	National Population
Gender *	Male	49.8	49.6
	Female	50.2	50.4
Age *	18–24	10.0	10.0
	25–34	16.5	16.1
	35–44	20.3	20.7
	45–54	23.5	23.4
	55–70	29.7	29.5
Current education level **	Post graduate specialization/PhD	5.4	*n.a.*
	Master's Degree	15.7	19.3
	Bachelor's Degree	10.4	
	High school diploma	58.0	61
	Other	10.6	*n.a.*
Occupation	Housewife/Househusband	12.7	*n.a.*
	Employee	21.3	*n.a.*
	Self-employed	11	*n.a.*
	Worker	11	*n.a.*
	Retired	14	*n.a.*
	Unemployed	12	*n.a.*
	Other	18	*n.a.*
Area of Residence *	Northwest	26.3	26.2
	Northeast	18.9	19
	Centre	22.5	22
	South	32.3	32.8

Table 1. *Cont.*

		Sample	National Population
Children <12	Yes	28	*n.a.*
	No	72	*n.a.*
Household economic status ***	Below national average	30.9	*n.a.*
	In line with national average	63.3	*n.a.*
	Above national average	5.8	*n.a.*
Special dietary needs	Yes	33	*n.a.*
	No	67	*n.a.*
Personal and/or household members' health problems ****	High blood pressure	26	*n.a.*
	High cholesterol	27	*n.a.*
	Cardiovascular problems	8.4	*n.a.*
	Osteoporosis	9.2	*n.a.*
	Dental caries	15.7	*n.a.*
	Other	5	*n.a.*
	No problems	43	*n.a.*

* Italian National Institute of Statistics official data on the resident population on 1 January 2017 by age (18–70), gender and geographical area [41]. ** Italian National Institute of Statistics official data for 2017 on the population between 25 and 64 years [42]. *** Average net annual income of Italian families according to Italian National Institute of Statistics official data was about €30,000 for 2016. **** Although the incidence of personal or family pathologies is not representative of the national reality, it is possible to find a correspondence between the main pathologies indicated by our sample and those detected by the 2017 ISTAT survey [43] according to which 40% of Italians suffer from a chronic disease. The most prevalent chronic disease is hypertension while osteoporosis accounts for 7.6%.

4.2. Motivation and Ability to Process NHCs

With respect to general health interest in food choices (Table 2), respondents consider themselves particular about the healthiness of food (3.4) and consider it important to follow a diet rich in vitamins and minerals (3.5). In addition, with reference to the need for health-related information, respondents seem sensitive to being informed if one or more components of the foods chosen reduces a risk factor of developing a disease (3.6). In terms of perceived ability to process NHCs, respondents tend to consider themselves quite knowledgeable (3) and are quite confident about their ability to understand the claims (3.2).

Table 2. Motivation and ability to process nutrition and health claims (NHCs).

Health Interest in Food Choices (Cronbach's Alpha = 0.834)	Mean	S.D.
I always follow a healthy and balanced diet	3.2	0.96659
It is important for me that my diet is low in fat	3.3	0.97366
It is important for me that my daily diet contains a lot of vitamins and minerals	3.5	0.95734
I am very particular about the healthiness of food I eat	3.4	0.96194
Need for health-related information (Cronbach's alpha = 0.838)		
It is necessary for me to know the nutrient content of food products	3.5	1.01235
It bothers me if health-related information is not available on food labels	3.3	1.11105
It is important for me to be informed if one or more components of the foods I choose reduces a risk factor in the development of a human disease	3.6	1.03137
Perceived ability to process health and nutrition claims(Cronbach's alpha = 0.863)	**Mean**	**S.D.**
Compared to most people, I am quite knowledgeable about health and nutrition claims	3.0	0.85479
Compared to most people, I am more confident in using health and nutrition claims in making food choices	3.1	0.85999
I am confident about my ability to understand health and nutrition claims	3.2	0.86439

4.3. Consumer Attention to, and Use and Evaluation of NHCs

With respect to the degree of attention and use of nutritional and health information on the label, 33% of respondents say they often pay attention to the nutritional panel while shopping. The level of attention is higher for NCs (40% often pay attention) but is lower in the case of HCs (29% often pay attention). Considering the level of NHC use, 36% of respondents state that they often buy products with an NC on the label, while this percentage decreases to 26% for HCs. Cross-tabulation shows that the high level of attention to nutritional panels is related to the high level of attention to, and use of, NHCs (p-value = 0.00).

As shown by Table 3, respondents generally stated a slightly higher level of familiarity with NCs than HCs. Pairwise comparisons t-test was performed in order to verify whether the differences between means are statistically significant. Among the NCs proposed the most familiar is "no sugar added" while among the HCs the most familiar is "Omega-3 fatty acids help to maintain a healthy cardiovascular system".

With regard to the credibility level our results show that the level of credibility is generally lower than that of familiarity. Respondents are more confident towards HCs, among which Omega-3 fatty acids obtained on average the highest level of credibility.

Table 3. Familiarity and credibility towards specific health and nutrition claims.

	Familiarity	**Credibility**
Nutrition Claims	**Mean (S.D.)**	**Mean (S.D.)**
Low in sodium	3.6 (0.956)	3.2 (0.846)
Reduced kcal	3.6 (0.963)	3 (0.892)
No sugar added	3.8 (0.949)	3.1 (0.961)
Health Claims		
Plant sterols have been shown to lower blood cholesterol levels. High cholesterol is a risk factor in the development of coronary heart disease	3.4 (1.096)	3.4 (0.930)
Omega-3 fatty acids help to maintain a healthy cardiovascular system	3.6 (1.091)	3.5 (0.899)
Chewing gum sweetened with 100% xylitol helps neutralize plaque acids. Plaque acids are a risk factor in the development of dental caries	3.1 (1.009)	3 (0.952)

All pairwise comparisons among familiarity and credibility means for nutrition claims (NCs) and health claims (HCs) are statistically significant (according to t-test p-value < 0.05) except for familiarity between Low in sodium/Omega-3 and Reduced kcal/Omega-3.

4.4. *Objective knowledge of NHCs*

In terms of general nutritional knowledge, our results show that the average value of the sample's nutritional knowledge index is quite low (2.37 mean value, S.D. 1.19). With reference to specific claims, many of the interviewees do not know the actual meaning of the selected claims. We constructed a normalized index using the scores obtained by each question used (assigning value 1 for a correct answer, otherwise value 0). The scores ranged from 0 to 6. The mean value of the ability to process index is quite low (2.7; S.D. 1.31).

However, contrary to our expectations, failure to know the actual meaning is more marked for nutrition than for health claims (Figure 1). Among the nutrition claims, the claim that consumers stated was most familiar, namely "no sugar added", proved the most difficult to interpret by consumers who were aware of the actual meaning of the claim, and in most cases considered that food reporting such claims contains no sugar at all or just contains less sugar (53%). Similarly but with less intensity "low in sodium" is misinterpreted by consumers, often indicating that products carrying this claim contain less sodium than other similar products or that the food was produced without sodium added (47%), while with reference to the claim "reduced in kcal content" the level of consumer knowledge is higher than the other two claims, with 61% of respondents indicating the correct answer.

For health claims respondents showed great difficulty in interpreting the claim "Omega-3 fatty acids help to maintain a healthy cardiovascular system", which is the claim considered most familiar. Almost half the sample opted for the response that it helps to reduce the risk of heart attack or the level of cholesterol in the blood (48%). The degree of objective knowledge of the plant sterol claim is higher: 65% of respondents indicated the correct answer, while the meaning of the HC related to "xylitol in chewing gum" was the best known. This could be due to the fact that most chewing gum on the market is advertised in relation to this substance.

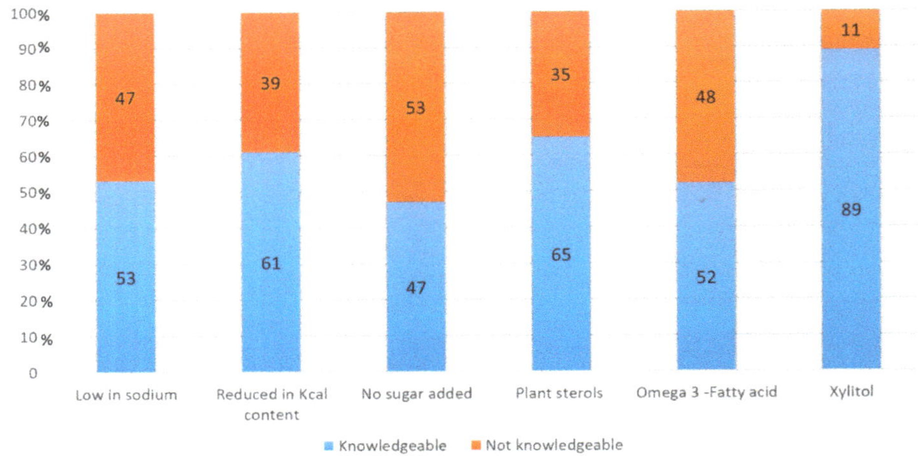

Figure 1. Share of interviewees who knew the actual meaning of the selected NHCs.

4.5. Consumer Segmentation

Applying K-means cluster analysis, three segments of consumers with different attention levels and use frequency of NHCs were identified. From the application of this method, the division into three segments was the optimal solution, where homogeneity is maximized within the individual clusters and minimized between them. In order to determine the optimal number of clusters the Calinski–Harabasz pseudo F-value was considered. The division into three clusters shows a higher value of the pseudo F′ test compared with other solutions tried. Differences between clusters in terms of attention and frequency of health and nutrition claims used are significant ($p = 0.000$).

The largest segment, namely *"Potential users"* in Cluster 1, accounted for 47% of respondents and included consumers who reported moderate levels of attention and use for both nutrition and health claims in line with the average sample.

Cluster 2, namely *"Claims users"*, accounted for 33% of respondents who reported the highest levels of attention and use of nutrition claims as well as health claims. Indeed, 35% of individuals in this cluster state they always pay attention to nutrition claims and 30% always use them. As for health claims 70% in this cluster state that they often pay attention to such claims and 68% often use them. The degree of attention towards nutritional panels is also higher in this cluster.

Cluster 3 accounted for 20% of respondents, showing the lowest levels of attention and use of nutrition claims and a very low level of attention and use of health claims. In this cluster, 49% rarely pay attention to nutrition claims and 20% never, which applies also to HC use. With regard to health claims, 41% never pay attention to them and 45% never use them while shopping. This cluster was therefore called *"Non-users"*.

The clusters were then profiled according to the degree of knowledge of specific claims and attitudinal determinants towards health and nutrition claims (Table 4) and socio–demographic characteristics (Table 5).

With respect to attitudinal determinants towards health and nutrition claims, Table 4 shows that the *Non-users* Cluster shows the lowest level of motivation to use NHCs, as confirmed by low levels of both general health interest in food choices and the need for health-related information. By contrast, the level of general health interest and need for health-related information is higher among *Claims users*, the cluster with the highest interest and use of nutrition and health claims. *Claims users* also consider themselves better able to interpret such claims.

With reference to specific claims, significant differences among clusters are revealed in terms of familiarity and credibility. Once again, *Claims users* includes consumers with the highest level of familiarity and credibility for both health and nutrition claims, and *Non-users* includes consumers with a lower degree of familiarity and credibility towards claims. Interestingly, for all three clusters HCs enjoy a higher degree of credibility than NCs.

As regards nutritional knowledge, consumers in the *Claims users* Cluster are more knowledgeable than consumers in the *Non-users* Cluster. Although the same result emerges for the NHC knowledge index, for the specific index for NCs actual knowledge is scant in all three clusters and no significant differences are noted. By contrast, as regards HC knowledge, consumers in the *Non-users* Cluster show a lower level compared to the other two clusters. No significant difference was detected in relation to objective knowledge of each specific claim used in the study.

Table 4. Cluster profiling in terms of motivations, evaluation and knowledge of NHCs.

	Potential Users	Claims Users	Non-Users	Total Sample	*p*-Value *
Attention to nutritional panels	3.29 ^{a}	4.08 ^{b}	2.12 ^{c}	3.33	0.000
Attention to NCs	3.49 ^{a}	4.25 ^{b}	2.14 ^{c}	3.48	0.004
Attention to HCs	2.82 ^{a}	4.19 ^{b}	1.71 ^{c}	3.06	0.000
Buying frequency of NC-labelled products	3.31 ^{a}	4.20 ^{b}	2.13 ^{c}	3.38	0.000
Buying frequency of HC-labelled products	2.72 ^{a}	4.02 ^{b}	1.75 ^{c}	2.97	0.000
General health interest **	3.26 ^{a}	3.84 ^{b}	2.70 ^{c}	3.34	0.000
Need for health-related information **	3.45 ^{a}	4.11 ^{b}	2.70 ^{c}	3.52	0.000
Perceived ability to process health and nutrition claims **	3.10 ^{a}	3.60 ^{b}	2.53 ^{c}	3.16	0.000
Familiarity with specific NCs ***	3.55 ^{a}	4.06 ^{b}	3.12 ^{c}	3.64	0.000
Credibility of specific NCs ***	3.04 ^{a}	3.51 ^{b}	2.69 ^{c}	3.13	0.003
Familiarity with specific HCs ***	3.30 ^{a}	3.74 ^{b}	2.92 ^{c}	3.37	0.000
Credibility of specific HCs ***	3.18 ^{a}	3.56 ^{a}	2.86 ^{b}	3.24	0.000
Nutritional Knowledge Index	2.38 ^{a}	2.54 ^{a}	2.04 ^{b}	2.37	0.004
NHC Knowledge Index	2.75 ^{a}	2.76 ^{a}	2.32 ^{b}	2.67	0.013
NC Knowledge Index	1.07	1.04	0.90	1.03	0.177
HC Knowledge Index	1.23 ^{a}	1.28 ^{a}	1.05 ^{b}	1.21	0.085

p-value are related to F test in one-way ANOVA. ** Based on the mean value of items used; *** Based on the mean value of familiarity and credibility of each nutrition (low in sodium, reduced kcal content and no sugar added) and health claims (related to plant sterols; omega-3 fatty acids and xylitol). Different subscripts indicate a significant difference at $p < 0.05$ using Tukey's HSD test.

With respect to socio-demographics, the *Non-users* cluster differs from the other two in the higher incidence of males with a lower level of education. Age does not differ significantly among groups, nor does the presence of children in the household, occupation, household economic status or area of residence. The incidence of individuals with personal and/or household members' health problems that influence food choices is significantly different among clusters. The *Claims user* cluster reports the highest incidence of individuals with personal and/or household members' health problems, which suggests that the presence of personal or family pathologies generates a greater need for information and greater attention to claims in this cluster.

Table 5. Cluster profiles based on socio-demographics.

		Potential Users	Claims Users	Non-Users	*p*-Value
Gender *	Male	47	46	61	0.032
	Female	53	54	39	
Mean age		44.76	46.20	45.45	0.606
Education Level *	Post-graduate specialization/PhD	5.1	8.3	1.0	0.044
	Master's degree	16.5	15.5	14.3	
	Bachelor's degree	14.0	7.1	7.1	
	High school diploma	55.1	58.9	63.3	
	Other	9.3	10.1	14.3	
Children <12	Yes	30	27	23	0.436
	No	70	73	77	
Special dietary needs	Yes	15	19	13	0.075
	No	85	81	87	
Personal and/or household members' health problems *	Yes	54	60	57	0.046
	No	46	40	43	

* p-value < 0.05 for chi-square test.

5. Discussion

This paper explores the interest and use of NHCs in Italy, providing results from a national representative sample. Overall our results showed that attention to, and use of, NHCs is not very widespread among Italian consumers. At the same time, although consumers consider themselves quite capable of understanding claims, when objective knowledge is detected, the level of understanding of the selected claims is quite low, with misinterpretation and confusion being generated about the real meaning of the claims for both nutrition and health.

The degree of attention and use is higher for NCs than for HCs, both on average and for the six specific claims used in our study. However, the degree of familiarity as well as credibility of NHCs varies according to the claim considered, confirming that consumers' responses to NHCs are strictly connected to the specific claim as reported elsewhere [15,29]. As for familiarity, among NCs consumers show the most familiarity with the "No sugar added" claim. This could be due to the fact that in recent years Italian consumers have shown particular sensitivity towards sugar-free products, which also represent a particularly growing market trend at this time [39,40]. Among HCs, consumers show that they have great familiarity as well as a higher level of confidence with the general function claim "Omega-3 fatty acids help to maintain a healthy cardiovascular system". This might be due to the fact that Omega 3 is a substance associated with various health benefits [30,44,45] and mostly coupled with a great variety of food products in everyday use (from dairy products to eggs). In addition, extensive marketing promotion and communication efforts have built a healthy reputation of the omega-3 concept in recent years and this might influence the credibility of health messages [46,47].

Our result on general function claims contrasts with previous evidence which showed that consumers perceive disease risk reduction claims as being more appealing than general function claims [12,13], but it is in line with other research findings that consumers usually prefer short claims with general mentions of health effects compared to claims with specific information relating to disease risk reduction or containing warnings [48,49].

As for knowledge, in contrast with our results on familiarity, consumers' objective knowledge of claims is higher for HCs than for NCs. In particular, although "No sugar added" is the most familiar claim, it is the most difficult to interpret. The same holds for the Omega 3 claim among HCs. With respect to no sugar added claims being difficult to interpret, similar results were found by Patterson et al. [50] showing that consumers could underestimate the actual nutritional value of products that carry sugar claims, while for Omega 3 the difficulty among consumers could be connected to the existence of various authorized claims for this substance present on the market that may create confusion.

What was found with reference to familiarity and knowledge leads to two different considerations. On the one hand, our results are in line with previous research findings, according to which consumers show a positive bias towards NHCs connected to potential halo effects of some claims (consumers tend to rate the product higher on attributes mentioned in the claim) and magic bullet effects that occur when consumers attribute inappropriate health benefits to the product [51].

These potential effects need to be taken into particular consideration by policy makers, in light of the strong influence that advertising campaigns may have on consumers' perception of the health properties of food products. In this regard, it is important to recall that some food companies have in the past conducted advertising campaigns using deceptive health claims, improperly leading individuals to attribute health properties to these products. Indeed, in Italy several companies have been sanctioned pursuant to Regulation 1924/2006 within the consumer protection activity carried out by the Italian Competition Authority [52,53].

In addition, even if consumers may be familiar with the nutrients mentioned in the claims, they may not understand the role that food products or nutrients play in their diets and overall health [6]. As a consequence, efforts should be made to simplify claims and make them more understandable to consumers, and above all to promote an understanding of the real beneficial effect on health. On the other hand, our results contrast with previous research suggesting that familiarity greatly affects consumer knowledge and understanding of NHCs [6].

The above evidence allows a critical issue to be highlighted: As consumers perceive products labelled with NHCs as healthier, their scant knowledge of the actual meaning of such claims could mislead their purchasing decisions [30]. This issue is also recognized by the EC Regulation on NHCs, which prescribes that claims should be used only if the average consumer can be expected to understand the beneficial effect. However, the level of consumer ability to understand claims may depend on several factors such as the use of scientific terms and the length of the claim [54]. Thus, in accordance with Hung et al. [9], it is important to increase consumer awareness and understanding of the actual meaning of NHCs available on the market, in order to protect the public from being misled while evaluating the healthiness of food. This suggests that policy makers as well as food marketers should also focus on facilitating understanding of NHCs, for example through adapting the wording or length of health claims.

Cluster analysis, in line with previous research [5,8,15,55], shows the existence of three consumer profiles with different levels of attention and use frequencies of NHCs. However, in contrast with Hung and Verbeke [8], our results show that the *non-users* segment represents the smallest cluster, while the largest comprises consumers quite interested in claims, which we called *potential users*.

It is worth pointing out that consumers who show a higher degree of familiarity and use of NHCs (the cluster called *claims users*) also pay more attention to nutritional panels on labels. This is an interesting result given that previous research found that the use of nutrition claims together with a detailed nutrition facts panel increases consumer utility beyond the increase provided by each label in isolation [35]. This suggests that nutritional panels and NHCs tend to reinforce each other and that the use and credibility of claims can be strengthened by educating consumers and promoting a more frequent use of nutritional panels.

Consistent with other studies our results show the key role of motivations in influencing NHC use and familiarity. Indeed, *Claims users* show on average the highest level of general health interest in food choices, while *Non-users* have the lowest. Thus, in accordance with Dean et al. [12] and Hung et al. [7], food products with NHCs are more appealing to consumers who are interested in healthy eating, who are also those that show the greatest need of health-related information. As a consequence, consumer attitudes to using NHCs could be encouraged by stimulating consumer interest in healthy eating.

Finally, with regard to the presence of personal and/or household members' health problems our results confirm previous research findings [14,20] that this condition could affect interest in and use of NHCs. Those in *Claims users* show the highest incidence of respondents who claim to be in this condition. However, also in the *Non-users* cluster a high incidence of individuals with personal and/or household members' health problems is revealed. Thus the incidence of personal or family experience with health issues on attention to, and use of, NHCs is controversial. In this regard, in accordance with Verbeke et al. [30], the link between diet and health is complex and consumers may react differently, also with regard to different NHCs. This complexity is confirmed by our results according to which, contrasting with other studies [8,12], the existence of special dietary needs does not differ significantly among clusters.

Another interesting difference that emerges among the clusters concerns the consumers' perceived ability to process NHCs. The *Non-users* cluster shows the lowest level of ability to process NHCs, which may influence their lower use of NHCs, since previous research shows that usage of claims is strictly connected to ability to process [7].

Furthermore, the degree of nutritional knowledge as well as the degree of objective knowledge of NHCs differs among clusters. This confirms that a higher level of nutritional knowledge may support the use of NHCs [15,35] and suggests that policy makers should focus on enhancing nutritional knowledge in the population by implementing public campaigns that allow consumers to process information contents. However, it has been shown elsewhere that the effects of nutrition knowledge on claims depend on the claim type involved [56].

Overall, in line with previous research our results show that consumers with lower education are less motivated towards NHCs [15,20,36], supporting the idea that better educated consumers are generally more likely to search for nutritional and health information [21] and that education is positively related to knowledge and understanding of nutritional information [57].

Finally, our results confirm that males are less motivated towards NHCs [15,20,36], but contrary to other research there are no significant differences in the identified clusters as regards age or the presence of children [9,15,24]. However, with reference to the presence of children we deliberately excluded claims related to children's development and health in order to prevent responses being affected by the different household composition of the sample.

This study has some limitations that constitute areas for further research. First of all, the influence of product-specific characteristics on NHC use, familiarity and credibility was not investigated. It would be worth ascertaining the influence of carrier products as well as their sensory profile on the use of NHCs. In addition, we considered only a limited number of specific claims, while it would be useful to consider a wider range of claims present on the national market. Moreover, even if we consider both nutrition and health claims, no interaction effects are investigated in the present analysis. Finally, a further limitation concerns the use of self-reported data that may be susceptible to social desirability bias. In this regard, future research should involve the carrying-out of experimental and observational studies.

6. Conclusions

The findings of this study suggest the existence of three segments of consumers in Italy, with different levels of attention and use frequency of NHCs, with specific profiles in terms of motivation, nutritional knowledge, and ability to process and understand NHCs. Hence, first of all, policy interventions as well as communication efforts targeting different consumer segments are required to support the use of NHCs and to avoid misleading interpretation and perception. According to our results it would be useful to adopt appropriate communication strategies to promote consumer awareness of the importance of healthy eating, such as public education campaigns, using both traditional and innovative tools. Specifically, new technologies, such as so-called smart labels or QR codes or other apps for mobiles, can make an important contribution, even if it has been suggested that such new labelling technologies could motivate consumers to access the information if combined with additional interventions [58]. Also, a nudge-based approach could represent a valid method to support consumers in making healthier food choices during their purchases, complementing other tools [9,59].

Furthermore, the results show a low level of objective knowledge of NHCs among Italian consumers. It should therefore be considered a priority for policy makers and marketers to inform consumers better about the actual of NHCs, in the context of a healthy diet, in order to improve their use and effectiveness, avoiding misleading interpretation and perception, such as overestimation of the benefit or attribution of inappropriate health benefits. Overall, but especially to reach the *Non-users* cluster, information should be provided simply and clearly, avoiding the use of complex scientific terminology. Finally, it should be emphasized that among the three clusters identified, that of *potential users* is numerically the largest. The communication and information policies outlined above could therefore have a significant impact on increasing understanding and use of NHCs.

Author Contributions: Conceptualization, A.A. and A.M.; Data curation, A.A.; Methodology, A.A.; Supervision, A.M.; Writing—original draft, A.A. and A.M.

Funding: This research was funded by the University of Naples Parthenope within the research project: "Analysis of the factors that determine the transition towards sustainable food consumption".

Conflicts of Interest: The authors declare no conflict of interest.

References

1. World Health Organization Europe. Better Food and Nutrition in Europe: A Progress Report Monitoring Policy Implementation in the WHO European Region. 2018. Available online: http://www.euro.who.int/en/health-topics/disease-prevention/nutrition/publications/2018/better-food-and-nutrition-in-europe-progress-report-2018 (accessed on 4 June 2019).
2. Brambila-Macias, J.; Shankar, B.; Capacci, S.; Mazzocchi, M.; Perez-Cueto, F.J.A.; Verbeke, W.; Traill, W.B. Policy interventions to promote healthy eating: A review of what works, what does not, and what is promising. *Food Nutr. Bull.* **2011**, *32*, 365–375. [CrossRef] [PubMed]
3. Verhagen, H.; van Loveren, H. Status of nutrition and health claims in Europe by mid 2015. *Trends Food Sci. Technol.* **2016**, *56*, 39–45. [CrossRef]
4. Santeramo, F.G.; Carlucci, D.; Devitiis, B.D.; Seccia, A.; Stasi, A.; Viscecchia, R.; Nardone, G. Emerging trends in European food, diets and food industry. *Food Res. Int.* **2018**, *104*, 39–47. [CrossRef] [PubMed]
5. Ballco, P.; de-Magistris, T.; Caputo, V. Consumer preferences for nutritional claims: An exploration of attention and choice based on an eye-tracking choice experiment. *Food Res. Int.* **2019**, *116*, 37–48. [CrossRef] [PubMed]
6. Lähteenmäki, L. Claiming health in food products. *Food Qual. Prefer.* **2013**, *27*, 196–201. [CrossRef]
7. Hung, Y.; Grunert, K.G.; Hoefkens, C.; Hieke, S.; Verbeke, W. Motivation outweighs ability in explaining European consumers' use of health claims. *Food Qual. Prefer.* **2017**, *58*, 34–44. [CrossRef]
8. Hung, Y.; Verbeke, W. Consumer evaluation, use and health relevance of health claims in the European Union. *Food Qual. Prefer.* **2019**, *74*, 88–99. [CrossRef]
9. Hung, Y.; Hieke, S.; Grunert, K.G.; Verbeke, W. Setting Policy Priorities for Front-of-Pack Health Claims and Symbols in the European Union: Expert Consensus Built by Using a Delphi Method. *Nutrients* **2019**, *11*, 403. [CrossRef] [PubMed]
10. Vassallo, M.; Saba, A.; Arvola, A.; Dean, M.; Messina, F.; Winkelmann, M.; Claupein, E.; Lähteenmäki, L.; Shepherd, R. Willingness to use functional breads. Applying the Health Belief Model across four European countries. *Appetite* **2009**, *52*, 452–460. [CrossRef] [PubMed]
11. Saba, A.; Vassallo, M.; Shepherd, R.; Lampila, P.; Arvola, A.; Dean, M.; Winkelmann, M.; Claupein, E.; Lähteenmäki, L. Country-wise differences in perception of health-related messages in cereal based food products. *Food Qual. Prefer.* **2010**, *21*, 385–393. [CrossRef]
12. Dean, M.; Lampila, P.; Shepherd, R.; Arvola, A.; Saba, A.; Vassallo, M.; Claupein, E.; Winkelmann, M.; Lähteenmäki, L. Perceived relevance and foods with health-related claims. *Food Qual. Prefer.* **2012**, *24*, 129–135. [CrossRef]
13. Contini, C.; Casini, L.; Stefan, V.; Romano, C.; Juhl, H.J.; Lähteenmäki, L.; Scozzafava, G.; Grunert, K.G. Some like it healthy: Can socio-demographic characteristics serve as predictors for a healthy food choice? *Food Qual. Prefer.* **2015**, *46*, 103–112. [CrossRef]

14. Cavaliere, A.; De Marchi, E.; Banterle, A. Does consumer health-orientation affect the use of nutrition facts panel and claims? An empirical analysis in Italy. *Food Qual. Prefer.* **2016**, *54*, 110–116. [CrossRef]
15. Cavaliere, A.; Ricci, E.C.; Banterle, A. Nutrition and health claims: Who is interested? An empirical analysis of consumer preferences in Italy. *Food Qual. Prefer.* **2015**, *41*, 44–51. [CrossRef]
16. Steinhauser, J.; Hamm, U. Consumer and product-specific characteristics influencing the effect of nutrition, health and risk reduction claims on preferences and purchase behavior—A systematic review. *Appetite* **2018**, *127*, 303–323. [CrossRef] [PubMed]
17. Andrews, J.C. Motivation, ability, and opportunity to process information: Conceptual and experimental manipulation issues. *NA Adv. Consum. Res.* **1988**, *15*, 219–225.
18. Moorman, C.; Matulich, E. A model of consumers' preventive health behaviours—The role of health motivation and health ability. *J. Consum. Res.* **1993**, *20*, 208–228. [CrossRef]
19. Visschers, V.H.M.; Hess, R.; Siegrist, M. Health motivation and product design determine consumers' visual attention to nutrition information on food products. *Public Health Nutr.* **2010**, *138*, 1099–1106. [CrossRef] [PubMed]
20. Verbeke, W. Consumer acceptance of functional foods: Sociodemographic, cognitive and attitudinal determinants. *Food Qual. Prefer.* **2005**, *16*, 45–57. [CrossRef]
21. Drichoutis, A.C.; Lazaridis, P.; Nayga, R.M., Jr.; Kapsokefalou, M.; Chryssochoidis, G. A theoretical and empirical investigation of nutritional labels use. *Eur. J. Health Econ.* **2008**, *9*, 293–304. [CrossRef] [PubMed]
22. Soederberg Miller, L.M.; Cassady, D.L. The effects of nutrition knowledge on food label use. A review of the literature. *Appetite* **2015**, *92*, 207–216. [CrossRef] [PubMed]
23. Grunert, K.G.; Wills, J.M.; Fernandez-Celemin, L. Nutrition knowledge, and use and understanding of nutrition information on food labels among consumers in the UK. *Appetite* **2010**, *55*, 177–189. [CrossRef] [PubMed]
24. Ares, G.; Giménez, A.; Gámbaro, A. Influence of nutritional knowledge on perceived healthiness and willingness to try functional foods. *Appetite* **2008**, *51*, 663–668. [CrossRef] [PubMed]
25. Lalor, F.; Madden, C.; McKenzie, K.; Wall, P.G. Health claims on foodstuffs: A focus group study of consumer attitudes. *J. Funct. Foods* **2011**, *3*, 56–59. [CrossRef]
26. Coleman, K.L.; Miah, E.M.; Morris, G.A.; Morris, C. Impact of health claims in prebiotic-enriched breads on purchase intent, emotional response and product liking. *Int. J. Food Sci. Nutr.* **2014**, *65*, 164–171. [CrossRef] [PubMed]
27. Strijbos, C.; Schluck, M.; Bisschop, J.; Bui, T.; de Jong, I.; van Leeuwen, M.; von Tottleben, M.; van Breda, S.G. Consumer awareness and credibility factors of health claims on innovative meat products in a cross-sectional population study in the Netherlands. *Food Qual. Prefer.* **2016**, *54*, 13–22. [CrossRef]
28. Williams, P. Consumer understanding and use of health claims for foods. *Nutr. Rev.* **2005**, *63*, 256–264. [CrossRef] [PubMed]
29. Van Trijp, H.C.; van der Lans, I.A. Consumer perceptions of nutrition and health claims. *Appetite* **2007**, *48*, 305–324. [CrossRef] [PubMed]
30. Verbeke, W.; Scholderer, J.; Lähteenmäki, L. Consumer appeal of nutrition and health claims in three existing product concepts. *Appetite* **2009**, *52*, 684–692. [CrossRef] [PubMed]
31. Grunert, K.G.; Scholderer, J.; Rogeaux, M. Determinants of consumer understanding of health claims. *Appetite* **2011**, *56*, 269–277. [CrossRef] [PubMed]
32. Verhagen, H.; Vos, E.; Francl, S.; Heinonen, M.; van Loveren, H. Status of nutrition and health claims in Europe. *Arch. Biochem. Biophys* **2010**, *501*, 6–15. [CrossRef] [PubMed]
33. Leathwood, P.D.; Richardson, D.P.; Strater, P.; Todd, P.M.; van Trijp, H.C.M. Consumer understanding of nutrition and health claims: Sources of evidence. *Br. J. Nutr.* **2007**, *98*, 474–484. [CrossRef] [PubMed]
34. Hodgkins, C.E.; Egan, B.; Peacock, M.; Klepacz, N.; Miklavec, K.; Pravst, I.; Pohar, J.; Gracia, A.; Groeppel-Klein, A.; Rayner, M.; et al. Understanding how consumers categorize health related claims on foods: A consumer-derived typology of health-related claims. *Nutrients* **2019**, *11*, 539. [CrossRef] [PubMed]
35. Barreiro-Hurlé, J.; Gracia, A.; De-Magistris, T. Does nutrition information on food products lead to healthier food choices? *Food Policy* **2010**, *35*, 221–229. [CrossRef]
36. Vecchio, R.; Van Loo, E.J.; Annunziata, A. Consumers' willingness to pay for conventional, organic and functional yogurt: Evidence from experimental auctions. *Int. J. Consum. Stud.* **2016**, *40*, 368–378. [CrossRef]
37. Roininen, K.; Lähteenmäki, L.; Tuorila, H. Quantification of consumer attitudes to health and hedonic characteristics of foods. *Appetite* **1999**, *33*, 71–88. [CrossRef] [PubMed]

38. Parmenter, K.; Wardle, J. Development of a general nutrition knowledge questionnaire for adults. *Eur. J. Clin. Nutr.* **1999**, *53*, 298–308. [CrossRef] [PubMed]
39. Nielsen. Osservatorio Immagino Nielsen GS1 Italy. Le Etichette dei Prodotti Raccontano i Consumi Degli Italiani. 2019. Available online: https://osservatorioimmagino.it/ (accessed on 5 September 2019).
40. Nielsen. Alimentazione e Salute: Italiani Sempre più Consapevoli e Attenti agli Ingredienti. 2016. Available online: https://www.nielsen.com/it/it/insights/report/2016/food-and-health-italian-ever-more-aware-and-look-to-the-ingredients (accessed on 10 October 2019).
41. Italian National Institute of Statistics. Official Data on Resident Population in the Italian Municipalities. Available online: http://demo.istat.it/pop2017/index.html (accessed on 15 May 2019).
42. Italian National Institute of Statistics. Annual Report. Livelli di Istruzione Della Popolazione e Ritorni Occupazionali: I Principali Indicatori. Available online: https://www.istat.it/it/files/2018/07/Indicatori-dellistruzione.pdf (accessed on 15 May 2019).
43. Italian National Institute of Statistics. Annuario Sullo Stato di Salute Della Popolazione nelle Regioni Italiane nel 2017. Available online: https://www.istat.it/it/files/2018/12/C04.pdf (accessed on 15 May 2019).
44. Sioen, I.; Bilau, M.; Verdonck, F.; Verbeke, W.; Willems, J.L.; De Henauw, S.; Van Camp, J. Probabilistic intake assessment of polybrominated diphenyl ethers and omega-3 fatty acids through fish consumption. *Mol. Nutr. Food Res.* **2008**, *52*, 250–257. [CrossRef] [PubMed]
45. Wills, J.M.; genannt Bonsmann, S.S.; Kolka, M.; Grunert, K.G. European consumers and health claims: Attitudes, understanding and purchasing behaviour. *Proc. Nutr. Soc.* **2012**, *71*, 229–236. [CrossRef] [PubMed]
46. Vecchio, R.; Lombardi, A.; Cembalo, L.; Caracciolo, F.; Cicia, G.; Masucci, F.; Di Francia, A. Consumers' willingness to pay and drivers of motivation to consume omega-3 enriched mozzarella cheese. *Br. Food J.* **2016**, *118*, 2404–2419. [CrossRef]
47. Annunziata, A.; Mariani, A.; Vecchio, R. Consumer understanding and use of health claims: The case of functional foods. *Recent Pat. Food Nutr. Agric.* **2014**, *6*, 113–126. [CrossRef] [PubMed]
48. Annunziata, A.; Vecchio, R. Consumer perception of functional foods: A conjoint analysis with probiotics. *Food Qual. Prefer.* **2013**, *28*, 348–355. [CrossRef]
49. Wansink, B.; Sonka, S.T.; Hasler, C.M. Front-label health claims: When less is more. *Food Policy* **2004**, *29*, 659–667. [CrossRef]
50. Patterson, N.J.; Sadler, M.J.; Cooper, J.M. Consumer understanding of sugars claims on food and drink products. *Nutr. Bull.* **2012**, *37*, 121–130. [CrossRef] [PubMed]
51. Nocella, G.; Kennedy, O. Food health claims—What consumers understand. *Food Policy* **2012**, *37*, 571–580. [CrossRef]
52. Passarini, P.; Cavicchi, A.; Santini, C.; Mazzantini, G. Deceptive advertising and unfair commercial practices in the agrifood sector: The role of the Italian competition authority. *Br. Food J.* **2017**, *119*, 1781–1800. [CrossRef]
53. Di Via, L.; Leone, P. Italian Antitrust Case Law on Unfair and Misleading Nutritional Claims. *Riv. Dirit. Aliment.* **2011**, *2*, 1–14.
54. Stancu, V.; Grunert, K.G.; Lähteenmäki, L. Consumer inferences from different versions of a beta-glucans health claim. *Food Qual. Prefer.* **2017**, *60*, 81–95. [CrossRef]
55. Jurado, F.; Gracia, A. Does the valuation of nutritional claims differ among consumers? Insights from Spain. *Nutrients* **2017**, *9*, 132. [CrossRef] [PubMed]
56. Petrovici, D.; Fearne, A.; Nayga, R.M., Jr.; Drolias, D. Nutritional knowledge, nutritional labels, and health claims on food: A study of supermarket shoppers in the South East of England. *Br. Food J.* **2012**, *114*, 768–783. [CrossRef]
57. Gracia, A.; Loureiro, M.; Nayga, R.M., Jr. Do consumers perceive benefits from the implementation of a EU mandatory nutritional labelling program? *Food policy* **2007**, *32*, 160–174. [CrossRef]
58. Li, T.; Messer, K.D. To Scan or Not to Scan: The Question of Consumer Behavior and QR Codes on Food Packages. *J. Agric. Resour. Econ.* **2019**, *44*, 311–327.
59. Vecchio, R.; Cavallo, C. Increasing healthy food choices through nudges: A systematic review. *Food Qual. Prefer.* **2019**, *78*, 103714. [CrossRef]

nutrients

MDPI

Article

Plant-Derived Extracts Feed-Addition and Packaging Type Influence Consumer Sensory Perception of Pork

Begoña Panea [1,2,*] and Guillermo Ripoll [1,2]

1 Centro de Investigación y Tecnología Agroalimentaria de Aragón (CITA), Avda. Montañana, 930, 50059 Zaragoza, Spain; gripoll@aragon.es

2 Instituto Agroalimentario de Aragón—IA2 (CITA-Universidad de Zaragoza), C/Miguel Servet, 177, 50013 Zaragoza, Spain

* Correspondence: bpanea@aragon.es

Received: 1 October 2019; Accepted: 31 October 2019; Published: 4 November 2019

Abstract: This paper investigates whether the combination of the addition of extracts derived from plants (plants derived extracts, PDE) to pork feedstuff and the meat conservation conditions (packaging and time exposure) affect consumers' perception of pork quality, studied by means of visual appraisal, purchase intention and a home test. The three PDE groups were control, garlic extract and blended oil composed by carvacrol, timol, cynamic aldehide and eugenol extracts. Meat was packed in film, vacuum or modified atmosphere (MAP) packaging. A visual test was designed comprising a four-day storage step followed by a four-day exposure step in a refrigerated island display case. All studied effects influenced visual appraisal scores, being time exposure and packaging effects more noticeable than PDE or pig-sex effects. Meat from MAP scored higher than the rest. Scores decreased as exposure time increased, but this evolution was less perceptible in vacuum packaging and was faster for meat from the garlic group. Only gender affected the visual appraisal scores, with women scoring higher than men. Neither PDE addition nor pig sex affected to purchase intention whereas both exposure time and packaging type did. A maximum of 2 days of exposure would be recommended. In the home-test, meat from male pigs obtained higher scores than meat from female pigs, and none of the consumer-related effects influenced the given scores.

Keywords: packaging; exposure time; plant extract; visual appraisal; consumer home-test

1. Introduction

To ensure the healthy, safe and high-quality food demanded by consumers, livestock must be in good sanitary condition. It has been demonstrated that some extracts derived from plants (PDEs) have antimicrobial properties [1,2] and therefore they have been tested to reinforce the control of zoonosis, such as salmonellosis, for example [3]. Nevertheless, animals' diet modifications can induce meat-quality modifications [4]. If a diet causes changes in the degree of saturation of intramuscular fat, meat could be prone to lipid oxidation during ageing, thus affecting the meat colour [5]. Colour has been reported as one of the most important sensory quality characteristics [6] because it affects freshness perception and consumer' purchase intention. To guarantee lipid stability, tocopherol and other synthetic antioxidants have been frequently used, but consumers reject these synthetic products; consequently, recent research has been focused on antioxidants from natural resources, including PDEs [7]. Different plants are used for almost all livestock species [8–13]. Moreover, since it is widely accepted that diet may affect meat sensory characteristic, there are several studies regarding the possibility of varying sensory profile by means of different plant materials, such as rosemary, oregano, ginger, garlic or chicory [14,15].

Sensory appraisal is a whole consumer experience, including both extrinsic and intrinsic quality cues [16]. Tenderness, juiciness or flavor are the most important intrinsic cues for consumers [17] and as meat is often retail packed in order to lengthen its shelf-life [18], packaging should preserve this intrinsic trait in the finest conditions until consumption [19]. When meat is packed with an oxygen-permeable film, it maintains its attractive red colour, but it is not protected against oxidation or microbial contamination. Vacuum packaging is frequently used in industry because vacuum packaging increases the shelf life of meat by reducing microbial growth [20], but the meat develops a dark-brown colour that is rejected by consumers. Finally, a modified atmosphere (MAP) maintains the meat's desirable red colour and prevents bacterial growth, but it promotes increased lipid oxidation during storage [21].

Thus, the aim of the paper was to investigate if the combination of the addition of PDE to pork feedstuff and the meat conservation conditions (packaging and time exposure) affect consumers' perception of pork quality, studied by means of visual appraisal, purchase intention and a home-test.

2. Materials and Methods

2.1. Animals and Handling

The procedures used in the trial followed the Spanish guidelines for experimental animal protection [22] and were approved by the Institutional Animal Care and Use Committee of the Research Centre (Procedure number 2011-03). A total of 150 Duroc × (Landrace × Large White) animals that were intended for Protected Geographic Indication (P.G.I.) "Jamón de Teruel" ham production were randomly allotted into three experimental groups: control, garlic and oil. The feed for all animals was a cereal mixture composed of corn, soya, wheat, barley and rapeseed that were given ad libitum. Pigs were housed in 80% slotted floor pens (3.50 m × 3.00 m) in a natural-environment barn and had free access to a pelleted diet and water throughout the trial. The diet was formulated to ensure the requirements of pigs of that age [23]. The compositions of the diets [24,25] are shown in Table 1.

Table 1. Composition of basic feedstuff that is used in the experiment.

	Initiation from 20 to 45 kg	Growth from 45–75 kg	Growth from 75 kg to Slaughter
		Ingredients composition	
Corn (%)	10.0	-	-
Sunflower undercoated (%)	-	3.7	-
Sunflower 28% (%)	-	-	2.0
Soya 44% (%)	15.9	7.6	1.8
Rapeseed (%)	4.0	10	12.0
Wheat (%)	10.0	15	16.0
Barley (%)	53.7	61	60.6
		Proximate composition	
Dry matter (%)	89.2	89.5	89.4
Crude protein (%)	16.0	15.0	14.0
Crude fibre (%)	4.6	5.0	5.5
Non digestible fibre (%)	16.8	18.3	19.7

Animals from the control group were fed only this diet, whereas the other two lots were fed with the diet added with each PDE from initiation to slaughter. The garlic group had 1 kg/Tm of a garlic (*Garlic sativum*) complex (Garlicon® Domca, S.A.U., Granada, Spain), resulting from a 25 g/Tm combination of propyl propane thiosulfonate /propyl propane thiosulfonate added to their feed. The oil group feed had 2 kg/Tm of a compound (Repaxol®, Molimen, Barcelona) formed by a mixture of carvacrol (from oregano), thymol (from thyme), cinnamic aldehyde (from cinnamon) and eugenol (from clove), which was microencapsulated by a lipidic matrix and added to the feed.

When the animals reached the weight described by P.G.I. "Jamón de Teruel", they were slaughtered at a slaughterhouse that was authorized by the European Union (EU) (Calamocha, Spain). The pigs were electrically stunned (225 to 380 V/0.5 A for 5 to 6 s), exsanguinated, scalded, skinned, eviscerated, and split down the midline according to standard commercial procedures. Carcasses were kept at 4 °C for 24 h. Then, 24 carcasses from each experimental group (12 barrows and 12 gilts) were randomly selected and subsequently, left loins from the 5th thoracic vertebra to the 6th lumbar vertebra were excised and transported to our laboratory. Once in the laboratory, the m. longissimus thoracis et lumborum were deboned and sliced to obtain the samples described below.

2.2. Sample Preparation and Packaging

(a) Visual Appraisal

Since the aim of the experiment was to determine the possibility of lengthening the meat shelf-life, visual test was designed comprising a four-days storage step in different packaging followed by a four-days exposure step in the same packaging.

Then, from each animal, three 2-cm steaks were used for visual appraisal. Steak 1 (FILM) was placed on a polystyrene tray overwrapped with a polyethylene low density (PE-LD) oxygen permeable film (Coimbra Pack, S.L., Spain) without contact with the meat surface and stored until 4th post-mortem day at 4 °C in darkness.

Steak 2 (VACUUM) was vacuum packed (MCOEX material, Coimbra Pack, S.L., Spain), stored until 4th post-mortem day at 4 °C in darkness and thereafter, it was extracted from the vacuum bag and placed on a polystyrene tray overwrapped with a PE-LD oxygen permeable film (Coimbra Pack, S.L., Spain) without contact with the meat surface.

Steak 3 (MAP) was packed under MAP (2:1 gas:meat ratio, commercial atmosphere of 70% O_2, 30% CO_2, Praxair España, Spain) with a cover film Cryovac 1825–50, (Cryovac Europe, Barcelona Spain, oxygen permeability of 14.8 $cm^3 \cdot m^{-2} \cdot 24\ h^{-1}$ at 1 atm and a water vapor permeability of 16 $g \cdot m^{-2} \cdot 24\ h^{-1}$) stored until 4th post-mortem day at 4 °C in darkness and, thereafter, it was extracted from the MAP and placed on a polystyrene tray overwrapped with a PE-LD oxygen permeable film (Coimbra Pack, S.L., Spain) without contact with the meat surface.

(b) Home Test

Two 2-cm steaks per animal were vacuum packed (MCOEX material, Coimbra Pack, S.L., Spain), at 1 day of ageing, frozen at −20 °C and destined for the consumers' home test.

2.3. Methodological Procedures

All the consumers participants in both experiments were recruited among students and workers, without relation with the current research, of Aula Dei Campus (Zaragoza, Spain). Personal data such as identification or electronic mail were not required and there was no financial compensation. Participants were clearly informed of the aim of the study and gave implicit consent for research use of the supplied information according to European regulations and the study was conducted in accordance with the Declaration of Helsinki.

(c) Visual Appraisal

Once placed on the polystyrene tray, all the samples were expose for 4 days in a Carrier Multinor 1540/80 refrigerated island display case (Carrier Refrigeración Ibérica SA, Madrid, Spain) with a display area of 1 m^2 (1.3m × 0.8m) at 0–2 °C, to simulate supermarket conditions. Samples were evaluated by the consumers each day of the storage time, and samples were available to the consumers from 08:00 to 16:00. During this time, samples were moved randomly three times to avoid possible presentation-order, first-order and carry-over effects. The lightning was provided by light-emitting diode (LED) bulbs with a luminous flux of 816 lumen, a color temperature of 4000 K, a color-rendering index >80 and a standard deviation color matching equal to MacAdam ellipses. The illuminance on the surface of the chops was approximately 1300 lx, ensuring the minimum level of illuminance in areas with high visual requirements.

Consumers was provided with a form on which they were asked about their gender and age. Regarding the steaks, the consumers were asked to evaluate the appearance of 5 samples using a continuous 10-point scale, from 1 (very bad) to 10 (very good). Additionally, they were asked about their purchase intention (yes/no).

(d) Home Test

Each consumer had a pack including 2 loin portions, one each from a different combination pig-sex*PDE. Test were performed following a complete balanced design. Samples of the packet were identified by a three-figure random number and they were accompanied by an evaluation questionnaire. Consumers were not given any information about the steaks other than the species and accurate guidelines for storage and cooking were specified on the questionnaire. Consumers evaluated each sample for preference for taste, juiciness and tenderness, using a continuous 10-point scale (from 1 = dislike very much, to 10 = like very much). Consumers were also asked their gender, age, general liking for meat (high, medium, low) and weekly meat consumption (1–2; 3–6; daily).

2.4. Statistics

Statistical analyses were performed using XLStat 17.03 software. Consumers were categorized into 4 categories of age.

Visual appraisal score was analyzed using two independent analysis of variance (ANOVA) procedures: the first with PDE, pig sex, exposure time and type of packaging as main effects; the second ANOVA with the consumer's gender and age as main effects. Means for visual appraisal scores were calculated. The Duncan test was used to compare means, and the level of significance was $p < 0.05$. Crosstabs for studied effects were carried out and the influence of the effects on purchase intention was investigated by the chi-square test.

For the consumers' home test, preference for juiciness, taste and tenderness scores were mean-centered. In this way, the effect of the use of the scale was eliminated. Then, we used two ANOVA procedures, the first with PDE and pig sex effects and the second with consumer's gender, consumer' age, liking for meat and meat-consumption frequency as main effects. The means for the corrected scores in function of significant effects were calculated. The Duncan test was used to compare means, and the level of significance was $p < 0.05$. Crosstabs for studied effects were carried out and differences for scores were investigated by the chi-square test.

3. Results

3.1. Visual Appraisal Scores and Purchase Intention

3.1.1. Effect of Extracts Derived from Plants (PDE), Pig Sex, Exposure Time and Packaging Type

A total of 179 people filled in the survey. All studied effects influenced significantly the visual appraisal scores (Table 2), being time exposure and packaging effects more noticeable than PDE or pig sex effects. Both time exposure and packaging type presented significant interactions between them and with PDE, whereas pig sex did not. None of the triple or quadruple interactions were significant. Then, we obviated sex effect and studied visual appraisal scores in function of PDE group, exposure time and packaging type, which were shown in Figure 1.

Global means for PDE groups were 6.42 for control group, 6.02 for garlic group and 5.73 for oil group. Global mean for males was 6.27 whereas global mean for females was 5.95.

In general, visual appraisal scores decreased as exposure time increased. Nevertheless, this evolution was less perceptible in VACUUM than in FILM or MAP. The effect of PDE disappeared from the 3rd exposure-day onward in FILM. At 1 day of exposure, the oil group obtained lower scores than the rest. The decrease for scores were faster for meat from garlic group than for meat from control group. Regarding VACUUM, meat without PDE shown the lowest scores at 1st day of exposure but from the second day of exposure onward no differences were found between the three different PDE groups. Finally, within MAP, the effect of PDE was noticeable, with oil group showing the lowest scores at any exposure time. In general, MAP scored higher than VACUUM or FILM.

Table 2. Values of p for studied effects (plant extract addition, pig sex, exposure time and packaging type) on consumer visual appraisal scores of pork meat.

Effect	p Value
PDE addition (D)	0.015
Pig sex (S)	0.001
Exposure time (T)	<0.000
Packaging type (P)	<0.000
D × S	0.590
D × T	0.005
D × P	0.011
S × T	0.293
S × P	0.265
T × P	<0.000

PDE—Plant derived extracts.

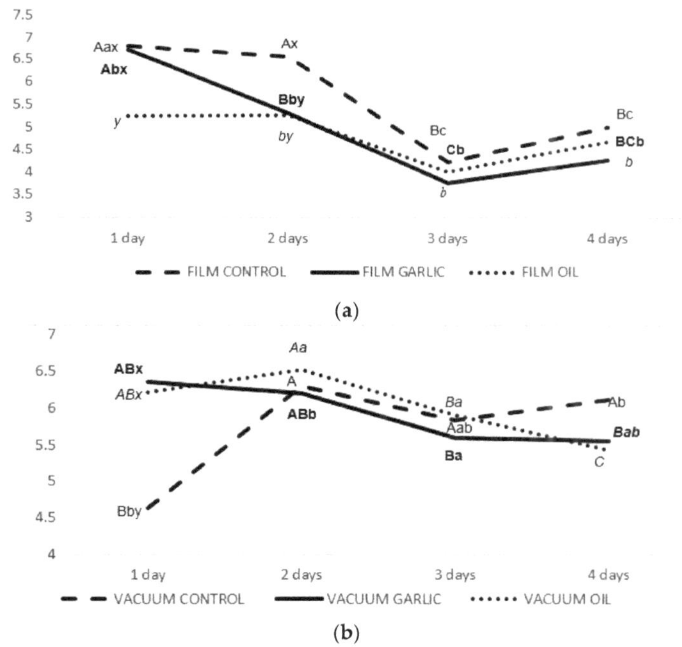

Figure 1. *Cont.*

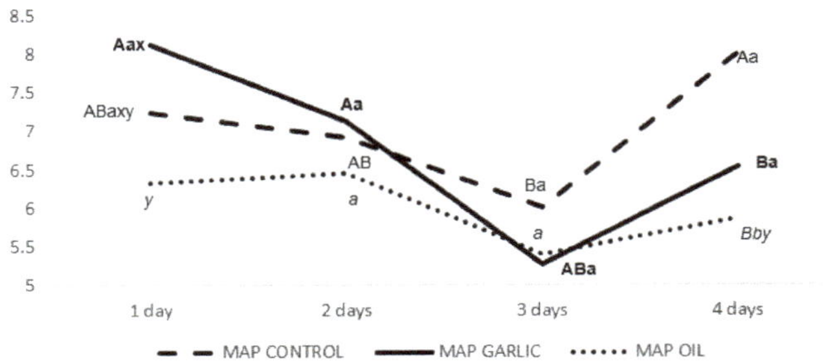

(c)

Figure 1. Visual appraisal scores of pork meat in function of PDE feed-addition, exposure time and packaging type. (**a**) Film packaging; (**b**) vacuum packaging; (**c**) MAP (modified atmosphere) packaging). a,b. Different letters means statistical differences ($p < 0.05$) between packaging types inside PDE group and exposure time. A,B. Different letters means statistical differences ($p < 0.05$) between exposure time inside PDE group and packaging type. x,y. Different letters mean statistical differences ($p < 0.05$) between PDE groups inside packaging type and exposure time. Bold letters-garlic group. *Italics*-oil group.

Regarding purchase intention, neither PDE addition ($p = 0.124$) nor pig sex ($p = 0.149$) affected purchase intention whereas both exposure time and packaging type did ($p < 0.000$). Figure 2 shows percentages of purchase intention in function of packaging type and exposure time. No differences between YES percentages and NO percentages were found for film whereas percentages of YES were higher than percentages of NO in vacuum packaging or in MAP packaging. Frequency of YES were higher than No frequencies at 1st exposure-day and at 2nd exposure-day whereas no differences between YES percentages and NO percentages were found at 3rd or at 4th exposure-day. Therefore, a maximum of two days of exposure is recommended.

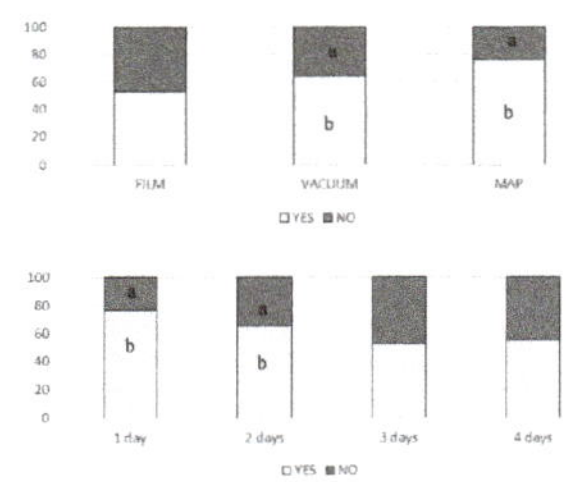

Figure 2. Percentages of purchase intention in function of packaging type and exposure time. a,b Different letters means significant p value ($p < 0.05$) for X^2 test inside a packaging type or inside a exposure time.

3.1.2. Effect of Consumer Profile

Females (Table 3) represented 58% of respondents and by ages, 7.3% were less than 25 years old, 30.5% were from 26 to 40 years old, and 48.4% were from 41 to 55 years old.

Table 3. Consumer profile for visual appraisal and purchase intention tests.

Consumers' Age Groups	Men	Women
≤25 years (%)	9.4	5.7
26–40 years (%)	26.6	33.3
41–55 years (%)	36.0	57.2
>55 years (%)	28.0	3.8

Only gender affected the visual appraisal scores ($p < 0.0001$) and no significant interactions were found between consumer's gender and consumer' age ($p = 0.071$). Table 4 shown the means and standard error for visual appraisal scores. In general, women gave higher scores (6.3 on average) than men (5.8 on average). Crosstabs shown that men scored 4 or 5 more frequently than expected whereas women scored 8 or 10 more frequently than expected ($X^2 < 0.000$). Men gave the highest scores to meat from the control group whereas women gave the lowest scores to meat from the oil group. All the respondents scored the meat packed in MAP packaging higher than meat packed in film or vacuum packaging, but women distinguished between film packaging and vacuum packaging whereas men did not. Similarly, women clearly penalized meat exposed for 3 or 4 days whereas men only discriminated clearly meat with 1 day of exposure.

Table 4. The means and standard error of studied effects (consumer' gender and age) on visual appraisal scores of pork meat.

	Treatment	Men	Women
PDE group	Control	6.1 a	6.6 a
	Garlic	5.6 b	6.3 a
	Oil	5.6 b	5.8 b
Packaging type	Film	5.5 b	5.8 c
	Vacuum	5.7 b	6.3 b
	MAP	6.5 a	7.0 a
Exposure time	1 day	6.5 a	6.9 a
	2 days	5.9 b	6.5 a
	3 days	5.1 c	5.6 b
	4 days	5.5 bc	5.9 b
Global mean		5.8	6.3
s.e.		0.09	0.08

a,b. Different letters in a column means significant p value ($p < 0.05$) inside PDE groups, packaging types or exposure times. MAP—modified atmosphere packaging.

Consumers' gender did not affect purchase intention ($p > 0.05$) whereas consumer' age showed a tendency ($p = 0.052$). Independently of the consumer age group (Table 5), percentages of YES (I'd buy it) were higher than percentages of NO. Crosstabs between consumers' ages and PDE, pig sex, packaging type or exposure time shown than people aged ≤25 preferred meat from 1 day-exposure time more frequently than expected ($p = 0.042$). Likewise, people aged 26–40 chosen film packaging less frequently than expected ($p < 0.000$) and preferred meat from 1 day-exposure time more frequently than expected ($p = 0.025$). People aged 41–55 also preferred meat from 1 day-exposure time ($p = 0.018$) and they chose meat from the control group instead of meat from the oil group ($p = 0.004$). Finally, people aged >55 chose film packaging less frequently than expected ($p = 0.008$) and preferred meat from 1 day-exposure time ($p = 0.007$) but they depreciated meat from garlic group and valorised meat from oil group ($p = 0.016$). A resume of these preferences is shown in Table 6.

Table 5. The percentages for purchase intention of pork from in function of consumer' age.

Consumer' Age	Purchase Intention	
	Yes (%)	No (%)
≤25	70.8 a	29.2 b
26–40	61.9 a	38.1 b
41–55	61.2 a	38.8 b
>55	73.2 a	26.8 b

a,b. Different letters in a row means significant differences between percentages ($p < 0.05$).

Table 6. Resume of consumer preferences derived from crosstabs between consumer' age and PDE, pig-sex, packaging type or exposure time.

Consumers' Age	Frequency < Than Expected	Frequency > Than Expected
≤25	Film packaging	Garlic group 1 day-exposure meat
26–40	Film packaging 3 days-exposure meat	1 day-exposure meat
41–55	Oil group	Control group 1 day-exposure meat
>55	Film packaging 3 days-exposure meat Garlic group	Oil group 1 day-exposure meat

3.2. Consumers' Home Test

Effect of PDE and Pig Sex

A total of 72 people filled the enquiry (Table 7). Sample was equally spread by gender and age. Most people like meat (73.6%) being higher the percentage for men (89.7%) than for women (57.6%). Independent of the gender, higher meat-frequency consumption is 3–6 times a week

Table 7. Consumers' profile for the home test. Percentages of valid responses.

Survey Question	Consumers' Age Groups	Men	Women
	≤25	6.9	6.1
Age	26–40	55.2	54.2
	41–55	24.1	24.2
	>55	13.8	15.2
Do you like meat?	Like very much	89.7 a	57.6 b
	Neither like nor dislike	10.3 b	42.4 a
	Not very much	0.0	0.0
How often a week do you eat meat?	1 or 2	17.2	27.3
	From 3 to 6	75.9	63.6
	Daily	6.9	9.1

a, b letters means statistical differences between genders in a X^2 test.

Table 8 shows the p values for PDE addition and pig-sex effects in a pork-meat consumer's home-test. PDE addition affected both taste (p = 0.012) and juiciness (p = 0.047) whereas sex only affected taste (p = 0.012). The interaction between effects was significant for both juiciness (p = 0.001) and tenderness (p = 0.002).

Table 8. The *p* values for PDE addition and pig-sex effects in a pork-meat consumers home test.

Effect	Taste	Juiciness	Tenderness
PDE (D)	0.012	0.047	0.107
Sex (S)	0.012	0.930	0.928
D × S	0.103	0.001	0.002

Figure 3 shows centered means for the taste variable as well as the *p* values for PDE addition and pig-sex effects. In this picture, when mean-centered value was positive, scores for the item were higher than the global mean. Meat from male pigs obtained higher scores than meat from female pigs. Regarding the PDE addition effect, garlic group obtained higher scores than the other two groups, with no statistical differences between them.

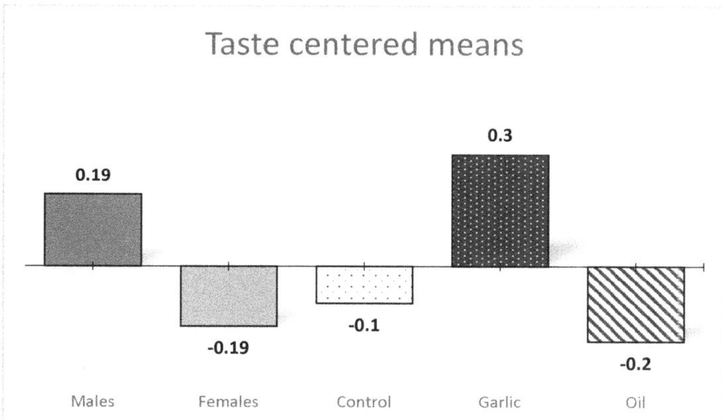

(a)

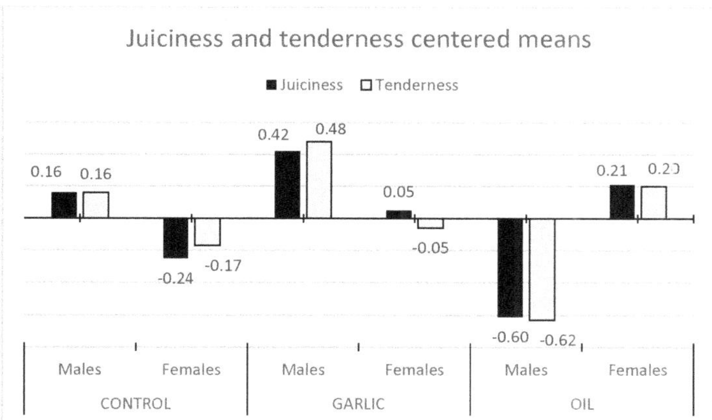

(b)

Figure 3. The centered means for sensory variables in a pork-meat consumers home test in function of pig sex and PDE addition. (**a**) centered mens for taste in function of pig-sex or PDE addition. (**b**) centered means for juiciness and tenderness in function of pig sex and PDE addition.

3.3. Effect of Consumer Profile

None of the consumer-related effects (gender, age, linking for meat or frequency of meat consumption) influenced the given scores ($p > 0.05$).

4. Discussion

4.1. Visual Scores and Purchase Intention

There are many studies about influence of the PDE on meat quality and most of them have focused on TBARs (thiobarbituric acid reactive substances), colour or flavour [15,26–29] but from the knowledge of the authors, almost none of them regard the effect on visual appraisal or on purchase intention. Ripoll et al. in a study with lambs reared in 4 different feed regimes found there was a significant relationship between feeding system and visual appraisal scores of M. rectus abdominis because of differences in colour [30]. This conclusion can explain the effect found for PDE addition on visual scores, since as reported in a previous paper with the same animal material used in the current experiment [31] differences were found in meat colour between animal groups. Colour was defined such as the most important factor influencing consumer perception, and more important than other factors as marbling or sample shape [32], because colour is used by consumer as an indicator of spoilage [33].

Borgogno et al. found, in packed goat meat, that the liking scores decreased significantly from the first day to the third day [34]. In a study with several bones, Grobbel et al. found an interaction bone–packaging–display time for visual appraisal since colour changes and they concluded that the darkening was more extensive for sample packaged in PVC (polyvinylchloride) and high-oxygen MAP than for those in ultra-low oxygen MAP [35]. In the current experiment, and independently on the packaging and PDE addition effect, colour hue values decreased from the 1st to the 4th day (61.5 and 58.5, respectively), which could to explain the effect of exposure time on visual scores. In current experiment, meat packed in MAP had in general higher visual scores than meat packed in vacuum or overwrapped and hue values for MAP were higher in MAP than in film or vacuum packaging at both 1st (63. 7 in MAP, 57.8 in film and 63.1 in vacuum) and 4th days of exposure (61.9 in MAP, 55.5 in film and 58.2 in vacuum). This result disagreed with those reported by other authors [36] who described that meat packed in PVC is preferred to meat packed in other packaging. Nevertheless, Carpenter et al. found a positive correlation between visual scores and purchase intention [36], which is in agreement with current results, since the higher percentages for purchase likelihood were found in MAP (76.2%; Table 3).

In the current experiment, a consumer's gender influenced visual appraisal scores, according with Jiang et al. who concluded that gender affects the emotion relevance to food in specific way [37]. Femininity and masculinity have been associated with the type of food that individuals eat [38]. It has been reported that females show higher preference for white meat because females are often suspicious about red meat and express less meat-eating satisfaction and this finding is much more intense in young females, whose expectation about red meat is linked to negative attitudes [32,39,40]. Consumer age is often considered in consumers studies and there are many studies focused on how socio-demographic variables influenced the likelihood of eating meat [41,42] the influence of extrinsic cues, such as health value or risk perception [43] or on consumer tasting perception of cooked meat [33,41,44] but scarce literature is available regarding age differences on visual appraisal and, when it exists, the clustering of consumers is often made based on marbling or colour [45–47]. For example, Brewer et al., in a visual appraisal experiment with pork of different marbling degree concluded that highly marbled chops appeared lighter coloured, had less acceptable appearance, and were less likely to be purchased than leaner chops [48]. Current results show that a consumer's age modified the visual perception, that is, the quality expectation. Most of the consumers expressed that film was the least preferable packaging and also, they concurred that time-exposure depreciate visual appearance of the meat, but age-related differences were found for the PDE group who chose this "as the most preferably" from a visual point of view. Lahucky et al. reported that addition of oregano prevents to oxidation and then, protects to meat color, which can to explain why oil group was chosen by older consumers [49].

4.2. Consumer Home Test

Some authors have reported that the eating quality of the cooked pork was unaffected by the plant extracts in diet [15,49] whereas other authors reported a slight effect of plant addition on sensory attributes [28], being juiciness higher in control pork than in supplemented meat and without differences in tenderness, which agree with present results [26]. As in the present study, meat from animals used by Rossi et al. [26] had no differences on chemical parameters which in turn make difficult the explanation of the results. In the current experiment, no differences in intramuscular fat content were found between groups and hence, no great differences in sensory characteristics were expected a priori but in fact we have a significant interaction between PDE group and pig sex ($p = 0.001$). Although it is generally accepted that an increased level of intramuscular fat has a positive influence on the sensory quality of pig meat, literature reveals contradictory results. For example, Flores et al. found a close relationship between intramuscular fat content and juiciness although this was lower with hardness [50] but Fernandez et al. found that in the case of Duroc crosses, such as was the case of the animals of the current experiment, sensory characteristics were only slightly affected by intramuscular fat level [51]. Several studies have proposed that a minimum of 2% of intramuscular fat is necessary for sensory acceptance of pork [52] but as summarized in Font i Furniols et al., there was not a consensus in literature about this percentage. The effect of intramuscular fat on consumer preferences has been associated with skatole concentration of fat tissue because skatole is responsible for boar taint [45]. In this sense, Leong et al. reported that there is an increase in skatole and indole concentrations in the fat tissue when garlic essential oil was included the diet [53] but the present result has shown that the garlic group obtained higher taste scores than control or oil groups.

Concerning pig sex, D'Souza et al. found that meat from barrows was more tender than meat from females, whereas the sex of the pigs did not influence either juiciness or overall acceptability of the pork [54]. Also, Elsbernd et al. reported no pig-sex effect on juiciness, tenderness, or chewiness of pork [55]. Nevertheless, our results have shown that meat presented a meat with a better taste and, in general, was more tender and more juicy. Font et al. shown that literature is not consistent in the effect of boar taint on meat acceptability and some studies show no differences in the acceptability of meat from entire male and female or castrates whereas others do [56].

Regarding consumers' profile, Ngapo et al. in a consumer study carried out in several countries, found that most Spanish consumers like pork (90% of respondents), the taste is the most important reason to eat it (81% of respondents), consumer eat pork once a week (67%) and they think that the pork they buy is always or almost always of good quality (90%) [40]. These results agree with those reported by other studies on Spanish consumers [56,57].

Some authors reported that the acceptability of the pig meat depends on the gender and age of consumers, in disagreement with our results [56]. Mathews et al. reported that women were more critical than men and that, in general, the oldest group of consumers had the lowest percentage of dislike scores, both for both flavor and odor [58]. In the same way, it has been reported that consumers over 30 and especially the middle-aged group (31–50) place more emphasis on tenderness than younger consumers [44].

Also, some authors have argued that consumers preferences depend on several factors, such as tradition, religion, age, education, gender, income, etc. [59,60] but other authors [16,61] have concluded that socio-demographic variables as well as some life-style related variables, such as frequency of consumption, were less significant than intrinsic cues in the consumer global appraisal.

5. Conclusions

All studied effects influenced visual appraisal scores, with time exposure and packaging effects being more noticeable than PDE or pig-sex effects. Scores decreased as exposure time increased, and the decrease was faster for meat from the garlic group than for meat from the control group. Meat from MAP scored higher than meat from vacuum packaging or from film packaging and a maximum of 2 days of exposure would be recommended. Neither PDE addition nor pig sex affected purchase intention whereas both exposure time and packaging type did. In the home test, meat from male pigs obtained higher scores than meat from female pigs and the garlic group obtained higher scores than the rest.

None of the consumer-related effects (gender, age, linking for meat or frequency of meat consumption) influenced the given scores in the home test and only gender affected visual appraisal scores.

Therefore, consumer profile was not an important factor but time exposure and packaging type definitely affect consumer appraisal.

Author Contributions: Conceptualization, B.P.; methodology, B.P. and G.R.; formal analysis, B.P. and G.R.; investigation, B.P. and G.R.; resources, B.P. and G.R.; data curation, B.P. and G.R.; writing—original draft preparation, B.P.; writing—review and editing, B.P. and G.R.; funding acquisition, B.P.

Funding: This research was funded by Teruel Investment Funds (FIT 2016-7).

Acknowledgments: Raúl Mainar, Carolina Albertí, Sara Andrés and ALAT of CITA. The authors are members of the MARCARNE network, funded by CYTED (116RT0503).

Conflicts of Interest: The authors declare no conflict of interest.

References

1. Govaris, A.; Solomakos, N.; Pexara, A.; Chatzopoulou, P. The antimicrobial effect of oregano essential oil, nisin and their combination against Salmonella Enteritidis in minced sheep meat during refrigerated storage. *Int. J. Food Microbiol.* **2010**, *137*, 175–180. [CrossRef]
2. De Oliveira, T.L.C.; de Araújo Soares, R.; Piccoli, R.H. A Weibull model to describe antimicrobial kinetics of oregano and lemongrass essential oils against Salmonella Enteritidis in ground beef during refrigerated storage. *Meat Sci.* **2013**, *93*, 645–651. [CrossRef] [PubMed]
3. Vico, J.P.; Jaime, R.C.M. Prevalencia de"salmonella" spp en cerdos de cebo en Aragón. In Proceedings of the XXXIX Jornadas de Estudio: XIII Jornadas sobre Producción Animal, Zaragoza, Spain, 12–13 May 2009; pp. 182–184.
4. Andersen, H.J.; Oksbjerg, N.; Young, J.F.; Therkildsen, M. Feeding and meat quality—A future approach. *Meat Sci.* **2005**, *70*, 543–554. [CrossRef] [PubMed]
5. Geay, Y.; Bauchart, D.; Hocquette, J.-F.; Culioli, J. Valeur diététique et qualités sensorielles des viandes de ruminants. Incidence de l'alimentation des animaux. *Prod. Anim.* **2002**, *1*, 37–52.
6. Barbut, S. Effect of illumination source on the appearance of fresh meat cuts. *Meat Sci.* **2001**, *59*, 187–191. [CrossRef]
7. Shah, M.A.; Bosco, S.J.D.; Mir, S.A. Plant extracts as natural antioxidants in meat and meat products. *Meat Sci.* **2014**, *98*, 21–33. [CrossRef]
8. Botsoglou, N.; Christaki, E.; Fletouris, D.; Florou-Paneri, P.; Spais, A. The effect of dietary oregano essential oil on lipid oxidation in raw and cooked chicken during refrigerated storage. *Meat Sci.* **2002**, *62*, 259–265. [CrossRef]
9. Luna, A.; Labaque, M.; Zygadlo, J.; Marin, R. Effects of thymol and carvacrol feed supplementation on lipid oxidation in broiler meat. *Poult. Sci.* **2010**, *89*, 366–370. [CrossRef]
10. Simitzis, P.; Deligeorgis, S.; Bizelis, J.; Dardamani, A.; Theodosiou, I.; Fegeros, K. Effect of dietary oregano oil supplementation on lamb meat characteristics. *Meat Sci.* **2008**, *79*, 217–223. [CrossRef]
11. Simitzis, P.; Symeon, G.; Charismiadou, M.; Bizelis, J.; Deligeorgis, S. The effects of dietary oregano oil supplementation on pig meat characteristics. *Meat Sci.* **2010**, *84*, 670–676. [CrossRef]
12. Han, J.; Rhee, K. Antioxidant properties of selected Oriental non-culinary/nutraceutical herb extracts as evaluated in raw and cooked meat. *Meat Sci.* **2005**, *70*, 25–33. [CrossRef] [PubMed]
13. Ortuño, J.; Serrano, R.; Jordán, M.J.; Bañón, S. Shelf life of meat from lambs given essential oil-free rosemary extract containing carnosic acid plus carnosol at 200 or 400 mg kg^{-1}. *Meat Sci.* **2014**, *96*, 1452–1459. [CrossRef] [PubMed]

14. Hansen, L.L.; Agerhem, H.; Rosenvold, K.; Jensen, M.T. Effect of Brussels sprouts and inulin/rape seed cake on the sensory profile of pork *M. longissimus dorsi*. *Meat Sci.* **2002**, *61*, 441–448. [CrossRef]
15. Janz, J.A.; Morel, P.C.; Wilkinson, B.H.; Purchas, R.W. Preliminary investigation of the effects of low-level dietary inclusion of fragrant essential oils and oleoresins on pig performance and pork quality. *Meat Sci.* **2007**, *75*, 350–355. [CrossRef] [PubMed]
16. Bernués, A.; Ripoll, G.; Panea, B. Consumer segmentation based on convenience orientation and attitudes towards quality attributes of lamb meat. *Food Qual. Prefer.* **2012**, *26*, 211–220. [CrossRef]
17. Acebrón, L.B.; Dopico, D.C. The importance of intrinsic and extrinsic cues to expected and experienced quality: An empirical application for beef. *Food Qual. Prefer.* **2000**, *11*, 229–238. [CrossRef]
18. Belcher, J.N. Industrial packaging developments for the global meat market. *Meat Sci.* **2006**, *74*, 143–148. [CrossRef]
19. Holman, B.W.B.; Kerry, J.P.; Hopkins, D.L. Meat packaging solutions to current industry challenges: A review. *Meat Sci.* **2018**, *144*, 159–168. [CrossRef]
20. Gill, C. Extending the storage life of raw chilled meats. *Meat Sci.* **1996**, *43*, 99–109. [CrossRef]
21. Renerre, M. La couleur de la viande et sa mesure. *Bull. Tech.* **1982**, *47*, 47–54.
22. Ministerio de la Presidencia. *Real Decreto 53/2013, de 1 de Febrero, por el que se Establecen las Normas Básicas Aplicables para la Protección de los Animales Utilizados en Experimentación y Otros Fines Científicos, Incluyendo la Docencia*; Boletín Oficial del Estado: Madrid, Spain, 2013; Volume 34, pp. 11370–11421.
23. National Academy of Sciences. Subcommittee on Swine Nutrition, CoAN, Board on Agriculture, National Research Council. In *Nutrient Requirements of Swine. Tenth Revised Edition*; National Academy Press: Washington, DC, USA, 1998.
24. De Blas, C.; Mateos, G.G.; García-Rebollar, P.G. *Tablas Fedna de Composición y Valor Nutritivo de Alimentos para la Fabricación de Piensos Compuestos*, 3rd ed.; Fundación Española para el Desarrollo de la Nutrición Animal: Madrid, Spain, 2010; p. 502.
25. The Association of Official Analytical Chemists. *Official Methods of Analysis*, 17th ed.; The Association of Official Analytical Chemists: Gaithersburg, MD, USA, 2000.
26. Rossi, R.; Pastorelli, G.; Cannata, S.; Tavaniello, S.; Maiorano, G.; Corino, C. Effect of long term dietary supplementation with plant extract on carcass characteristics meat quality and oxidative stability in pork. *Meat Sci.* **2013**, *95*, 542–548. [CrossRef] [PubMed]
27. O'Grady, M.N.; Carpenter, R.; Lynch, P.B.; O'Brien, N.M.; Kerry, J.P. Addition of grape seed extract and bearberry to porcine diets: Influence on quality attributes of raw and cooked pork. *Meat Sci.* **2008**, *78*, 438–446. [CrossRef] [PubMed]
28. Ranucci, D.; Beghelli, D.; Trabalza-Marinucci, M.; Branciari, R.; Forte, C.; Olivieri, O.; Badillo Pazmay, G.V.; Cavallucci, C.; Acuti, G. Dietary effects of a mix derived from oregano (*Origanum vulgare* L.) essential oil and sweet chestnut (*Castanea sativa* Mill.) wood extract on pig performance, oxidative status and pork quality traits. *Meat Sci.* **2015**, *100*, 319–326. [CrossRef] [PubMed]
29. Zhang, J.; Wang, Y.; Pan, D.D.; Cao, J.X.; Shao, X.F.; Chen, Y.J.; Sun, Y.Y.; Ou, C.R. Effect of black pepper essential oil on the quality of fresh pork during storage. *Meat Sci.* **2016**, *117*, 130–136. [CrossRef] [PubMed]
30. Ripoll, G.; Alberti, P.; Joy, M. Influence of alfalfa grazing-based feeding systems on carcass fat colour and meat quality of light lambs. *Meat Sci.* **2012**, *90*, 457–464. [CrossRef]
31. Panea, B.; Ripoll, G. Influence of feed-added plant derived extracts on the shelf-life in pork stored in three packaging types. 2019; Manuscript under review.
32. Ngapo, T.M.; Martin, J.F.; Dransfield, E. International preferences for pork appearance: I. Consumer choices. *Food Qual. Prefer.* **2007**, *18*, 26–36. [CrossRef]
33. Font-i-Furnols, M.; Guerrero, L. Consumer preference, behavior and perception about meat and meat products: An overview. *Meat Sci.* **2014**, *98*, 361–371. [CrossRef]
34. Borgogno, M.; Corazzin, M.; Sacca, E.; Bovolenta, S.; Piasentier, E. Influence of familiarity with goat meat on liking and preference for capretto and chevon. *Meat Sci.* **2015**, *106*, 69–77. [CrossRef]
35. Grobbel, J.P.; Dikeman, M.E.; Yancey, E.J.; Smith, J.S.; Kropf, D.H.; Milliken, G.A. Effects of ascorbic acid, rosemary, and Origanox in preventing bone marrow discoloration in beef lumbar vertebrae in aerobic and anaerobic packaging systems. *Meat Sci.* **2006**, *72*, 47–56. [CrossRef]
36. Carpenter, C.E.; Cornforth, D.P.; Whittler, D. Consumer preferences for beef color and packaging did not affect eating satisfaction. *Meat Sci.* **2001**, *57*, 359–363. [CrossRef]

37. Jiang, Y.; King, J.M.; Prinyawiwatkul, W. A review of measurement and relationships between food, eating behaviour and emotion. *Trends Food Sci. Technol.* **2014**, *36*, 15–28. [CrossRef]
38. Vartanian, L.R.; Herman, C.P.; Polivy, J. Consumption stereotypes and impression management: How you are what you eat. *Appetite* **2007**, *48*, 265–277. [CrossRef] [PubMed]
39. Kubberod, E.; Ueland, O.; Rodbotten, M.; Westad, F.; Risvik, E. Gender specific preferences and attitudes towards meat. *Food Qual. Prefer.* **2002**, *13*, 285–294. [CrossRef]
40. Ngapo, T.M.; Martin, J.F.; Dransfield, E. International preferences for pork appearance: II. Factors influencing consumer choice. *Food Qual. Prefer.* **2007**, *18*, 139–151. [CrossRef]
41. Verbeke, W.; De Smet, S.; Vackier, I.; Van Oeckel, M.J.; Warnants, N.; Van Kenhove, P. Role of intrinsic search cues in the formation of consumer preferences and choice for pork chops. *Meat Sci.* **2005**, *69*, 343–354. [CrossRef]
42. Ngapo, T.M.; Dransfield, E.; Martin, J.F.; Magnusson, M.; Bredahl, L.; Nute, G.R. Consumer perceptions: Pork and pig production. Insights from France, England, Sweden and Denmark. *Meat Sci.* **2003**, *66*, 125–134. [CrossRef]
43. Verbeke, W.; Frewer, L.J.; Scholderer, J.; De Brabander, H.F. Why consumers behave as they do with respect to food safety and risk information. *Anal. Chim. Acta* **2007**, *586*, 2–7. [CrossRef]
44. Aaslyng, M.D.; Oksama, M.; Olsen, E.V.; Bejerholm, C.; Baltzer, M.; Andersen, G.; Bredie, W.L.; Byrne, D.V.; Gabrielsen, G. The impact of sensory quality of pork on consumer preference. *Meat Sci.* **2007**, *76*, 61–73. [CrossRef]
45. Font-i-Furnols, M.; Tous, N.; Esteve-Garcia, E.; Gispert, M. Do all the consumers accept marbling in the same way? The relationship between eating and visual acceptability of pork with different intramuscular fat content. *Meat Sci.* **2012**, *91*, 448–453. [CrossRef]
46. Fortomaris, P.; Arsenos, G.; Georgiadis, M.; Banos, G.; Stamataris, C.; Zygoyiannis, D. Effect of meat appearance on consumer preferences for pork chops in Greece and Cyprus. *Meat Sci.* **2006**, *72*, 688–696. [CrossRef]
47. Dransfield, E.; Ngapo, T.M.; Nielsen, N.A.; Bredahl, L.; Sjoden, P.O.; Magnusson, M.; Campo, M.M.; Nute, G.R. Consumer choice and suggested price for pork as influenced by its appearance, taste and information concerning country of origin and organic pig production. *Meat Sci.* **2005**, *69*, 61–70. [CrossRef] [PubMed]
48. Brewer, M.; Zhu, L.; McKeith, F. Marbling effects on quality characteristics of pork loin chops: Consumer purchase intent, visual and sensory characteristics. *Meat Sci.* **2001**, *59*, 153–163. [CrossRef]
49. Lahucky, R.; Nuernberg, K.; Kovac, L.; Bucko, O.; Nuernberg, G. Assessment of the antioxidant potential of selected plant extracts—In vitro and in vivo experiments on pork. *Meat Sci.* **2010**, *85*, 779–784. [CrossRef] [PubMed]
50. Flores, M.; Armero, E.; Aristoy, M.C.; Toldra, F. Sensory characteristics of cooked pork loin as affected by nucleotide content and post-mortem meat quality. *Meat Sci.* **1999**, *51*, 53–59. [CrossRef]
51. Fernandez, X.; Monin, G.; Talmant, A.; Mourot, J.; Lebret, B. Influence of intramuscular fat content on the quality of pig meat—1. Composition of the lipid fraction and sensory characteristics of *M. longissimus lumborum*. *Meat Sci.* **1999**, *53*, 59–65. [CrossRef]
52. Bidner, B.S.; Ellis, M.; Witte, D.P.; Carr, S.N.; McKeith, F.K. Influence of dietary lysine level, pre-slaughter fasting, and rendement napole genotype on fresh pork quality. *Meat Sci.* **2004**, *68*, 53–60. [CrossRef]
53. Leong, J.; Morel, P.C.; Purchas, R.W.; Wilkinson, B.H. Effects of dietary components including garlic on concentrations of skatole and indole in subcutaneous fat of female pigs. *Meat Sci.* **2011**, *88*, 45–50. [CrossRef]
54. D'Souza, D.N.; Mullan, B.P. The effect of genotype, sex and management strategy on the eating quality of pork. *Meat Sci.* **2002**, *60*, 95–101. [CrossRef]
55. Elsbernd, A.J.; Patience, J.F.; Prusa, K.J. A comparison of the quality of fresh and frozen pork from immunologically castrated males versus gilts, physical castrates, and entire males. *Meat Sci.* **2016**, *111*, 110–115. [CrossRef]
56. Font, I.F.M.; Gispert, M.; Guerrero, L.; Velarde, A.; Tibau, J.; Soler, J.; Hortos, M.; Garcia-Regueiro, J.A.; Perez, J.; Suarez, P.; et al. Consumers' sensory acceptability of pork from immunocastrated male pigs. *Meat Sci.* **2008**, *80*, 1013–1018. [CrossRef]
57. Font-i-Furnols, M. Consumer studies on sensory acceptability of boar taint: A review. *Meat Sci.* **2012**, *92*, 319–329. [CrossRef] [PubMed]

58. Matthews, K.; Homer, D.; Punter, P.; Béague, M.-P.; Gispert, M.; Kempster, A.; Agerhem, H.; Claudi-Magnussen, C.; Fischer, K.; Siret, F. An international study on the importance of androstenone and skatole for boar taint: III. Consumer survey in seven European countries. *Meat Sci.* **2000**, *54*, 271–283. [CrossRef]
59. Verbeke, W.; Perez-Cueto, F.J.; Barcellos, M.D.; Krystallis, A.; Grunert, K.G. European citizen and consumer attitudes and preferences regarding beef and pork. *Meat Sci.* **2010**, *84*, 284–292. [CrossRef] [PubMed]
60. Ngapo, T.M. Consumer preferences for pork chops in five Canadian provinces. *Meat Sci.* **2017**, *129*, 102–110. [CrossRef]
61. Ripoll, G.; Albertí, P.; Panea, B. Consumer Segmentation Based on Food-Related Lifestyles and Perception of Chicken Breast. *Int. J. Poult. Sci.* **2015**, *14*, 262–275. [CrossRef]

nutrients

MDPI

Article

Role of Food Neophobia and Allergen Content in Food Choices for a Polish Cohort of Young Women

Dominika Guzek [1,*], Joanna Pęska [1] and Dominika Głąbska [2]

1 Department of Organization and Consumption Economics, Faculty of Human Nutrition and Consumer Sciences, Warsaw University of Life Sciences (SGGW–WULS), 159C Nowoursynowska Street, 02-787 Warsaw, Poland; joanna_peska@sggw.pl

2 Department of Dietetics, Faculty of Human Nutrition and Consumer Sciences, Warsaw University of Life Sciences (SGGW–WULS), 159C Nowoursynowska Street, 02-787 Warsaw, Poland; dominika_glabska@sggw.pl

* Correspondence: dominika_guzek@sggw.pl; Tel.: +48-22-593-71-34

Received: 29 September 2019; Accepted: 28 October 2019; Published: 1 November 2019

Abstract: Young women are vulnerable to a number of factors which influence their food choices, including beliefs about food products, or information about nutritional value, while information, that product is free from specific component generates consumer perceptions of its healthfulness. Among the factors which may influence such perception, there is food neophobia (FN). The aim of this study was to determine the influence of FN and information about allergens on the food product choices in the Polish cohort of young women, in the choice experiment when given a model restaurant menu. The web-based choice experiment, in a group of 600 women, aged 18–30 years, with no food allergies diagnosed, was conducted using a mock Italian-style restaurant menu. For 2 starters, 2 soups, 3 main courses and 3 desserts that were included, the allergen content, neophobic potential and perceived lack of healthiness, for a Polish population, were defined. Each respondent randomly received the version containing only a description of dishes, or a description accompanied by the allergens listed. The FN was assessed using the Food Neophobia Scale (FNS). The type of menu (with or without allergens listed) did not influence the choices of dishes. The highest FN level was observed for the women being inhabitants of villages (median of 32). The respondents characterized by a high level of FN less commonly chose dishes characterized by neophobic potential as a starter (Carpaccio), main course (Risotto ai frutti di mare) and dessert (Zabaglione). At the same time, the highest FN level was observed for respondents who chose dishes with no neophobic potential (median of 34.5). However, for allergen content and perceived lack of healthiness, no association with FN was observed, so it may be stated that for neophobic respondents, only neophobic potential is a factor limiting the choice of dishes. It may be concluded that food neophobia in young women may limit the consumption of dishes with unknown food products, and the influence is observed independently of other features of a dish, such as allergen content or perceived healthiness. The problem may appear especially for inhabitants of villages, who are characterized by the highest level of FN.

Keywords: Food Neophobia Scale (FNS); food neophobia; consumer choice; choice experiment; allergen; menu

1. Introduction

The food choices of male and female individuals differ. This variance is due to sex-related differences in their preferences and brain responses to food products [1]. Women, especially young ones, are more vulnerable to factors other than hunger or preferences, as was observed by Wardle et al. [2] in a cohort aged 17–30, from 23 countries. Information acquired about the nutritional value of various food products, their dieting, and their beliefs about food products also influence their food choices [2]. These factors may have an important influence on their nutritional behaviors [3]. The factors gain more importance as women, compared with men, have a higher frequency of dieting [4], or even disordered eating habits [5]. But it is women who are commonly responsible for the nutrition of their whole families [6], even though their own diet is often not properly balanced [2]. This state is mainly true for young women. The situation was emphasized by a consensus statement of the International Summit on the Nutrition of Adolescent Girls and Young Women to be one requiring a high level of international priority [7]. To this end, it must be highlighted that knowledge about the food choices determinants may be necessary to conduct the dedicated education needed to influence young women's nutritional behaviors [8].

Younger individuals are believed to be characterized by a higher sensory-specific satiety than older ones. Sensory-specific satiety is defined as the reduction of the pleasantness of a specific food product as it is consumed [9] that may promote the individual to consume a variety of products [10]. However, at the same time, young women are very prone to restrained eating [11], including weight control [12], even if they are already malnourished [13]. While combining this predilection with common nutritional beliefs, which are not always true [14], as well as with the growing popularity of a number of diets which include the elimination of specific food products, even with no medical reasons to do so [15], it may generate a problem in this group. Among popular diets, which are commonly applied even if not justified, are gluten-free diets [16] or wheat elimination diets [17], which may have a negative impact on general health if there is no medical need to apply such diet [18].

Currently, it has been observed that products with declared information on their packaging that they are free from specific components generate consumer perceptions of their healthfulness [19], even if sometimes it is just a misconception, rather than an objective assessment of the products nutritional value [20]. In this context, it should be emphasized that in spite of the fact that there are sex-dependent differences in food allergy development [21], resulting in a higher frequency of food-related symptoms of allergies and intolerances in women than men [22], more women than men follow the allergen-free diet, but a number of women following such diet even have no adverse reaction diagnosed by physicians [22]. In the study by Zysk et al. [23] on a group of women following the gluten-free diet with no justified medical reason, another factor was also observed, namely, quite a high level of food neophobia, which may cause reluctance in the individual to try new food products and as a result, cause a lower variety of consumed food products [24]. Taking this factor into account, it is also important to understand the influence of allergen content, and information about it, as well as of the food neophobia on the food product choices; hence, the aim of the present study was to determine the effect of these factors in a Polish cohort of young women in the choice experiment when given a model restaurant menu.

2. Materials and Methods

2.1. Ethics Approval Statement

The study was conducted according to the guidelines laid down in the Declaration of Helsinki. The study was approved by the Ethics Committee of the Faculty of Human Nutrition and Consumer Sciences of the Warsaw University of Life Sciences WULS-SGGW (No. 20/2017; 19.06.2017). All the participants provided their informed consent to participate in the study.

2.2. Study Participants

The study was conducted in a group of young Polish women that were recruited via social media using the CAWI (Computer-Assisted Web Interview) method to participate in a web-based choice experiment. The inclusion criteria were as follows: Caucasian individuals, women, age 18–30 years and declared to have no food allergy diagnosed by a physician. The exclusion criteria were as follows: not providing informed consent to participate and any missing data in the Food Neophobia Scale (FNS) questionnaire. There were no additional exclusion criteria associated with socio-economic factors, this was in order to include a wide range of participants being representative of the Polish young women.

The participants were asked about their age (open-ended question), place of residence (closed-ended question with the following options to choose: village, town or city of <500,000 inhabitants, and city of >500,000 inhabitants), and economic status (closed-ended question with the following options to choose: very bad, bad, average, good, very good; while for the analysis very bad and bad, as well as good and very good options were combined). It was decided to ask respondents about their subjective economic status, as it was proven for Caucasians as significantly related to household income [25].

2.3. Choice Experiment

The web-based choice experiment was conducted using a mock Italian-style restaurant menu. Such a type of restaurant was chosen as they are very popular in Poland. In the annual research conducted in a representative sample of adult Polish respondents, Italian cuisine is always preferred by more than half of Polish consumers, as was stated also in the recent study in 2018 [26]. In some years, this type of cuisine was even graded higher than national Polish cuisine, as was observed in 2017 [27]. This predilection has resulted in a growing number of Italian and Italian-style restaurants [28].

The mock Italian-style menu was developed by a professional dietitian educated also in the field of gastronomy with practice in restaurants, including those serving Italian-style dishes. Afterwards, the menu was verified by 2 other dietitians, as well as reviewed, discussed and corrected, if needed, for its understandability and to obtain dishes being within the defined categories (for allergen content, neophobic potential and perceived lack of healthiness, as indicated in Table 1). The menu was not branded with any specific name of a restaurant, or its logotype, and was planned to be as simple as possible. The only presented information were the offered dishes in 4 categories (starters, soups, main courses, desserts) with additional information about the ingredients and recipe, including allergens, if needed. The menu was developed in two identical versions: presenting names of dishes with information about the ingredients and recipe (menu A) and presenting names of dishes with information about the ingredients and recipe, as well as for each dish, as an additional information, the allergens listed, based on Regulation (EU) No 1169/2011 [29] (menu B). For the developed menu, no prices were presented and no other information about visual appearance of dishes were provided (no photographs of dishes and no other additional information were included).

Table 1. The dishes and descriptions provided in a developed mock Italian-style menu, as well as defined categories of dishes for allergen content, neophobic potential and perceived lack of healthiness based on the ingredients.

Dishes and Descriptions Provided in a Mock Italian-Style Menu			Defined Categories of Dishes Based on the Ingredients		
Course	**Dish**	**Provided Description**	**Content of Ingredients that Are Among 14 Major Food Allergens Listed in Regulation (EU) No 1169/2011** [29]	**Content of Ingredients that for the Consumers May Cause the Neophobic Response**	**Content of Ingredients that for the Consumers May Be Perceived as not Healthy**
Starter	Carpaccio	Paper-thin raw beef slices	–	Raw meat	Meat
	Caprese	Slices of tomatoes and mozzarella	Milk and products thereof (including lactose) (annex 2, pt. 7)	–	–
Soup	Minestrone	Vegetable soup with zucchini, green beans, carrot, spinach and fennel	–	Spinach and fennel	–
	Pappa al pomodoro	Tomato bread soup	Cereals containing gluten (annex 2, pt. 1)	–	Bread
Main course	Risotto ai frutti di mare	Seafood risotto with parmesan	Crustaceans and products thereof (annex 2, pt. 2); milk and products thereof (including lactose) (annex 2, pt. 7); molluscs and products thereof (annex 2, pt. 14);	Seafood	–
	Spaghetti alla siciliana	Spaghetti with tomatoes and roasted eggplant	Cereals containing gluten (annex 2, pt. 1)	–	Pasta
	Bistecca alla fiorentina	Beef steak served with rosemary roasted potatoes	–	–	Meat
Dessert	Tiramisu	Espresso-dipped ladyfingers with mascarpone-egg cream	Cereals containing gluten (annex 2, pt. 1); eggs and products thereof (including lactose) (annex 2, pt. 3); milk and products thereof (including lactose) (annex 2, pt. 7)	Egg cream	–
	Zabaglione	Egg yolk dessert with alcohol	Eggs and products thereof (including lactose) (annex 2, pt. 3)	Egg dessert	Alcohol
	Granita	Frozen fruit dessert	–	–	–

While developing the menu, an excessive number of dishes were not planned—for each course, at least two but not more than three dishes were proposed (2 starters, 2 soups, 3 main courses and 3 desserts) and no beverages were included. For each course, the dishes were characterized by a comparable energy value. At the same time, for each course, it was planned to have diverse dishes in terms of the following:

- Allergen content—defined as a content of ingredients that are among 14 major food allergens listed in Regulation (EU) No 1169/2011 [29]—it was planned to have in each category at least one dish with the noticeable allergen (as one of the main components of the dish) and at least one dish with no allergens at all.
- Neophobic potential—defined as a content of ingredients that for the Polish consumers may cause a neophobic response, due to the fact that they are not commonly consumed and are not typical of the Polish diet (the following were included: raw meat [30], seafood [31], spinach and fennel [32], and egg desserts [33] —these items are, in general, rejected by neophobic individuals)—it was planned to have in each category at least one dish with the noticeable component that may be perceived as unknown/not consumed (as one of the main components of dish) and at least one dish with no such components at all (all components neutral).
- Perceived lack of healthiness—defined as a content of ingredients that for the Polish consumers may be perceived as not healthy (the following were included: meat [34], bread [35], pasta [36], and alcohol [37] — these items are perceived by some consumers as not healthy)—it was planned to have in each category at least one dish with the noticeable component that may be perceived as unhealthy (as one of the main components of dish) and at least one dish with no such components at all (all components neutral).

The dishes and descriptions provided in a developed mock Italian-style menu for both menu A and menu B are presented in Table 1, with the defined categories of dishes for allergen content, neophobic potential and perceived lack of healthiness based on the ingredients, that were not presented in menus, except for the allergens, based on Regulation (EU) No 1169/2011 [29] that were listed for menu B.

As for the planned dishes, there were three variables included (allergen content, neophobic potential, and perceived lack of healthiness) and each of them had two options (dish defined as within criterion or not). There were 8 possibilities (2 × 2 × 2) to define a dish within the criteria. Taking it into account, it was planned to have at least one dish for one possibility (being a setup of variables). Moreover, if more than one dish was planned within the same setup of variables, they were to be dishes for various courses. The planned dishes distributed for the included variables are presented in Table 2.

Table 2. The planned dishes distributed for the included variables (allergen content, neophobic potential and perceived lack of healthiness).

Variables			Planned Dish	Course
Allergen Content	Neophobic Potential	Perceived Lack of Healthiness		
–	–	–	Granita	Dessert
+	–	–	Caprese	Starter
–	+	–	Minestrone	Soup
–	–	+	Bistecca alla fiorentina	Main course
+	+	+	Zabaglione	Dessert
+	+	–	Risotto ai frutti di mare Tiramisu	Main course Dessert
+	–	+	Pappa al pomodoro Spaghetti alla siciliana	Soup Main course
–	+	+	Carpaccio	Starter

The obtained mock Italian-style restaurant menu was applied in the web-based choice experiment. Each participant saw only one version of the menu (menu A with no allergens listed or menu B with allergens listed). For each individual, the system randomly chose the version and participants did not receive any information that there was more than one possible version. Each respondent was asked a simple question about her choice of dishes—she was asked to indicate the dishes that she would order while eating in this restaurant. The respondents were informed that they may choose whatever they want and they were asked to imagine a situation when they may afford whatever they want (the prices were not presented). Moreover, they were allowed to not choose a dish within each course, as they were informed that they should indicate only those dishes that they would order to consume (it was assumed that for some young women, a four-courses meal may be too large to consume).

2.4. Assessment of Food Neophobia

To measure the level of food neophobia, the Food Neophobia Scale (FNS) by Pliner and Hobden [38] was applied. The FNS is a 10-item scale with each item rated using a 7-point Likert scale (from 1 = strongly disagree, to 7 = strongly agree). For the presented study, the obtained answers for each question were attributed to a number of points (from 1 to 7). For 5 positive items, they were reversed for scoring and for 5 negative ones, they were not reversed. The resultant FNS ranged from 10 to 70. For the analyzed respondents, the internal consistency was assessed using Cronbach's alpha (0.79). The result revealed good internal consistency, as it was above 0.7 [39].

While interpreting the obtained results of the FNS, the obtained scores were analyzed as a continuous variable (obtained score) and in categories (of low, medium or high levels of food neophobia). In order to stratify respondents into categories of FNS, their results were grouped according to the group mean and standard deviation (SD) values, as is commonly applied [40–43]:

- Low food neophobia level—FNS score < mean value minus SD value.
- Average food neophobia level—FNS score within the range from mean value minus SD to mean value plus SD.
- High food neophobia level—FNS score > mean value plus SD value.

2.5. Statistical Analysis

To verify the normality of distribution, the Shapiro–Wilk test was used. For the internal consistency testing conducted for the FNS, Cronbach's alpha coefficient was applied.

Three stages of analysis of data were applied, as follows:

- The influence of the information with the allergens listed on the consumer choices of dishes was verified (menu A versus menu B)—as no differences were stated, the results obtained for menu A and menu B may have been combined.
- The influence of food neophobia (FNS score) on the consumer choices of dishes was verified—as the influence was stated, a deepened analysis was conducted taking into account the general characteristics of the chosen dishes.
- The influence of food neophobia (FNS score) on the characteristics of chosen dishes—for this analysis, in terms of allergen content, neophobic potential and the perceived lack of healthiness, the choices of each respondent for each course (Table 2) were stratified into the following groups:

  (a) Choice of dishes with no allergen content; choice of 1–3 dishes with allergen content; choice of all dishes (4 of them) with allergen content.

  (b) Choice of dishes with no neophobic potential; choice of 1–3 dishes with neophobic potential; choice of all dishes (4 of them) with neophobic potential.

  (c) Choice of dishes with no perceived lack of healthiness; choice of 1–3 dishes with perceived lack of healthiness; choice of perceived lack of healthiness (4 of them) with perceived lack of healthiness.

To compare the frequency in sub-groups, the chi-squared test was used. To identify the differences between groups, due to the non-parametric distributions, the U Mann–Whitney test was used (for 2 groups) or Kruskal–Wallis Analysis of Variance (ANOVA) with multiple comparisons of results (for more than 2 groups). The $p \leq 0.05$ was accepted as the level of significance. Statistical analysis was performed using Statistica software version 8.0 (StatSoft Inc., Tulsa, OK, USA).

3. Results

3.1. Characteristics of the Studied Group

The basic characteristics of the studied group of 600 women is presented in Table 3. The majority of recruited participants (aged 18–30) were inhabitants of big cities (over 60%) and assessed their economic situation as either good or very good (55%).

Table 3. Basic characteristics of the studied group of women (n = 600).

Characteristics		Results
Age (years)	Mean ± SD	21.2 ± 1.9
	Median (range)	21 * (18–30)
Place of residence	Village	72 (12.0%)
	Towns and cities <500,000 inhabitants	153 (25.5%)
	Cities >500,000 inhabitants	375 (62.5%)
Economic status	Very bad and bad	36 (6.0%)
	Average	234 (39.0%)
	Good and very good	330 (55.0%)

* non-parametric distribution (verified using Shapiro–Wilk test: $p \leq 0.05$). SD: Standard deviation.

The FNS scores in the studied group of women are presented in Table 4. It was observed that the median score in the group was 29, while for the sub-groups of low, average and high food neophobia level they were 16 (n = 104), 30 (n = 407) and 44 (n = 89), respectively.

Table 4. Food Neophobia Scale (FNS) scores in the studied group of women (n = 600).

FNS Score	Total (n = 600)	Food Neophobia Level		
		Low [1] (n = 104)	Average [2] (n = 407)	High [3] (n = 89)
Mean ± SD	30.2 ± 10.1	16.1 ± 2.4	30.3 ± 5.7	46.3 ± 5.9
95% CI	29.4–31.0	15.6–16.5	29.8–30.9	45.1–47.5
Median	29 *	16 *	30 *	44 *
Range	10–70	10–19	20–40	41–70
25th–75th	23–37	14–18	26–35	42–49

* non-parametric distribution (verified using Shapiro–Wilk test: $p \leq 0.05$); [1] low level (< mean value − SD); [2] average level (the range from mean value − SD to mean value + SD); [3] high level (> mean value − SD) [40–43]. CI: Confidence intervals.

The FNS scores in the studied group of women stratified by place of residence and economic status are presented in Table 5. It was observed that the place of residence influenced the observed FNS score, as the highest food neophobia level was observed for the women being inhabitants of villages (p = 0.0290).

Table 5. Food Neophobia Scale (FNS) scores in the studied group of women ($n = 600$) stratified by place of residence and economic status.

Characteristics		Mean ± SD	Median (range)	*P*-Value
Place of residence	Village	33.5 ± 9.9	32 (15–56) [a]	0.0070
	Towns and cities <500,000 inhabitants	30.2 ± 9.6	29 (10–56) [b]	
	Cities >500,000 inhabitants	29.6 ± 10.2	29 * (11–70) [b]	
Economic status	Good and very good	29.9 ± 10.4	29 *(10–70)	0.1775
	Average	31.0 ± 9.6	31 (11–58)	
	Very bad and bad	28.2 ± 10.3	30 (13–50)	

* non-parametric distribution (verified using Shapiro–Wilk test: $p \leq 0.05$); [a,b] different letters indicate significant differences between groups ($p < 0.05$).

3.2. Influence of the Information with the Allergens Listed on the Consumer Choices

The comparison of the choices of dishes stratified by information about allergen content (type of menu) is presented in Table 6. It was observed that for none of the courses did the type of menu influence the choices of dishes, as the frequency of choosing a specific dish within a course was similar, independently from the received menu ($p > 0.05$).

Table 6. Comparison of the choices of dishes stratified by information about allergen content (type of menu).

Course	Dish	Total	Menu A [1] ($n = 300$)	Menu B [2] ($n = 300$)	*p*-Value
Starters	Carpaccio	193 (34.8)	97 (35.3)	96 (34.4)	0.8302
	Caprese	361 (65.2)	178 (64.7)	183 (65.6)	
Soup	Minestrone	178 (33.9)	92 (35.0)	86 (32.8)	0.6671
	Pappa al pomodoro	347 (66.1)	171 (65.0)	176 (67.2)	
Main course	Risotto ai frutti di mare	149 (24.2)	70 (21.8)	79 (23.5)	0.4499
	Spaghetti alla siciliana	281 (45.8)	131 (40.8)	150 (44.7)	0.1408
	Bistecca alla fiorentina	184 (30.0)	120 (37.4)	107 (31.8)	0.3125
Dessert	Tiramisu	377 (55.1)	182 (53.8)	195 (56.4)	0.3106
	Zabaglione	90 (13.2)	39 (11.5)	51 (14.7)	0.2085
	Granita	217 (31.7)	117 (34.6)	100 (28.9)	0.1740

[1] menu A—standard menu with information about the ingredients and recipe; [2] menu B—standard menu with information about the ingredients and recipe followed by allergens listed.

3.3. Influence of the Food Neophobia on the Consumer Choices

As it was observed that the information about allergen content (type of menu) does not influence the choices of dish, the respondents were analyzed combined, to assess the influence of the FNS score.

The choice of dishes compared between groups stratified by the FNS scores of respondents is presented in Table 7. It was observed, that for starters, main courses and desserts, the FNS scores influenced choices. The respondents characterized by a high level of food neophobia less commonly chose Carpaccio as a starter ($p = 0.0058$), Risotto ai frutti di mare as a main course ($p < 0.0001$) and Zabaglione as a dessert ($p < 0.0001$). All of them were defined as dishes characterized by neophobic potential, while for Carpaccio and Zabaglione, perceived lack of healthiness was also defined and for Risotto ai frutti di mare and Zabaglione, allergen content was also indicated.

Table 7. Choice of dishes compared between groups stratified by Food Neophobia Scale (FNS) scores of respondents.

Course	Dish	Total	Level of Food Neophobia			p-Value
			Low [1]	Average [2]	High [3]	
Starters	Carpaccio	193 (34.8)	49 (48.0)	123 (32.7)	21 (27.6)	0.0058
	Caprese	361 (65.2)	53 (52.0)	253 (67.3)	55 (72.4)	
Soup	Minestrone	178 (33.9)	27 (31.0)	125 (34.9)	26 (32.5)	0.7581
	Pappa al pomodoro	347 (66.1)	60 (69.0)	233 (65.1)	54 (67.5)	
Main course	Risotto ai frutti di mare	149 (22.7)	49 (40.8)	92 (20.6)	8 (8.9)	<0.0001
	Spaghetti alla siciliana	281 (42.7)	36 (30.0)	200 (44.7)	45 (50.0)	
	Bistecca alla fiorentina	227 (34.6)	35 (29.2)	155 (34.7)	37 (41.1)	
Dessert	Tiramisu	377 (55.1)	61 (51.7)	264 (57.0)	52 (50.5)	<0.0001
	Zabaglione	90 (13.2)	19 (16.1)	63 (13.6)	8 (7.8)	
	Granita	217 (31.7)	38 (32.2)	136 (29.4)	43 (41.7)	

[1] low level (< mean value − SD); [2] average level (the range from mean value − SD to mean value + SD); [3] high level (> mean value + SD) [40–43].

The FNS scores of respondents compared between groups stratified by choice of dishes in the categories of neophobic potential, allergen content and perceived lack of healthiness are presented in Figure 1.

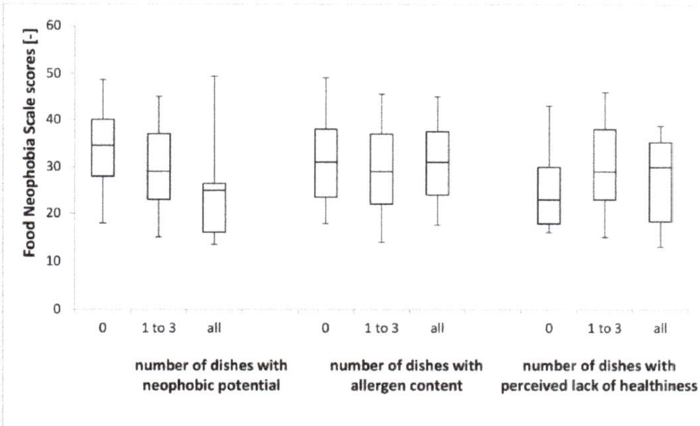

Figure 1. Food Neophobia Scale (FNS) scores of respondents compared between groups stratified by choice of dishes in the categories of neophobic potential, allergen content and perceived lack of healthiness.

The highest FNS scores (highest food neophobia level) were noted for respondents who chose dishes with no neophobic potential ($p = 0.0051$), confirming the previous analysis. However, while comparing participants who chose 1–3 dishes with neophobic potential and all of them with neophobic potential, it was stated that there were no differences between them. At the same time, for allergen content ($p = 0.8443$) and perceived lack of healthiness ($p = 0.1206$), there were no significant differences.

4. Discussion

Many popular diets are restricted ones that focus on avoiding some food products or their components, which results in misleading information being promoted by the media that some products, e.g., gluten, lactose, fat, are harmful always and for everybody. A survey by Watson [44] found that 65% of American adults think that gluten-free products are healthier than regular ones, and over 27% choose gluten-free products to lose weight. This situation is driven by several factors, not only media messages but also consumer-directed marketing strategies by manufacturers and press reports of the clinical benefits of gluten avoidance influence people [45]. As a result, it is observed that food products labeled as 'free from' any ingredient can be perceived as better for health, in comparison with products with no such information [19]. However, following a restricted diet, based on avoiding some type of products, could result in a deficiency in healthy individuals without health issues, which was indicated as a possible health risk in a national Polish study in a representative cohort of girls and young women [46]. For example, gluten-free processed grain products are often lower in fiber, iron, zinc, and potassium than normal grain products [47]. Even if the consumers believe that this trend of restricted diets is healthy and that they have an optimal diet, they commonly do not meet the requirements of the dietary guidelines [48].

Taking all these factors into account, in the present study, the influence of information about allergen content was assessed. It was analyzed to determine if this information had any impact on the consumer choice. For the applied mock Italian-style restaurant menu, women receiving menus with allergens listed made comparable choices to women receiving the menus with no allergens listed. This may be interpreted as a lack of influence of allergen content on the choices of dishes in the assessed group of young women, but it cannot be stated for a fact that this is the general approach in the studied group. The similarity in choices may be attributed to the growing trend of so-called cheat-meals, being not in accordance with the applied general diets, which are often consumed while being in restaurants, or eating out [49]. At the same time, such lack of differences for the applied menus with and without allergens listed may result from the fact that for a person avoiding specific food products or components, the name of the product may be sufficient information to indicate that it may be a source of a specific allergen, as it is observed that for consumers with and without food allergies, the level of knowledge about allergens is similar [50]. It is also associated with the fact that consumers avoiding some allergens, who are not able to identify products with allergen and without, would commonly rather avoid those allergens and products from the same group as these are perceived as suspicious products. This was observed for tree nut allergic individuals [50].

The additional fact that may be indicated here is a general lack of trust in food allergen labels declared by consumers [51]. Moreover, they report that some ingredient lists include too much information, as well as too-technical terms, so they are hard for them to understand [52]. As a result, consumers often declare that they are not satisfied with the current allergen labelling practices, which results in their personal stress and feelings of insecurity [53].

While analyzing the indicated approaches, it may be supposed, that the studied women with no food allergies diagnosed may not avoid any allergens from their diet, but based on literature, this is a rare situation. For example, in the United States of America, it was observed that 30% of inhabitants are interested in avoiding gluten [16]. At the same time, it may be supposed that in the studied group, respondents either do not avoid allergens in a restrictive way and while eating out, they do not follow their individual restriction, or that they preventively avoid some products which they suspect to contain allergens, even based on the limited information. Both approaches for individuals with no food allergies cannot be interpreted as improper, if only they were not associated with a decreased quality of diet due to overconsumption in social situations [54], excessive food products avoiding that may result in reduced variety of food products consumed [55].

The other factor influencing consumer food choices is food neophobia which was, in the present study, the major determinant of choice of dishes from the applied mock Italian-style restaurant menu. It was stated that women defined as neophobic more than others chose dishes without components of neophobic potential, and that women choosing dishes without components of neophobic potential have the higher FNS scores. However, it was not associated with allergen content and perceived healthiness of dish, so this may indicate, that for neophobic ones, specifically neophobic potential of food products was a determinant of choice, but not other factors. Taking this into account, it must be indicated that in spite of the fact that food neophobia is a rejection of novel or unknown food products that occurs mainly in children [56] and should decrease with age [57], for adults, it also significantly determines food choices. This is in agreement with the stable level of food neophobia from adolescence (approximately 13 years old) through adulthood suggested by some researchers [58]. At the same time, it should be emphasized that Tuorila et al. [40] observed that food neophobia may increase in older people, so food neophobia may not be so stable for adults as was previously supposed.

As a food neophobia may influence dietary habits, it may result in lower quality of diet, inadequate dietary intake [59] and consequent deficiencies [60]. Some studies indicated that food neophobia was related to reduced vegetable intake [42,60], or vegetable and fruit intake [60,61], while it was more strongly associated with vegetable than with fruit intake [59]. However, so far, there is no study published that analyzes the influence of food neophobia on choice of dishes from the restaurant menu. The study of Camarena et al. [62] analyzed the influence of ethnocentrism and food neophobia on the ethnic food consumption in Spain, but for the Polish population, Italian cuisine, which is analyzed in the present study, should not be considered as an ethnic one. This type of cuisine was introduced in Poland in the XVI century and is nowadays quite popular in all the regions of the country [25]. Hence, the reason for rejecting some dishes by neophobic respondents must have been associated with a low level of familiarity with those specific dishes (novel food) or with the specific ingredients (unknown food products).

As was mentioned previously, some ingredients may cause a neophobic response due to the fact that they are not commonly consumed and are not typical of the Polish diet such as raw meat [30], seafood [31], spinach and fennel [32], and egg desserts [33]. Therefore, the dishes that included those ingredients may be rejected by neophobic individuals. Such a situation can be easily explained by analyzing the basic mechanisms; however, individual differences must also be taken into account.

The reduced acceptance of meat and fish products was stated in several studies. An evolutionary explanation was presented in that the reluctance to use these products was due to a protective mechanism against the possibility of accidental poisoning by raw or perishable food products [30,33]. It is in general indicated that animal food products are a more likely source of food poisoning than non-animal foods, so this mechanism may be also attributed to seafood and eggs [33]. At the same time, in the case of the vegetables, the bitter taste may be rejected due to the natural subconscious mechanism of defense of the organism from poisoning, as most potential poisons in nature have a bitter taste [63]. Taking all these factors into account, the dishes defined as those of neophobic potential may be justified and their avoidance by neophobic respondents may be explained.

However, avoidance of some products by respondents with food neophobia may lead to a reduced variety of products consumed. The resultant reduced quality of the respondent's diets accompanied by possible inadequate dietary intake of some nutrients may lead to some deficiencies. In the analyzed group, such risk may be observed especially for respondents from villages, as the place of residence influences consumers to a higher food neophobia level. At the same time, the economic status was not a determinant of food neophobia in the studied group, which allows to conclude that women from villages are especially prone to have a reduced quality of diet due to their food neophobia.

In spite of the fact that the presented study revealed interesting new observations from a large homogenic sample of respondents, associated with the influence of food neophobia in young women, which may limit the consumption of dishes with unknown food products, some limitations of the study should be also listed. The study was conducted only in a group of young women, so the situation in other groups is still unknown, and it was conducted only using one mock Italian-style menu, so the observations should be verified for other menus, including not only Italian-style, but also the other ones. Moreover, the influence of broad interfering factors should also be included in further studies.

5. Conclusions

It may be concluded that food neophobia in young women may limit the consumption of dishes with unknown food products, and the influence is observed independently of other features of the dish, such as allergen content or perceived healthiness. It may cause not properly balanced diets, due to avoiding some foods and a reduced diversity of consumed products. The problem may appear greater especially for inhabitants of villages, who are characterized by the highest level of food neophobia. This problem should be taken into account in the education programs and public health policy.

Author Contributions: Conceptualization, D.G. (Dominika Guzek) J.P., and D.G. (Dominika Głąbska); Formal analysis, D.G. (Dominika Guzek), D.G. (Dominika Głąbska); Investigation, J.P.; Methodology, D.G. (Dominika Guzek); Writing—original draft, D.G. (Dominika Guzek), J.P., and D.G. (Dominika Głąbska); Writing—review and editing, D.G. (Dominika Guzek), J.P., and D.G. (Dominika Głąbska).

Funding: This research is financed by the Polish Ministry of Science and Higher Education within funds of the Faculty of Human Nutrition and Consumer Sciences, Warsaw University of Life Sciences (WULS), for scientific research.

Conflicts of Interest: The authors declare no conflict of interest.

References

1. Manippa, V.; Padulo, C.; van der Laan, L.N.; Brancucci, A. Gender Differences in Food Choice: Effects of Superior Temporal Sulcus Stimulation. *Front. Hum. Neurosci.* **2017**, *11*, 597. [CrossRef] [PubMed]
2. Wardle, J.; Haase, A.M.; Steptoe, A.; Nillapun, M.; Jonwutiwes, K.; Bellisle, F. Gender differences in food choice: The contribution of health beliefs and dieting. *Ann. Behav. Med.* **2004**, *27*, 107–116. [CrossRef] [PubMed]
3. Hardcastle, S.J.; Thøgersen-Ntoumani, C.; Chatzisarantis, N.L. Food Choice and Nutrition: A Social Psychological Perspective. *Nutrients* **2015**, *7*, 8712–8715. [CrossRef] [PubMed]
4. Kiefer, I.; Rathmanner, T.; Kunze, M. Eating and dieting differences in men and women. *J. Men's Health Gend.* **2005**, *2*, 194–201. [CrossRef]
5. Murnen, S.K.; Smolak, L. Femininity, masculinity, and disordered eating: A meta-analytic review. *Int. J. Eat Disord.* **1997**, *22*, 231–242. [CrossRef]
6. Charles, N.; Kerr, M. *Women, Food and Families*; Manchester University Press: Manchester, UK, 1990. [CrossRef]
7. Krebs, N.; Bagby, S.; Bhutta, Z.A.; Dewey, K.; Fall, C.; Gregory, F.; Hay, W., Jr.; Rhuman, L.; Caldwell, C.W.; Thornburg, K.L. International summit on the nutrition of adolescent girls and young women: Consensus statement. *Ann. N. Y. Acad. Sci.* **2017**, *1400*, 3–7. [CrossRef]
8. Lawrence, W.; Barker, M. A review of factors affecting the food choices of disadvantaged women. *Proc. Nutr. Soc.* **2009**, *68*, 189–194. [CrossRef]
9. Westenhoefer, J. Age and gender dependent profile of food choice. *Forum Nutr.* **2005**, *57*, 44–51.
10. Wilkinson, L.L.; Brunstrom, J.M. Sensory specific satiety: More than just habituation? *Appetite* **2016**, *103*, 221–228. [CrossRef]
11. Dong, D.; Wang, Y.; Jackson, T.; Chen, S.; Wang, Y.; Zhou, F.; Chen, H. Impulse control and restrained eating among young women: Evidence for compensatory cortical activation during a chocolate-specific delayed discounting task. *Appetite* **2016**, *105*, 477–486. [CrossRef]
12. Field, A.E.; Haines, J.; Rosner, B.; Willett, W.C. Weight-control behaviors and subsequent weight change among adolescents and young adult females. *Am. J. Clin. Nutr.* **2010**, *91*, 147–153. [CrossRef] [PubMed]
13. Zhang, L.; Qian, H.; Fu, H. To be thin but not healthy—The body-image dilemma may affect health among female university students in China. *PLoS ONE* **2018**, *13*, 0205282. [CrossRef] [PubMed]
14. Casazza, K.; Brown, A.; Astrup, A.; Bertz, F.; Baum, C.; Brown, M.B.; Dawson, J.; Durant, N.; Dutton, G.; Fields, D.A.; et al. Weighing the Evidence of Common Beliefs in Obesity Research. *Crit. Rev. Food Sci. Nutr.* **2015**, *55*, 2014–2053. [CrossRef] [PubMed]
15. Ramachandran, D.; Kite, J.; Vassallo, A.J.; Chau, J.Y.; Partridge, S.; Freeman, B.; Gill, T. Food Trends and Popular Nutrition Advice Online-Implications for Public Health. *Online J. Public Health Inf.* **2018**, *10*, 213. [CrossRef] [PubMed]
16. Jones, A.L. The Gluten-Free Diet: Fad or Necessity? *Diabetes Spectr.* **2017**, *30*, 118–123. [CrossRef] [PubMed]
17. Nash, D.T.; Slutzky, A.R. Gluten sensitivity: New epidemic or new myth? *Procedings (Bayl. Univ. Med. Cent)* **2014**, *27*, 377–378. [CrossRef]
18. Lebwohl, B.; Cao, Y.; Zong, G.; Hu, F.B.; Green, P.H.R.; Neugut, A.I.; Rimm, E.B.; Sampson, L.; Dougherty, L.W.; Giovannucci, E.; et al. Long term gluten consumption in adults without celiac disease and risk of coronary heart disease: Prospective cohort study. *BMJ* **2017**, *357*, j1892. [CrossRef]

19. Priven, M.; Baum, J.; Vieira, E.; Fung, T.; Herbold, N. The Influence of a Factitious Free-From Food Product Label on Consumer Perceptions of Healthfulness. *J. Acad. Nutr. Diet.* **2015**, *115*, 1808–1814. [CrossRef]
20. Asioli, D.; Aschemann-Witzel, J.; Caputo, V.; Vecchio, R.; Annunziata, A.; Næs, T.; Varela, P. Making sense of the "clean label" trends: A review of consumer food choice behavior and discussion of industry implications. *Food Res. Int.* **2017**, *99*, 58–71. [CrossRef]
21. Pali-Schöll, I.; Jensen-Jarolim, E. Gender aspects in food allergy. *Curr. Opin. Allergy Clin. Immunol.* **2019**, *19*, 249–255. [CrossRef]
22. Afify, S.M.; Pali-Schöll, I. Adverse reactions to food: The female dominance-A secondary publication and update. *World Allergy Organ J.* **2017**, *10*, 43. [CrossRef] [PubMed]
23. Zysk, W.; Głąbska, D.; Guzek, D. Food Neophobia in Celiac Disease and Other Gluten-Free Diet Individuals. *Nutrients* **2019**, *11*, 1762. [CrossRef] [PubMed]
24. Ostrove, J.M.; Adler, N.E.; Kuppermann, M.; Washington, A.E. Objective and subjective assessments of socioeconomic status and their relationship to self-rated health in an ethnically diverse sample of pregnant women. *Health Psychol.* **2000**, *19*, 613–618. [CrossRef] [PubMed]
25. van den Heuvel, E.; Newbury, A.; Appleton, K.M. The Psychology of Nutrition with Advancing Age: Focus on Food Neophobia. *Nutrients* **2019**, *11*, 151. [CrossRef]
26. Poland on Plate 2018: More and More Poles Use Gastronomic Facility (Report). Available online: http://www.portalspozywczy.pl/horeca/wiadomosci/polska-na-talerzu-2018-coraz-wiecej-polakow-korzysta-z-lokali-gastronomicznych-raport,159812_2.html (accessed on 29 August 2019).
27. Poland on Plate 2018: Sour Flavors and Italian Restaurants are the Culinary Choices of Poles (Report). Available online: http://www.portalspozywczy.pl/horeca/wiadomosci/kwasne-smaki-i-restauracje-wloskie-kulinarnymi-wyborami-polakow,150282.html (accessed on 29 August 2019).
28. Głuchowski, A.; Rasińska, E.; Czarniecka-Skubina, E. Rynek usług gastronomicznych w Polsce na przykładzie Warszawy [The Catering Services Market in Poland on the Example of Warsaw]. *Handel Wew.* **2017**, *4*, 118–133.
29. EEA relevance. Regulation (EU) No 1169/2011 of the European Parliament and of the Council of 25 October 2011 on the provision of food information to consumers, amending Regulations (EC) No 1924/2006 and (EC) No 1925/2006 of the European Parliament and of the Council, and repealing Commission Directive 87/250/EEC, Council Directive 90/496/EEC, Commission Directive 1999/10/EC, Directive 2000/13/EC of the European Parliament and of the Council, Commission Directives 2002/67/EC and 2008/5/EC and Commission Regulation (EC) No 608/2004. *Off. J. Eur. Union* **2011**, *20*, 18–63.
30. Cooke, L.; Carnell, S.; Wardle, J. Food neophobia and mealtime food consumption in 4–5 year old children. *Int. J. Behav. Nutr. Phys. Act.* **2006**, *6*, 14. [CrossRef]
31. Knaapila, A.; Silventoinen, K.; Broms, U.; Rose, R.J.; Perola, M.; Kaprio, J.; Tuorila, H.M. Food neophobia in young adults: Genetic architecture and relation to personality, pleasantness and use frequency of foods, and body mass index—A twin study. *Behav. Genet.* **2011**, *41*, 512–521. [CrossRef]
32. Nekitsing, C.; Hetherington, M.M.; Blundell-Birtill, P. Developing Healthy Food Preferences in Preschool Children Through Taste Exposure, Sensory Learning, and Nutrition Education. *Curr. Obes. Rep.* **2018**, *7*, 60–67. [CrossRef]
33. Cooke, L.; Wardle, J.; Gibson, E.L. Relationship between parental report of food neophobia and everyday food consumption in 2-6-year-old children. *Appetite* **2003**, *41*, 205–206. [CrossRef]
34. Frewer, L.; Van Trijp, H. *Understanding Consumers of Food Products*, 1st ed.; Woodhead Publishing: Cambridge, UK, 2007; p. 696.
35. Gellynck, X.; Kühne, B.; Van Bockstaele, F.; Van de Walle, D.; Dewettinck, K. Consumer perception of bread quality. *Appetite* **2009**, *53*, 16–23. [CrossRef] [PubMed]
36. U.S. International Trade Commission. *Industry and Trade Summary: Pasta*; U.S. International Trade Commission: Washington, DC, USA, 2003; p. 47.
37. Rosenberg, G.; Bauld, L.; Hooper, L.; Buykx, P.; Holmes, J.; Vohra, J. New national alcohol guidelines in the UK: Public awareness, understanding and behavioural intentions. *J. Public Health* **2018**, *40*, 549–556. [CrossRef] [PubMed]
38. Pliner, P.; Hobden, K. The development of a scale to measure the trait of food neophobia in humans. *Appetite* **1992**, *19*, 105–120. [CrossRef]
39. DeVellis, R.F. *Scale Development*; Sage Publications: Newbury Park, NJ, USA, 1991.

40. Tuorila, H.M.; Lähteenmäki, L.; Pohjalainen, L.; Lotti, L. Food neophobia among the Finns and related responses to familiar and unfamiliar foods. *Food Qual. Prefer.* **2001**, *12*, 29–37. [CrossRef]
41. Falciglia, G.A.; Couch, S.C.; Gribble, L.S.; Pabst, S.M.; Frank, R.A. Food neophobia in childhood affects dietary variety. *J. Am. Diet. Assoc.* **2000**, *100*, 1474–1481. [CrossRef]
42. Guzek, D.; Głąbska, D.; Lange, E.; Jezewska-Zychowicz, M. A Polish study on the influence of food neophobia in children (10–12 years old) on the intake of vegetables and fruits. *Nutrients* **2017**, *9*, 563. [CrossRef]
43. Guzek, D.; Głąbska, D.; Mellová, B.; Zadka, K.; Żywczyk, K.; Gutkowska, K. Influence of Food Neophobia Level on Fruit and Vegetable Intake and Its Association with Urban Area of Residence and Physical Activity in a Nationwide Case-Control Study of Polish Adolescents. *Nutrients* **2018**, *13*, 897. [CrossRef]
44. Watson, E. Health/Weight Conscious Consumers are Driving the Gluten-Free Market, Not Celiac, Says Mintel. Available online: http://www.foodnavigator-usa.com/Markets/Health-weight-consciousconsumers-are-driving-the-gluten-freemarket-not-celiacs-says-Mintel (accessed on 24 September 2019).
45. Niland, B.; Cash, B.D. Health Benefits and Adverse Effects of a Gluten-Free Diet in Non-Celiac Disease Patients. *Gastroenterol. Hepatol.* **2018**, *14*, 82–91.
46. Czarnocińska, J.; Jeżewska-Zychowicz, M.; Babicz-Zielińska, E.; Kowalkowska, J.; Wądołowska, L. *Postawy Względem Żywności, Żywienia i Zdrowia a Zachowania Żywieniowe Dziewcząt i Młodych Kobiet w Polsce*; Wydawnictwo UWM: Olsztyn, Poland, 2013; pp. 1–114.
47. Missbach, B.; Schwingshackl, L.; Billmann, A.; Mystek, A.; Hickelsberger, M.; Bauer, G.; König, J. Gluten-free food database: The nutritional quality and cost of packaged gluten-free foods. *PeerJ* **2015**, *22*, 1337. [CrossRef]
48. Bailey, R.L.; Denby, N.; Haycock, B.; Sherif, K.; Steinbaum, S.; von Schacky, C. Perceptions of a Healthy Diet: Insights From a 3-Country Survey. *Nutr. Today* **2015**, *50*, 282–287. [CrossRef]
49. Pila, E.; Mond, J.M.; Griffiths, S.; Mitchison, D.; Murray, S.B. A thematic content analysis of #cheatmeal images on social media: Characterizing an emerging dietary trend. *Int. J. Eat. Disord.* **2017**, *50*, 698–706. [CrossRef] [PubMed]
50. Choi, Y.; Ju, S.; Chang, H. Food allergy knowledge, perception of food allergy labeling, and level of dietary practice: A comparison between children with and without food allergy experience. *Nutr. Res. Pr.* **2015**, *9*, 92–98. [CrossRef] [PubMed]
51. Chow, Y.L.B. Everybody Else Got to Have This Cookie: The Effects of Food Allergen Labels on the Well-Being of Canadians. Master's Thesis, McMaster University, Hamilton, ON, Canada, 2011. Available online: http://hdl.handle.net/11375/11078 (accessed on 22 October 2019).
52. Ju, S.Y.; Park, J.H.; Kwak, T.K.; Kim, K.E. Attitudes and preferences of consumers toward food allergy labeling practices by diagnosis of food allergies. *Nutr. Res. Pr.* **2015**, *9*, 517–522. [CrossRef] [PubMed]
53. Cornelisse-Vermaat, J.R.; Voordouw, J.; Yiakoumaki, V.; Theodoridis, G.; Frewer, L.J. Food-allergic consumers' labelling preferences: A cross-cultural comparison. *Eur. J. Public Health* **2008**, *18*, 115–120. [CrossRef]
54. Institute of Medicine (US). Committee on Military Nutrition Research. In *Not Eating Enough: Overcoming Underconsumption of Military Operational Rations*; Marriott, B.M., Ed.; Eating Situations, Food Appropriateness, and Consumption; National Academies Press (US): Washington, DC, USA, 1995; p. 18. Available online: https://www.ncbi.nlm.nih.gov/books/NBK232460 (accessed on 24 September 2019).
55. Whitelock, E.; Ensaff, H. On Your Own: Older Adults' Food Choice and Dietary Habits. *Nutrients* **2018**, *210*, 413. [CrossRef]
56. Dovey, T.M.; Staples, P.A.; Gibson, E.L.; Halford, J.C. Food neophobia and picky/fussy eating in children: A review. *Appetite* **2008**, *50*, 181–193. [CrossRef]
57. Koivisto-Hursti, U.-K.; Sjödén, P.O. Food and general neophobia and their relationship with self-reported food choice: Familial resemblance in Swedish families with children of ages 7–17 years. *Appetite* **1997**, *29*, 89–103. [CrossRef]
58. Nicklaus, S.; Boggio, V.; Chababnet, C.; Issanchou, S. Prospective study of food variety seeking in childhood, adolescence and early adult life. *Appetite* **2005**, *44*, 289–297. [CrossRef]
59. Knaapila, A.J.; Sandell, M.A.; Vaarno, J.; Hoppu, U.; Puolimatka, T.; Kaljonen, A.; Lagström, H. Food neophobia associates with lower dietary quality and higher BMI in Finnish adults. *Public Health Nutr.* **2015**, *18*, 2161–2171. [CrossRef]
60. Perry, R.A.; Mallan, K.M.; Koo, J.; Mauch, C.E.; Daniels, L.A.; Magarey, A.M. Food neophobia and its association with diet quality and weight in children aged 24 months: A cross sectional study. *Int. J. Behav. Nutr. Phys. Act.* **2015**, *13*, 13. [CrossRef]

61. Cooke, L.J.; Wardle, J.; Gibson, E.L.; Sapochnik, M.; Sheiham, A.; Lawson, M. Demographic, familial and trait predictors of fruit and vegetable consumption by pre-school children. *Public Health Nutr.* **2004**, *7*, 295–302. [CrossRef] [PubMed]
62. Camarena, D.; Sanjuán, A.; Philippidis, G. Influence of ethnocentrism and neo-phobia on ethnic food consumption in Spain. *Appetite* **2011**, *57*, 121–130. [CrossRef] [PubMed]
63. Martins, Y.; Pliner, P. Human food choices: An examination of the factors underlying acceptance/rejection of novel and familiar animal and nonanimal foods. *Appetite* **2005**, *45*, 214–224. [CrossRef] [PubMed]

nutrients

MDPI

Article

To Raise or Not to Raise the Level of Ingredients in Yoghurts: Polish Consumer Preferences Regarding Dairy Products

Marta Sajdakowska * and Agnieszka Tekień

Department of Food Market and Consumer Research, Institute of Human Nutrition Sciences, Warsaw University of Life Sciences (SGGW-WULS), 159C Nowoursynowska Street, 02-787 Warsaw, Poland; agnieszka_tekien@sggw.pl

* Correspondence: marta_sajdakowska@sggw.pl; Tel.: +48-225-937-145

Received: 19 September 2019; Accepted: 18 October 2019; Published: 19 October 2019

check for updates

Abstract: Modern consumers are becoming increasingly aware of the perceived health benefits of food. As a result, they are in search of various types of information, for example, information on the packaging of food products that could confirm to what extent the purchased product will meet their expectations regarding the proper composition, that is, nutritional value, or perceived health values earlier mentioned. Furthermore, consumers increasingly seek new dairy products with additional health benefits and, therefore, it is essential to explore which attributes are important drivers of food choices and how producers can better respond to shifting consumer values and needs in each dairy product category. Therefore, the aims of our research was twofold: (1) To determine different segments of consumers based on their preferences towards food and nutrition, including opinion on new food products with a particular emphasis on a dairy market as well as (2) to study the importance of some statements related to nutrition presented on the yoghurt label with a precise focus on aspects of the increased and decreased content of some ingredients. The data were collected using a CAPI (Computer Assisted Personal Interview) survey on a sample of 489 adult Polish consumers. Respondents provided answers to questions and took part in a discrete choice-based experiment. The obtained data were analysed using the clustering method. The segmentation was performed using a hierarchical Ward's method. As a result, four segments were identified Quality-oriented, Involved, Quality Enthusiasts, and Neutral. The results indicated that in relation to the features that are important in the case of yoghurts, the following were indicated above all: Beneficial effects on health, its sensory values, as well as its availability on the market and production by traditional methods. Consumers belonging to Quality Enthusiasts seemed to be the most promising segment due to their openness to new products, as well as positive feedback on yoghurt. From the perspective of taking action on the food market, Involved may also be interesting, as it showed their openness to new products available on the food market. However, due to the relatively lower, compared to other segments, assessment on the beneficial effect of yoghurt on health, their taste, aroma, availability, as well as the importance of information on care for the proper method of breeding animals, this segment can pose a special challenge to entrepreneurs. Moreover, Involved seemed to be more demanding and critical towards some projects undertaken on the market by policy makers and marketing practitioners.

Keywords: consumer; yoghurt; label; a discrete choice-based experiment

1. Introduction

In the last years, consumer demand for health-enhancing food products has risen rapidly [1–4]. On the one hand, the changing lifestyles and growing health concerns towards the negative impact of certain food products, e.g., saturated fatty acids originating from animals has increased consumer preferences for dairy-alternative products [5].

On the other hand, the review of available scientific evidence conducted by Thorning et al. (2016) supported the fact that the intake of milk and dairy products contributes to meeting nutrient recommendations and may protect against the most prevalent, chronic non-communicable diseases [3]. Tapsell (2015) stated that a combination of evidence is still necessary and more research is needed across various regions, but indications remain that fermented dairy foods such as yoghurt and cheese are an integral part of diets that are protective against cardiovascular disease (CVD) [4].

Results of other studies indicate that the suggestion to restrict or eliminate full-fat dairy from the diet may not be the optimal strategy for reducing cardiometabolic disease risk and should be re-evaluated in light of recent evidence [6].

Results of studies of Mazidi et al. (2019) showed that higher total dairy consumption was associated with lower total and cerebrovascular mortality, while higher milk consumption was associated with higher risk of coronary heart diseases (CHD). These results do not support public health advice to reduce total dairy fat consumption, although the association between milk consumption and CHD mortality requires further study [7].

Moreover, there is evidence for the impact of fermented foods and beverages (e.g., yoghurt), produced or preserved by the action of microorganisms, on general health, namely their significance on the gut microbiota balance and brain functionality [2].

In addition to this, when it comes to yoghurt, it is the most frequently consumed healthy and nutritious food around the world. Therefore, it has potential in conveying nutritious ingredients to the human diet. Most people in developing or underdeveloped countries suffer from micronutrient deficiencies and enriched food products can reduce nutritional diseases. A study shows that food enrichment can prevent most diseases and it has a significant impact in improving the health of the community [8]. Furthermore, yoghurt consumption is associated with a lower body mass index, lower body weight, smaller waist circumference, and lower body fat in epidemiological studies [9]. Moreover, one should also not forget about the indirect economic benefits associated with the consumption of yoghurt. Increasing yoghurt consumption in the adult population in the UK by 100g per day could generate substantial cost savings to the UK National Health Service as well as significant benefits for a patient through reductions in the incidence of type 2 diabetes (T2D) [10].

Apart from the aspect related to the beneficial effects of food on health, information on the product's label also plays a major role in consumer decisions [11]. Moreover, there is some empirical evidence for the existence of a positive relationship between nutritional label use and healthy food choices [12]. In general, food labelling is regulated in order to both help consumers make informed choices regarding the food they consume and to prevent any practices that may mislead them [13]. Furthermore, differences in nutritional knowledge related to nutritional recommendations and the links between nutrient consumption and health have been the main reason for differences in healthiness perception and willingness to try functional foods among consumers [14]. With regards to yoghurt, sweetness and information on sugar content have had significant effects on liking and purchase probability [15].

Other research findings showed that the response to labels differ, both with regard to the way information is presented (facts or claims) and with the type of information (nutrition or health) offered. Consumer utility increases when nutrition and health information labels in food products are present and tends to provide higher utility than facts panel only in the case of the less healthy product [16]. While consumers generally consider the nutrition composition of yoghurt to be more important than the tested claims, some groups of consumers are more sensitive to the use of health-related statements [13].

Since some consumers do not understand health and nutrition claims and even express doubts over the claimed effects of food, the importance of information about the content of a nutrient in the product should also be examined in order to find out which attribute is more convincing for consumers when the purchase decisions are being made [17].

Apart from the health-related aspects of dairy products and the role of food information labels in consumer purchasing decisions, generally, consumer knowledge of the relationship between diet and health is important for innovation in the food sector, therefore, producers must find a way to reduce unhealthy ingredients and enhance healthy ones in food products [18].

Producers have focused on reducing sodium, fat, cholesterol contents in food products [19] as well as in reducing sugar contents, and even some recommendations have been proposed by policy makers [20].

Therefore, the aims of our research was in twofold: (1) To determine different segments of consumers based on their preferences towards food and nutrition including opinion on new food products with a particular emphasis on the dairy market as well as (2) to study the importance of some statements related to nutrition presented on the yoghurt label with a precise focus on aspects of the increased and decreased content of some ingredients.

2. Material and Methods

2.1. Design of the Experiment

In our research, a discrete choice-based experiment was conducted in order to elicit consumer preferences referring to different product profiles with some levels of attributes [21], so study participants made a discrete choice from a set of presented alternatives which contained a number of attributes with different levels, combined within choice sets [22]. It was analysed with discrete choice models [23] in order to get utilities important to respondents from a situation where consumers choose a particular product among available products [24]. This kind of experiment was selected instead of rating-based conjoint due to its greater similarity with market behaviour, i.e., more similar to what a consumer really does when buying food [25].

2.2. Data Collection Process

This paper presents some of the findings from a larger study [26]. The sample was drawn from the Social Security addresses database and was representative in terms of age and gender. The survey was conducted in each of the 16 voivodships in Poland. After drawing the starting addresses, the method of random route was used in the selection of the sample [27]. The interviews were conducted face-to-face at respondents' homes by a professional market research agency, respecting the ESOMAR (European Society for Opinion and Marketing Research) code of conduct using the CAPI (Computer Assisted Personal Interview) technique. All respondents were aged 21 years and over. Only respondents who met the recruitment criteria, i.e., made their own or cooperative food purchase and declared dairy products consumption, participated in the study. Those who declared the purchase and consumption of yoghurts among consumers from the total sample, took part in the research (N = 489) (Figure 1).

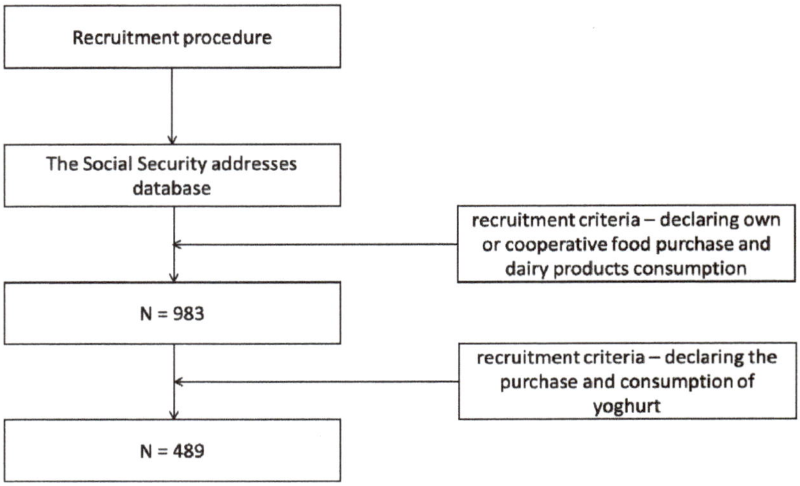

Figure 1. Participants' inclusion in the study.

2.3. Consumer Attributes

Taking into account various determinants that are important for consumers during the purchasing choice of yoghurt [25], we decided to select the following for the experiment attributes: Increased content of ingredients (8 levels), reduced content of ingredients (5 levels), additional claims (4 levels), and price (12 levels). Attributes and their levels are presented in Table 1.

Table 1. Attributes and levels used in the discrete choice-based experiment design.

Attribute	Attribute Level
Increased content of ingredients	With an increased amount of vitamins and minerals With an increased amount of fiber With an increased amount of live bacteria cultures With an increased amount of cholesterol lowering ingredients With an increased amount of omega 3 acids With an increased amount of coenzyme Q10 With an increased amount of ingredients positively affecting body shape, complexion, nails No information
Reduced content of ingredients	No added sugar Low energy content Low salt content Reduced sugar content No information
Additional claims	High nutritional value High health values High quality No information
Price	12 levels (from PLN * 1.00 to PLN 15.99)

* PLN = Polish new zloty (approx. 1 PLN = EUR 0.23).

During testing, the respondents were presented with screens displaying the full characteristics of each product (every product was described on each of the attributes). As shown in Table 2, 3 product configurations were presented on the screen and the respondent had to indicate the preferred alternative. In this study, a "no-choice" option was not included although it has been pointed out that sometimes this option may give a better market penetration prediction [23]. As a result, it was a situation of a forced choice where the respondent had to opt for one configuration in order to choose the preferred product (out of 3) to buy. Each respondent had to choose one out of 3 products. The task was repeated 12 times for each respondent. During the research, consumers were asked the following question: "Which yoghurt do you want to buy?"

Table 2. Example of the discrete choice-based experiment screen used in the study.

Option 1	Option 2	Option 3
	Yoghurt	
with high fiber content	with an increased amount of live bacteria cultures	with an increased amount of cholesterol-lowering ingredients
with no added sugar	with law salt content	with reduced sugar content
with high health values	of high quality	of high nutritional value
PLN 12.99	PLN 3.59	PLN 2.29

2.4. Data Analysis

The Hierarchical Bayesian (HB) network model was applied in Sawtooth SMRT (Sawtooth Software Market Research Tool). This is a dedicated software for a discrete choice-based experiment, which enables the estimation of coefficients for the individual utilities of each attribute level as well as an estimate of individual-level part-worth values. The HB algorithm has the ability to borrow information from other respondents in order to stabilise part-worth estimation for each individual. It is a valuable feature for this method. The hierarchy of the Bayesian network results from the fact that two levels can be distinguished [24]. At a higher level, the assumption is that each individual's part-worths are described by a multivariate normal distribution. Such a distribution is characterised by a vector of means and a covariance matrix. At a lower level, it is assumed that given the individual's part-worths, their probabilities of choosing particular alternatives are governed by a multinomial logit model [28]. After modelling with the HB network, part-worth utilities for all product attributes and levels were obtained and, in further studies, an analysis in IBM SPSS Statistics version 23PL (IBM Corp. in Armonk, NY, USA) was carried out to study the data more closely. The script for this discrete choice-based experiment was prepared in the Sawtooth SMRT software v. 4.22 (Sawtooth Software, Provo, UT, USA). A "full profile" option was used as a rotation scheme of used product variants. This means that all of the attributes of each product were always shown on screen. The profiles were generated using the complete enumeration method, where each presented product profile differed from the adjacent product in each of the presented attributes.

In order to obtain an in-depth analysis, the sample of respondents who took part in the discrete choice-based study was subjected to segmentation. The following questions from the survey questionnaire were selected as the basis for segmentation: (1) Please specify how much you agree or disagree with the following statements: "I like to buy new dairy products", where 1 means I definitely disagree and 7 means I definitely agree, (2) "How important is it for you that the yoghurt you eat does not contain artificial additives", where 1 means definitely not important and 7 means definitely important, as well as (3) "How important is it for you that the yoghurt you eat is organic", where 1 means definitely not important and 7 means definitely important.

Segmentation was performed in SPSS v.23 using the hierarchical Ward's method. Using Ward's method [28,29] for each cluster, the means were calculated for all variables. In the next step, the squared Euclidean distance to the cluster means was calculated for each considered case. Then, all these distances were summed for all of the cases. At each step, the two clusters to be merged were those that resulted in the smallest increase in the overall sum of the squared within-cluster distances. The coefficient in the agglomeration schedule is the within-cluster sum of squares at that step. It is not the distance at which clusters are joined.

3. Results

3.1. Socio-Demographic Profile of the Total Sample and the Clusters Identified

As a result of the segmentation analysis, four cardinality clusters (segments) were identified from the following multiplicity: Cluster 1 (N = 78, 16% of total sample), cluster 2 (N = 123, 25%), cluster 3 (N = 233, 48%), and cluster 4 (N = 55, 11%) (Table 3). In the next step, the data from the discrete choice-based experiment module was calculated for all the clusters.

Table 3. Socio-demographic profile of the total sample and the clusters identified (%) (N = 489).

Variable	Total Sample (n = 489)	Cluster 1 (n = 78, 16%)	Cluster 2 (n = 123, 25%)	Cluster 3 (n = 233, 48%)	Cluster 4 (n = 55, 11%)	Sig.
Gender						0.040
Female	59%	47%	58%	37%	56%	
Male	41%	53%	42%	63%	44%	
Education						0.484
Primary and junior high school	8%	13%	6%	6%	15%	
Vocational	29%	23%	32%	30%	29%	
Secondary	41%	38%	46%	43%	27%	
Higher	20%	26%	14%	19%	29%	
Refusal of answer	1%	0%	2%	2%	0%	
Age						0.008
34 and below	37%	28%	40%	43%	22%	
35–54	39%	40%	37%	37%	53%	
55–64	15%	18%	15%	14%	11%	
65 and over	9%	14%	8%	6%	15%	
Place of residence						0.000
Country	34%	37%	29%	29%	47%	
Up to 20,000	13%	13%	9%	15%	7%	
20,000–100,000	21%	26%	24%	20%	20%	
100,000–500,000	18%	15%	17%	24%	11%	
Over 500,000	14%	9%	21%	11%	15%	
Children						0.219
Yes	49%	46%	46%	48%	62%	
No	34%	28%	35%	38%	24%	
Refusal of answer	1%	1%	2%	1%	0%	
N.A.	16%	24%	18%	13%	15%	
Financial status						0.898
Sufficient budget without necessity to economise	10%	10%	11%	10%	11%	
We live frugally and have enough money to buy what we need	40%	47%	41%	38%	38%	
We live very frugally to save money for major purchases	28%	22%	24%	31%	31%	
We have enough money for the cheapest food or clothing	10%	9%	15%	9%	5%	
We have enough money for the cheapest food only, there is not enough money for clothing	2%	4%	0%	2%	4%	
There is not enough money even for the cheapest food or clothing	1%	0%	1%	1%	0%	
I don't know/hard to say	7%	4%	7%	8%	9%	
Refusal of answer	2%	4%	2%	1%	2%	

In the examined general sample (Table 3), women dominated (59%). Taking age into account, it shows that the majority of respondents were aged 35–54 years and up to 34 (39% and 37%, respectively). Over 2/5 of respondents (41%) had secondary education and almost 1/3 had vocational education (29%). Taking into account the place of residence indicates that over 1/3 of respondents (34%) live in the country, every fifth respondent live in cities with a population of 20,000–100,000 inhabitants (21%), and cities of 100,000–500,000 inhabitants (18%). Almost half of the respondents declared having children (49%) and almost 70% of the surveyed assessed that they lived frugally and had enough money to buy what they needed, and that they lived very frugally to save money for major purchases (40% and 28%, respectively). In segment 1 (16%), men (53%) dominated as well as people aged 35–54 years (40%) as well as people with a secondary (38%) and higher education level (26%), live in the country (37%) and cities of 20,000–100,000 residents (26%), and have children (46%). Almost half of the respondents in segment 1 declared that they lived frugally and had enough money for everything (47%). In segment 2 (25%), women (58%) dominated as well as those up to 34 years and those aged 35–54 years (40% and 37%, respectively), have secondary education (46%), reside in a village and town of 20,000–100,000 residents respectively (29% and 24%). Almost half of the people in this segment declared that they have children (46%), and more than 2/5 said that they live frugally and have enough money for everything (41%). In segment 3 (48%), men (63%), people up to 34 years of age and 35–54 years (43% and 37%, respectively), as well as people with secondary education (43%) dominated. Segment 3 comprised of people living in the country and large cities of 100,000–500,000 inhabitants (29% and 24% respectively), have children (48%), and declare that they live frugally and have enough for everything (38%), and that they live very frugally to save money for major purchases (31%).

In segment 4 (11%), women (56%), people aged 35–54 years (53%), live in the country (47%), and have children (62%) dominated. Similarly to segment 3, those belonging to segment 4 declared that they live frugally and have money for everything (38%) or that they live very frugally to save money for major purchases (31%) (Table 3).

In order to characterise the segments, several questions from the questionnaire were used. The first (1) of them concerned opinions on the quality of dairy products: "Please indicate how much you agree with the following statements". Please provide answers on a scale of 1–7, where 1 meant "definitely disagree" and 7 meant "definitely agree". Agreement was assessed in relation to the following statements: Quality is important to me when choosing dairy products, I buy high-quality dairy products because they have a good effect on my children's health, I buy high-quality dairy products because they have a positive effect on my body shape, I buy high-quality dairy products for family members who have health issues. These statements referring to the importance of food quality were also used as the basis for naming individual segments: Quality-oriented (cluster 1), Involved (cluster 2), Quality Enthusiasts (cluster 3), and Neutral (cluster 4) (Table 4).

The next questions that were used in segment profiling were: (2) "Please say how much you agree or disagree with the statements about dairy products", where 1 meant strongly disagree and 7 meant strongly agree and they referred to new dairy products available on the market. Table 4 presents the items used in the question. The next question concerned selected features of yoghurts and was: (3) "Which of the following features of yoghurts which you usually consume are important for you?", where 1 meant definitely not important and 7 meant definitely important. The last question that was used in profiling (4) related to information that refers to the presence of various ingredients in yoghurts and was: "Please specify if you would be interested in the following types of yoghurts". Please use a scale from 1–7, where 1 meant definitely not and 7 meant definitely yes.

Analysis of the data presented in Table 4 indicates that the quality of dairy products played a major role in the purchasing decisions of the respondents. This thesis is confirmed by relatively high levels of agreement with the statement: "Quality is important to me when choosing dairy products". The respondents also declared a high level of agreement with the statement: "I buy high-quality dairy products because they have a good impact on the health of my children" (5.11), but no differences at a statistically significant level were found between individual clusters. At a medium level, there was an agreement with the statement: "I buy high-quality dairy products because they have a positive effect on my body shape", while respondents in clusters 2 and 3 appreciated the impact on the body shape to be significantly higher compared to respondents in segments 1 and 4.

Table 4. Clusters' profile on attitudinal questionnaire referring to the importance of dairy products quality and openness to buy new products as well as changes in the market including dairy products ($N = 489$).

Statements	Total Sample	Cluster 1 QUALITY ORIENTED	Cluster 2 INVOLVED	Cluster 3 QUALITY ENTHUSIASTS	Cluster 4 NEUTRAL	Sig.
Quality matters to me while choosing a dairy products	5.96	6.28 a	5.40 b	6.15 a	5.96 a	0.000
I buy high-quality dairy products because they have a beneficial influence on the health of my children	5.11	4.91 a	4.93 a	5.23 a	5.24 a	0.375
I buy high-quality dairy products because they have a beneficial influence on my body shape	4.48	3.79 b	4.72 a	4.71 a	3.91 b	0.000
I buy high-quality dairy products only for those family members who have health issues	3.35	3.06 b,c	3.94 a	3.11 b,c	3.40 a,c	0.000
I am very particular about the new dairy products I will eat	4.79	4.49 a	4.63 a	4.98 a	4.76 a	0.139
Ethnic food deprived of dairy products looks too weird to eat (e.g., Asian cuisine)	4.58	4.88 a	4.46 a	4.49 a	4.76 a	0.954
If I do not know what is in a dairy product, I will not try it	4.54	4.71 a	4.37 a	4.51 a	4.82 a	0.514
New dairy products arouse my curiosity	4.39	3.04 b	4.78 a	5.00 a	2.23 b	0.000
I do not trust new foods [1]	4.32	2.91 b	4.56 a	4.89 a	3.36 b	0.000
I like to buy new and various dairy products	4.26	1.71 c	4.92 a	5.24 a	2.24 b	0.000
I will eat virtually everything	4.16	3.69 b	4.52 a	4.30 a	3.38 b	0.000
At dinner parties, I will try new dishes based on dairy products	4.09	2.88 b	4.77 a	4.49 a	2.56 b	0.000
I like dairy products from cuisines of different countries	3.39	2.67 c	4.70 a	4.02 b	2.35 c	0.000
I constantly try new and varied dairy products	3.83	2.12 b	4.49 a	4.40 a	2.35 b	0.000
I am usually amongst the first ones to try new dairy products	3.74	1.95 b	4.59 a	4.27 a	2.11 b	0.000
I am afraid to eat dairy products I have never tried before	3.74	4.47 a	4.03 a	3.34 b,c	3.73 a,c	0.000
I know more than others about the latest dairy products	3.40	2.32 b	4.02 a	3.75 a	2.06 b	0.000
Among friends I am usually the first person to try new dairy products	3.30	1.81 b	4.20 a	3.64 a	2.02 b	0.000
I like to try ethnic restaurants (e.g., Asian cuisine)	3.27	2.51 b	4.26 a	3.19 b	2.49 b	0.000
I look for information about what new dairy products appear on the market	3.09	1.92 c	3.96 a	3.27 b	2.09 c	0.000

[1] The item was reversed. a, b, c means with the same letter are not significantly different; ANOVA (Analysis of Variance) Tukey's post hoc test.

Regarding the agreement with statements on openness to buy new products and changes in the market including dairy products, the respondents agreed most strongly with the following statements: "I am very particular about the new dairy products I will eat (4.79)", "Ethnic food deprived of dairy products looks too weird to eat (e.g., Asian cuisine)" (4.58), and "If I do not know what is in a dairy product, I will not try it" (4.54).

In relation to the above statements, no statistically significant differences were noted between individual segments. Further statements about new dairy products with average scores of 4 and above, i.e., "New dairy products arouse my curiosity" (4.39), "I trust new dairy products", "I like to buy new and various dairy products" (4.26), "I will eat virtually everything" (4.16), and "At dinner parties, I will try new dishes based on dairy products" (4.09) had a significantly higher level of agreement in segments 2 and 3 compared to segments 1 and 4, which may indicate a greater openness of consumers in segments 2 and 3 to new products on the food/dairy market compared to those in segments 1 and 4.

3.2. Attitudinal Questionnaire

Analysis of the data presented in Table 5 indicates that in relation to the features that are important in the case of yoghurts, the following were indicated above all: Beneficial effects on health, its sensory values (statements about taste and smell), as well as its availability on the market and production by traditional methods. Furthermore, the Polish origin of yoghurt and the origin of farms where the producer cares in a special way for the proper way of breeding animals. Respondents from segment 2 assessed the following aspects lower than in other segments: The significance of beneficial effects on health, flavour, availability of yoghurts, and information on Polish origin. Compared to other segments, these surveyed features also showed the least appreciation for the importance of the origin of yoghurt from farms where the producer pays special attention to the proper way of breeding animals.

Table 5. Clusters' profile on attitudinal questionnaire referring to the importance of some attributes of yoghurt ($N = 489$).

Statements Referring to Attributes of Yoghurt that Is Usually Consumed	Total Sample	Cluster 1 QUALITY ORIENTED	Cluster 2 INVOLVED	Cluster 3 QUALITY ENTHUSIASTS	Cluster 4 NEUTRAL	Sig.
It was good for health	6.47	6.85 a	5.57 b	6.76 a	6.76 a	0.000
It was tasty	6.44	6.67 a	5.64 b	6.76 a	6.58 a	0.000
It was easily accessible	6.13	6.50 a	5.26 b	6.44 a	6.20 a	0.000
It had a pleasant smell	6.09	6.32 a	5.37 b	6.42 a	6.00 a	0.000
It was made in a traditional way	6.06	6.63 a	5.20 b	6.42 a	5.65 b	0.000
It came from farms where the producer pays special attention to the proper way of breeding animals	5.99	6.68 a	4.92 c	6.40 a	5.65 b	0.000
It was of Polish origin	5.97	6.55 a	5.05 b	6.33 a	5.69 c	0.000
It contained a lot of vitamins and minerals	5.91	6.27 a	5.09 b	6.31 a	5.55 b	0.000
It contained fiber	5.67	6.09 a	4.80 b	6.09 a	5.24 b	0.000
It was low in sugar	5.34	5.65 a	4.76 b	5.70 a	4.73 b	0.000
It was low in fat	5.10	5.28 a,c	4.73 b,c	5.43 a	4.25 b	0.000
It was low in calories	5.08	5.35 a	4.74 a,c	5.34 a	4.40 b,c	0.000
It contributed to maintaining a slim body	5.06	4.56 b	4.85 b	5.52 a	4.31 b	0.000
The packaging was appealing	4.95	4.55 b,c	4.69 c	5.48 a	3.85 b	0.000

a, b, c means with the same letter are not significantly different; ANOVA Tukey's post hoc test.

Analysis of the data presented in Table 6 indicates that in relation to specific examples of information that would accompany the availability of yoghurt on the market, the respondents most often declared interest in yoghurt "with live bacteria cultures" and "produced using raw materials from organic farms". Next, interest was indicated in yoghurts with increased levels of substances beneficial to health with reduced levels of some ingredients, e.g., salt or sugar, in order to prevent various diseases, e.g., obesity, hypertension, diabetes, followed by yoghurts with higher levels of certain vitamins and minerals in order to prevent nutritional deficiencies. It was noted that respondents in segments 1 and

3 were significantly more interested in yoghurts with live bacteria cultures and yoghurts produced using raw materials from organic farms than those in segments 2 and 4. Respondents from segment 3 obtained the highest average scores compared to other segments in relation to yoghurts with live bacteria cultures with increased levels of substances beneficial to health, yoghurts with higher levels of some vitamins and minerals, and with reduced levels of some ingredients, e.g., salt or sugar, to prevent various diseases such as obesity, hypertension, and diabetes. On the other hand, respondents from segment 1 compared to other surveyed features obtained the highest rating in relation to yoghurts produced using raw materials from organic farms. In the case of segment 2, relatively low scores were obtained in comparison to other segments with regard to yoghurts with live bacteria cultures made from raw materials from organic farms with reduced levels of some ingredients, e.g., salt or sugar, in order to prevent various diseases such as obesity, hypertension, and diabetes, as well as an increased level of substances with beneficial effects on health.

Table 6. Profile on attitudinal questionnaire referring to purchase of various yoghurt types (*N* = 489).

Statements	Total Sample	Cluster 1 QUALITY ORIENTED	Cluster 2 INVOLVED	Cluster 3 QUALITY ENTHUSIASTS	Cluster 4 NEUTRAL	Sig.
With live bacteria cultures	5.89	6.19 a	5.07 c	6.23 a	5.87 b	0.000
Produced using raw materials from organic farms	5.81	6.36 a	4.98 b	6.17 a	5.36 b	0.000
With an increased level of substances beneficial to health	5.56	5.62 a,c	5.11 b,c	5.87 a	5.16 a,c	0.000
With reduced levels of some ingredients, e.g., salt or sugar, to prevent various diseases e.g., obesity, hypertension, diabetes	5.41	5.56 a,c	4.88 b	5.75 a	4.98 b,c	0.000
With higher levels of some vitamins and minerals to prevent nutritional deficiencies	5.40	5.41 b,c	4.98 b	5.73 a,c	4.89 b	0.000

a, b, c means with the same letter are not significantly different; ANOVA Tukey's post hoc test.

As indicated in the previous part of the study, in the next part of the survey the respondents were asked to answer the question: "Which yogurt would you like to buy?" using the discrete choice-based experiment method. The subjects were offered several attributes of yogurt: (1) Increased level of ingredients, (2) reduced level of ingredients, (3) additional information, and (4) price.

3.3. The Relative Importance of Attributes

An analysis of the values (mean relatives) presented in Table 7 indicates that consumers perceived the price and information on increasing selected ingredients in yoghurts as the two most important attributes prompting them to buy food products/yoghurts (70.4% and 15.7%, respectively), and the following "additional information" and information on reducing selected ingredients (8.5% and 5.4%, respectively) were mentioned. Along with the increase in price, a decrease in utility/lower utility was noted. Regarding ingredients with increased levels, the most accepted information was "With an increased number of live bacteria cultures", "With an increased amount of vitamins and minerals", and "With an increased amount of fiber" (0.939, 0.546, 0.453, respectively).

Table 7. The part-worth utilities and relative importance of attributes for total sample and four identified clusters (*N* = 489).

Attribute	Attribute Level	Total Sample	Cluster 1 QUALITY ORIENTED	Cluster 2 INVOLVED	Cluster 3 QUALITY ENTHUSIASTS	Cluster 4 NEUTRAL	Sig.
Increased level of ingredients	Relative importance (%)	15.7	19.4	13.0	16.0	16.3	
	With an increased amount of vitamins and minerals	0.546	0.594	0.582	0.540	0.411	0.517
	With an increased amount of fiber	0.453	0.523	0.338	0.489	0.420	0.189
	With an increased number of live bacteria cultures	0.939	1.160	0.645	0.951	1.209	0.001
	With an increased amount of cholesterol lowering ingredients	0.133	0.188	0.062	0.160	0.118	0.751
	With an increased amount of omega3 acid	−0.232	−0.185	−0.292	−0.170	−0.412	0.427
	With an increased amount of coenzyme Q10	−0.774	−0.834	−0.552	−0.837	−0.860	0.073
	With an increased amount of ingredients that have a beneficial effect on the body shape, complexion and nails	−0.081	−0.360	0.191	−0.109	−0.149	0.005
	Lack of information	−0.985	−1.085	−0.973	−1.024	−0.737	0.262
Reduced level of ingredients	Relative importance (%)	5.4	8.5	4.2	6.2	3.6	
	No added sugar	0.323	0.497	0.147	0.382	0.241	0.032
	Low energy content	−0.338	−0.485	−0.206	−0.382	−0.221	0.008
	Low salt content	−0.038	−0.166	0.165	−0.085	−0.083	0.002
	With reduced sugar content	0.270	0.426	0.207	0.279	0.151	0.055
	Lack of information	−0.218	−0.272	−0.312	−0.195	−0.087	0.331
Additional information	Relative importance (%)	8.5	8.4	7.4	9.2	8.1	
	On high nutritional value	0.257	0.197	0.224	0.317	0.164	0.169
	On high health values	0.315	0.352	0.274	0.320	0.345	0.717
	On high quality	0.155	0.076	0.154	0.173	0.177	0.495
	Lack of information	−0.727	−0.625	−0.652	−0.811	−0.686	0.079
Price	Relative importance (%)	70.4	63.7	75.4	68.7	71.9	
	PLN * 1.00	5.064	4.251	5.540	4.978	5.441	0.301
	PLN 1.29	4.547	3.805	5.000	4.477	4.822	0.230
	PLN 1.69	3.670	3.147	4.034	3.586	3.882	0.140
	PLN 2.19	3.118	2.661	3.436	3.085	3.126	0.094
	PLN 2.79	1.634	1.549	1.721	1.617	1.594	0.580
	PLN 3.59	−0.535	−0.316	−0.705	−0.481	−0.663	0.089
	PLN 4.59	−1.568	−1.249	−1.766	−1.540	−1.667	0.165
	PLN 5.89	−2.502	−2.099	−2.699	−2.490	−2.625	0.353
	PLN 7.59	−3.108	−2.691	−3.392	−3.059	−3.220	0.233
	PLN 9.79	−3.288	−2.879	−3.568	−3.238	−3.403	0.198
	PLN 12.99	−3.468	−3.046	−3.754	−3.417	−3.598	0.191
	PLN 15.99	−3.564	−3.133	−3.848	−3.518	−3.687	0.164

* PLN = Polish new zloty (approx. 1 PLN = EUR 0.23).

With regard to "additional information", the highest level of preferences was noted for information: On high health values (0.315), weight is attached to the role of health in general compared to information on high nutritional value and high quality (0.257 and 0.155, respectively). When the information was not available (No information), negative usability was noted (−0.727).

For information relating to reduced levels of ingredients: A higher level of utility was obtained for the information "No added sugar" (0.332) compared to the information "Reduced sugar" (0.270). Other levels of attributes have negative usability. For respondents in segment 1, the most important factors taken into account when choosing yoghurt was information on ingredients whose amount in the product was increased (19.4%) and information on ingredients whose amount was reduced (8.5%). On the other hand, the significance of price was the lowest, compared to the opinions of people in other segments (63.7%). For consumers in segment 2, the price was the most important factor during the purchase of yoghurt (75.4%) compared to the sequence of factors for other segments and the remaining attributes played a relatively less significant role. In the case of segment 3, respondents attached a lot of importance to "additional information" compared to other segments (9.2%). The significance of the price was relatively lower for segment 3 (68.7%) compared to segment 2 (75.4%), as well as compared to segment 4 (71.9%). As noted, the respondents in segment 4 belonged to the group of people for whom the price was relatively important, but they also attached attention to information on increasing selected components (16.3%). For both segment 1 and segment 4, for ingredients whose increase was most preferred, the highest part-worth utilities were recorded for information "With increased number of live bacteria cultures" (1.160 and 1.209, respectively).

4. Discussion

The study presents the results of a consumer survey using a questionnaire among Polish consumers. The analysis of the results obtained indicated that the quality of dairy products is of great importance to Polish consumers. This is confirmed by the results of studies by other authors, in which it was emphasised that the perception of food quality by consumers affect their purchase decisions and dietary patterns [29]. In addition, the survey found that some consumers were open to food and nutrition news. This concerned two consumer segments (2 and 3). These two segments compared to 1 and 4 also contained, apart from those aged over 35, relatively more younger consumers, i.e., 34 and below, which may have affected the acceptance of new/innovative products as indicated by the results of other surveys among consumers, including those relating to Polish consumers [30].

4.1. The Most Important Yoghurt Features

Regarding yoghurts, among the most important features characterising them, beneficial effects on health, flavour, and availability on the market were indicated. The literature confirms the obtained research results refering to flavour and availability [15,31,32], as well as the importance of the perceived beneficial impact of food on health was indicated in other consumer studies [26,30,33–36].

Moreover, as consumers demand healthy foods with a pleasant taste, in recent years, some functional dairy products have been produced by means of enrichment and fortification. Consequently, yoghurt has begun to attract new consumer groups due to its pleasant taste and increased health benefits [37].

The beneficial effect on health in the described own study was also confirmed by relatively high assessments regarding the presence of certain ingredients in food. Augmenting the ingredients whose increased quantity positively associates with health effects e.g., fiber, vitamins, and minerals, was strongly accepted and the reduced amount of ingredients that consumers perceive as negatively affecting health e.g., sugar or fat, was accepted. This is confirmed by the acceptance of changes in food of animal origin by Polish consumers [38]. In addition, the literature indicates that for individual ingredients that have been raised or lowered, some consumers accept fiber-enriched yoghurt [39]. Moreover, when it comes to vitamins and minerals, due to the high consumption rate of dairy products such as yoghurt, the fortification of these products will effectively reduce or prevent diseases associated with nutritional deficiencies [8]. When it comes to reducing the level of sugar and fat in dairy products, the results of research among Polish consumers confirm the preferences in this category of products [38].

In the case of this research, the method of producing yoghurt is also worth emphasising, in particular referring to traditional production methods and involvement in animal welfare. These two aspects are particularly important because of the perceived quality of dairy products by consumers, including their freshness and taste [31,40–42]. This is also confirmed by the research of other authors of the source, in which attention is drawn to the importance of traditional production methods [43] and the growing importance of caring for animals in consumer statements [44,45]. However, the importance of the aspect related to caring for animals is presented differently, depending on the study [46,47].

The discrete choice-based study also noted the importance of price as a factor in food, which is confirmed by the studies of other authors [33,48–50]. Price was ranked as the most important food choice factor in five countries (Spain, Greece, Ireland, Portugal, and the Netherlands), sensory appeal was ranked first for three countries (Norway, Germany, and the UK), while natural content was ranked as the most important factor in Poland. Familiarity and ethical concern were consistently ranked as the least important in all countries [51]. The results of the review made by Roman et al. (2017) clearly showed that for the majority of consumers in developed countries, naturalness in food products is important. This finding could be observed across countries and in the different years when the studies were conducted [52]. On the other hand, Aschemann-Witzel [53] indicated that the consumer's role and consumption aspect of the supply chain were identified to be crucial in improving healthy choices and achieving sustainability goals.

As for the importance of live bacteria cultures, it was also important for the subjects tested. This may be associated with the generally positive perception of yoghurt through the prism of the content of live bacterial cultures, which is confirmed by the studies of other authors [54,55]. Research shows that yoghurt still plays an important role in the human diet today due to its pleasant taste and health benefits. To meet consumer demands for healthier options, manufacturers are making low fat and non-fat versions of their most popular flavours. Unfortunately, when adding prebiotics to yoghurt, a negative impact was observed on the sensory characteristics of the yoghurt turning off consumers [37].

4.2. *Perspectives of Information on the Yoghurt's Labels*

Although the health motive seems to be one of the most important factors affecting functional foods, its effect partially depends on the consumer's knowledge of a particular health-enhancing effect, as well as on the barriers and benefits they perceive from the use of nutritional labels [56]. In general, nutritional claim (NC) requirements on food packages are among the most important and influential EU policy measures related to diet, which has the capacity to promote healthy eating. The results of other research indicate that the low-sugar NC was the least preferred claim. Overall, the presence of NCs generally increases visual attention, which may be linked to an increased likelihood of affecting the final decision to purchase yogurts with NCs [57]. In summary, yoghurt has always been one of the vital players in the spectrum of fermented food products that has transformed science and technology into health and wellness through diet. Considering the fast evolution of functional yoghurts either at a research stage or marketplace, further development should demand an accurate measurement of quality, safety, and efficacy to meet consumer expectations on quality and claimable health benefits [54]. Studies show that the acceptance of products with health claims is influenced by several factors. Familiarity with the product, health claim, or functional ingredient used plus personal relevance appear as the most important determinants. The choice of a carrier product can determine to what extent people trust a health claim or are willing to try the respective product. Furthermore, consumers like simple wording, but they may also demand detailed explanations. However, more research is needed on consumer understanding of health claims in order to maximise the potential for functional foods in order to contribute to healthy, balanced diets [58].

5. Conclusions

An analysis of the available results of our research indicated that consumers belonging to segment 3 (Quality Enthusiasts) seemed to be the most promising segment due to their openness to new products, as well as positive feedback on yoghurt. From the perspective of taking action on the food market, segment 2 (Involved) may also be interesting, as it showed openness to new products available on the food market. However, due to the relatively low, compared to other segments, assessment on the beneficial effect of yoghurt on health, their taste and aroma, availability, as well as the importance of information on the proper way of breeding animals, this segment could be a special challenge for entrepreneurs. Segment 2 seemed to be more demanding and critical (or perhaps more cautious) regarding some projects undertaken on the market.

Author Contributions: M.S. developed the concept of the study and supervised the experiments. A.T. analysed the data. M.S. and A.T. interpreted the data. M.S. wrote the original draft of the manuscript. M.S. wrote the manuscript. M.S. and A.T. were involved in critically revising the manuscript. Both authors read and approved the final manuscript.

Funding: Research financed by Polish Ministry of Science and Higher Education within funds of Institute of Human Nutrition Sciences, Faculty of Human Nutrition, Warsaw University of Life Sciences (WULS), for scientific research.

Acknowledgments: The survey was a part of the "BIOFOOD—Innovative, Functional Products of Animal Origin" project that was co-financed by the European Union from the European Regional Development Fund within the Innovative Economy Operational Programme 2007–2013.

Conflicts of Interest: The authors declare no conflict of interest.

References

1. Jezewska-Zychowicz, M.; Królak, M. Do Consumers ' Attitudes Towards Food Technologies and Motives of Food Choice In fl uence Willingness to Eat Cereal Products Forti fi ed with Fibre? *Pol. J. Food Nutr. Sci.* **2015**, *65*, 281–292. [CrossRef]
2. Bell, V.; Ferrão, J.; Pimente, L.; Pintado, M.; Fernandes, T. One Health, Fermented Foods, and Gut Microbiota. *Foods* **2018**, *7*, 195. [CrossRef] [PubMed]
3. Thorning, T.K.; Raben, A.; Tholstrup, T.; Soedamah-Muthu, S.S.; Givens, I.; Astrup, A. Milk and dairy products: good or bad for human health? An assessment of the totality of scientific evidence. *Food Nutr. Res.* **2016**, *60*, 32527. [CrossRef] [PubMed]
4. Tapsell, L.C. Fermented dairy food and CVD risk. *Br. J. Nutr.* **2015**, *113*, S131–S135. [CrossRef] [PubMed]
5. Laassal, M.; Kallas, Z. Consumers Preferences for Dairy-Alternative Beverage Using Home-Scan Data in Catalonia. *Beverages* **2019**, *5*, 55. [CrossRef]
6. Astrup, A.; Geiker, N.R.W.; Magkos, F. Effects of Full-Fat and Fermented Dairy Products on Cardiometabolic Disease: Food Is More Than the Sum of Its Parts. *Adv. Nutr.* **2019**, *10*, 924S–930S. [CrossRef]
7. Mazidi, M.; Mikhailidis, D.P.; Sattar, N.; Howard, G.; Graham, I.; Banach, M. Consumption of dairy product and its association with total and cause specific mortality—A population-based cohort study and meta-analysis. *Clin. Nutr.* **2019**, in press. [CrossRef]
8. Gahruie, H.H.; Eskandari, M.H.; Mesbahi, G.; Hanifpour, M.A.; Hadi, H.G.; Hadi, E.M.; Gholamreza, M.; Amin, H.M. Scientific and technical aspects of yogurt fortification: A review. *Food Sci. Hum. Wellness* **2015**, *4*, 1–8. [CrossRef]
9. Eales, J.; Lenoir-Wijnkoop, I.; King, S.; Wood, H.; Kok, F.J.; Shamir, R.; Prentice, A.; Edwards, M.; Glanville, J.; Atkinson, R.L. Is consuming yoghurt associated with weight management outcomes? Results from a systematic review. *Int. J. Obes.* **2016**, *40*, 731–746. [CrossRef]
10. Lenoir-Wijnkoop, I.; Mahon, J.; Claxton, L.; Wooding, A.; Prentice, A.; Finer, N. An economic model for the use of yoghurt in type 2 diabetes risk reduction in the UK. *BMC Nutr.* **2016**, *2*, 1–10. [CrossRef]
11. Graham, D.J.; Jeffery, R.W. Predictors of nutrition label viewing during food purchase decision making: An eye tracking investigation. *Public Health Nutr.* **2012**, *15*, 189–197. [CrossRef] [PubMed]
12. Barreiro-Hurle, J.; Gracia, A.; De-Magistris, T. Does nutrition information on food products lead to healthier food choices? *Food Policy* **2010**, *35*, 221–229. [CrossRef]
13. Miklavec, K.; Pravst, I.; Grunert, K.G.; Klopčič, M.; Pohar, J. The influence of health claims and nutritional composition on consumers' yoghurt preferences. *Food Qual. Prefer.* **2015**, *43*, 26–33. [CrossRef]
14. Ares, G.; Giménez, A.; Gámbaro, A. Influence of nutritional knowledge on perceived healthiness and willingness to try functional foods. *Appetite* **2008**, *51*, 663–668. [CrossRef] [PubMed]
15. Johansen, S.B.; Næs, T.; Øyaas, J.; Hersleth, M. Acceptance of calorie-reduced yoghurt: Effects of sensory characteristics and product information. *Food Qual. Prefer.* **2010**, *21*, 13–21. [CrossRef]

16. Barreiro-Hurle, J.; Gracia, A.; Barreiro-Hurle, J.; De-Magistris, T.; Barreiro-Hurlé, J.; De-Magistris, T. The Effects of Multiple Health and Nutrition Labels on Consumer Food Choices. *J. Agric. Econ.* **2010**, *61*, 426–443. [CrossRef]
17. Svederberg, E.; Wendin, K. Swedish consumers' cognitive approaches to nutrition claims and health claims. *Food Nutr. Res.* **2011**, *55*, 5929. [CrossRef]
18. Grunert, K.G.; Wills, J.M. A review of European research on consumer response to nutrition information on food labels. *J. Public Heal.* **2007**, *15*, 385–399. [CrossRef]
19. Chamorro, A.; Miranda, F.J.; Rubio, S.; Valero, V.; Chamorro-Mera, A. Innovations and trends in meat consumption: An application of the Delphi method in Spain. *Meat Sci.* **2012**, *92*, 816–822. [CrossRef]
20. Erickson, J.; Slavin, J. Are restrictive guidelines for added sugars science based? *Nutr. J.* **2015**, *14*, 124. [CrossRef]
21. Louviere, J.J.; Flynn, T.N.; Carson, R.T. Discrete Choice Experiments Are Not Conjoint Analysis. *J. Choice Model.* **2010**, *3*, 57–72. [CrossRef]
22. Van Wezemael, L.; Caputo, V.; Nayga, R.M.; Chryssochoidis, G.; Verbeke, W. European consumer preferences for beef with nutrition and health claims: A multi-country investigation using discrete choice experiments. *Food Policy* **2014**, *44*, 167–176. [CrossRef]
23. Enneking, U.; Neumann, C.; Henneberg, S. How important intrinsic and extrinsic product attributes affect purchase decision. *Food Qual. Prefer.* **2007**, *18*, 133–138. [CrossRef]
24. Orme, B. Sawtooth Software, Research Paper Series, Hierarchical Bayes: Why All the Attention? 2001, Copyright 2000–2001, Sawtooth Software, Inc. Available online: https://www.sawtoothsoftware.com/support/technical-papers/hierarchical-bayes-estimation/hierarchical-bayes-why-all-the-attention-2000 (accessed on 18 October 2019).
25. Ares, G.; Giménez, A.; Deliza, R. Influence of three non-sensory factors on consumer choice of functional yogurts over regular ones. *Food Qual. Prefer.* **2010**, *21*, 361–367. [CrossRef]
26. Sajdakowska, M.; Gębski, J.; Gutkowska, K.; Żakowska-Biemans, S. Importance of Health Aspects in Polish Consumer Choices of Dairy Products. *Nutrients* **2018**, *10*, 1007. [CrossRef]
27. Bauer, J.J. Selection Errors of Random Route Samples. *Sociol. Methods Res.* **2014**, *43*, 519–544. [CrossRef]
28. Sawtooth Software, Technical Paper Series: The CBC/HB System for Hierarchical Bayes Estimation Version 5.0 Technical Paper; 2009, Copyright 2000-2009, Sawtooth Software, Inc. Available online: https://www.sawtoothsoftware.com/support/technical-papers/hierarchical-bayes-estimation/cbc-hb-technical-paper-2009 (accessed on 18 October 2019).
29. Mascarello, G.; Pinto, A.; Parise, N.; Crovato, S.; Ravarotto, L. The perception of food quality. Profiling Italian consumers. *Appetite* **2015**, *89*, 175–182. [CrossRef]
30. Sajdakowska, M.; Jankowski, P.; Gutkowska, K.; Guzek, D.; Żakowska-Biemans, S.; Ozimek, I. Consumer acceptance of innovations in food: A survey among Polish consumers. *J. Consum. Behav.* **2018**, *17*, 253–267. [CrossRef]
31. Bayarri, S.; Carbonell, I.; Barrios, E.; Costell, E. Impact of sensory differences on consumer acceptability of yoghurt and yoghurt-like products. *Int. Dairy J.* **2011**, *21*, 111–118. [CrossRef]
32. Domanski, T.; Bryla, P. The fragile strength of a leading Polish yoghurt company (case study of Bakoma). *Br. Food J.* **2013**, *114*, 618–635. [CrossRef]
33. Czarnacka-Szymani, J.; Jezewska-Zychowicz, M. Impact of nutritional information on consumers' acceptance of cheese with reduced sodium chloride content. *Int. Dairy J.* **2015**, *40*, 47–53. [CrossRef]
34. Kraus, A. Development of functional food with the participation of the consumer. Motivators for consumption of functional products. *Int. J. Consum. Stud.* **2015**, *39*, 2–11. [CrossRef]
35. Pohjanheimo, T.; Sandell, M. Explaining the liking for drinking yoghurt: The role of sensory quality, food choice motives, health concern and product information. *Int. Dairy J.* **2009**, *19*, 459–466. [CrossRef]
36. Rahnama, H.; Rajabpour, S. Factors for consumer choice of dairy products in Iran. *Appetite* **2017**, *111*, 46–55. [CrossRef] [PubMed]
37. Das, K.; Choudhary, R.; Thompson-Witrick, K.A. Effects of new technology on the current manufacturing process of yogurt-to increase the overall marketability of yogurt. *LWT* **2019**, *108*, 69–80. [CrossRef]
38. Gutkowska, K.; Sajdakowska, M.; Zakowska-Biemans, S.; Kowalczuk, I.; Kozłowska, A.; Olewnik-Mikołajewska, A. Poziom Akceptacji Zmian Na Rynku Żywności Pochodzenia Zwierzęcego W Opinii Konsumentów. *Zywn. Nauk. Technol. Jakosc Food. Sci. Technol. Qual.* **2012**, *19*, 187–202.

39. Tomic, N.; Dojnov, B.; Miocinovic, J.; Tomasevic, I.; Smigic, N.; Djekic, I.; Vujcic, Z. Enrichment of yoghurt with insoluble dietary fiber from triticale—A sensory perspective. *LWT* **2017**, *80*, 59–66. [CrossRef]
40. Argın, S.; Eskinazi, B.; Tavlı, D. A Consumer Perception Study on Functional Dairy Products Among Consumers in Istanbul, Turkey. *Turk. J. Agric. Food Sci. Technol.* **2019**, *7*, 963. [CrossRef]
41. Ates, H.C.; Ceylan, M. Effects of socio-economic factors on the consumption of milk, yoghurt, and cheese: Insights from Turkey. *Br. Food J.* **2010**, *112*, 234–250. [CrossRef]
42. Lefèvre, M. Do Consumers Pay More for What They Value More? The Case of Local Milk-hased Dairy Products in Senegal. *Agric. Resour. Econ. Rev.* **2014**, *43*, 158–177. [CrossRef]
43. Almli, V.L.; Næs, T.; Enderli, G.; Sulmont-Rossé, C.; Issanchou, S.; Hersleth, M. Consumers' acceptance of innovations in traditional cheese. A comparative study in France and Norway. *Appetite* **2011**, *57*, 110–120. [CrossRef] [PubMed]
44. Napolitano, F.; Girolami, A.; Braghieri, A. Consumer liking and willingness to pay for high welfare animal-based products. *Trends Food Sci. Technol.* **2010**, *21*, 537–543. [CrossRef]
45. Olewnik-Mikołajewska, A.; Guzek, D.; Głąbska, D.; Sajdakowska, M.; Gutkowska, K. Fodder enrichment and sustaining animal well-being as methods of improving quality of animal-derived food products, in the aspect of consumer perception and acceptance. *Anim. Sci. Pap. Rep.* **2016**, *34*, 361–372.
46. Vanhonacker, F.; Verbeke, W. Public and Consumer Policies for Higher Welfare Food Products: Challenges and Opportunities. *J. Agric. Environ. Ethics* **2014**, *27*, 153–171. [CrossRef]
47. EFSA Special Eurobarometer Wave EB91.3 Food safety in the EU Report Fieldwork. 2019; ISBN 9789294990822.
48. Font-I-Furnols, M.; Guerrero, L. Consumer preference, behavior and perception about meat and meat products: An overview. *Meat Sci.* **2014**, *98*, 361–371. [CrossRef]
49. Shan, L.C.; De Brún, A.; Henchion, M.; Li, C.; Murrin, C.; Wall, P.G.; Monahan, F.J. Consumer evaluations of processed meat products reformulated to be healthier—A conjoint analysis study. *Meat Sci.* **2017**, *131*, 82–89. [CrossRef]
50. Żakowska-Biemans, S.; Sajdakowska, M.; Issanchou, S. Impact of Innovation on Consumers Liking and Willingness to Pay for Traditional Sausages. *Pol. J. Food Nutr. Sci.* **2016**, *66*, 119–127. [CrossRef]
51. Markovina, J.; Stewart-Knox, B.J.; Rankin, A.; Gibney, M.; De Almeida, M.D.V.; Fischer, A.R.; Kuznesof, S.A.; Poínhos, R.; Panzone, L.; Frewer, L.J. Food4Me study: Validity and reliability of Food Choice Questionnaire in 9 European countries. *Food Qual. Prefer.* **2015**, *45*, 26–32. [CrossRef]
52. Román, S.; Sánchez-Siles, L.M.; Siegrist, M. The importance of food naturalness for consumers: Results of a systematic review. *Trends Food Sci. Technol.* **2017**, *67*, 44–57. [CrossRef]
53. Aschemann-Witzel, J. Consumer perception and trends about health and sustainability: trade-offs and synergies of two pivotal issues. *Curr. Opin. Food Sci.* **2015**, *3*, 6–10. [CrossRef]
54. Fazilah, N.F.; Ariff, A.B.; Khayat, M.E.; Rios-Solis, L.; Halim, M. Influence of probiotics, prebiotics, synbiotics and bioactive phytochemicals on the formulation of functional yogurt. *J. Funct. Foods* **2018**, *48*, 387–399. [CrossRef]
55. Nowak, A.; Ślizewska, K.; Libudzisz, Z. Probiotyki—Historia i mechanizmy działania. *Zywn. Nauk. Technol. Jakosc Food. Sci. Technol. Qual.* **2010**, *17*, 5–19.
56. Žeželj, I.; Milošević, J.; Stojanovic, Z.; Ognjanov, G. The motivational and informational basis of attitudes toward foods with health claims. *Appetite* **2012**, *59*, 960–967. [CrossRef] [PubMed]
57. Ballco, P.; De-Magistris, T.; Caputo, V. Consumer preferences for nutritional claims: An exploration of attention and choice based on an eye-tracking choice experiment. *Food Res. Int.* **2019**, *116*, 37–48. [CrossRef] [PubMed]
58. Wills, J.M.; Bonsmann, S.S.G.; Kolka, M.; Grunert, K.G. European consumers and health claims: attitudes, understanding and purchasing behaviour. *Proc. Nutr. Soc.* **2012**, *71*, 229–236. [CrossRef] [PubMed]

nutrients

MDPI

Article

The Role of Health Information in Consumers' Willingness to Pay for Canned Crushed Tomatoes Enriched with Lycopene

Fabio Verneau [1], Francesco La Barbera [1] and Marilena Furno [2,*]

[1] Department of Political Science, University of Naples Federico II, 80138 Naples, Italy; verneau@unina.it (F.V.); francesco.labarbera@unina.it (F.L.B.)
[2] Department of Agricultural Sciences, University of Naples Federico II, 80055 Naples, Italy
* Correspondence: marfurno@unina.it

Received: 16 July 2019; Accepted: 5 September 2019; Published: 10 September 2019

check for updates

Abstract: The paper investigated whether information about the health benefit produced by lycopene could influence consumers' willingness to pay (WTP) for canned crushed tomatoes enriched with lycopene. An additional aim was to determine whether the main socio-demographic variables, such as sex, age, income and selected attitudinal factors, affect WTP. To this end, a non-hypothetical experimental auction was carried on with five repeated rounds. Results show a relevant impact of information on WTP in the case of lycopene-enriched products, whereas no difference in bids emerges for the conventional product, either on average or at the quantiles. Previous knowledge seems to have a modest influence upon WTP, but it shows a significant interaction with the information shock provided during the experiment, so that the effect of the latter is more pronounced when previous knowledge is low. In addition, age, sex, food technology neophobia, trust in science and implicit attitudes towards food technology significantly affect participants' WTP.

Keywords: quantile regression; experimental auction; lycopene; information

---

1. Introduction

Functional products are among the most interesting food categories in terms of market opportunities, and one of the faster growing food sectors. This global growth is strongly related to the increased consumers' awareness about the role of food in wellbeing, either to help prevent chronic diseases, or to optimize health [1–3]. Moreover, due to the economic costs of the medical care of chronic diseases, authorities have developed actions and policies to promote the consumption of functional food [4]. Despite the considerable success and interest, there is not yet a universally accepted definition for this group of food [5]. On a very general level, they could be defined as food products providing an added health benefit over and above the traditional food nutritional value [6,7].

From the production point of view, functional food is an opportunity to face the increasing global competition and to boost a progressively more saturated food demand [8]. However, even if functional food could be a good firm strategy by creating differentiated, value-added products, many of those innovative products have failed, probably due to a lack of cooperation between the disciplines involved in the new product development process, especially those focusing on consumers' attitudes and knowledge [9]. As a matter of fact, at least in Europe, the performance of new functional food launched on the market has not always been in line with the optimistic expectations of the producers, and the results in terms of profitability have been sometime disappointing. A controversial attitude towards functional products has been observed in Europe, where people are more reluctant in adopting functional food than in Asia or North America [8,10].

The last survey of the Spanish Government revealed that consumers attach more importance to the origin of the product (i.e., those produced in their own region, or with a Protected Denomination of Origin), than to functionality [11]. Moreover, Backstrom and colleagues [12], analyzing focus group discussions, report that functional food is associated with a medicine-like representation, evoking quite negative impressions among participants. In a similar fashion, Frewer, Scholderer, and Lambert [6] emphasize that consumers' risk perceptions may have an important role in reducing any willingness to accept functional food. Finally, Furno et al. [13] performed experimental auctions on between samples in order to study hypothetical bias., they used as a product canned tomatoes enriched in lycopene, and highlighted ing the role of explicit and implicit attitudes towards food technology.

Due to the controversial attitudes of European consumers, the development of consumer-oriented new products becomes a crucial element to recover the required up-front investments, and more generally to generate adequate levels of profitability and the growth rate of companies. Indeed, the development of new functional components and their relative technological solutions can be very challenging and expensive. To avoid major failures in investments, it is therefore important to monitor consumers' attitudes to ensure that new functional products fulfil consumer expectations, and that any health-related information is honestly and attractively communicated. In fact, knowledge and information have been demonstrated to exert a strong effect on consumers' decision-making towards functional food. In their 1999 report of a qualitative study, the International Food Information Council (IFIC) [14] indicated that lack of knowledge is the major reason for not consuming functional food. Also, de-Magistris and colleagues [15] highlighted the crucial role of knowledge about nutrition in their study on the influence of body image on consumers' willingness to pay (WTP) for potato chips. Several other studies indicated information and knowledge, both subjective and objective, as relevant antecedents of functional products' consumption [16–18]. In addition to the intervening factors already emerged in previous studies, Temesi et al. [19] have found a new factor called the Perceived Correspondence of Health that plays an important role in consumers' perception of functional ingredients.

However, most of the research on consumers' acceptance of functional food, dealing with information and its effects on the willingness to pay (WTP) or accept, is limited to investigating whether information and/or different types of information may alter this WTP [20–23]. To the best of our knowledge, the study of the main factors influencing the change in the willingness to pay due to the administration of information is a topic still understudied.

The products chosen for the present study are canned tomatoes, one of the most important products of the food industry in southern Italy [24]. Tomatoes naturally contain lycopene, a carotenoid responsible for the red color which has received considerable attention due to its various biological activities [25]. As a matter of fact, several studies indicate lycopene among the main natural components capable of inhibiting the proliferation of cancer cells in some of the most common types of cancer [26,27]. Yet, it should also be noted that the European Food Safety Authority (EFSA) has not yet approved any health claims related to lycopene considering the scientific evidence still insufficient [28]. Given this not yet well-defined scenario on health claims related to lycopene, the focus of the present study is to investigate whether information about the possible health benefits produced by lycopene could influence consumers' WTP for canned crushed tomatoes enriched with lycopene.

Our specific objectives are:

1. To determine whether information on the benefit of lycopene affects the consumers' WTP both on canned tomatoes enriched with lycopene and conventional canned tomatoes (not enriched).
2. To analyze whether and how several selected variables, namely socio-demographics, attitudinal constructs and control variables, affect WTP. More in detail, the impact of these variables on WTP will be investigated also in the tails of the bid distribution;
3. To explore whether the selected variables affect the possible change in prices, after the informative shock; that is, assessing the moderator role of the variables on the WTP.

We undertook this study using data from an experimental laboratory auction conducted in Italy. Two treatments differing in the information provided to the participants were taken into consideration. This study expands the literature on consumers' preferences for functional food products using the experimental auction method. Currently, the use of a non-hypothetical auction method has gained popularity in estimating the WTP for product attributes or new products [29], alongside a real Choice Experiment [15,30]. A major reason for the increasing popularity of experimental auctions is their incentive compatibility property, that is, subjects have the dominant strategy to submit bids equal to the true value of the goods. The experimental auction would then be demand-revealing, and hence the participants would provide truthful bids [31]. Indeed, in accordance with Chang, Lusk and Norwood [32], the WTP values from experimental auctions can be assumed to be the true values corresponding to the actual payments in the marketplace (i.e., scanner data from shopping), thus, they are a better approximation of the true preferences.

2. Materials and Methods

2.1. Participants

A total of 100 participants were recruited for the auctions, selected among the college students from several departments of the University of Naples. All of the various steps of the experiment were computer administered; moreover some phases of the experiment, such as the Single Category Implicit Association Test (SC-IAT) text, require promptness and great attention. It is very likely that students have a greater ability to perform such tasks. In addition, Depositario et al. (2009) pointed out that prices and opinions on products issued by students are in line with those declared by responsible purchasers. [33] Therefore, the students' answers in this experiment should not generate significant distortions. Main statistics on socio-demographic characteristics are presented in Table 1. Age is a numerical variable expressing the number of years of the participant's life. Gender is a binary variable which assumes value 1 in the case of the female gender, and 0 in the case of the male gender. Income is a categorical variable which measures net monthly family income and assumes value 1 for incomes up to 1000 Euros; 2 for incomes between 1000 and 2000 Euros, 3 for incomes between 2000 and 3000 Euros, 4 for incomes over 3000 Euros.

Table 1. Summary statistics of the explanatory variables of the model (n = 100).

Socio-Demographic	Mean	Std. Dev.	q25	q50	q75
Age	23.88	4.00	21.0	24.0	25.5
Gender	0.560	0.49	0.00	1.00	1.00
Income	2.300	0.87	2.00	2.00	3.00

2.2. Measures

In the late 1980s and early 1990s, as a response to widespread criticism against self-report measures, new instruments began to appear in social psychology for measuring implicit attitudes, that is, attitudes that are not completely conscious, and/or that are automatically activated. Research showed the relevant predictive validity of these new measures [34,35], also in the food domain [36–38]. Among the most widely used instruments for measuring implicit attitudes are the Implicit Association Test (IAT) [39], and its development, the Single Category Implicit Association Test (SC-IAT) [40]. The latter was administered in the current study to measure the participants' implicit attitude towards food technology. Participants were asked to categorize stimuli belonging to the target category (food technology) and stimuli belonging to two opposite attribute categories (Positive and Negative), using the keyboard keys "A" and "L." In two subsequent steps, the target category and one attribute category (e.g., Positive) share the same response key (step 1); then the target category and the other attribute category (e.g., Negative) share the same response key (step 2). A longer reaction time indicates that for the respondent it is more difficult to associate the target and attribute category (e.g., food technology and positive), whereas a shorter reaction time means that the two categories are easily associated.

The stimuli in the attribute category were obtained by a pilot study in which 50 undergraduate students were asked to associate five adjectives to "food technology." The ten most frequent adjectives were associated with their antonyms and used in the SC-IAT as the stimuli of the attribute category (e.g., "moral" and "immoral," "useful" and "useless"). For the target category, the stimuli used were the following words: GMO, Nanotechnology, Coloring, Additive, Preservative, Sterilizing, Thickening. The presentation of the combination of target and attribute categories was counterbalanced, so that half of the participants were presented with Food technology and Positive first, and the other half with Food technology and Negative first. The final index was obtained by subtracting the participants' time latency in the Food technology and Negative step from the participants' time latency in the Food technology and Positive step (correcting for individual standard deviation; see [41]), so that positive values indicate a positive implicit attitude about food technology, whereas negative values indicate a negative implicit attitude. Tested for reliability, the SC-IAT proved adequate ($\alpha = 0.71$).

The Food Technology Neophobia Scale (FTNS) was used to measure attitudes toward food technology [42–44]. In our study, the following subscales were used: 1. New food technologies are unnecessary (FTNS1); 2. Perception of risk (FTNS2); 3. (FTNS3) Healthy choices. Participants indicated their agreement with the statements on a 7-point scale from 1 (totally disagree) to 7 (totally agree). The three subscales proved to have adequate internal consistency ($\alpha_{FTNS1} = 0.80$; $\alpha_{FTNS2} = 0.71$; $\varrho_{FTNS3} = 0.73$). Items were averaged in single scores. For FTNS1, the higher the score, the more negative the attitude; in the case of FTNS2 the higher the score, the more negative the perception of risk, and finally, for FTNS3 the higher the score, the more positive the perception of benefit.

Participants also completed the Trust in Science Scale (TISS), a six items scale that focuses on public attitudes toward controversial scientific research and technologies [45]. Participants indicated their agreement on a scale from 1 to 5 ($\alpha = 0.69$). Items were averaged in a single score (the higher the score, the higher the trust in science).

To measure social desirability (SD), the short form of the Marlowe-Crowne Social Desirability scale [46] was used. Answers were collected on a 1 to 6 scale ($\alpha = 0.67$). Items were averaged in a single score (the higher the score, the higher the social desirability).

Participants' self-reported consumption frequency of tomato products (FREQ) was measured by a single item, with the following response format: 1 = rarely or never; 2 = 1 or 2 times a week; 3 = several times a week; 4 = daily.

Participants were also asked to self-report their level of knowledge about lycopene (KNOW) through a single item, with a 1 to 6 response scale. The higher the score, the higher the perceived level of knowledge about lycopene. Finally, SHOCK is a dummy variable which assumes value 0 in the treatment without information and value 1 in the following treatment with information about lycopene. Table 2 provides the summary statistics.

Table 2. Summary statistics of the Control variables and measures.

Control Variables	Mean	Std. Dev.	q25	q50	q75
Consumption Frequency (FREQ)	2.880	0.58	3.00	3.00	3.00
Lycopene Knowledge (KNOW)	2.810	2.87	0.00	2.50	6.00
Measures	**Mean**	**Std. Dev.**	**q25**	**q50**	**q75**
SC_IAT	−0.068	0.41	−0.40	−0.09	0.25
Social Desirability (SD)	3.975	0.65	3.61	4.00	4.33
FTNS Unnecessary (FTNS1)	3.445	1.12	2.50	3.33	4.33
FTNS Risks (FTNS2)	4.040	1.19	3.37	4.00	4.75
FTNS Benefits (FTNS3)	3.185	1.28	2.50	3.00	4.00
Trust in Science (TISS)	2.31	0.49	1.80	2.20	2.40
Treatment Variable	**Mean**	**Std. Dev.**	**q25**	**q50**	**q75**
SHOCK	0.500	0.50	0.00	0.50	1.00

2.3. Experimental Design and Procedure

Several sessions of experimental auctions were conducted in the computer lab of the Department of Agricultural Sciences in Portici (Naples) in order to assess the willingness to pay (WTP) for a specific functional product (crushed tomatoes enriched with lycopene). The experimental design was computer administered in order to accelerate data acquisition and to minimize the possibility of error in the collecting of the data. The software were:

- *Z-tree* [47], for the collection of bids in auctions;
- *Google Drive*, to administer the questionnaires;
- *Millisecond Inquisit*, for the administration of SC-IAT.

For this experiment, the fifth-price mechanism with a full bidding process was employed. Currently, a fifth-price auction is considered an incentive-compatible mechanism [29]. In addition, the fifth-price mechanism represents an attempt to combine the advantages of the second-price mechanism and the random n^{th}-price mechanism. Therefore, using the fifth-price mechanism could engage bidders with values on both tails of the value distribution [23]. We used the full bidding process instead of endowment bidding to eliminate any aversion to loss and risk exchanging of the participants [29,48]. Following Drichoutis, Lazaridis and Nayga [49], Bernard and He [50], and Hellyer, Fraser and Haddock-Fraser [48], we did not use the reference price, since we were aware of the possibility of bid affiliation. No price feedback among multiple rounds was reported [51].

The experiment is divided into several stages. During the experiment each participant is asked to use an ID (identifier) in order to trace the source computer of the data, thereby preserving complete anonymity.

At first, the SC-IAT is administered. After that, the experimenter provides participants with all of the information on the auction mechanism. Any communication between the subjects during the auction is prevented. The subjects are informed about the dominant strategy to reveal their true value for the products offered. To understand the bidding behavior and the mechanism, two training rounds are conducted using three different candy bars.

The products in the auction are two packs of three 400-gram cans of crushed tomatoes: Conventional crushed tomatoes, and crushed tomatoes enriched in lycopene (50% more). During the auction, each participant is asked to submit simultaneously a bid for each of the two crushed tomato products. The bids are collected, and the step is repeated in four additional rounds. In order to avoid any ordering effect, the products are simultaneously shown to the participants, and the order of the bids for each product is randomly assigned.

At the end of the five rounds, participants are informed that a second part of the experiment would start in a few minutes. Each participant is given a brief note on the nature of lycopene and the potential positive effects on human health (see Appendix A). The information note is written considering the results produced by scientific research.

After reading the text, that is the informative shock, the participants take part in a further non-hypothetical auction that takes place in the same conditions as the previous one.

Immediately after the post-information auctions, participants complete the Food Technology Neophobia Scale (FTNS), the Trust in Science Scale (TISS), the Marlowe-Crowne Social Desirability scale, and the items about the frequency with which they consume tomato products and about their knowledge of lycopene.

Finally, a first random draw determines which auction is chosen (with or without informative shock), a second draw determines the selected round; finally, a random draw defines which of the crushed tomato products is selected. The top four bidders on the bidding product of the selected auction and round purchase the crushed tomato package, paying a price equivalent to the fifth-highest bid for the product.

Basic socio-demographic characteristics are collected before thanking and de-briefing participants.

3. Statistical Analysis

3.1. Quantile Regression

The OLS analysis is complemented by the estimates of the regression at various quantiles. While OLS computes the regression at the mean, the quantile regression estimator [52,53] allows the impact of the explanatory variables to be investigated at different levels of the dependent variable: Not only at the center, the mean or the median, but also in the tails of the distribution, such as the first and third quartiles of the conditional distribution of bids. This is particularly relevant in the case of asymmetry, as is the case in the bid distributions.

For instance, in the linear regression model $y_i = x_i{'}\beta + e_i$, where x_i is the row vector including the ith observation for all the explanatory variables of the model, the quantile regression objective function at the selected quantile θ is:

$$\sum_i \{\theta - 1(y_i \le x_i\beta)\}\left|y_i - x_i{'}\beta\right|$$

where the absolute value of the regression errors, $|e_i| = \left|y_i - x_i{'}\beta\right|$, is asymmetrically weighted by θ or $(1-\theta)$. The weights set the position of the estimated line and allow the estimated line to move away from the mean of the conditional distribution. For instance, the 10th quantile regression partitions the residuals 10% below and 90% above the estimated regression, based on weights of $\theta = 0.10$ assigned to the positive residuals and $(1-\theta) = 0.90$ to the negative ones. This determines the prevalence of negative errors, thereby locating the regression line in the lower tail, precisely at the 10th quantile. At the median the weights are symmetric, $\theta = 1 - \theta = 0.5$ and the objective function simplifies into $0.5\sum\left|y_i - x_i{'}\beta\right|$. Just as the sample median equally divides the observations, the median regression equally splits the regression residuals half above and half below the estimated equation [53].

3.2. Decomposition Analysis

To analyze a data set split in two different subsets, each identified by an index assuming values 0, the bid experiments before information on lycopene, and 1 otherwise, a decomposition approach can be implemented. On average, the Oaxaca [54] and Blinder [55] decomposition writes the difference between these subsets as:

$$\begin{gathered} E\,(y_1 - y_0) = E[\beta_1\, x_1 - \beta_0\, x_0\,] = E[\beta_1\, x_1 - \beta_0\, x_1 + \beta_0\, x_1 - \beta_0\, x_0] = \\ E[(\beta_1 - \beta_0)\, x_1 + \beta_0\,(x_1 - x_0)] = E(y_{1/1} - y_{0/1} + y_{0/1} - y_{0/0}) \end{gathered} \quad (1)$$

where $E(y_{1/1}) = \beta_1 E(x_1)$ and $E(y_{0/0}) = \beta_0 E(x_0)$ are the realizations of the dependent variable within each subset, while the term $E(y_{0/1}) = \beta_0 E(x_1)$, defined by the covariates of subset 1 evaluated at the coefficients of subset 0, represents the counterfactual, and cannot be observed. Counterfactual distributions are the result of a change in covariates, or a change in their relationship with the dependent variable, the regression coefficients. In the bids example $y_{1/1}$ coincides with bids after the shock and $y_{0/0}$ coincides with bids before the shock. They are respectively the observed bids in each subset, while $y_{0/1}$ provides the subset 1 covariate, after shock, multiplied by subset 0 coefficients, pre-shock. The first term of the decomposition, $y_{1/1} - y_{0/1}$, measures the difference in bids due to changes in the regression coefficients, $(\beta_1 - \beta_0)$. The second term, instead, looks at the difference in bids due to changes in the covariates between subsets, and provides a measure of the composition effect. In this analysis the focus is on the coefficients effect since the bids are provided by the same individuals, so there is no difference in the covariates. The terms in the decomposition are computed at their average values,

$$Ey_1 - Ey_0 = (\beta_1 - \beta_0)\, Ex_1 + \beta_0\,(Ex_1 - Ex_0\,) \quad (2)$$

where expectations are replaced by their sample average and the parameters β_1, β_0 by their OLS estimates. The result is an average measure of bids difference between the two subsets. We focus on the behavior of $(\beta_1 - \beta_0)$ Ex_1, that would measure for each coefficient the impact of the informative shock on the bids for enriched and for conventional tomatoes.

3.3. Censoring

Since WTP cannot be negative, we also consider the Powell [56] censored least absolute deviations estimator (CLAD) and the bootstrap estimates of its sampling variance. The CLAD estimator is a generalization of the least absolute deviations (LAD) estimator, that is, the median regression. Unlike the standard estimators of the censored regression model, such as Tobit or other maximum likelihood approaches, the CLAD estimator is robust to heteroscedasticity and is consistent and asymptotically normal for a wide class of error distributions. Bootstrap estimates of the sampling variance are provided using a simple random design. The objective function of the Powell estimator is given by $\Sigma|y_i - \max(0; x_i'\beta)|$ and is iteratively estimated.

4. Results

4.1. The Model

The model compares the WTP for the enriched and the conventional tomatoes as a function of three main groups of variables: Socio-demographic, attitudinal and control variables. The summary statistics of all the study variables are collected in Tables 1–3.

Table 3. Summary statistics of the dependent variables.

Dependent Variable	Mean	Std. Dev.	q25	q50	q75
BIDL50 (n = 1000)	2.161	1.58	1.20	1.90	2.80
BIDCONV (n = 1000)	1.378	0.87	0.80	1.20	1.80
BIDL50, SHOCK = 0 (n = 500)	1.977	1.39	1.05	1.80	2.70
BIDCONV, SHOCK = 0 (n = 500)	1.367	0.88	0.80	1.20	2.70
BIDL50, SHOCK = 1 (n = 500)	2.346	1.74	1.31	2.10	3.00
BIDCONV, SHOCK = 1 (n = 500)	1.390	0.87	0.80	1.20	1.80

Note. BIDL50 = bids for the enriched product; BIDCONV = bids for the conventional product; SHOCK = treatment variable that assumes value 0 before the information shock, and value 1 after the information shock; n = number of observations.

In order to evaluate if information on the benefit of lycopene affects the consumers' WTP both on canned tomatoes enriched with lycopene and conventional canned tomatoes (not enriched), we started with the comparison of the patterns of the bids before and after the informative shock (SHOCK in the tables) across the five rounds and for each product, as reported in Table 4 and Figure 1.

Table 4. Summary of Mean and Median (in parenthesis) willingness to pay (WTP) by Products and Treatment.

	Pre-Shock		Post Shock	
	Conventional	Enriched	Conventional	Enriched
1st Round, n = 100	1.36 (1.20)	2.00 (1.75)	1.31 (1.20)	2.31 (2.05)
2nd Round, n = 100	1.37 (1.22)	1.97 (1.80)	1.41 (1.20)	2.35 (2.10)
3rd Round, n = 100	1.38 (1.20)	1.92 (1.79)	1.37 (1.20)	2.38 (2.05)
4th Round, n = 100	1.35 (1.20)	1.99 (1.90)	1.48 (1.30)	2.39 (2.10)
5th Round, n = 100	1.38 (1.20)	2.01(1.85)	1.38 (1.20)	2.30 (2.00)

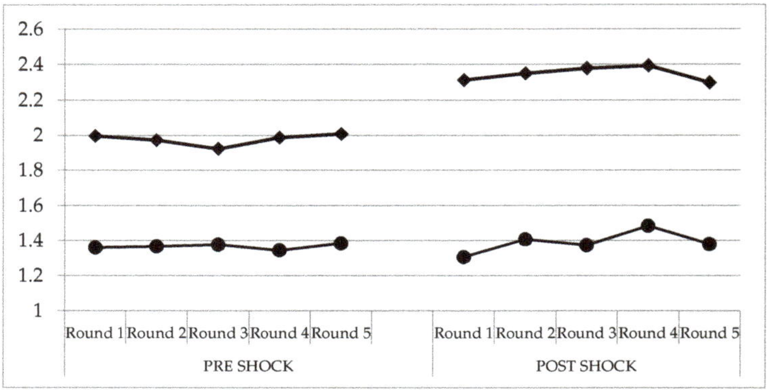

Figure 1. The Effect of Information on Mean Bids by Product and by Round. Note: the y axis is expressed in Euros (€).

Informative shock increases the bids for the enriched product, whereas it does not affect their level (mean) for the conventional one, although it affects the dispersion of the average bids. Figure 2 displays the histograms of the dependent variables, the bids for the enriched and the conventional product. The left graph is more dispersed than the one depicting the conventional product, and its sample mean is higher. Both distributions are characterized by skewness, that can be related to the presence of a few large values in the upper tail of the distributions, particularly in the bids for the enriched product.

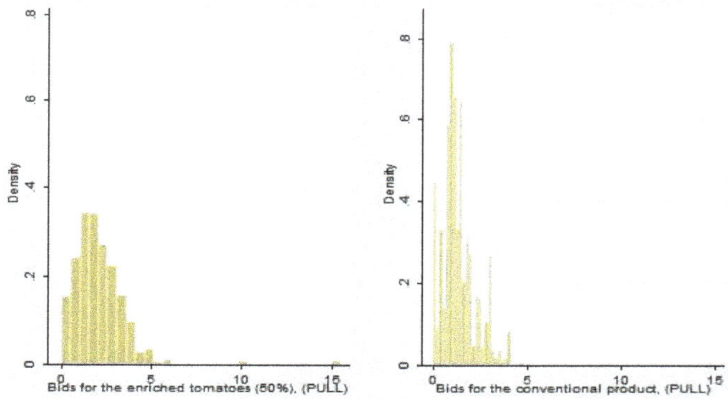

Figure 2. Empirical distributions of bids for the enriched and the conventional product.

The second specific objective of the present research was to evaluate whether and how several selected variables, namely socio-demographics, attitudinal constructs and control variables, affect WTP. To pursue the objective two quantile regressions were estimated both for the conventional and the enriched products. The estimates of the regression coefficients are reported in Table 5 for the enriched product and in Table 6 for the conventional tomatoes. The last two columns of Tables 5 and 6 report the OLS estimated coefficients explaining average bids for the enriched and the conventional tomatoes as a function of the selected variables.

Table 5. OLS and quantile regressions for the enriched tomatoes (n = 1000).

	q25		q50		q75		OLS	
Variables	**B**	**SE**	**B**	**SE**	**B**	**SE**	**B**	**SE**
AGE	−0.018	0.013	0.036 **	0.012	0.021	0.021	0.099 **	0.011
GENDER	0.457 **	0.116	0.398 **	0.105	0.432 *	0.188	0.722 **	0.102
INCOME	−0.014	0.066	−0.020	0.060	0.025	0.108	−0.004	0.059
FREQ	0.143	0.092	0.285 **	0.084	0.586 **	0.150	0.498 **	0.082
FTNS1	−0.041	0.057	0.024	0.051	−0.149	0.092	−0.064	0.050
FTNS2	−0.145 **	0.056	−0.247 **	0.051	−0.234 *	0.092	−0.137 **	0.050
FTNS3	−0.003	0.044	0.146 **	0.040	0.326 **	0.072	0.241 **	0.039
SD	0.041	0.085	0.065	0.077	0.243	0.138	0.182 *	0.075
TISS	0.172	0.114	0.234 *	0.104	0.627 **	0.186	0.362 **	0.101
SC-IAT	0.223	0.131	0.583 **	0.119	0.381	0.214	0.418 **	0.116
SHOCK	0.301 **	0.101	0.2 *	0.092	0.403 *	0.164	0.369 **	0.089
KNOW	−0.004	0.018	0.017	0.016	−0.002	0.030	−0.038 *	0.016
constant	1.02	0.618	−0.298	0.561	−1.78	1.00	−3.37 **	0.549

Note: Dependent variable: Bids for the enriched product. B = estimated coefficients; SE = Standard Error. * = $p < 0.05$; ** = $p < 0.01$.

Table 6. Estimated regressions for the conventional product (n = 1000).

	q25		q50		q75		OLS	
Variables	**B**	**SE**	**B**	**SE**	**B**	**SE**	**B**	**SE**
AGE	0.004	0.008	0.006	0.007	−0.011	0.014	−0.004	0.007
GENDER	0.421 **	0.074	0.245 **	0.065	0.498 **	0.124	0.377 **	0.060
INCOME	−0.070	0.043	−0.120 **	0.037	0.101	0.071	−0.000	0.035
FREQ	0.040	0.059	0.007	0.052	0.185	0.099	0.124 **	0.048
FTNS1	0.015	0.036	0.002	0.032	0.052	0.061	−0.018	0.029
FTNS2	−0.028	0.036	−0.048	0.032	−0.157 *	0.060	−0.083 **	0.029
FTNS3	0.067 **	0.028	0.134 **	0.025	0.216 **	0.047	0.093 **	0.023
SD	−0.002	0.054	0.085	0.048	0.081	0.091	0.030	0.044
TISS	−0.182 *	0.073	−0.182 **	0.065	0.182	0.123	0.072	0.060
SC-IAT	0.147	0.084	0.119	0.075	0.269	0.141	0.214 **	0.069
SHOCK	−0.016	0.065	−0.004	0.057	0.030	0.108	0.023	0.053
KNOW	0.024 *	0.011	0.055 **	0.010	0.038 *	0.019	0.025 **	0.009
constant	0.702	0.398	0.908 *	0.351	−0.046	0.662	0.661 *	0.324

Note: Dependent variable: Bids for the enriched product. B = estimated coefficients; SE = Standard Error. * = $p < 0.05$; ** = $p < 0.01$.

The results for the first, second and third quartile regression are reported in the first three columns of Tables 5 and 6. Across quartiles the estimated coefficients do change, and depending on the selected quantile, the explanatory variables have a different impact on the dependent variable, a different explanatory power.

The results of the two regressions offer further evidence about the positive effect of the information about the benefit of lycopene on consumers' WTP. Indeed, the informative shock (SHOCK) is statistically relevant at all quantiles and at the conditional mean/OLS in the regression for the enriched product of Table 5. Vice versa, for the conventional product this same variable is not statistically different from zero, as reported in Table 6. With regard to the role played by the selected socio-demographic variables, gender exerts a positive impact on bids both for the enriched and conventional product, while age and income are mostly not significant.

Overall, attitude toward food technologies exerts a key role in determining bids. The results of the two regressions for the conventional and the enriched canned tomatoes (Tables 5 and 6) outline the different role of technophobia in the case of more or less familiar technology. In fact, in the case of conventional crushed tomatoes, which represent a very familiar food product generated by a mature technology, the perception of risks (FTNS2) is significant only at the 75th percentile. In the case of enriched crushed tomatoes, that is an innovative and less familiar food product, the perception of risks is always statistically significant at all quantiles (see also [44]). Moreover, although canned tomatoes are considered by consumers as the result of mature and well-known technologies, it is confirmed that the implicit attitude (SC-IAT) reduces WTP for canned tomato pulp. Finally, the two regressions seem to reserve a minor role for self-reported previous knowledge. In fact, previous knowledge does not seem to play any role in the case of enriched products [57].

Both explicit (FTNS) and implicit attitudes (SC-IAT) are significant predictors, particularly at the upper quartile. As expected, the perception of risks linked to technologies decreases WTP while the perception of benefits increases it. The effect of implicit attitude is statistically significant and increases across quantiles. Although significant in the conventional product, the self-reported knowledge provides small estimated coefficients.

From a marketing point of view, this set of results highlights the existence of different consumers' clusters. The clustering suggests the possibility of a fruitful communication and price strategy to build a market segmentation.

4.2. Results of the Decomposition Analysis

The third and final specific objective of the paper was to explore if and how the selected variables affect the possible change in prices after the informative shock; that is, assessing the moderator role of the variables on the WTP.

To this end a decomposition approach is implemented and next the relevant variables signaled by the decomposition are added to the main model as interaction terms. This allows us to measure the relevance of their additional impact.

The variable SHOCK is a dummy, assuming a zero value before, and unit value after the informative talk on the qualities of lycopene. The dummy measures the additional explanatory power of information on bids, net of the effect of the informative talk on all the other variables of the equation. It is however possible that the informative talk has affected bids not only directly, but also indirectly, through its impact on some/all the other explanatory variables. In order to check if this is the case, we implemented a decomposition analysis.

Table 7 reports the results of the decomposition. In this table the left columns report the decomposition of the enriched product and the right columns those of the conventional product. The top section of the table reports the overall average decomposition, while the bottom section of the table considers the impact of the informative shock on each regression coefficient as computed on average, by OLS. In the top section, the comparison between bids before and after the informative shock in the enriched product, as computed on average, yields a total difference $Ey_1 - Ey_0 = 0.369$ that is statistically different from zero. The latter is split into the endowment effect ($Ex_1 - Ex_0$), that is equal to zero, since people characteristics do not change after the informative shock, and the coefficient effect that is equal to the total difference, $(\beta_1 - \beta_0)\ Ex_1 = 0.369$. Therefore, there is a significant difference between bids before and after the shock in the enriched product case that cannot be explained by the covariates of the model but must be credited to information. For the conventional product instead, the overall average difference in bids before and after the shock is not statistically relevant.

Table 7. Oaxaca-Blinder decomposition of the total difference due to information (n = 1000).

Functional			Conventional		
mean group_0	1.98	0.062	mean group_0	1.37	0.040
mean group_1	2.35	0.078	mean group_1	1.39	0.039
mean difference	0.369 **	0.100	mean difference	0.023	0.056
in endowments	0	0.054	in endowments	0	0.044
in coefficients	0.369 **	0.091	in coefficients	0.023	0.035
Variable	**B**	**SE**	**Variable**	**B**	**SE**
AGE	1.74 **	0.56	AGE	0.007	0.34
GENDER	0.094	0.11	GENDER	−0.065	0.06
INCOME	−0.018	0.26	INCOME	−0.040	0.16
FREQ	0.114	0.46	FREQ	−0.226	0.28
KNOW	−0.241 **	0.09	KNOW	−0.004	0.05
FTNS1	0.350	0.34	FTNS1	0.056	0.20
FTNS2	−0.226	0.40	FTNS2	−0.148	0.24
FTNS3	0.290	0.25	FTNS3	−0.068	0.15
SD	0.559	0.59	SD	0.315	0.35
TISS	−0.182	0.42	TISS	−0.105	0.25
SC_IAT	0.005	0.01	SC_IAT	0.001	0.01
Constant	−2.112	1.08	Constant	0.299	0.64

Note: Dependent variable: Bids for the enriched and conventional products. B = estimated coefficients; SE = Standard Error, ** = $p < 0.01$.

Next the impact of the shock on each explanatory variable is reported in the lower section of this table. For the enriched product the informative shock has a relevant impact on the age coefficient, and on the knowledge of the qualities of lycopene. After the briefing on lycopene qualities, the age coefficient increases, while the knowledge coefficient decreases the impact on bids. Vice versa, in the conventional product the knowledge of lycopene properties and age have no relevant impact on any of the variables explaining bids.

4.3. Censoring

Unlike the standard estimators of the censored regression model such as Tobit or other maximum likelihood approaches, the CLAD estimator is robust to heteroscedasticity and is consistent and asymptotically normal for a wide class of error distributions. Bootstrap estimates of the sampling variance are provided using a simple random design. The objective function of the Powell estimator, given by $\Sigma|y_i - \max(0; x_i'\beta)|$, is iteratively estimated. Table 8 reports the results. For comparison's sake, the table reports the results of the uncensored median regression (LAD in the table) and of the mean regression (OLS in the table) estimators.

We observe that, adding the interactions Shock*Age and Shock*Know, the main effect of the shock variable drops in significance. The negative sign of the know*shock coefficient means that individuals with a low level of previous knowledge are characterized by higher bids as a result of the informative shock. The relevant role played by attitude variables towards technologies, which already emerged in the quantile regression analysis on functionalized products, is fully confirmed.

The LAD estimated coefficients are numerically identical to CLAD results, the sole difference is in the estimated standard errors. The OLS results are not numerically identical to the CLAD estimates, however they fully confirm the CLAD findings.

Table 8. Censored and uncensored regression with (Model 1) and without (Model 2) interactions (n = 1000).

	Model 1						Model 2					
Variable	**LAD**	**SE**	**CLAD**	**SE**	**OLS**	**SE**	**LAD**	**SE**	**CLAD**	**SE**	**OLS**	**SE**
SHOCK	0.262	0.555	0.262	0.518	−1.079	1.266	0.200 *	0.092	0.200 *	0.080	0.369 **	0.089
INCOME	−0.006	0.060	−0.006	0.125	−0.004	0.052	−0.020	0.060	−0.020	0.124	−0.004	0.051
AGE	0.035 *	0.017	0.035 *	0.015	0.063 *	0.031	0.036 **	0.012	0.036 **	0.013	0.099 **	0.029
GENDER	0.392 **	0.105	0.392 **	0.015	0.722 **	0.109	0.398 **	0.105	0.398 *	0.157	0.722 **	0.110
FREQ	0.299 **	0.084	0.300 **	0.071	0.498 **	0.085	0.285 **	0.084	0.285 **	0.081	0.498 **	0.086
FTNS1	0.005	0.051	0.005	0.073	−0.064	0.043	0.024	0.051	0.024	0.075	−0.064	0.043
FTNS2	−0.252 **	0.051	−0.252 **	0.049	−0.186 **	0.043	−0.247 **	0.051	−0.247 **	0.049	−0.186 **	0.043
FTNS3	0.175 **	0.040	0.175 **	0.052	0.241 **	0.053	0.146 **	0.040	0.146 **	0.044	0.241 **	0.054
TISS	0.329 **	0.104	0.330	0.226	0.362 **	0.096	0.234 *	0.104	0.234	0.210	0.362 **	0.095
SD	0.169*	0.077	0.170	0.148	0.181 *	0.073	0.065	0.077	0.065	0.152	0.181 *	0.074
SC-IAT	0.661 **	0.119	0.661 **	0.193	0.417 **	0.112	0.583 **	0.119	0.583 **	0.175	0.417 **	0.112
KNOW	0.062 **	0.023	0.062 *	0.029	0.009	0.022	0.017	0.017	0.017	0.024	−0.038 *	0.016
Shock*Know	−0.089 **	0.032	−0.089 **	0.023	−0.094 *	0.036						
Shock*Age	0.008	0.023	0.008	0.023	0.072	0.057						

Note: Dependent variable: Bids for the enriched product. B = estimated coefficients; SE = Standard Error. * = $p < 0.05$; ** = $p < 0.01$.

5. Discussion

The present paper examined the following research questions. First, we evaluated whether information on the benefit of lycopene affects the consumers' WTP both on canned tomatoes enriched with lycopene and conventional canned tomatoes (not enriched). Secondly, we studied whether and how several selected variables, namely socio-demographics, attitudinal constructs and control variables, affect WTP. Finally, we explored whether the selected variables affect the possible change in prices after the informative shock, which means assessing the moderator role of the variables on the WTP.

5.1. Premium Price for Lycopene-enriched Canned Tomato

From our results emerges the higher willingness to pay for the enriched product compared to the conventional one. This premium price that participants are willing to pay for the functionalized product has been confirmed in all five rounds.

This means that, even without any specific information about the beneficial properties of lycopene, the WTP changes, due to the label on the package of the enriched product highlighting the higher content of lycopene.

The higher WTP for the functionalized products confirms previous results obtained in the case of other food products such as bread and yogurt [48,57]. However, it is important to stress that, differently from previous research, in our study the functional attribute is naturally held both in the conventional and in the enriched product. The higher quantity of lycopene of the functionalized product is determined by the use of specific tomato varieties identified by genomic research and characterized by a higher content of lycopene. Second, previous research used claims focusing on the beneficial health effect of the functionalized product. Therefore, participants' evaluation of the functionalization itself, and of the healthiness of the attribute in question, are not distinguishable. On the contrary, our study was designed to allow the investigation of participants' attitudes and behavioral reaction to functionalization itself, and the relevance of the attribute is particularly related to the functionalization, which was communicated through an information shock. However, the effect of the information is limited only to the case of lycopene-enriched products, whereas no difference in bids emerges for the conventional product either on average or at the quantiles. A possible explanation of these results could be that the label of the functionalized product, which communicates the higher content in lycopene, increased the salience of this attribute.

5.2. *Effect of Attitudinal and Socio-demographic Variables*

The second issue discusses the premium price that participants are willing to pay for the enriched product as influenced by the explanatory variables. First of all, a significant role of socio-demographic variables emerges. As shown in Tables 5 and 6, women are characterized by a higher WTP both in the conventional and lycopene-enriched case. Income is not significant at any of the quantiles considered. The influence of age is never significant for the conventional crushed tomatoes, whereas in the case of the enriched products it shows a positive and significant coefficient both on average and at the median. Finally, a higher frequency of consumption is associated with a higher WTP.

The quantile regression on functionalized products shows the importance of attitudinal variables. The set of variables used to measure the degree of aversion/favor towards food technologies contribute to determine the WTP for the functionalized product. While FTNS1, which measures the perceived un-usefulness of new food technology, is never statistically significant, the perception of the risks associated with the use of food-related technologies (FTNS2) has a negative effect on WTP, which is more pronounced in the case of the enriched products. The role played by the perception of benefits (FTNS3) becomes more relevant in the case of higher WTP, since the coefficient increases moving from the median to the 75th percentile. The role of the FTNS three dimensions is in line with previous results in the literature considering different food products and food categories [42]. The importance of the attitude towards technologies is also confirmed by the significant role played by implicit attitude, measured through SC-IAT [13,38]. Finally, to complete the set of indicators used to capture the effects of technophobia, the degree of trust towards science also affects the WTP for enriched tomatoes: Individuals with higher confidence in scientific research show a higher WTP. TISS does not exert any effect in the case of more familiar products like the conventional canned tomatoes. To the best of our knowledge, this is the first study that shows the relevance of explicit and implicit attitudes towards food technology on people's WTP in a non-hypothetical experiment. Findings also suggest that, at least in the case of the cluster of consumers with the higher WTP, it may be reasonable to build communicative messages and market segmentation drawing on health benefits associated with functionalized food products [58].

5.3. *The Role of Information and Knowledge on WTP for Lycopene-Enriched Canned Tomato*

Referring to the last issue, assessing the moderator role of the variables on the WTP, the Oaxaca decomposition suggests a potential role of moderator for previous knowledge and age: When the level of knowledge is low, the informative shock produces greater effects in terms of premium price, while the informative shock is associated with higher bids for the enriched product as people grow older. The CLAD regression allows the effects of the informative shock on bidding to be analyzed in a censoring context to account for 0 bids (Table 8). The CLAD regression is also carried out considering the interactions terms Shock*Knowledge and Shock*Age suggested by the results of the Oaxaca decomposition. The shock variable is no longer significant when adding interactions; therefore, the variables selected to explain the divergence in the bid before and after the informative shock completely explain the phenomenon. The role of information is already well-known, and recently confirmed also in the case of very familiar functionalized food products characterized by an image of naturalness such as apples [59]. However, to the best of our knowledge, this research is the first attempt to identify the factors that are involved in the formation of the premium price resulting from an information shock. The CLAD regression carried out without interaction confirms the results already discussed for the regression with interactions, because in this case the dummy variable on the treatment is significant, confirming the importance of the level of previous knowledge as a moderator of the effect of the informative shock on bids. The OLS regressions confirm these results. Finally, since we can interpret the administration of the information shock as an increase in nutritional knowledge, our results also confirm the evidence recently produced by Steinhauser and Hamm [60] on the positive effect of nutritional knowledge on the intention to purchase.

5.4. Limitation of the Study and Further Research

Turning to the implementation of the experimental auction, there are some limitations that have to be acknowledged. First the nature of the sample, composed by university students who may have a starting degree of knowledge about lycopene different from that of the general population, and they maybe also have a different attitude towards food technologies. It is up to future research to widen the study on the topic of food functionalization involving more heterogeneous samples. Second, the within treatment, which allows us to trace the effect of the information in reference to each participant, could also generate in the participants anchoring effects, and could amplify problems of social desirability, thus introducing some bias in the estimates. Therefore, another intriguing path for further research may be the study of WTP for functionalized food products with between-subject treatments. Finally, the explicit and implicit attitudinal variables we use in the study, which showed a significant role, are cognitive in nature. An important development of the study about functionalization should include the emotional dimension of evaluation [61,62].

6. Conclusions

In the above analysis we investigated three main issues: Is information on the benefit of lycopene relevant to determine consumers' WTP? What are the main factors affecting the WTP? Are there variables assuming a moderator role on the WTP?

—The informative shock is found relevant in defining consumer willingness to pay both on average and in the tails of WTP distribution.

—WTP for functionalized canned tomatoes is strongly affected by attitudes toward technology. Indeed, both explicit and implicit attitudes have a relevant and significant impact on WTP, together with the trust in science.

—Finally, the moderator role of knowledge on WTP has been ascertained. People with little knowledge on lycopene properties increase their willingness to pay after the informative shock.

The quantile regression approach implemented throughout the paper has pointed out a changing impact of each variable along the WTP distribution. From a marketing point of view this analysis shows the existence of different consumers' clusters. This suggests implementing different communication and price strategies to build a market segmentation.

Author Contributions: Although this work is the result of a common effort, F.V. was overall responsible for the supervision and conceptualization of the work and took the lead in organizing the structure of the manuscript. F.V.; F.L.B. conceived and wrote Section 1. F.L.B has contributed in the choice of the attitude variables and the relative measures, in the data collection and analysis, and he wrote Section 2. M.F. has contributed to the selection and implementation of statistical models together with the result interpretation, she wrote Sections 3 and 4. F.V. is responsible for Section 5. All authors have contributed in critically reviewing and editing the manuscript. All authors read and approved the final version.

Funding: This research was funded by Ministry of University and Research (MIUR), Grant/Award Number: MIUR-PON02-GenoPOMpro.

Conflicts of Interest: The authors declare no conflict of interest.

Appendix A. Informative Shock

Lycopene is a natural element which can be found in some plant origin food. The major source of lycopene is represented by the tomato. It has been shown that lycopene has a high antioxidant capacity that neutralizes free radicals, protecting cells from the damage that these reactive elements can cause. Lycopene also reduces the oxidation of lipids, and in particular acts against the oxidation of cholesterol. Another possible application of lycopene is represented by the protective action it exerts on the skin in case of long exposure to UV rays. Finally, it has been highlighted that lycopene can inhibit the growth of cancer cells and the most recent studies show that the weekly consumption of seven or more portions of tomatoes leads to a reduction in cancer risk.

References

1. Regmi, A.; Gehlhar, M.J. *New Directions in Global Food Markets*; US Department of Agriculture: Washington, DC, USA, 2005.
2. Sadler, J. *Innovation in Functional Food and Drinks*; Business Insights Ltd.: London, UK, 2005.
3. Szakály, Z.; Szente, V.; Kövér, G.; Polereczki, Z.; Szigeti, O. The influence of lifestyle on health behavior and preference for functional foods. *Appetite* **2012**, *58*, 406–413. [CrossRef] [PubMed]
4. Peake, H.; Stockely, M.; Frost, G. What nutritional support literature do hospital nursing staff require? *J. Hum. Nutr. Diet.* **2001**, *14*, 225–230. [CrossRef] [PubMed]
5. Alzamora, S.M.; Salvatori, D.; Tapia, M.S.; López-Malo, A.; Welti-Chanes, J.; Fito, P. Novel functional foods from vegetable matrices impregnated with biologically active compounds. *J. Food Eng.* **2005**, *67*, 205–214. [CrossRef]
6. Frewer, L.; Scholderer, J.; Lambert, N. Consumer acceptance of functional foods: Issues for the future. *Br. Food J.* **2003**, *105*, 714–731. [CrossRef]
7. Heasman, M.; Mellentin, J. *The Functional Foods Revolution: Healthy People. Healthy Profits*; Routledge: London, UK, 2001.
8. Grunert, K.G. European consumers' acceptance of functional foods. *Ann. NY Acad. Sci.* **2010**, *1190*, 166–173. [CrossRef]
9. Jousse, F. Modeling to improve the efficiency of product and process development. *Compr. Rev. Food Sci. Food Saf.* **2008**, *7*, 175–181. [CrossRef]
10. Siro, I.; Kápolna, E.; Kápolna, B.; Lugasi, A. Functional food. Product development, marketing and consumer acceptance—A review. *Appetite* **2008**, *51*, 456–467. [CrossRef]
11. OCDA. Estudio de Mercado. In *Informe Histórico Observatorio del Consumo y la Distribución Alimentaria*; Spanish Ministry of Agriculture, Food and Environment: Madrid, Spain, 2001.
12. Bäckström, A.; Pirttilä-Backman, A.M.; Tuorila, H. Dimensions of novelty: A social representation approach to new foods. *Appetite* **2003**, *40*, 299–307. [CrossRef]
13. Furno, M.; Verneau, F.; Sannino, G. Assessing hypothetical bias: An analysis beyond the mean of functional food. *Food Qual. Prefer.* **2016**, *50*, 15–26. [CrossRef]
14. IFIC. *Functional Foods: Attitudinal Research (1996–1999)*; International Food Information Council Foundation: Washington, DC, USA, 1999.
15. de-Magistris, T.; López-Galán, B.; Caputo, V. The impact of body image on the WTP values for reduced-fat and low-salt content potato chips among obese and non-obese consumers. *Nutrients* **2016**, *8*, 830. [CrossRef]
16. Verbeke, W. Consumer acceptance of functional foods: Socio-demographic, cognitive and attitudinal determinants. *Food Qual. Prefer.* **2005**, *16*, 45–57. [CrossRef]
17. Verbeke, W. Functional foods: Consumer willingness to compromise on taste for health? *Food Qual Prefer.* **2006**, *17*, 126–131. [CrossRef]
18. La Barbera, F.; Amato, M.; Sannino, G. Understanding consumers' intention and behaviour towards functionalised food: The role of knowledge and food technology neophobia. *Br. Food J.* **2016**, *118*, 885–895. [CrossRef]
19. Temesi, Á.; Bacsó, Á.; Grunert, K.G.; Lakner, Z. Perceived Correspondence of Health Effects as a New Determinant Influencing Purchase Intention for Functional Food. *Nutrients* **2019**, *11*, 740. [CrossRef] [PubMed]
20. de-Magistris, T.; Del Giudice, T.; Verneau, F. The effect of information on willingness to pay for canned tuna fish with different corporate social responsibility (CSR) certification: A pilot study. *J. Consum. Aff.* **2015**, *49*, 457–471. [CrossRef]
21. Dillaway, R.; Messer, K.D.; Bernard, J.C.; Kaiser, H.M. Do consumer responses to media food safety information last? *Appl. Econ. Perspect. Policy* **2011**, *33*, 363–383. [CrossRef]
22. De Steur, H.; Buysse, J.; Feng, S.; Gellynck, X. Role of Information on Consumers' Willingness-to-pay for Genetically-modified Rice with Health Benefits: An Application to China. *Asian Econ. J.* **2013**, *27*, 391–408. [CrossRef]
23. Lusk, J.L.; House, L.O.; Valli, C.; Jaeger, S.R.; Moore, M.; Morrow, J.L.; Traill, W.B. Effect of information about benefits of biotechnology on consumer acceptance of genetically modified food: Evidence from experimental auctions in the United States, England, and France. *Eur. Rev. Agric. Econ.* **2004**, *31*, 179–204. [CrossRef]

24. ISMEA. *I Numeri della Filiera del Pomodoro da Industria*; Istituto di Servizi per il Mercato Agricolo Nazionale: Rome, Italy, 2017.
25. Poojary, M.M.; Passamonti, P. Extraction of lycopene from tomato processing waste: Kinetics and modelling. *Food Chem.* **2015**, *173*, 943–950. [CrossRef]
26. Takeshima, M.; Ono, M.; Higuchi, T.; Chen, C.; Hara, T.; Nakano, S. Anti-proliferative and apoptosis-inducing activity of lycopene against three subtypes of human breast cancer cell lines. *Cancer Sci.* **2014**, *105*, 252–257. [CrossRef]
27. Kaur, G.; Verma, N. Nature curing cancer–review on structural modification studies with natural active compounds having anti-tumor efficiency. *Biotechnol. Rep.* **2015**, *6*, 64–78. [CrossRef] [PubMed]
28. EFSA Panel on Dietetic Products, Nutrition and Allergies (NDA). Scientific Opinion on the substantiation of health claims related to lycopene and protection of DNA, proteins and lipids from oxidative damage (ID1608, 1609, 1611, 1662, 1663, 1664, 1899, 1942, 2081, 2082, 2142, 2374), protection of the skin from UV-induced (including photo-oxidative) damage (ID1259, 1607, 1665, 2143, 2262, 2373), contribution to normal cardiac function (ID1610, 2372), and maintenance of normal vision (ID1827) pursuant to Article 13(1) of Regulation (EC) No 1924/2006. *EFSA J.* **2011**, *9*, 1–28.
29. Lusk, J.L.; Shogren, J. *Experimental Auctions: Methods and Applications in Economic and Marketing Research*; Cambridge University Press: Cambridge, UK, 2007.
30. Ballco, P.; de-Magistris, T.; Caputo, V. Consumer preferences for nutritional claims: An exploration of attention and choice based on an eye-tracking choice experiment. *Food Res. Int.* **2019**, *116*, 37–48. [CrossRef] [PubMed]
31. Corrigan, J.R.; Rousu, M.C. Testing whether field auction experiments are demand revealing in practice. *J. Agric. Res. Econ.* **2008**, 290–301. [CrossRef]
32. Chang, J.B.; Lusk, L.J.; Norwood, B.F. How Closely Do Hypothetical Surveys and Laboratory Experiments Predict Field Behaviour? *Am. J. Agric. Econ.* **2009**, *91*, 518–534. [CrossRef]
33. Depositario, D.P.; Nayga, R.; Wu, X.; Laude, T.P. Should students be used as subjects in experimental auctions? *Econ. Lett.* **2009**, *102*, 122–124. [CrossRef]
34. Greenwald, A.G.; Banaji, M.R. Implicit social cognition: Attitudes, self-esteem, and stereotypes. *Psychol. Rev.* **1995**, *102*, 4. [CrossRef]
35. Greenwald, A.G.; Poehlman, T.A.; Uhlmann, E.L.; Banaji, M.R. Understanding and using the Implicit Association Test: III. Meta-analysis of predictive validity. *J. Personal. Soc. Psychol.* **2009**, *97*, 17. [CrossRef]
36. La Barbera, F.; Verneau, F.; Amato, M.; Grunert, K.G. Understanding Westerners' disgust for the eating of insects: The role of food neophobia and implicit associations. *Food Qual. Prefer.* **2018**, *64*, 120–125. [CrossRef]
37. Verneau, F.; La Barbera, F.; Kolle, S.; Amato, M.; Del Giudice, T.; Grunert, K.G. The effect of communication and implicit associations on consuming insects: An experiment in Denmark and Italy. *Appetite* **2016**, *106*, 30–36. [CrossRef]
38. Verneau, F.; La Barbera, F.; Del Giudice, T. The role of implicit associations in the hypothetical bias. *J. Consum. Aff.* **2017**, *51*, 312–328. [CrossRef]
39. Greenwald, A.G.; McGhee, D.E.; Schwartz, J.L. Measuring individual differences in implicit cognition: The implicit association test. *J. Personal. Soc. Psychol.* **1998**, *74*, 1464. [CrossRef]
40. Karpinski, A.; Steinman, R.B. The single category implicit association test as a measure of implicit social cognition. *J. Personal. Soc. Psychol.* **2006**, *91*, 16. [CrossRef] [PubMed]
41. Greenwald, A.G.; Nosek, B.A.; Banaji, M.R. Understanding and using the Implicit Association Test: I. An improved scoring algorithm: Correction to Greenwald et al. (2003). *J. Personal. Soc. Psychol.* **2003**, *85*, 481. [CrossRef]
42. Cox, D.N.; Evans, G. Construction and validation of a psychometric scale to measure consumers' fears of novel food technologies: The food technology neophobia scale. *Food Qual. Prefer.* **2008**, *19*, 704–710. [CrossRef]
43. Coppola, A.; Verneau, F.; Caracciolo, F. Neophobia in food consumption: An empirical application of the FTNS scale in southern Italy. *Ital. J. Food Sci.* **2014**, *26*, 81–90.
44. Verneau, F.; Caracciolo, F.; Coppola, A.; Lombardi, P. Consumer fears and familiarity of processed food. The value of information provided by the FTNS. *Appetite* **2014**, *73*, 140–146. [CrossRef]
45. Bak, H.J. Education and public attitudes toward science: Implications for the "deficit model" of education and support for science and technology. *Soc. Sci. Q.* **2001**, *82*, 779–795. [CrossRef]

46. Reynolds, W.M. Development of reliable and valid short forms of the Marlowe-Crowne Social Desirability Scale. *J. Clin. Psychol.* **1982**, *38*, 119–125. [CrossRef]
47. Fischbacher, U. z-Tree: Zurich toolbox for ready-made economic experiments. *Exp. Econ.* **2007**, *10*, 171–178. [CrossRef]
48. Hellyer, N.E.; Fraser, I.; Haddock-Fraser, J. Food Choice, Health Information and Functional Ingredients: An Experimental Auction Employing Bread. *Food Policy* **2012**, *37*, 232–245. [CrossRef]
49. Drichoutis, A.C.; Lazaridis, P.; Nayga, R.M. The role of reference prices in experimental auctions. *Econ. Lett.* **2008**, *99*, 446–448. [CrossRef]
50. Bernard, J.C.; He, N. Confounded by the field: Bidding in food auctions when field prices are increasing. *Agric. Resour. Econ. Rev.* **2010**, *39*, 275–287. [CrossRef]
51. Corrigan, J.R.; Drichoutis, A.; Lusk, J.; Nayga, R.; Rousu, M.C. Repeated Rounds with Price Feedback in Experimental Auction Valuation: An Adversarial Collaboration. *Am. J. Agric. Econ.* **2012**, *94*, 97–115. [CrossRef]
52. Koenker, R.; Bassett, G. Regression quantiles. *Econometrica* **1978**, *46*, 33–50. [CrossRef]
53. Koenker, R. *Quantile Regression*; Cambridge University Press: Cambridge, UK, 2005.
54. Oaxaca, R. Male-female wage differentials in urban labor markets. *Int. Econ. Rev.* **1973**, *14*, 693–709.
55. Blinder, A.S. Wage discrimination: Reduced form and structural estimates. *J. Hum. Resour.* **1973**, *8*, 436–455. [CrossRef]
56. Powell, J.L. Least absolute deviations estimation for the censored regression model. *J. Econom.* **1984**, *25*, 303–325. [CrossRef]
57. Vecchio, R.; Van Loo, E.J.; Annunziata, A. Consumers' willingness to pay for conventional, organic and functional yogurt: Evidence from experimental auctions. *Int. J. Consum. Stud.* **2016**, *40*, 368–378. [CrossRef]
58. Coppola, A.; La Barbera, F.; Verneau, F. Fair trade products' consumption: A market segmentation by personal values. *Qual. Access Success* **2015**, *16* (Suppl. S1), 23.
59. Wortmann, L.; Enneking, U.; Daum, D. German consumers' attitude towards selenium-biofortified apples and acceptance of related nutrition and health claims. *Nutrients* **2018**, *10*, 190. [CrossRef] [PubMed]
60. Steinhauser, J.; Hamm, U. Consumer and product-specific characteristics influencing the effect of nutrition, health and risk reduction claims on preferences and purchase behavior–A systematic review. *Appetite* **2018**, *127*, 303–323. [CrossRef] [PubMed]
61. Amato, M.; Fasanelli, R.; Riverso, R. Emotional profiling for segmenting consumers: The case of household food waste. *Qual. Access Success* **2019**, *20* (Suppl. S2), 27–32.
62. La Barbera, F.; Riverso, R.; Verneau, F. Understanding beliefs underpinning food waste in the framework of the theory of planned behaviour. *Qual. Access Success* **2016**, *17* (Suppl. S1), 130.

nutrients

Article

Trans Fat Labeling Information on Brazilian Packaged Foods

Camila Zancheta Ricardo [1,*], Isabela Mateus Peroseni [2], Laís Amaral Mais [3], Ana Paula Bortoletto Martins [1,3] and Ana Clara Duran [1,2]

1 Center for Epidemiological Research in Nutrition and Public Health, University of Sao Paulo, Sao Paulo 01246-904, Brazil
2 Center for Food Studies, University of Campinas, Campinas 13083-970, Brazil
3 Brazilian Institute for Consumer Defense, Sao Paulo 05002-050, Brazil
* Correspondence: ca.zancheta@gmail.com

Received: 28 June 2019; Accepted: 3 September 2019; Published: 6 September 2019

Abstract: Although the adverse effects of trans fat consumption are well documented, industrially-produced trans fats are still used in a variety of food products. Our objective was to investigate the presence of trans fat information on the nutrition facts panel, in the list of ingredients, and the use of trans fat claims in packaged food and beverages marketed in Brazil. This was a cross-sectional study that used data from packaged food and beverages available in the five supermarket chains with the largest market share in Brazil. Of the 11,434 products that were analyzed, 81.3% did not present a source of trans fats in the list of ingredients. The percentages of products with specific (hydrogenated fats or oils) and unspecific trans fat terms (margarine, vegetable fat, and vegetable cream) in the list of ingredients were 4.1% and 14.6%, respectively. Bakery products, cookies and crackers, candies and desserts, snacks, and convenience foods had the highest percentages of trans fat claims. We also found claims in products with ingredients that are sources of trans fats. In conclusion, trans fat ingredients were found in almost one-fifth of the Brazilian packaged foods. The current Brazilian legislation is not sufficient to inform consumers about the content of trans fats in packaged foods. Along with measures to restrict the use of industrially-produced trans fats, improvements in nutritional labeling are also needed.

Keywords: trans-fatty acids; ultra-processed food products; food labels; consumer; non-communicable disease

1. Introduction

Trans fats are unsaturated fatty acids with at least one double bond in the trans configuration. They occur naturally at low levels in meat and dairy products from ruminants and are industrially produced by the partial hydrogenation of vegetable oils [1]. Trans fats are attractive to the food industry because they increase the shelf life of foods and oils and improve taste and texture at a low cost. Therefore, a variety of ultra-processed food products (UPP), such as bakery products, cookies and crackers, snacks, sweets, ice cream, and ready-to-eat and frozen foods, contain trans fats [2–5].

The adverse health effects of trans fat consumption, particularly the association with increased risk of coronary heart disease (CHD), are well documented in the literature [1,6,7]. Although there is a clear relationship between the intake of trans fats and the risk of CHD, ruminant trans fats have no apparent health effects, which is likely due to the low content in the diet of the general population [8]. Furthermore, no benefits for health or "safe level" of trans fat intake has been identified. Compounding this issue, the consumption of trans fats exceeding the maximum limit proposed by the World Health Organization (WHO) (1% of total energy intake) [9] occurs mainly because of the availability of industrially-produced trans fats [10].

Over the past decades, voluntary and regulatory measures to lower the content of industrially-produced trans fats, including improvements in food labeling, food reformulation, and restrictions in use, have been adopted around the world [11]. The most effective measures are policy interventions aimed at eliminating industrially-produced trans fats [4]. Denmark was the first country to restrict the use of trans fats by imposing a limit of two grams of trans fats per 100 g of fat or oil, which has virtually eliminated the consumption of trans fats in the country [12]. The restriction on the use of trans fat ingredients have been discussed in many countries worldwide and the WHO recommends the elimination of industrially-produced trans fats to reduce the burden of non-communicable diseases (NCDs) as one of their priority targets, which has guided the WHO's strategic plan from 2019–2023 [13].

In 2003, the Brazilian Health Regulatory Agency (Agência Nacional de Vigilância Sanitária—Anvisa) included the information regarding the content of trans fats as a mandatory item on the nutrition facts panel of packaged foods and beverages labels in the country. However, the current legislation fails to provide clear and adequate information related to the content and presence of trans fats in packaged foods and beverages. First, trans fat content below 0.2 g per serving is considered insignificant and must be declared as zero on the nutrition facts panel [14]. Still, the portion sizes for different food groups are determined by legislation based on a recommended consumption, with small portion sizes for high caloric foods, such as biscuits and chocolates, [15], usually minor in comparison to the real portion people in Brazil eat [16]. Moreover, Brazilian regulations allow a variation of 20% in the nutrient content [14] and of 30% in the portion sizes [15] declared on the packages. Second, although the use of nutrition claims about trans fats are regulated in the country, it is possible that products with trans fat ingredients state a message such as "zero trans fats" or "trans fat free" if they present less than 0.1 g of trans fats per portion and also a low content of saturated fat (a maximum of 1.5 g of trans and saturated fats in a portion or per 100 g of prepared dishes and with less than 10% of energy from saturated fat) [17]. In addition to labeling rules, in 2007, the Brazilian Ministry of Health signed a voluntary agreement with food industry representatives to reduce trans fats in processed foods [18]. Despite these efforts, the consumption of trans fats in Brazil exceeds the limit proposed by WHO [10].

A review that investigated the consumption of trans fats in 29 countries worldwide found that in the majority of the studied countries, the trans fat intake is under the limit of 1% of total energy intake, and the main source of the trans fats consumed is from animal products [10]. However, data from the last survey about food consumption in Brazil conducted during the period 2008–2009 showed that the average trans fat intake in the country was equivalent to 1.4% of the total energy intake [19], and the main source of trans fats was industrial (e.g., oils and fats, biscuits, pizza, grains, seeds, nuts, chocolate, soups, savory snacks, meals, and restaurant foods) [10]. The scenario in the country is concerning since Brazil is facing changes in dietary pattern, with a crescent consumption of UPP [20], and the consumption of those products is associated with higher trans fat intake [21]. Analysis of a national survey showed that the quintile of Brazilians with lower UPP consumption was in line with the recommendation of trans fat (0.8% of total energy intake) and for those in the quintile with highest UPP consumption the trans fat intake was 1.9% of total energy intake, double that of those with lower UPP consumption and almost two times the limit established by WHO [19]. Moreover, in Brazil cardiovascular disease (CVD) is the leading cause of deaths [22], and it was estimated that more than 10% of the CHD deaths were due to high trans fat intake [23], equivalent to at least 10,000 deaths in the country in 2010 [24].

Considering the challenges of developing and implementing effective regulatory measures for the prevention and control of NCDs related to unhealthy diets in Brazil, the objective of the present study was to evaluate the presence of trans fat information and claims regarding trans fats in packaged foods and beverages marketed in Brazil.

2. Materials and Methods

2.1. Data Collection

This cross-sectional study used data from a sample of packaged foods and beverages that are sold in Brazilian supermarkets. Supermarkets were selected as the source for the data collection because they are the main places in which food is purchased in the country [25]. The five supermarket chains with the largest market share in Brazil were identified using food retail annual sales data from Euromonitor International in 2016 [26]. The primary study area was São Paulo, the largest Brazilian city, which is located in the Southeastern region of the country. Because one of the supermarket chains only had stores in the Northeastern region of the country, we collected data in Salvador, which is considered the largest city in the region. We obtained formal permission from all supermarket chains included in this study.

Trained fieldworkers photographed all sides of the packages of the products available in the stores between April and July 2017, according to the methods proposed by Kanter et al. (2017) [27]. We included approximately 13,000 different items. The product name, brand, flavor, package size, nutrition facts panel, and list of ingredients were entered by trained nutritionists in an online platform using a template developed by researchers from the Institute of Nutrition and Food Technology, University of Chile (Instituto de Nutrición y Tecnología de los Alimentos—INTA, Universidad de Chile, Santiago, Chile) and the University of North Carolina at Chapel Hill (UNC, Chapel Hill, NC, USA). We adapted and translated the template for this research.

We excluded duplicate records, products available in more than one package size, multipacks with different items, products without a nutrition facts panel, products without a list of ingredients, and products with missing values for portion size and/or calories. The final database was composed of 11,434 products. All products were included in the analysis of the nutrition facts panel and the list of ingredients.

Due to budget and time constraints, we performed the collection of claims information in a random subsample of 30% of the products ($n = 3491$). This was the first project aimed to identify the prevalence of nutrition claims in Brazil, so we guaranteed that our sample size was large enough to estimate even a small prevalence ($p = 0.01$) and its 95% confidence intervals (CI), considering a precision of 0.005 [28]. Claims collection and classification were based on the International Network for Food and Obesity/NCD Research, Monitoring and Action Support (INFORMAS) protocol [29]. In the present study, we considered the presence of trans fat claims in any part of the packages.

For analysis, we classified the food and beverages products in 25 categories presented in Appendix A.

2.2. Data Quality

For data on nutritional composition, 10% of the sample was entered twice by the same person, and 10% of the sample was entered by a second person for intra- and interrater reliability analysis. For data on claims, the entire subsample was entered twice: 50% for intra- and 50% for interrater reliability analysis. Intraclass correlation coefficients (ICC) were calculated for the content of trans fats, and Cohen's kappa coefficients were calculated for the presence of claims. The ICC values were 0.905 and 0.949, and the kappa values were 0.949 and 0.893 for intra- and interrater, respectively. These analyses confirmed the reliability of the data [30]. We compared the proportion of food groups and the average nutrients in the database and the subsample, using chi-square and *t*-test, and found no differences.

2.3. Search for Terms in the List of Ingredients

Table 1 shows the keywords that were used to identify products with trans fat ingredients. Specific terms clearly identify ingredients that are a source of industrially-produced trans fats, but unspecific terms may or may not indicate a source of trans fats. Therefore, the latter terms are imprecise in determining the presence of industrially-produced trans fats in a product.

Table 1. Keywords used to identify products with trans fats in the list of ingredients.

Type of Term	Searched Items
Specific	Partially hydrogenated fat, hydrogenated vegetable oil, hydrogenated.
Unspecific	Vegetable fat, margarine, vegetable cream.

Adapted from Silveira et al., 2013 [3].

Based on the information of trans fats on the nutrition facts panel and in the list of ingredients, the foods were divided into four categories:

- With trans fats: foods with information indicating a content of trans fats greater than zero on the nutrition facts panel;
- False negatives: foods with no trans fats declared on the nutrition facts panel but with specific sources of trans fats in the list of ingredients;
- Possible false negatives: foods with no trans fats declared on the nutrition facts panel but with unspecific sources of trans fats in the list of ingredients;
- Without trans fats: foods with no specific or unspecific terms related to trans fats in the list of ingredients.

2.4. Statistical Analysis

We calculated the proportion and 95% confidence intervals (95% CI) of foods with information of trans fats on the nutrition fact panel, in the list of ingredients and with trans fat claims, by food categories. Additionally, we analyzed the proportion of the four aforementioned categories for the presence of trans fats in foods and beverages with claims.

3. Results

Of the 11,434 food products, 81.3% did not include ingredients that are sources of trans fats in the list of ingredients. The percentages of products with specific and unspecific ingredient sources of trans fats in the list of ingredients were 4.1% and 14.6%, respectively. Bakery products (20.7%), cookies and crackers (15.1%), snacks (12.6%), candies and desserts (8.2%), and convenience foods (6.0%) had the highest proportion of products with specific trans fat ingredients. Unspecific keywords in the list of ingredients were more prevalent in the same five food groups (Table 2).

Over 90% of the products reported zero or an insignificant amount of trans fats on the nutrition facts panel. Bakery products (17.6%), cheeses and cheese spreads (17.3%), convenience foods (16.7%), and cookies and crackers (13.0%) had the highest proportions of items with a listed content of trans fats above zero grams. Only a small portion of the products (1.4%) did not report the content of trans fats on the nutrition facts panel, and these products were mostly packaged fruits and vegetables, which are not a source of trans fats (Table 2).

Figure 1 shows the proportion of foods and beverages based on the presence of trans fats. Approximately 10% of the cookies and crackers, bakery products, and snacks were false negatives (i.e., they had specific sources of trans fats in the list of ingredients but declared no trans fat content on the nutrition facts panel). More than half of the cookies and crackers and approximately one-third of the bakery products and snacks were possible false negatives (i.e., they listed possible sources of trans fats in the list of ingredients and declared being free of trans fats on the nutrition facts panel).

Table 2. Number and proportion of packaged foods with sources of trans fat in the list of ingredients and with trans fats declared on the nutrition facts panel.

Food Category	n	List of Ingredients						Nutrition Facts Panel					
		None		Specific		Unspecific		Zero or Insignificant		Above Zero		Not Informed	
		n	%	n	%	n	%	n	%	n	%	n	%
Breakfast cereals and granola bars	308	225	73.1	2	0.6	81	26.3	304	98.7	4	1.3	0	0.0
Bakery products	595	244	41.0	123	20.7	228	38.3	490	82.4	105	17.6	0	0.0
Convenience foods	795	477	60.0	48	6.0	270	34.0	651	81.9	133	16.7	11	1.4
Unsweetened dairy beverages	181	181	100.0	0	0.0	0	0.0	179	98.9	2	1.1	0	0.0
Sweetened dairy beverages	483	472	97.7	0	0.0	11	2.3	476	98.6	4	0.8	3	0.6
Snacks	356	203	57.0	45	12.6	108	30.3	340	95.5	14	3.9	2	0.6
Cookies and crackers	747	185	24.8	113	15.1	449	60.1	638	85.4	97	13.0	12	1.6
Canned vegetables	345	345	100.0	0	0.0	0	0.0	330	95.7	2	0.6	13	3.8
Oils and fats	351	338	96.3	7	2.0	6	1.7	325	92.6	24	6.8	2	0.6
Sauces, herbs, and dressings	801	720	89.9	18	2.2	63	7.9	785	98.0	9	1.1	7	0.9
Coffee and tea	94	94	100.0	0	0.0	0	0.0	84	89.4	0	0.0	10	10.6
Candies and desserts	1220	765	62.7	100	8.2	355	29.1	1132	92.8	83	6.8	5	0.4
Cereals, beans, other grain products	735	698	95.0	2	0.3	35	4.8	717	97.6	10	1.4	8	1.1
Packaged fruits and vegetables	907	907	100.0	0	0.0	0	0.0	841	92.7	0	0.0	66	7.3
Meat, poultry, seafood, and eggs	577	577	100.0	0	0.0	0	0.0	518	89.8	55	9.5	4	0.7
Sugar and other nonnutritive sweeteners	106	106	100.0	0	0.0	0	0.0	106	100.0	0	0.0	0	0.0
Processed meats	810	759	93.7	1	0.1	50	6.2	769	94.9	36	4.4	5	0.6
Fruit juices	150	150	100.0	0	0.0	0	0.0	150	100.0	0	0.0	0	0.0
Nectars	160	160	100.0	0	0.0	0	0.0	160	100.0	0	0.0	0	0.0
Fruit-flavored drinks	220	220	100.0	0	0.0	0	0.0	220	100.0	0	0.0	0	0.0
Carbonated beverages	106	106	100.0	0	0.0	0	0.0	106	100.0	0	0.0	0	0.0
Other beverages	286	285	99.7	0	0.0	1	0.3	281	98.3	0	0.0	5	1.7
Nuts and seeds	80	74	92.5	2	2.5	4	5.0	79	98.8	0	0.0	1	1.3
Cheeses and cheese spreads	607	594	97.9	11	1.8	2	0.3	502	82.7	105	17.3	0	0.0
Fruit preserves	414	412	99.5	0	0.0	2	0.5	410	99.0	0	0.0	4	1.0
Total	**11,434**	**9297**	**81.3**	**472**	**4.1**	**1665**	**14.6**	**10,593**	**92.6**	**683**	**6.0**	**158**	**1.4**

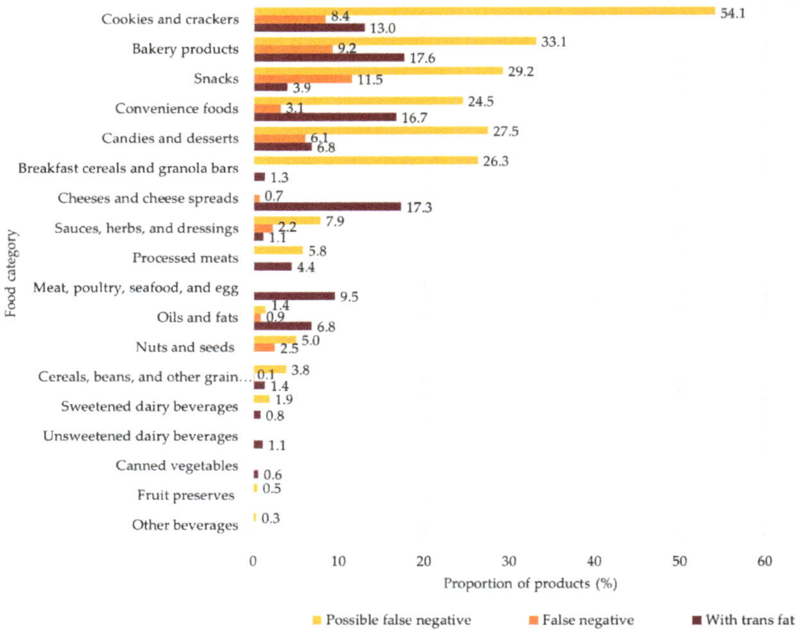

Figure 1. The proportion of products with trans fat content declared on the nutrition facts panel, false negatives, and possible false negatives by food category. Note: All the products of the sample were classified in the four groups related to the presence of trans fats (n = 11,434). Food categories in which all items were classified in the "without trans fats" group were not shown in this figure.

Nutrition claims regarding trans fats were found in 4.6% of the assessed products. Bakery products had the highest proportion (21.1%) of foods with claims of the absence of trans fats, and in four other groups, claims were presented in around 10% of the products: breakfast cereal and granola bars (11.6%), convenience foods (9.2%), snacks (11.1%), and cookies and crackers (12.0%) (Table 3). Most of the foods with claims had no ingredients that were possible sources of trans fats in the list of ingredients. However, we found trans fat ingredients in 24.4% of the 160 products with claims. This proportion was higher in cookies and crackers and bakery products, in which almost half of the products could present some trans fat ingredient (Figure 2).

Table 3. Number and proportion of packaged foods with trans fat claims.

Food Category	n Subsample	Presence of Claims			
		n	%	95% CI	
Bakery products	175	37	21.1	15.7	27.8
Cookies and crackers	225	27	12.0	8.4	16.9
Breakfast cereals and granola bars	95	11	11.6	6.5	19.8
Snacks	108	12	11.1	6.4	18.6
Convenience foods	238	22	9.2	6.2	13.7
Fruit preserves	136	7	5.1	2.5	10.4
Candies and desserts	365	13	3.6	2.1	6.0
Oils and fats	106	3	2.8	0.9	8.5
Sauces, herbs, and dressings	262	7	2.7	1.3	5.5
Cereals, beans, other grain products	230	6	2.6	1.2	5.7

Table 3. *Cont.*

Food Category	n Subsample	Presence of Claims			
		n	%	95% CI	
Processed meats	257	6	2.3	1.1	5.1
Nectars	50	1	2.0	0.3	13.1
Packaged fruits and vegetables	261	5	1.9	0.8	4.5
Other beverages	77	1	1.3	0.2	8.8
Meat, poultry, seafood, and eggs	171	2	1.2	0.3	4.6
Unsweetened dairy beverages	56	0	0.0	0.0	0.0
Sweetened dairy beverages	149	0	0.0	0.0	0.0
Canned vegetables	110	0	0.0	0.0	0.0
Coffee and tea	30	0	0.0	0.0	0.0
Sugar and other nonnutritive sweeteners	37	0	0.0	0.0	0.0
Fruit juices	53	0	0.0	0.0	0.0
Fruit-flavored drinks	63	0	0.0	0.0	0.0
Carbonated beverages	35	0	0.0	0.0	0.0
Nuts and seeds	22	0	0.0	0.0	0.0
Cheeses and cheese spreads	180	0	0.0	0.0	0.0
Total	**3491**	**160**	**4.6**	**3.9**	**5.3**

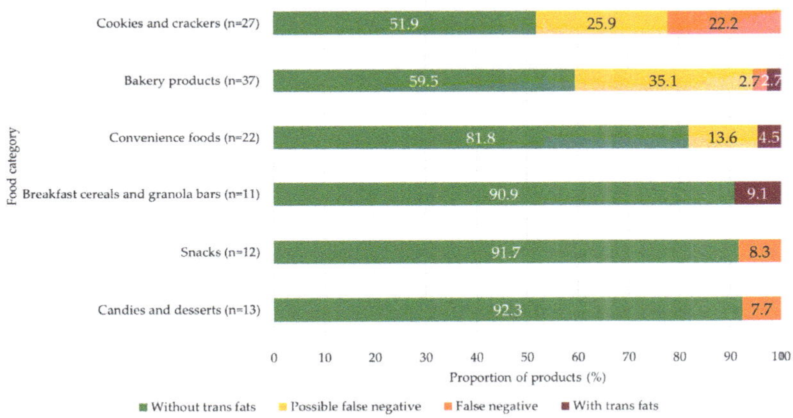

Figure 2. The proportion of products in the four categories related to the trans fat presence among the products displaying trans fat claims. Note: We displayed the six food categories with the highest proportion of trans fat claims in this figure. In other food categories, products with trans fat claims did not have trans fat ingredients. The exceptions were two products in cereal products (possible false negative) and two products in oils and fats (with trans fats). However, we decided not to show those groups in the figure since the total number of products with claims was very low.

4. Discussion

Most of the Brazilian packaged foods and beverages did not report the presence of trans fats in the list of ingredients or on the nutrition facts panel. However, our results showed that industrially-produced trans fats may still be present in almost one-fifth of the products. Four percent of the evaluated products presented hydrogenated fats or oils. Furthermore, 15% of food and beverages had at least one ingredient that might be a source of trans fats, but that do not allow for a definitive assessment of their presence. Specific and unspecific sources of trans fats were also identified in foods with claims about their absence.

Although the content of trans fats in Brazilian food products has been shown to be reduced after the implementation of legislation on food labeling [31] and voluntary reduction agreements in the food industry [32], more recent evidence on this topic is lacking in the country. A 2010 study of food labels of products available in a supermarket in Santa Catarina, located in the Southern region of the country, found trans fat terms in more than half of the products. However, only 20% of the products declared trans fats on the nutrition facts panel [3]. Our sample included a larger number of products and food categories, but we found similar results for some categories, such as bakery products and cookies and crackers. Another study, conducted during the period 2014–2015 in Rio de Janeiro, in the Southeastern region of Brazil, evaluated some of the most consumed processed foods, including vegetable oils, margarine, biscuits, snacks, and ice cream, and found a drop in trans fat content in those products compared with similar research made in 2003, prior to the regulation. However, this drop in trans fats has not been observed in all products, and one of the samples of biscuits presented values as high as 12.9 g per 100 g [31].

Systematic reviews about the effectiveness of policies for reducing trans fats found that mandatory food labeling and voluntary self-regulation, the two measures adopted in Brazil, resulted in a decrease of trans fat levels in the food supply, but they are not as effective as national bans [33,34]. While the restriction led to elimination of trans fat in Denmark, in Canada, where mandatory labeling and voluntary limits were also implemented, some categories like bakery products and restaurant food remained with trans fats in a portion of the products [34]. Voluntary agreements depend on the extent to which the industry is willing to agree to and comply with the arrangement [35]. In Brazil, we have no official data on the progress obtained with the measure. Also, the information available on the total amount of trans fats withdrawn from the market was presented by the association of the food industries, without details on the applied methodology [32]. In the current Brazilian scenario, some products with trans fats continue to be sold and they usually are cheaper than a trans fat free option [36,37], which represents a disadvantage for lower-income consumers. Food labeling, on the other hand, depends on health literacy for consumers to use and understand it. Recent work conducted in Brazil showed that consumers use food labels to check nutrient content and ingredient information, but the format and the technicality of the labels often made the information inaccessible, particularly for those of low socioeconomic status [38]. Moreover, while it is an effective effort to influence consumer purchasing and product reformulation, mandatory labeling seems to have a limited long-term effect [39].

The scientific basis for the establishment of the cutoffs for the inclusion of trans fat content on nutrition facts panel and nutrition claims are not declared in the Brazilian legislation [14,17] and this is an issue that should be discussed. The current legislation allows food packages displaying zero content in products with 0.2 g of trans fats per serving. However, this amount corresponds to approximately 10% of the daily limit recommended by WHO (less than 1% of the total energy intake is equal to less than 2.2 g per day for a 2000-calorie diet) [9]. In comparison, the same cutoff of 0.2 g per serving is adopted for saturated fats, but this limit represents only 1% of the 22 g considered as a reference daily value for a 2000-calorie diet [14]. In addition, for some UPP, particularly those with trans fat ingredients, such as biscuits, chocolates, and snacks, the consumed serving sizes by Brazilian population are more than three times the declared serving size [16]. This means that the consumption of trans fats can exceed the upper recommended limit even when consumers choose products declared as trans fat free. Biochemical analyses of breads, cookies, oils, margarine, sauces, snacks, chocolates, cakes, and popcorn purchased in supermarkets located in a city of São Paulo found that most of the products complied with the legislation. However, only a small portion of the products that were declared as being free of trans fats on the nutrition facts panel did not have any trans fats in their composition (11 products out of 189—or 6%) [40].

Another important aspect of food labeling is the lack of standardization in the terms used in the list of ingredients, which may create confusion among consumers. Silveira et al. (2013) [3] identified that more than 20 ingredients may be a source of trans fats in Brazilian food products, and approximately half of those ingredients do not allow consumers to determine if trans fats are contained within the product. In this study, conducted in 2010, the majority of products with ingredients that are potential sources of trans fats presented imprecise terms, such as vegetable fat, vegetable cream, or margarine [3]. We found similar results which reinforce the need to establish and impose standardized names to be used in food labeling in the country to guarantee the right to information for consumers.

Regarding nutrition claims, as mentioned previously, the current Brazilian rules permit the use of messages that claim that the product is free of trans fats on products with trans fat content below 0.1 g per serving [17]. However, the presence of these claims on food products should be evaluated with caution. Positive messages about nutrition are used as marketing strategies by the food industry and have a selective focus, ignoring the presence of other unhealthy nutrients and encouraging the consumption of products with poor nutritional quality [41]. In this study, even foods with ingredients that are sources of trans fats showed trans fat free claims. Moreover, bakery products, cookies and crackers, candies and desserts, snacks, and convenience foods displayed the majority of trans fat claims. Many of the items in these food groups are classified as UPP [42] and should be avoided, according to the recommendations of the Brazilian Dietary Guidelines [43].

Imposing limits on the use of industrially-produced trans fats is the measure that has had the greatest impact on trans fat consumption in many countries [34], and it is currently under discussion in the regulatory process at the Brazilian agency [44]. This intervention has the advantage of not depending on individual actions and has the potential to positively impact the entire population [11]. Many countries have adopted measures restricting the use of trans fats in food production, such as Denmark, Hungary, and Norway, which have resulted in a drastic reduction in the use of such fats, revealing that these measures are feasible and effective public policies [45]. In Denmark, the first country in which restriction on the use of industrial trans fat was implemented, there was an estimated decrease in mortality attributable to CVD on average by about 14.2 deaths per 100,000 people per year in the three years that followed the policy [46]. Taking into account all the available evidence, recently, in April 2019, the European Union adopted a new regulation that must be effective until 2021, which limits the amount of industrially-produced trans fats in all foods to a maximum of two grams per 100 g of fat [47].

In 2008, the Pan American Health Organization (PAHO) launched the "Trans Fat Free Americas", which recommended a limit of 5% of total fats from trans fats in industrialized food products and pointed to the need for legislative measures [48]. Countries in the Latin American region, such as Chile and Argentina, have already imposed a limit on trans fat use [49]. In 2015, in the United States (USA), the Food and Drug Administration (FDA) put partially hydrogenated oils, the main source of industrially-produced trans fats, on the list of substances that are not generally recognized as safe for use in food. Three years later, in 2018, nationwide bans were implemented on the use of this ingredient for the majority of the products in the USA [50]. The elimination of industrially-produced trans fats is a priority for WHO, which has recently launched a guide with strategic actions to be implemented in countries, including low- and middle-income countries in which the legislation regarding the use of industrially-produced trans fats is weaker [51].

Implementing policies that aim to restrict trans fats in the food supply results in the need for product reformulation, and the employed substitutes can be concerning as trans fats could be replaced with saturated fats to maintain the required or preferred solid fat content. However, a study conducted in 14 countries that aimed to compare foods with high and low content of trans fats showed that french fries, cookies, cakes, and microwave popcorn with low content of trans fatty acids had a higher content of saturated fats but also higher content of monounsaturated and polyunsaturated fatty acids [52]. The results of the aforementioned study indicated that, in different countries, trans fats were replaced with a combination of fats [52]. A separate study which investigated changes in the levels of trans and saturated fats in the United States from 1993 to 2009 found lower levels of both fats after reformulation in more than 90% of evaluated products [53]. Although any measure adopted to remove industrial trans fats would be expected to produce health benefits, WHO recommends the use of mono- and polyunsaturated fats as a replacement for trans fats rather than using animal fats and tropical oils that are high in saturated fat [9,51]. We suggest that the use of other industrially-produced fats, as interesterified fats, should be considered with caution because their health effects are not clear [54,55].

It is important to highlight that, initially, trans fat use by the food industry was considered a healthier option in relation to animal fat, due to the presence of unsaturation and its vegetable origin, but it is currently known that its consumption negatively impacts lipid metabolism, causing detrimental effects on the heart health [56]. The use of interesterified oils makes it possible to obtain similar sensory characteristics in food products without trans fat content, but the process results in a modification in the arrangement of fatty acids on the glycerol backbone and its influence on lipid metabolism is not understood [54,55]. Given the lack of scientific evidence on the health effect of consumption of interesterified fat, there is a need for research that contributes to the understanding of the effects of such fats on health to ensure the safety of its use and clarify the risks related to its consumption. The focus on the reformulation of products should not be the final goal in the discussion to promote healthier habits since recommendations made based only on the nutrients have been shown ineffective in dealing with current nutritional problems [56,57].

This study has some limitations. Our data source was the labels of food products marketed in the country and, to classify the product by the presence of trans fats, we used the information available in the list of ingredients and on the nutrition facts panel without conducting laboratory analyses, which may have introduced bias in our findings. Another limitation was the use of a random subsample to evaluate nutrition claims. Nonetheless, we were cautious in selecting our subsample, we included a large number of products, and we did not find any statistically significant differences in the proportion of foods in each food category or on the average content of nutrients, including trans fats, when we compared our subsample with the total sample of the assessed Brazilian packaged foods (data not shown). As a strength, we attempted to include all available packaged foods and beverages in the five largest food retailer chains in the country. We also took advantage of the INFORMAS claims taxonomy, which was developed to standardize the classification of the different health-related labeling components in different countries, to assess the presence of claims [29].

In conclusion, despite advances in recent years regarding the use of trans fats in Brazil, important issues concerning the presence and information about trans fats remain. In food labeling, there is a possibility that products are declared as trans fat free even with the presence of ingredients that are sources of trans fats. Another related issue is the deficiency of standardized terms in the list of ingredients. The lack of accurate information on packages can misinform consumers and lead to excessive consumption of trans fats without knowing. Food labeling with clear and adequate messages is a citizen's right, as was established by the Brazilian Consumer Defense Code (Código de Defesa do Consumidor) [58]. In the case of trans fats, which are proven to cause significant health damage, the discussion on regulation must advance towards the restriction of industrial use. Trans fat limits should be imposed from the raw material used in the food industry, and no oils or fats with a high content of trans fats should be available to be used in Brazilian food products. This measure could benefit the entire population and aid inspection by competent agents. We believe our paper contributes to advancing the topic on the use of trans fats in Brazil and could be used to inform policymakers who are discussing the new regulation in the country, and also as a source of information for media and the general public about the products available in our supermarkets, aiming to increase public awareness of the need for policy options that supports better food environments.

Author Contributions: Conceptualization, C.Z.R. and A.C.D.; methodology, C.Z.R.; formal analysis, C.Z.R.; writing—original draft preparation, C.Z.R.; writing—review and editing, C.Z.R., I.M.P.; L.A.M.; A.P.B.M., and A.C.D.; project administration, A.C.D. and A.P.B.M.; funding acquisition, A.C.D. and A.P.B.M.

Funding: This research was funded by Bloomberg Philanthropies through a sub award agreement between the University of North Carolina at Chapel Hill and the Medicine Faculty Foundation (FFM), grant number 5104695, and by the International Development Research Center (IDRC), grant number 108166-001.

Acknowledgments: We thank North Carolina TraCS Institute, University of North Carolina, Chapel Hill, NC, USA UL1RR025747 for support for the REDCap electronic data capture tool.

Conflicts of Interest: The authors declare no conflict of interest. The funders had no role in the design of the study; in the collection, analyses, or interpretation of data; in the writing of the manuscript, or in the decision to publish the results.

Appendix A

Table A1. Description of food categories.

Food Category	Foods Included
Breakfast cereals and granola bars	Corn flakes, flavored oats, infant cereals, granolas, mueslis, granola bars, porridges, mix of cereals and fruits.
Bakery products	Breads, toasts, and cakes, including powders.
Convenience foods	Ready meals, French fries to frying, instant rice, instant noodles, instant soups, instant mashed potatoes, filled pastas, pizzas, pies, sandwiches, baby foods.
Unsweetened dairy products	Natural yogurts, milks, milk powders, milk compounds. Only products without sugar or nonnutritive sweeteners.
Sweetened dairy products	Yogurts, flavored milks, fermented milk, milk compounds, dairy beverages.
Snacks	Salted peanuts, salty snacks, potato chips, potato sticks, microwave popcorn (all flavors).
Cookies and crackers	Sweet and salty biscuits and cookies.
Canned vegetables	Canned beans and vegetables.
Oils and fats	Oils, margarines, butters, milk creams, vegetal fats.
Sauces, herbs, and dressings	Sauces, mayonnaises, herbs, catchup, salad dressings.
Coffee and tea	Coffees and herbs to prepare tea.
Candies and desserts	Chocolates, boiled sweets, flavored milk powder, jellies, cocoa powder, syrups, chocolate spreads, chewing gum, condensed milk, marshmallows, ice creams. Including versions with nonnutritive sweeteners.
Cereals, beans, other grain products	Beans, flours, rice, corns, pastas.
Packaged fruits and vegetables	Fruits and vegetables in natura, frozen fruit pulp, frozen vegetables.
Meat, poultry, seafood, and eggs	Meat, poultry, seafood, and egg, including chilled and frozen products.
Sugar and other nonnutritive sweeteners	Sugar, honey, nonnutritive sweeteners.
Processed meats	Burgers, sausages, canned fish, smoked meats, seasoned meats, salted meats, hams, salami, meat pates.
Fruit juices	Products declared as juices, fruit juices without water or added sugar, coconut water.
Nectars	Nectars made with juice and added water and/or sugar.
Fruit-flavored drinks	Fruit drink powder, fruit punch concentrate, fruit-flavored beverages.
Carbonated beverages	Sodas. Including diet and light versions.
Other beverages	Plant-based beverages, ready-to-drink teas, isotonic drinks, coconut milk.
Nuts and seeds	Nuts and seeds, including salted nuts.
Cheeses and cream cheeses	Cheeses, and cream cheeses.
Fruits preserves	Fruit jellies, fruit preserves, canned fruits, dried fruits, fruits sorbet.

References

1. Mozaffarian, D.; Katan, M.B.; Ascherio, A.; Stampfer, M.J.; Willett, W.C. Trans fatty acids and cardiovascular disease. *N. Engl. J. Med.* **2006**, *354*, 1601–1613. [CrossRef] [PubMed]
2. Albers, M.J.; Harnack, L.J.; Steffen, L.M.; Jacobs, D.R. 2006 marketplace survey of trans-fatty acid content of margarines and butters, cookies and snack cakes, and savory snacks. *J. Am. Diet. Assoc.* **2008**, *108*, 367–370. [CrossRef] [PubMed]
3. Silveira, B.M.; Gonzalez-Chica, D.A.; da Costa Proença, R.P. Reporting of trans-fat on labels of Brazilian food products. *Public Health Nutr.* **2013**, *16*, 2146–2153. [CrossRef] [PubMed]
4. Mozaffarian, D.; Stampfer, M.J. Removing industrial trans fat from foods. *BMJ* **2010**, *340*, c1826. [CrossRef] [PubMed]
5. Roe, M.; Pinchen, H.; Church, S.; Elahi, S.; Walker, M.; Farron-Wilson, M.; Buttriss, J.; Finglas, P. Trans fatty acids in a range of UK processed foods. *Food Chem.* **2013**, *140*, 427–431. [CrossRef] [PubMed]
6. Mozaffarian, D.; Aro, A.; Willett, W.C. Health effects of trans-fatty acids: Experimental and observational evidence. *Eur. J. Clin. Nutr.* **2009**, *63*, S5–S21. [CrossRef] [PubMed]
7. Teegala, S.M.; Willett, W.C.; Mozaffarian, D. Consumption and health effects of trans fatty acids: A review. *J. AOAC Int.* **2009**, *92*, 1250–1257. [PubMed]
8. Bendsen, N.T.; Christensen, R.; Bartels, E.M.; Astrup, A. Consumption of industrial and ruminant trans fatty acids and risk of coronary heart disease: A systematic review and meta-analysis of cohort studies. *Eur. J. Clin. Nutr.* **2011**, *65*, 773–783. [CrossRef] [PubMed]
9. WHO. World Health Organization. *Draft Guidelines: Saturated Fatty Acid and Trans-Fatty Acid Intake for Adults and Children*; WHO: Geneva, Switzerland, 2018.
10. Wanders, A.J.; Zock, P.L.; Brouwer, I.A. Trans Fat Intake and Its Dietary Sources in General Populations Worldwide: A Systematic Review. *Nutrients* **2017**, *9*, 840. [CrossRef]
11. Hyseni, L.; Bromley, H.; Kypridemos, C.; O'Flaherty, M.; Lloyd-Williams, F.; Guzman-Castillo, M.; Pearson-Stuttard, J.; Capewell, S. Systematic review of dietary trans-fat reduction interventions. *Bull. World Health Organ.* **2017**, *95*, 821–830G. [CrossRef]
12. L'abbé, M.R.; Stender, S.; Skeaff, C.M.; Tavella, M. Approaches to removing trans fats from the food supply in industrialized and developing countries. *Eur. J. Clin. Nutr.* **2009**, *63*, S50–S67.
13. WHO. World Health Organization. *Thirteenth General Programme of Work, 2019–2023: Promote Health, Keep the World Safe, Serve the Vulnerable*; WHO: Geneva, Switzerland, 2019.
14. Brasil Agência Nacional de Vigilância Sanitária. *Resolução-RDC Anvisa n°. 360, de 23 de Dezembro de 2003—Aprova o Regulamento Técnico Sobre Rotulagem Nutricional de Produtos Embalados*; Brasil Agência Nacional de Vigilância Sanitária: Brasília, Brazil, 2003.
15. Brasil Agência Nacional de Vigilância Sanitária. *Resolução-RDC Anvisa n°. 359, de 23 de Dezembro de 2003*; Brasil Agência Nacional de Vigilância Sanitária: Brasília, Brazil, 2003.
16. Kraemer, M.V.D.S.; Machado, P.P.; Kliemann, N.; Chica, D.A.G.; Proença, R.P.D.C. The Brazilian population consumes larger serving sizes than those informed on labels. *Br. Food J.* **2015**, *117*, 719–730. [CrossRef]
17. Brasil Agência Nacional de Vigilância Sanitária. *Resolução da Diretoria Colegiada—RDC N° 54, de 12 de Novembro de 2012. Dispõe Sobre o Regulamento Técnico Sobre Informação Nutricional Complementar*; Brasil Agência Nacional de Vigilância Sanitária: Brasília, Brazil, 2012.
18. Brasil Ministério da Saúde. *Nota técnica. Ações do Governo Brasileiro Sobre as Gorduras Trans*; Ministério da Saúde: Rio de Janeiro, Brazil, 2008. Available online: http://189.28.128.100/dab/docs/portaldab/documentos/nota_imprensa_gorduras_trans.pdf (accessed on 17 June 2019).
19. Costa Louzada, M.L.; Martins, A.P.; Canella, D.S.; Baraldi, L.G.; Levy, R.B.; Claro, R.M.; Moubarac, J.C.; Cannon, G.; Monteiro, C.A. Ultra-processed foods and the nutritional dietary profile in Brazil. *Rev. Saude Publica* **2015**, *49*, 38. [CrossRef] [PubMed]
20. Martins, A.P.B.; Levy, R.B.; Claro, R.M.; Moubarac, J.C.; Monteiro, C.A. Participação crescente de produtos ultraprocessados na dieta brasileira (1987–2009). *Rev. Saúde Pública* **2013**, *47*, 656–665. [CrossRef] [PubMed]
21. Da Costa Louzada, M.L.; Ricardo, C.Z.; Steele, E.M.; Levy, R.B.; Cannon, G.; Monteiro, C.A. The share of ultra-processed foods determines the overall nutritional quality of diets in Brazil. *Public Health Nutr.* **2018**, *21*, 94–102. [CrossRef] [PubMed]

22. Collaborators, G.B. Burden of disease in Brazil, 1990–2016: A systematic subnational analysis for the Global Burden of Disease Study 2016. *Lancet* **2018**. [CrossRef]
23. Wang, Q.; Afshin, A.; Yakoob, M.Y.; Singh, G.M.; Rehm, C.D.; Khatibzadeh, S.; Micha, R.; Shi, P.; Mozaffarian, D. Global Burden of Diseases Nutrition and Chronic Diseases Expert Group (NutriCoDE). Impact of Nonoptimal Intakes of Saturated, Polyunsaturated, and Trans Fat on Global Burdens of Coronary Heart Disease. *J. Am. Heart Assoc.* **2016**, *5*. [CrossRef]
24. Baena, C.P.; Chowdhury, R.; Schio, N.A.; Sabbag, A.E.; Guarita-Souza, L.C.; Olandoski, M.; Franco, O.H.; Faria-Neto, J.R. Ischaemic heart disease deaths in Brazil: Current trends, regional disparities and future projections. *Heart* **2013**, *99*, 1359–1364. [CrossRef] [PubMed]
25. Machado, P.P.; Claro, R.M.; Canella, D.S.; Sarti, F.M.; Levy, R.B. Price and convenience: The influence of supermarkets on consumption of ultra-processed foods and beverages in Brazil. *Appetite* **2017**, *116*, 381–388. [CrossRef]
26. Euromonitor International. Grocery Retailers in Brazil. 2016. Available online: https://www.euromonitor.com/grocery-retailers-in-brazil/report (accessed on 15 March 2017).
27. Kanter, R.; Reyes, M.; Corvalán, C. Photographic Methods for Measuring Packaged Food and Beverage Products in Supermarkets. *Curr. Dev. Nutr.* **2017**, *1*. [CrossRef]
28. Naing, L.; Winn, T.; Rusli, B. Practical Issues in Calculating the Sample Size for Prevalence Studies. *Arch. Orofac. Sci.* **2006**, *1*, 9–14.
29. Mike, R.; Stefanie, V. Informas Protocol: Food Labelling Module. 2015. Available online: https://figshare.com/articles/INFORMAS_Protocol_Food_Labelling_Module/5673643 (accessed on 17 June 2019).
30. Hallgren, K.A. Computing Inter-Rater Reliability for Observational Data: An Overview and Tutorial. *Tutor. Quant. Methods Psychol.* **2012**, *8*, 23–34. [CrossRef] [PubMed]
31. Dias, F.D.S.L.; Lima, M.F.; Velasco, P.C.D.; Salles-Costa, R.; Sardinha, F.L.D.C.; Carmo, M.D.G.T.D. Were policies in Brazil effective to reducing trans fat from industrial origin in foods? *Rev. Saúde Pública* **2018**, *52*. [CrossRef] [PubMed]
32. ABIA. Associação Brasileira das Indústrias de Alimentação. Relatório Anual 2010. 2010. Available online: https://pt.calameo.com/read/000182377a98ff937e76a (accessed on 17 June 2019).
33. Downs, S.M.; Thow, A.M.; Leeder, S.R. The effectiveness of policies for reducing dietary trans fat: A systematic review of the evidence. *Bull. World Health Organ.* **2013**, *91*, 262–269H. [CrossRef] [PubMed]
34. Downs, S.M.; Bloem, M.Z.; Zheng, M.; Catterall, E.; Thomas, B.; Veerman, L.; Wu, J.H. The Impact of Policies to Reduce trans Fat Consumption: A Systematic Review of the Evidence. *Curr. Dev. Nutr* **2017**, 1. [CrossRef] [PubMed]
35. European Commission. *Report from the Commission to the European Parliament and the Council Regarding Trans Fats in Foods and in the Overall Diet of the Union Population*; European Commission: Brussels, Belgium, 2015.
36. Ricciuto, L.; Lin, K.; Tarasuk, V. A comparison of the fat composition and prices of margarines between 2002 and 2006, when new Canadian labelling regulations came into effect. *Public Health Nutr.* **2009**, *12*, 1270–1275. [CrossRef] [PubMed]
37. Galdino, T.P.; Antunes, A.R.; Lamas, R.C.; Zingano, M.A.; Cruzat, V.F.; Coutinho, V.F.; Chagas, P. Biscoitos recheados: Quanto mais baratos maior teor de gordura trans? *Sci. Med.* **2010**, *20*, 270–276.
38. De Morais Sato, P.; Mais, L.A.; Khandpur, N.; Ulian, M.D.; Bortoletto Martins, A.P.; Garcia, M.T.; Spinillo, C.G.; Urquizar Rojas, C.F.; Jaime, P.C.; Scagliusi, F.B. Consumers' opinions on warning labels on food packages: A qualitative study in Brazil. *PLoS ONE* **2019**, *16*, e0218813. [CrossRef]
39. Wang, E.Y.; Wei, H.; Caswell, J.A. The impact of mandatory trans fat labeling on product mix and consumer choice: A longitudinal analysis of the U.S. Market for margarine and spreads. *Food Policy* **2016**, *64*, 63–81. [CrossRef]
40. Pinto, A.L.D.; Miranda, T.L.S.; Ferraz, V.P.; Athayde, D.D.; Salum, A. Determinação e verificação de como a gordura trans é notificada nos rótulos de alimentos, em especial naqueles expressos" 0% gordura trans"/Determination and verification of how" trans" fat is notified on food labels, particularly those expressed as being" 0% trans fat". *Braz. J. Food Technol.* **2016**, *19*, 1.
41. Nestle, M.; Ludwig, D.S. Front-of-Package Food Labels Public Health or Propaganda? *JAMA* **2010**, *303*, 2. [CrossRef] [PubMed]

42. Monteiro, C.A.; Cannon, G.; Moubarac, J.C.; Levy, R.B.; Louzada, M.L.C.; Jaime, P.C. The UN Decade of Nutrition, the NOVA food classification and the trouble with ultra-processing. *Public Health Nutr.* **2018**, *21*, 5–17. [CrossRef] [PubMed]
43. Ministry of Health of Brazil. *Dietary Guidelines for the Brazilian Population*; Ministry of Health of Brazil: Brasília, Brazil, 2015.
44. Brasil Agência Nacional de Vigilância Sanitária. *Relatório Preliminar de Análise de Impacto Regulatório sobre Rotulagem Nutricional*; Agencia Nacional de Vigilância Sanitária: Brasília, Brasil, 2018.
45. WHO. World Health Organization. *Eliminating Trans Fats in Europe: A Policy Brief*; WHO: Copenhagen, Denmark, 2015.
46. Restrepo, B.J.; Rieger, M. Denmark's Policy on Artificial Trans Fat and Cardiovascular Disease. *Am. J. Prev Med.* **2016**, *50*, 69–76. [CrossRef] [PubMed]
47. Commission Regulation (EU) 2019/649 Amending Annex III to Regulation (EC) No 1925/2006 of the European Parliament and of the Council as Regards Trans Fat, Other than Trans Fat Naturally Occurring in Fat of Animal Origin. Available online: https://eur-lex.europa.eu/legal-content/EN/TXT/PDF/?uri=CELEX:32019R0649&from=EN (accessed on 17 June 2019).
48. PAHO. Pan American Health Organization. *Trans Fat Free Americas*; Pan American Health Organization: Washington, DC, USA, 2008.
49. Colón-Ramos, U.; Monge-Rojas, R.; Campos, H. Impact of WHO recommendations to eliminate industrial trans-fatty acids from the food supply in Latin America and the Caribbean. *Health Policy Plan* **2014**, *29*, 529–541. [CrossRef] [PubMed]
50. FDA. Food and Drug Administration. Final Determination Regarding Partially Hydrogenated Oils (Removing Trans Fat). Available online: https://www.fda.gov/food/ingredientspackaginglabeling/foodadditivesingredients/ucm449162.htm (accessed on 17 June 2019).
51. WHO. World Health Organization. REPLACE Trans Fat: An Action Package to Eliminate Industrially-Produced Trans-Fatty Acids. 2018. Available online: https://www.who.int/docs/default-source/documents/replace-transfats/replace-action-package.pdf (accessed on 17 June 2019).
52. Stender, S.; Astrup, A.; Dyerberg, J. What went in when trans went out? *N. Engl. J. Med.* **2009**, *361*, 314–316. [CrossRef] [PubMed]
53. Mozaffarian, D.; Jacobson, M.F.; Greenstein, J.S. Food reformulations to reduce trans fatty acids. *N. Engl. J. Med.* **2010**, *362*, 2037–2039. [CrossRef] [PubMed]
54. Mensink, R.P.; Sanders, T.A.; Baer, D.J.; Hayes, K.C.; Howles, P.N.; Marangoni, A. The Increasing Use of Interesterified Lipids in the Food Supply and Their Effects on Health Parameters. *Adv. Nutr.* **2016**, *7*, 719–729. [CrossRef] [PubMed]
55. Mills, C.E.; Hall, W.L.; Berry, S.E.E. What are interesterified fats and should we be worried about them in our diet? *Nutr. Bull.* **2017**, *42*, 153–158. [CrossRef]
56. Astrup, A.; Bertram, H.C.; Bonjour, J.P.; de Groot, L.C.; de Oliveira Otto, M.C.; Feeney, E.L.; Garg, M.L.; Givens, I.; Kok, F.J.; Krauss, R.M.; et al. WHO draft guidelines on dietary saturated and trans fatty acids: Time for a new approach? *BMJ* **2019**, *366*, l4137. [CrossRef]
57. Scrinis, G.; Monteiro, C.A. Ultra-processed foods and the limits of product reformulation. *Public Health Nutr.* **2018**, *21*, 247–252. [CrossRef] [PubMed]
58. Presidência da República. *Lei N° 8.078, de 11 de setembro de 1990. Dispõe Sobre A Proteção do Consumidor E dá Outras Providências*; Presidência da República: Brasília, Brasil, 1990.

nutrients

Article

Are the Claims to Blame? A Qualitative Study to Understand the Effects of Nutrition and Health Claims on Perceptions and Consumption of Food

Tony Benson [1], Fiona Lavelle [1], Amanda McCloat [2], Elaine Mooney [2], Tamara Bucher [3,4], Bernadette Egan [5] and Moira Dean [1,3,4,*]

1 Institute for Global Food Security, School of Biological Sciences, Queen's University Belfast, Belfast BT9 5AG, UK
2 Department of Home Economics, St. Angela's College, Sligo F91 C634, Ireland
3 School of Health Sciences, Faculty of Health and Medicine, The University of Newcastle, Callaghan, NSW 2308, Australia
4 Priority Research Centre for Physical Activity and Nutrition, The University of Newcastle, Callaghan, NSW 2308, Australia
5 Food, Consumer Behaviour and Health Research Centre, University of Surrey, Guildford GU2 7XH, UK
* Correspondence: moira.dean@qub.ac.uk; Tel.: +44-289-097-6561

Received: 8 July 2019; Accepted: 27 August 2019; Published: 2 September 2019

Abstract: Nutrition and Health Claims (NHCs) have been found to influence perceptions of food and consumption behaviour. While previous quantitative research has identified factors that may explain these effects, the current study aimed to address the dearth of in-depth exploration as to the underlying reasons why and how claims may impact upon perceptions and behaviour and the relationships between key factors. Seventy-eight participants took part in 10 focus groups. Discussions were transcribed verbatim and Nvivo 11 was used for thematic analysis. Six themes were developed from the data: 1. Target populations for NHCs; 2. Influence of NHCs on purchasing behaviour; 3. Characteristics/perceptions of products displaying NHCs; 4. Believability of NHCs; 5. Superior yet superficial knowledge; 6. Consumption of products displaying NHCs. Knowledge was a key factor influencing how much individuals believe claims (Believability of NHCs) and their perceptions (Characteristics/perceptions of products displaying NHCs). These perceptions and the characteristics of products displaying claims also impacted believability, as well as purchasing behaviour and consumption. Future research should be cognisant of the role of knowledge and characteristics or perceptions of products in the relationship between NHCs and consumer behaviour, and modelling of these relationships would allow their relative strength to be identified.

Keywords: nutrition claims; health claims; portion size; perceptions; health halo; food labelling; consumer; nudging; qualitative; focus groups

1. Introduction

Nutrition and health claims (NHCs) are text or images on food or drinks products which highlight their nutritional or energy content and (for health claims) their impact on health. Specifically, a nutrition claim is a claim on a food or drink product which indicates that a food has beneficial nutritional properties due to the energy it provides or does not provide or the nutrients or other substances it contains or does not contain [1], for example "low in fat". A health claim is a statement that suggests or implies that a relationship exists between food and health [1] such as " … contains calcium which helps maintain healthy bones".

Nutrition and health claims may help consumers to make informed decisions about their purchase and intake of food and drink, which in turn may lead to an improvement in diet quality [2]. However, it has also been suggested (somewhat paradoxically) that NHCs may also lead to an increase in consumption, with a recent systematic review and meta-analysis finding that NHCs increase consumption and/or purchasing behaviour of food and drink [3]. The increase in consumption associated with NHCs may be due to a "health halo" effect [4] whereby an individual generalises from a nutrition or health claim that a product is healthier or has more favourable attributes than it actually does. NHCs may also impact consumers' perceptions of food products, with several studies showing impacts on perceived healthiness, tastiness, fillingness, naturalness, and attractiveness [5–8]. Specifically, claims have been found to increase perceived healthiness, decrease tastiness, attractiveness and naturalness perceptions and both increase or decrease fillingness perceptions, depending on the product. This research is further supported by a recent systematic review which found that overall, although based on low quality evidence, NHCs did influence consumers' perceptions of products [9].

The use of nutrition and health claims varies across countries due to cultural differences in regulation and understanding [10,11]. For example, 50% of individuals in Sweden check nutrition information on labels at least occasionally compared with 65% of individuals in Ireland. Specifically in relation to nutrition and health claims, a supermarket audit in Ireland found prevalence rates of 47% for nutrition claims and 18% for health claims [12]. This compares with European rates of 21% and 11% respectively [13] and an Australian rate of 14% for any claim [14]. While these studies are not directly comparable due to differences in methods and years of study (and interim changes in legislation, such as the establishment of the European Union (EU) register of nutrition and health claims), these results suggest that Ireland has a relatively high prevalence of NHCs. Despite these cross-country differences and the introduction of regulations in this area over 10 years ago, little research examining nutrition and health claims has been conducted on the island of Ireland.

A plethora of studies have investigated NHCs, yet scope for further research remains. For example, recent systematic reviews have highlighted the need for further studies on the effects of claims on dietary choices and consumption [3,9]. While the field consists mainly of quantitative studies, qualitative methods have also been used to examine NHCs. For example, a study using focus groups in the USA found that NHCs increased parents' healthiness perceptions of children's foods [15]. Similarly, Chan, Patch, and Williams [16] found that some Australians admitted that the presence of a low or reduced fat claim increased their consumption. Other focus group studies have found that claims could influence older women to purchase carrier products in Ireland [17], general acceptance of the potential for enhanced flavonoid content in fruit [18] and trust/distrust and relevance of claims as key themes [19]. While the outlined studies provide useful information on consumer attitudes towards and processing of claims, qualitative studies in this area have tended to investigate consumers' general perceptions and understanding of claims rather than the influence of claims on perceptions and consumption. From these studies, several common findings have emerged, showing that consumers lack understanding of some claims, NHCs may influence consumption and purchasing behaviours, and that consumers can be skeptical about claims. However, less well known is the structure and relationships between these factors and how NHCs influence consumer perceptions and behaviour. What are the pathways or underlying mechanisms by which claims influence the consumer? It is known that consumer characteristics such as sociodemographics and knowledge play a role in the link between claims and behaviour [3,5], but less well known is the relationship between these factors and their possible interactions and how these might mediate the effects of claims on perceptions and consumption behaviour. For example, it is known that NHCs impact consumption [3] but less attention has been given to why this happens or the psychological explanations for such effects. Therefore, these pathways should be explored.

The current research aimed to address the aforementioned limitations by using qualitative research to understand the relationships and pathways between NHCs and consumers' attitudes and behaviours and understand how and why NHCs influence perceptions and consumption behaviour. The current study also built upon the limited amount of previous qualitative research in this area conducted on the island of Ireland (includes Northern Ireland (NI) and the Republic of Ireland (ROI)) by using a mixed sample (in contrast to a young female sample previously used).

2. Materials and Methods

Ethical approval (07/16/BensonT) was granted by the Queen's University Belfast School of Biological Sciences Research Ethics Committee and written informed consent was obtained from all participants. The study was conducted in accordance with the Declaration of Helsinki.

2.1. Participants

Participants were recruited using convenience and snowball sampling. Given their potential to bias the results, individuals who were currently studying or had ever undertaken a degree in food, nutrition, or dietetics were excluded from participation. Similarly, those (or those with anyone in their household) working in any of the following areas were also excluded: advertising, marketing or market research, the food industry, diet, or nutrition. In total, 78 participants (see Table 1 for characteristics of participants) were recruited to ten focus groups (five held at a location in NI, five held at a location in ROI). Given the subject matter and the fact that the presence of the opposite sex can influence eating behaviours [20,21], groups were single sex (five exclusively male groups and five exclusively female groups), consisting of between six and nine participants and either younger (18–35 years old) or older (36–64 years old) with an overall mean age of 36.6 (SD = 13.2) years.

Table 1. Characteristics of focus group participants (n = 78).

Characteristic	n (%)
Location	n = 78
Northern Ireland	37 (47%)
Republic of Ireland	41 (53%)
Sex	
Male	38 (49%)
Female	40 (51%)
Education	
Primary school or less	1 (1%)
Secondary school (to age 15/16)	7 (9%)
Secondary school (to age 17/18)	16 (21%)
Additional training	13 (17%)
Undergraduate	17 (22%)
Postgraduate	24 (31%)
Socioeconomic status [1]	
Higher (ABC1)	61 (78%)
Lower (C2DE)	14 (18%)
Unknown	3 (4%)
Motivation to process NHCs [2]	
Mean (Standard Deviation)	3.24 (0.83)
Health or weight issues	
Overweight/obesity	20 (26%)
Irritable Bowel Syndrome or other digestive problems	14 (18%)
High blood cholesterol levels	7 (9%)
Hypertension	5 (6%)
Cancers (any type)	2 (3%)
Cardiovascular/heart disease	1 (1%)

Table 1. *Cont.*

Characteristic	n (%)
Other chronic conditions/diseases	4 (5%)
Current diet status	
Self-chosen slimming diet	10 (13%)
Slimming diet prescribed by a health professional	3 (4%)
Cholesterol lowering diet	1 (1%)
Diabetic diet	1 (1%)
Other medical diet	2 (3%)

[1] Socioeconomic status based on occupation status of the highest income earner in the household. Higher (ABC1) = higher and intermediate managerial or professional occupations, lower (C2DE) = unskilled, semi-skilled, skilled occupations and those unemployed. [2] Mean of three items on a scale from 1 to 5, with higher score indicating greater motivation to process NHCs.

2.2. Materials

2.2.1. Elicitation Materials

Several products and product images displaying NHCs were used as prompts in the discussion (see Table 2 for an overview of the materials used). To assess the spontaneous awareness and recognition of NHCs, participants were shown a chocolate bar displaying the regulated nutrition claim "no added sugar", as well as phrases which may be perceived as NHCs, for example, "gluten & nut free". This bar was also chosen as it was classified as "less healthy" according to the Office of Communications (Ofcom) nutrient profiling system [22]. Furthermore, recent research has called for examination of the effects of claims on hedonic products [23]. Therefore, it was felt that it would be interesting to gauge participants' opinions of the apparent juxtaposition of a "less healthy" product (typically viewed as a treat food) displaying nutrition and health messages.

Table 2. Overview of materials used in focus groups.

Product	Format	Claim(s) Displayed
Chocolate bar	Physical product	Regulated claim "no added sugar" as well as the claims "Good chocolate", "Gluten & nut free", "With live cultures", "Over a billion lactobacillus & bifidobacterium", and "63 calories per bar"
Breakfast cereal	Physical product	Regulated claims "Low in fat" and "Source of vitamin D"
Yoghurt	Photograph still from TV advertisement	Regulated claim "... contains calcium which helps maintain healthy bones"

The most common nutrition and health claims in the UK and Ireland relate to fat [12,24]. Therefore, the second product used as a prompt was breakfast cereal labelled with the claim "low in fat" as well as "source of vitamin D". A still image from a yoghurt advertisement was also used as a discussion prompt. This contained the health claim "... contains calcium which helps maintain healthy bones".

2.2.2. Questionnaire

Participants completed a short questionnaire (approximately ten minutes in duration) which collected sociodemographic information, such as age, sex, education, and socioeconomic status. Socioeconomic status was categorised as higher (ABC1—higher and intermediate managerial or professional occupations) or lower (C2DE—unskilled, semi-skilled, skilled occupations and those unemployed). Other factors which may influence perceptions of nutrition and health claims [25] were also assessed, such as current dieting and health issues and motivation to process (use of) nutrition and health claims [26].

2.3. Procedure

Following a review of the relevant literature, a semi-structured topic guide was created to explore attitudes towards and perceptions of nutrition and health claims. This was piloted with a group of eight female participants, with subsequent minor amendments made prior to full implementation, such as the addition of show cards with the definitions of "nutrition claim" and "health claim". Each focus group discussion was guided by an experienced moderator (TB) using the final topic guide (see Table 3). The groups began by outlining the purpose of the group and use of recording equipment. The order of the topic guide was designed to allow spontaneous mentions of NHC, followed by specific questions focusing on perceptions (tastiness, healthiness, and satiety) and consumption behaviour in relation to NHCs. Participants were also asked about their familiarity with and knowledge of NHC. Following the discussion, participants completed a questionnaire (see Section 2.2.2 Questionnaire). Each group session was audio recorded and lasted approximately 60 minutes. Upon completion, participants were paid an honorarium of £20 (NI) or €25 (ROI) to compensate for time and travel costs.

Table 3. Overview of topic guide used in focus groups.

Section	Question/Topics
Introduction	Explanation of purpose and format of focus groups Ice-breaker e.g., favourite foods
General thoughts on food packaging	What do you think should be displayed on food packaging? Are there certain foods for which you are more likely to look at the label?
Rapid elicitation	Participants asked for quick initial thoughts on chocolate bar
Perceptions of products with nutrition claims (chocolate bar followed by breakfast cereal)	Do you think this product would taste good or taste bad? Do you think this product is healthy or unhealthy? Do you think this product would "fill you up" or leave you hungry?
Awareness of nutrition claims	Can you give me examples of nutrition claims? On which types of products are nutrition claims typically displayed?
Knowledge of nutrition claims	What does "low in fat" mean? Are there any advantages or disadvantages to having nutrition claims displayed on products?
Use of nutrition claims	Do you look for nutrition claims on packaging before eating a product? Has there ever been an occasion where a nutrition claim on a product has: stopped you from eating a product? Made you eat more of a product? Made you eat less of a product? How believable is this claim?
Perceptions of products with health claims (still from yoghurt advertisement)	Do you think this product would taste good or taste bad? Do you think this product is healthy or unhealthy? Do you think this product would "fill you up" or leave you hungry?
Awareness of health claims	Can you give me examples of health claims? On which types of products are health claims typically displayed?
Knowledge of health claims	What does "... contains calcium which helps maintain healthy bones" mean? Are there any advantages or disadvantages to having health claims displayed on products?
Use of health claims	Do you look for health claims on packaging before eating a product? Has there ever been an occasion where a health claim on a product has: stopped you from eating a product? Made you eat more of a product? Made you eat less of a product? How believable is this claim?
Finish	Do you have anything further about nutrition and health claims that we have not mentioned today that you would like to add? Summarise and clarify key points from discussion

2.4. Data Analysis

Audio recordings from the groups were professionally transcribed and imported into NVivo 11 (QSR International Pty Ltd, Doncaster, Victoria, Australia) for analysis. Inductive thematic analysis was used to interpret the data, following the procedures of Braun and Clarke [27]. All the transcripts were read at least twice (by TB, psychologist in health) to achieve familiarity with and immersion in the data, with any prominent concepts in the data noted. Initial codes in the data were developed and subsequently arranged into themes. These themes were reviewed and checked by a second researcher (FL, a sport and health scientist) and any necessary refinements were made. Following this, the themes were named and defined and any links or relationships were established. Finally, transcripts were read again to ensure that the themes accurately represented the data and that there were no further applicable themes. Appropriate quotations were then chosen to represent each theme.

3. Results

Six themes were developed from the analysis: 1) Target populations for NHCs; 2) Influence of NHCs on purchasing behaviours; 3) Characteristics/perceptions of products displaying NHCs; 4) Believability of NHCs; 5) Superior yet superficial knowledge; 6) Consumption of products displaying NHCs. The meaning and content of these themes are explored in more detail below, with relevant quotations provided.

3.1. Target Populations for NHCs

There was general recognition amongst the groups that certain types of individuals or groups were most likely to benefit from using products with NHCs. It was commonly mentioned that those on diets or managing their weight, as well as those eating healthily, might be more likely to use these products:

> *"Yeah, I just joined Slimming World, so that's a big thing for me. I'm looking for 0% fat on things like dairy products, fromage frais, that kind of thing. Or, Diet Coke, I'd go to that, because it's no sugar compared to Coke ... "* (Group 4, female, 18–35 years old)

Those with specific illnesses or conditions were also mentioned as possible target populations for products with NHCs. In addition, parents were identified as purchasing products with claims for their children:

> *" ... you do want them (children) to grow up and they need healthy bones so the parents would get the stuff that contains calcium."* (Group 10, male, 18–35 years old)

3.2. Influence of NHCs on Purchasing Behaviour

Participants recognised that NHCs could sometimes influence their purchasing behaviours such that claims encouraged them to buy a product:

> *" ... that's a big factor, isn't it, you know, saying "diet" on it or "light". You would think 'okay I'll go for that option."* (Group 1, male, 36–64 years old)

In contrast, other participants remarked that claims did not influence their purchasing behaviour and that they did not look for these when shopping:

> *" ... I wouldn't purposely go out and say 'I'm only buying yogurts that are fortified with calcium ... '"* (Group 7, female, 18–35 years old)

When probed as to when or how NHCs might influence purchase, a subset of those who acknowledged that their purchasing behaviour might be influenced mentioned the type of product or the brand was a factor:

> *"A lot of the breakfast choices I'm influenced, like I would eat oats because I'm influenced by the cholesterol thing (claim) ... "* (Group 2, female, 36–64 years old)

There was also some acknowledgement that while NHC on packaging may not directly impact upon purchasing, claims sometimes made them more interested in a product and invited closer inspection.

3.3. Characteristics/Perceptions of Products Displaying NHCs

The presence of a nutrition or health claim on a product affects consumer perceptions of the product. Participants mentioned that there was often a price premium to be paid for the apparent benefits that come from consuming a product displaying claims:

> *"They make it more expensive, I suppose. They put up the price of it ... making it that bit more expensive than the competitors."* (Group 10, male, 18–35 years old)

Participants also understood the presence of an NHC to mean that ingredients had been substituted in order to achieve the claim. The example of a low-fat claim meaning that the reduction in fat had led to an increase in sugar was commonly cited. Somewhat linked to this substitution of ingredients, it was commented that this could lead to a change in the taste of products. Participants particularly mentioned that products with NHCs might be bland or poor in taste:

> *"You know, you just see "low in fat" and you think it's missing something, something's been pulled out of the process and it's not going to taste as good."* (Group 6, male 36–64 years old)

Also linked to the substitution of ingredients, a few participants mentioned that products with claims may be unhealthy, for example due to being higher in sugar. However, overall, participants viewed products with NHCs as being healthier than products without claims:

> *"Yes, if something has a health claim, like anti-oxidants, things like that, you do tend to think of them as healthy things or good things."* (Group 2, female, 36–64 years old)

Interestingly, the presence of claims on the health food shop chocolate bar seemed to confuse participants and prompted discussion as to whether the bar was 'healthy'. Despite being classified as a less healthy food according to its nutrient profile score, a sizeable proportion of participants felt that the bar was healthy. This suggests that consumers' typical healthiness perceptions of a food may be strongly influenced or changed by claims and that nutritionally less optimal products might be viewed as healthy. Finally, there was a perception amongst some participants that NHCs impact upon the portion size of products. Specifically, it was felt that products bearing NHCs have smaller serving and/or portion sizes in order to achieve their claims:

> *" ... it says it's low in fat but it weighs less than something beside it which doesn't say low in fat."* (Group 4, female, 18–35 years old)

3.4. Believability of NHCs

There were a number of different factors which impacted whether participants believed a claim or not. Overall, there was a strong awareness that NHCs are controlled by law, either in terms of regulations or with regards to advertising standards. This led to a general belief that claims must be true:

> *"The only reason I would believe it is because if it's on TV there's regulations, so it has to be true."* (Group 3, male, 18–35 years old)

Despite this awareness of regulations, consumers were still skeptical:

> *"I would generally be naïve enough to believe it, yes. I would take them at their face value, because they should be basing it on the nutritional facts, but ... how often do they check to make sure those facts are right?"* (Group 1, male, 36–64 years old)

The brand of a product was another factor which influenced the believability of NHCs:

> *"Before we would have tended to trust the big names, like McVitie's or Kellogg's, but now ... we don't really trust them and the whole thing has lost its credibility. It's just a series of numbers and percentages and grams and claims and people generally don't really trust them."* (Group 2, female, 36–64 years old)

The type of product and its suitability for certain claims was mentioned with differing opinions. A few felt that claims on healthier products such as cereal were more believable than claims on less healthy products, while others felt that there should be a link between the claim and the ingredients for it to be believable. However, when the claim was 'linked' with the ingredients or type of product, some felt that this was stating the obvious:

> *"That's kind of common sense, to me, like it's a yoghurt, it's made from … it came from cream, it comes from milk, you know what I mean?"* (Group 8, male, 18–35 years old)

The fact that some claims might be stating the obvious led to discussions around skepticism and whether NHCs are simply a marketing ploy designed to encourage consumers to buy products, or whether they are useful signposts to the consumer which highlight beneficial nutritional properties:

> *"There's certain foods you eat and they stick these labels on, low in fat or low in sugar, but the food is naturally going to be low in fat. It's just marketing, to make you think."* (Group 1, male, 36–64 years old)

Some participants felt that claims were believable in that while they may not help everyone, they may help certain individuals such as the elderly or those with certain conditions.

3.5. Superior yet Superficial Knowledge

Overall, participants were aware of NHCs and showed some general awareness and knowledge. When asked, participants were able to provide examples of claims and examples of the types of products on which claims were typically displayed. There was a basic understanding of what claims mean:

> *"It means that per 100 grams, or whatever the measure is, that there's a certain level where... there's categories, low, medium and high, it will be under the, obviously low, on the low spectrum to do with fat."* (Group 1, male, 36–64 years old)

There was also some evidence of superiority bias, with participants feeling that their knowledge was better than others (the general public) and that they would not be 'fooled' by NHCs:

> *"Well, I hope somebody doesn't eat one and says 'oh God, that's making my bones so much easier' because you would have to do it over a matter of time, you'd have to maintain it."* (Group 5, female, 36–64 years old)

Overall, however, the knowledge shown by participants was basic and superficial. All groups were asked for the exact meaning of the term "low in fat" and none of the participants in any of the groups were aware that this refers to a product having 3 g of fat or less per 100 g. Participants were also unable to distinguish between nutrition claims and health claims. When asked for examples of both separately, participants cited nutrition claims and health claims together as well as unofficial or unregulated claims, such as "gluten free" and "organic". This poor knowledge may be linked to the feeling of confusion or overload that some participants experience:

> *"It's almost like there is too much information and so many different claims, and then you hear these claims are false."* (Group 2, female, 36–64 years old)

3.6. Consumption of Products Displaying NHCs

As well as claims influencing purchasing behaviour, participants also recognised that NHCs can sometimes influence their consumption, with many stating that claims had or would increase their consumption. The claim on the product appeared to license further consumption:

> *"I think low calorie, as I was saying, the WeightWatchers thing, you end up having two or three packets, because you think it's very low calorie."* (Group 1, male, 36–64 years old)

There was also mention of compensating intake after eating a claim product, typically altering or increasing their intake later, having earlier eaten a product carrying NHCs:

"Those breakfast biscuits, because they are supposed to fill you, but an hour later you're saying 'it was only a light one anyway, I'm grand,' and you have a bowl of cereal, and then you have a slice of wholemeal bread because I've had low fat stuff this morning . . . "(Group 7, female, 18–35 years old)

Perhaps somewhat linked to participants' views that products with claims can taste bland or poor, a few participants mentioned that NHCs might decrease consumption due to the taste of products carrying these claims:

"The taste was just like 'I can't eat this.' It might be healthy but it's rotten" (Group 1, male, 36–64 years old)

4. Discussion

Numerous studies have investigated prevalence and understanding of and attitudes towards NHCs. It is generally well established that claims can influence consumer perceptions and consumption/purchasing behaviour of food and drink [3,9]. Specifically, claims are more likely to increase purchasing or consumption behaviour as well as increase healthiness perceptions of foods. It is also known that consumers may find NHCs useful but can also be confused and skeptical of their statements [15,16]. Less attention has been paid to the links between these findings and the possible underlying reasons why claims can impact the consumer. Using a qualitative approach, the current study aimed to identify and examine these links and reasoning. Six key themes were developed using thematic analysis: Target populations for NHCs, Influence of NHCs on purchasing behaviours, Characteristics/perceptions of products displaying NHCs, Believability of NHCs, Superior yet superficial knowledge, and Consumption of products displaying NHCs. While the themes were distinct, potential relationships between the themes were apparent, as displayed in Figure 1 and discussed in more detail below.

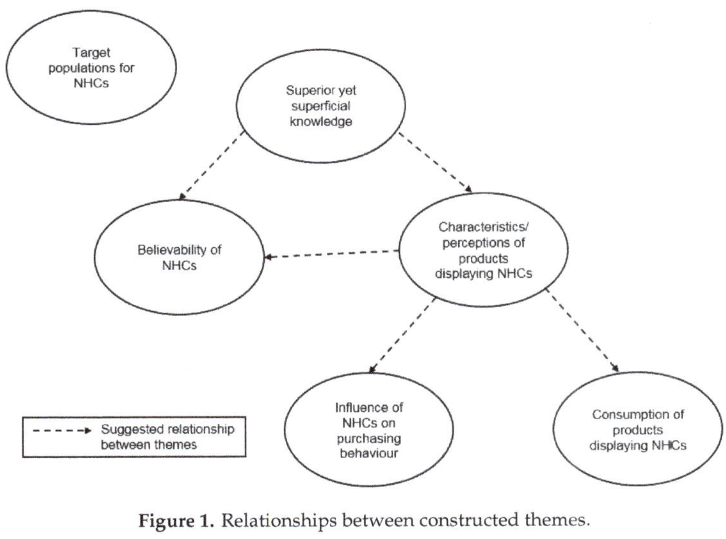

Figure 1. Relationships between constructed themes.

The 'Target populations for NHCs' was a simple, standalone theme. Participants recognised that products displaying nutrition and health claims were likely to be most beneficial (perceived relevance) to individuals with certain health issues or conditions, or specific groups within the population. While some participants recognized themselves as being part of those who would benefit from products with claims (such as those trying to lose weight), in general, it was mentioned that other groups would benefit. This may be due to participants not having relevant health conditions or a reluctance to disclose health conditions in a group setting. In addition, some participants may not have identified as being part of the target group even though they fit the profile of the target group in terms of age, parenthood, and those attempting weight management. Pregnant women, the elderly, those with diabetes, and those attempting to lose weight or gain muscle were all mentioned as potential key users of products with claims.

Participants' knowledge of NHCs appeared to be limited, lacking depth. While it is already known that nutrition knowledge impacts upon perceptions of claims [23], fewer studies have specifically looked at NHC knowledge. One study found that nutrition and health claim knowledge may be relatively low [5]. The superiority bias shown in the current study (individual's beliefs that they would not be misled by claims but others might) provides greater insight into an optimistic bias which has been found [5], whereby consumers believe that they have greater knowledge than others but actual knowledge is relatively low. Knowledge appeared to impact upon believability in the current study. If participants had knowledge or awareness of the ingredients, then they were more likely to believe the claim. For example, participants mentioned that they knew that calcium helps bones and therefore, a claim linking the two was believable and plausible. Lalor and colleagues [28] found that, overall, nutrition knowledge was not associated with perceived credibility of health claims, yet education level has been linked with belief in NHCs [29]. A link between knowledge and believability was seen in the current study as participants appeared unable to distinguish between different types of claims, for example, nutrition and health claims and unregulated and regulated claims. This may have impacted the believability; if participants are unaware that claims are regulated and there are stipulations for different claims, this may lead to low levels of belief and skepticism. Belief may also be linked with the wider concept of trust, which has previously been found to influence perceived benefits of claims [30]. In particular, those who have more trust in information sources tended to view claims as more beneficial than those with less trust, and food and health authorities were the most trusted sources. In addition, trust has been found to be a significant predictor of purchase intentions for functional foods (chocolate, yoghurt, and soup) [31]. Together, these findings might suggest that the type of knowledge, for example, knowledge in general, nutrition knowledge, or health knowledge is important as to whether knowledge influences believability. Future research should empirically investigate believability and different knowledge domains to determine their significance in the processing of claims.

Knowledge of NHCs (superior yet superficial knowledge) also had an impact on 'Characteristics/perceptions of products carrying NHCs'. For example, a few participants noted that a claim means that the ingredients in a product have been affected, such as the substitution of fat with sugar. This also appears in the wider literature [32]. This relationship reflects the findings in the literature that knowledge influences the acceptance of products with NHCs. Bimbo and colleagues [33] found that a higher diet/health related knowledge was linked with greater acceptance of dairy functional products, while Drolias and colleagues [34] found that nutrition knowledge was associated with a reported use of health claims but not nutrition claims.

The theme 'Characteristics/perceptions of products carrying NHCs' also has a relationship with the 'Believability of NHCs' theme. Current research in this area has identified participants' belief in claims (believability) as a possible key factor or moderator of the relationship between NHCs and perceptions or behaviour [5,35]. However, unlike other studies in which believability has impacted upon perceptions of products [5,36], in the current study, belief was mentioned more in relation to skepticism, with little or no explicit mention of perceptions or consumption behaviour. This may be due to a true lack of connection between believability, perception, and behaviour, or participants simply not bringing these connections to mind in the focus groups. As previously mentioned, believability is a key factor which merits further investigation in relation to nutrition and health claims [5].

Nutrition and health claims influenced purchasing behaviour for some participants, and while others were not influenced, there was acknowledgement that NHCs captured their attention and led them to inspect a product. As a few participants stated that NHCs only influenced their purchasing behaviour for certain types of products, this theme was linked with the theme 'characteristics/perception of products displaying NHCs'. In particular, breakfast and dairy products such as cereal, milk, and bread were mentioned as products where claims may be more likely to influence purchasing. This aligns with past research which has found that the type of product is perhaps one of the most important factors which affects consumer perceptions of products displaying NHCs [37,38]. For example, NHCs have been found to influence perceptions of cookies [8] but not those of cereal [39]. If certain types of products displaying NHCs are more likely to influence consumers, it may be necessary to ensure that only the products which are most nutritionally beneficial and suitable are authorised to display NHCs. Research in this area has examined the nutrient profile of foods carrying claims, finding that a sizeable proportion of foods with claims are not qualified to make a claim when subjected to a nutrient profiling system [40–42]. While Australia and New Zealand currently use a nutrient profiling system to approve claims on foods, such a system is yet to be established within the EU. The European Commission (EC) is currently conducting a re-evaluation to assess whether nutrient profiling for foods displaying NHCs is still necessary, the reasons preventing the introduction of profiles, and the costs and benefits of the absence of profiles [42]. Any creation of regulation in this area should be mindful of potential adverse effects. A recent study which modelled the potential impact of nutrient profile scoring in the UK found that while a nutrient profiling system might positively impact mortality rates if additional action is taken, such as the reformulation of products, the introduction of profiling without subsequent reformulation could lead to an increase in mortality rates as a result of less healthy diets [43]. However, as acknowledged, these are preliminary findings with large uncertainty intervals and further investigation is required.

Finally, there was acknowledgement from participants that their consumption behaviour could or had been affected by claims. There was the potential for a licensing effect of claims, whereby individuals reasoned that they could eat (more) of a product, with claims that it is healthy [44]. However, this was typically discussed in the context of specific brands or types of food, such as breakfast bars, cereals, and crisps. Furthermore, some participants noted that the pack size of products displaying claims can be smaller in order to meet the requirements for a claim. For example, lower calorie products may have less calories compared with standard products due to containing less food rather than reformulation. While previous qualitative research found that participants ate additional packs to compensate for a reduction in pack size [32], to our knowledge, this has not been investigated in more detail. Empirical examination of the effects of claims controlling for pack size would be interesting to understand if a "health halo" effect still persisted. Together, these findings suggest a potential relationship between the themes 'Consumption of products displaying NHCs' and 'Characteristics/perceptions of products displaying NHCs'. This is supported by a systematic review and meta-analysis of the effects of claims which found a preference for or consumption of foods with claims differed depending on the type of product [3]. As previously mentioned with regards to purchasing, if product characteristics also influence consumption behaviour, then it is important that the product has a beneficial nutritional profile and is suitable to display a claim. While several studies have examined the prevalence of food and drink products carrying claims and their nutrient profile [40–42], few have demonstrated the real-world implications of applying nutrient profiling to products with NHCs. Despite Australia having introduced a nutrient profile model for health claims, the impact of this does not appear to have been evaluated. Future work should examine or model the benefits and consequences (in areas such as public health and economics) of profiling products with claims, particularly as the development of an EU nutrient profiling model is pending.

Previous research in this area has identified key factors in the relationship between NHCs, perceptions, and behaviour. The current study advances research in this area by identifying the potential relationships between these factors and reasons underlying these pathways by using a qualitative approach. This investigation allowed for the creation of a model which can be taken forward in future research. The results suggest that knowledge (in particular familiarity and NHC knowledge) and characteristics/perceptions of products with NHCs are key influences on the purchasing and consumption of products displaying NHCs. The pathways and strength of the associations between factors should now be empirically validated. In addition, other factors from elsewhere in the literature and other potential relationships between factors which were not identified in the current study might also be tested. For example, there may be links between knowledge and target populations through perceived relevance if individuals believe that certain substances or nutrients identified by claims are most effective for certain population groups. A further strength of this study relates to the sample. While the use of a representative sample is unnecessary for qualitative research, this study used a relatively large and mixed group of participants, which allowed for a diverse range of views to be gathered. However, a greater proportion of participants were of a higher socioeconomic status and this should be considered in the interpretation of the results.

5. Conclusions

This study presents qualitative findings which provide further insight into the underlying reasons why nutrition and health claims may influence consumer perceptions and consumption behaviour. Knowledge was found to be a key influence on how claims are perceived. How much individuals believed claims appeared to be affected by the type and characteristics of products displaying claims. These characteristics alongside individuals' perceptions of claims may in turn impact upon purchasing and consumption behaviours.

Author Contributions: Conceptualization, T.B. (Tony Benson), A.M., E.M., T.B. (Tamara Bucher), B.E. and M.D.; Data curation, T.B. (Tony Benson); Formal analysis, T.B. (Tony Benson), F.L. and M.D.; Funding acquisition, A.M., E.M., T.B. (Tamara Bucher) and M.D.; Investigation, T.B. (Tony Benson), F.L., A.M., E.M. and M.D.; Methodology, T.B. (Tony Benson), A.M., E.M., T.B. (Tamara Bucher), B.E. and M.D.; Project administration, T.B. (Tony Benson), A.M., E.M. and M.D.; Resources, T.B. (Tony Benson), A.M., E.M. and M.D.; Supervision, M.D.; Writing—original draft, T.B. (Tony Benson); Writing—review & editing, T.B. (Tony Benson), F.L., A.M., E.M., T.B. (Tamara Bucher), B.E. and M.D.

Funding: This material is based upon work supported by safefood, The Food Safety Promotion Board, under Grant No. 09-2015.

Conflicts of Interest: The authors declare no conflict of interest. The founding sponsors had no role in the design of the study; in the collection, analyses, or interpretation of data; in the writing of the manuscript, and in the decision to publish the results.

References

1. Regulation (EC) No 1924/2006 of the European Parliament and of the Council of 20 December 2006 on Nutrition and Health Claims Made on Foods. 2006. Available online: https://eur-lex.europa.eu/LexUriServ/LexUriServ.do?uri=OJ:L:2006:404:0009:0025:EN:PDF (accessed on 18 June 2019).
2. Williams, P. Consumer understanding and use of health claims for foods. *Nutr. Rev.* **2005**, *63*, 256–264. [CrossRef] [PubMed]
3. Kaur, A.; Scarborough, P.; Rayner, M. A systematic review, and meta-analyses, of the impact of health-related claims on dietary choices. *Int. J. Behav. Nutr. Phys. Act.* **2017**, *14*, 93. [CrossRef] [PubMed]
4. Wansink, B.; Chandon, P. Can "low-fat" nutrition labels lead to obesity? *J. Mark. Res.* **2006**, *43*, 605–617. [CrossRef]
5. Benson, T.; Lavelle, F.; Bucher, T.; McCloat, A.; Mooney, E.; Egan, B.; Collins, C.E.; Dean, M. The Impact of Nutrition and Health Claims on Consumer Perceptions and Portion Size Selection: Results from a Nationally Representative Survey. *Nutrients* **2018**, *10*, 656. [CrossRef] [PubMed]
6. Wang, Q.; Oostindjer, M.; Amdam, G.V.; Egelandsdal, B. Snacks with Nutrition Labels: Tastiness Perception, Healthiness Perception, and Willingness to Pay by Norwegian Adolescents. *J. Nutr. Educ. Behav.* **2016**, *48*, 104–111.e1. [CrossRef] [PubMed]
7. Lähteenmäki, L.; Lampila, P.; Grunert, K.; Boztug, Y.; Ueland, Ø.; Åström, A.; Martinsdóttir, E. Impact of health-related claims on the perception of other product attributes. *Food Policy* **2010**, *35*, 230–239. [CrossRef]

8. Gravel, K.; Doucet, É.; Herman, C.P.; Pomerleau, S.; Bourlaud, A.-S.; Provencher, V. "Healthy," "diet," or "hedonic". How nutrition claims affect food-related perceptions and intake? *Appetite* **2012**, *59*, 877–884. [CrossRef]
9. Shemilt, I.; Hendry, V.; Marteau, T.M. What do we know about the effects of exposure to 'Low alcohol' and equivalent product labelling on the amounts of alcohol, food and tobacco people select and consume? A systematic review. *BMC Public Health* **2017**, *17*, 29. [CrossRef]
10. Landström, E.; Hursti, U.-K.K.; Magnusson, M. "Functional foods compensate for an unhealthy lifestyle". Some Swedish consumers' impressions and perceived need of functional foods. *Appetite* **2009**, *53*, 34–43. [CrossRef]
11. Masson, E.; Debucquet, G.; Fischler, C.; Merdji, M. French consumers' perceptions of nutrition and health claims: A psychosocial-anthropological approach. *Appetite* **2016**, *105*, 618–629. [CrossRef]
12. Lalor, F.; Kennedy, J.; Flynn, M.A.T.; Wall, P.G. A study of nutrition and health claims—A snapshot of what's on the Irish market. *Public Health Nutr.* **2010**, *13*, 704–711. [CrossRef] [PubMed]
13. Hieke, S.; Kuljanic, N.; Pravst, I.; Miklavec, K.; Kaur, A.; Brown, A.K.; Egan, M.B.; Pfeifer, K.; Gracia, A.; Rayner, M. Prevalence of Nutrition and Health-Related Claims on Pre-Packaged Foods: A Five-Country Study in Europe. *Nutrients* **2016**, *8*, 137. [CrossRef] [PubMed]
14. Sobierajski, M.; Spratt, B.; Williams, P.; Houston, A.; Roesler, A.; Ridges, L.; Rafferty, J.; Yeatman, H.R. Nutrition function, health and related claims on packaged Australian food products-prevalence and compliance with regulations. *Asia Pac. J. Clin. Nutr.* **2006**, *15*, 10.
15. Abrams, K.M.; Evans, C.; Duff, B.R.L. Ignorance is bliss. How parents of preschool children make sense of front-of-package visuals and claims on food. *Appetite* **2015**, *87*, 20–29. [CrossRef] [PubMed]
16. Chan, C.; Patch, C.; Williams, P. Australian consumers are sceptical about but influenced by claims about fat on food labels. *Eur. J. Clin. Nutr.* **2005**, *59*, 148. [CrossRef] [PubMed]
17. Lalor, F.; Madden, C.; McKenzie, K.; Wall, P.G. Health claims on foodstuffs: A focus group study of consumer attitudes. *J. Funct. Foods* **2011**, *3*, 56–59. [CrossRef]
18. Lampila, P.; van Lieshout, M.; Gremmen, B.; Lähteenmäki, L. Consumer attitudes towards enhanced flavonoid content in fruit. *Food Res. Int.* **2009**, *42*, 122–129. [CrossRef]
19. Coleman, K.L.; Miah, E.M.; Morris, G.A.; Morris, C. Impact of health claims in prebiotic-enriched breads on purchase intent, emotional response and product liking. *Int. J. Food Sci. Nutr.* **2014**, *65*, 164–171. [CrossRef]
20. Otterbring, T. Healthy or wealthy? Attractive individuals induce sex-specific food preferences. *Food Qual. Prefer.* **2018**, *70*, 11–20. [CrossRef]
21. Allen-O'Donnell, M.; Cottingham, M.D.; Nowak, T.C.; Snyder, K.A. Impact of Group Settings and Gender on Meals Purchased by College Students. *J. Appl. Soc. Psychol.* **2011**, *41*, 2268–2283. [CrossRef]
22. Department of Health. Nutrient Profiling Technical Guidance. 2011. Available online: https://assets.publishing.service.gov.uk/government/uploads/system/uploads/attachment_data/file/216094/dh_123492.pdf (accessed on 18 June 2019).
23. Steinhauser, J.; Hamm, U. Consumer and product-specific characteristics influencing the effect of nutrition, health and risk reduction claims on preferences and purchase behavior—A systematic review. *Appetite* **2018**, *127*, 303–323. [CrossRef] [PubMed]
24. Kaur, A.; Scarborough, P.; Matthews, A.; Payne, S.; Mizdrak, A.; Rayner, M. How many foods in the UK carry health and nutrition claims, and are they healthier than those that do not? *Public Health Nutr.* **2016**, *19*, 988–997. [CrossRef] [PubMed]
25. Dean, M.; Lampila, P.; Shepherd, R.; Arvola, A.; Saba, A.; Vassallo, M.; Claupein, E.; Winkelmann, M.; Lähteenmäki, L. Perceived relevance and foods with health-related claims. *Food Qual. Prefer.* **2012**, *24*, 129–135. [CrossRef]
26. Hung, Y.; Grunert, K.G.; Hoefkens, C.; Hieke, S.; Verbeke, W. Motivation outweighs ability in explaining European consumers' use of health claims. *Food Qual. Prefer.* **2017**, *58*, 34–44. [CrossRef]
27. Braun, V.; Clarke, V. Using thematic analysis in psychology. *Qual. Res. Psychol.* **2006**, *3*, 77–101. [CrossRef]
28. Lalor, F.; Kennedy, J.; Wall, P.G. Impact of nutrition knowledge on behaviour towards health claims on foodstuffs. *Br. Food J.* **2011**, *113*, 753–765. [CrossRef]
29. Lynam, A.-M.; McKevitt, A.; Gibney, M.J. Irish consumers' use and perception of nutrition and health claims. *Public Health Nutr.* **2011**, *14*, 2213–2219. [CrossRef] [PubMed]

30. Urala, N.; Arvola, A.; Lahteenmaki, L. Strength of health-related claims and their perceived advantage. *Int. J. Food Sci. Technol.* **2003**, *38*, 815–826. [CrossRef]
31. Siegrist, M.; Stampfli, N.; Kastenholz, H. Consumers' willingness to buy functional foods. The influence of carrier, benefit and trust. *Appetite* **2008**, *51*, 526–529. [CrossRef]
32. Spence, M.; Livingstone, M.; Hollywood, L.E.; Gibney, E.R.; O'Brien, S.A.; Pourshahidi, L.; Dean, M. A qualitative study of psychological, social and behavioral barriers to appropriate food portion size control. *Int. J. Behav. Nutr. Phys. Act.* **2013**, *10*, 92. [CrossRef]
33. Bimbo, F.; Bonanno, A.; Nocella, G.; Viscecchia, R.; Nardone, G.; De Devitiis, B.; Carlucci, D. Consumers' acceptance and preferences for nutrition-modified and functional dairy products: A systematic review. *Appetite* **2017**, *113*, 141–154. [CrossRef] [PubMed]
34. Petrovici, D.; Fearne, A.; Nayga, R.M.; Drolias, D. Nutritional knowledge, nutritional labels, and health claims on food: A study of supermarket shoppers in the South East of England. *Br. Food J.* **2012**, *114*, 768–783. [CrossRef]
35. Banks, A.P.; Egan, B.; Hodgkins, C.E.; Peacock, M.; Raats, M.M. The role of causal models and beliefs in interpreting health claims. *Br. J. Health Psychol.* **2018**, *23*, 933–948. [CrossRef] [PubMed]
36. Wansink, B.; van Ittersum, K.; Painter, J.E. How diet and health labels influence taste and satiation. *J. Food Sci.* **2004**, *69*, S340–S346. [CrossRef]
37. Bech-Larsen, T.; Grunert, K.G. The perceived healthiness of functional foods: A conjoint study of Danish, Finnish and American consumers' perception of functional foods. *Appetite* **2003**, *40*, 9–14. [CrossRef]
38. Williams, P.; Ridges, L.; Batterham, M.; Ripper, B.; Hung, M.C. Australian consumer attitudes to health claim–food product compatibility for functional foods. *Food Policy* **2008**, *33*, 640–643. [CrossRef]
39. Roberto, C.A.; Shivaram, M.; Martinez, O.; Boles, C.; Harris, J.L.; Brownell, K.D. The Smart Choices front-of-package nutrition label. *Appetite* **2012**, *58*, 651–657. [CrossRef] [PubMed]
40. Franco-Arellano, B.; Labonté, M.-È.; Bernstein, J.; L'Abbé, M. Examining the Nutritional Quality of Canadian Packaged Foods and Beverages with and without Nutrition Claims. *Nutrients* **2018**, *10*, 832. [CrossRef]
41. Kaur, A.; Scarborough, P.; Hieke, S.; Kusar, A.; Pravst, I.; Raats, M.; Rayner, M. The nutritional quality of foods carrying health-related claims in Germany, The Netherlands, Spain, Slovenia and the United Kingdom. *Eur. J. Clin. Nutr.* **2016**, *12*, 1388–1395. [CrossRef]
42. Pivk Kupirovič, U.; Miklavec, K.; Hribar, M.; Kušar, A.; Žmitek, K.; Pravst, I. Nutrient Profiling Is Needed to Improve the Nutritional Quality of the Foods Labelled with Health-Related Claims. *Nutrients* **2019**, *11*, 287. [CrossRef]
43. Kaur, A.; Scarborough, P.; Rayner, M. Regulating health and nutrition claims in the UK using a nutrient profile model: an explorative modelled health impact assessment. *Int. J. Behav. Nutr. Phys. Act.* **2019**, *16*, 18. [CrossRef] [PubMed]
44. Prinsen, S.; Evers, C.; de Ridder, D.T.D. Justified indulgence: self-licensing effects on caloric consumption. *Psychol. Health* **2019**, *34*, 24–43. [CrossRef] [PubMed]

nutrients

Article

Testing Emotional Eating Style in Relation to Willingness to Pay for Nutritional Claims

Belinda López-Galán [1] and Tiziana de-Magistris [1,2,*]

[1] Centro de Investigación y Tecnología Agroalimentaria de Aragón (CITA), Unidad de Economía Agroalimentaria y de los Recursos Naturales, Saragossa 50013, Spain
[2] Instituto Agroalimentario de Aragón (IA2), CITA-Universidad de Zaragoza, Saragossa 50059, Spain
* Correspondence: tmagistris@aragon.es; Tel.: +34-976-71-6352

Received: 25 June 2019; Accepted: 30 July 2019; Published: 1 August 2019

Abstract: In face of the high prevalence of non-communicable diseases, nutritional claims represent a useful tool to help people to make healthier food choices. However, recent research notes that when some people experience an intense emotional state, they increase their food consumption, particularly of energy-dense and sweet foods. In consequence, this study aims to assess whether emotional eating (EE) style influences the purchase of food products carrying these claims. To this end, a real choice experiment (RCE) was conducted with 306 participants who were asked to evaluate different types of toast. An error component random parameter logit (ECRPL) was used to analyze their preferences for reduced-fat and low-salt claims toast and the effects of the variation of the EE score on individual preferences. Findings of this study suggest that emotional eating negatively impacts purchasing behavior related to nutritional claims. In particular, a decrease of the willingness to pay between 9% and 16% for every unit of toast with nutritional claims was noted when an increase of EE individual score was registered. In this regard, to increase the effectiveness of the nutritional claims, policymakers and private sectors should consider the management of individuals' emotional states in designing public health policies and marketing strategies, respectively.

Keywords: emotional eating; food choices; nutritional claims

1. Introduction

The increase in non-communicable diseases (NCDs) is related to the poor quality of the human diet [1]. The World Health Organization (WHO) reports that cardiovascular diseases, cancer, respiratory diseases, and type II diabetes are the main causes of premature death for over 40 million people worldwide [2]. This poor state of public health [2] has had a negative impact on public health expenditure [3,4] and labor productivity [5]. As a consequence, policymakers have designed instruments to help people make better food choices [6].

One such instrument is the nutritional claim [7], a statement informing consumers if a food product has been reformulated through reduction or addition of one or more nutrients (e.g., reduced fat, low salt, high fiber) [7]. This reformulation must be in line with scientific evidence confirming that high consumption of a specific nutrient can be harmful to health (e.g., sugar or fat) or, conversely, that the food product contains a nutrient that is vital to a healthy diet (e.g., fiber). Hence, the objective of policymakers is to improve consumers' decision-making by increasing their awareness through the provision of scientifically proven nutritional information, thereby reducing the effects of the poor nutritional knowledge or impulsivity [7].

Preference studies suggest that nutritional claims are useful to consumers in choosing food products [8–12]. However, Prieto-Castillo et al. [13] calculated that, while 54% of Spanish consumers seem to understand nutritional labeling, only 20% use it in their decision-making. In this regard, evidence suggests that personal factors such as familiarity with the claim [8], taste [14], certificated quality [11], health concerns from past events [9,12], and psychological factors—such as body image [15] and emotional intelligence [16]—all influence the purchase of food products with nutritional claims.

In general, the influence of emotions on eating has been well-studied in terms of the behavioral [17,18] and the physiologic [19] components. However, this effect is too complex to predict due to the variability of emotional intensity among individuals [20]. Despite this, there is general consensus in research that emotions can negatively influence eating behavior when individuals present either poor ability to manage their emotions or high desire for eating [21].

Behavioral approaches identify a consistent construct that distinguishes people who eat more in response to emotions [22] because they are cue-reactive to external stimuli and those who do so when they cannot manage their emotions [21]. The latter behavior is known as "emotional eating", characterized by an overconsumption of high-density foods as a coping strategy to stress [23] or intense emotions, both negative [22] and positive [24]. Emotional eaters seem to display this pattern when an stressor appears in their everyday life, triggering an increase in food consumption, particularly that of sweet-fat (i.e., energy-dense) food products, leading to an improvement in their emotional state (Figure 1) [22,24].

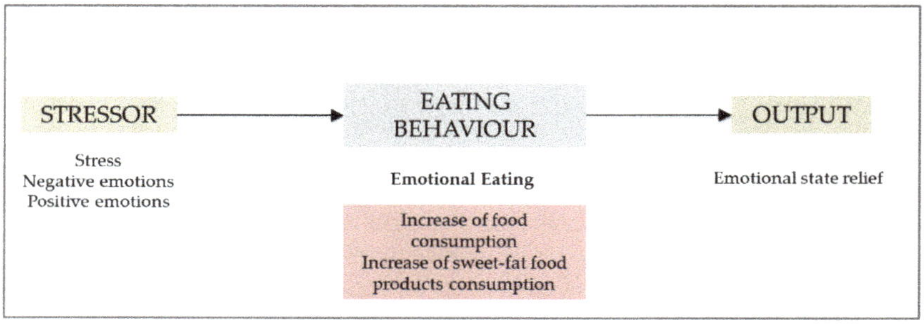

Figure 1. Emotional eating mechanism.

Studies that found that high levels of stress appear to modify existing eating behavior [23,25–28]. For example, Roberts [23] noted that healthy women increase their consumption of saturated fat (26%), carbohydrates (30%), and alcohol (33%) when they experience a significant increase in cortisol secretion (a physiologic parameter of stress). Groez et al. [27] indicated that perceived stress in women—irrespective of body mass index (BMI)—relates to a significant decrease in healthy food consumption and an increase in consumption of chips, hamburgers, and soft drinks.

Negative emotions are another stressor relevant to emotional eating behavior [29–33]. Van Strien et al. [30] reported that when sadness was induced in women, those who scored high on emotional eating increased their consumption of snacks. Additionally, the authors suggested that these participants use their eating behavior as a coping strategy to manage their emotional state. In another study, Van Strien et al. [33] noted that low emotional eaters did not increase their food intake when they experienced sadness or happiness, but high emotional eaters did so after the induction of a negative or positive emotional state. Only sweet foods present an increase of consumption.

Although most studies tie the overconsumption of energy-dense food products to negative emotions, recent research demonstrated that inducing positive emotions can also increase food consumption and/or that food intake improves people's emotional state [24,34]. In particular, Bongers et al. [24] found that self-reported female emotional eaters consumed more hedonic food (i.e., chips and chocolate) when they experienced positive emotions compared to neutral ones. The meta-analysis of Cardi et al. [34] similarly concluded that positive mood induced an increase in the eating of both sweet and savory foods. In addition, even healthy people can present emotional eating behaviors in social events that highlight the hedonic nature of their food intake.

However, some studies did not find negative or positive emotions to trigger emotional eating behavior [21,35,36]. Bongers et al. [35] found that negative emotions do not determine increase or decrease in appetite, and they consider negative and neutral emotional states to have equal probability in influencing appetite. Evers et al.'s [36] meta-analysis did not find any impact of negative or positive emotions on the consumption behaviors of self-reported emotional eaters. The authors agreed with Bongers et al. [21] that one reason for these results is that scales used to identify emotional eaters do not properly capture changes in their eating behavior.

Despite this evidence, Evers et al. [36] and Cardi et al. [34] reported great heterogeneity in the existing research findings, which they included in their meta-analyses. They noted that this could be explained by differences in the emotion induction procedure, sociodemographic characteristics of the samples, types of food offered, or the emotional eating scales used in the studies. In light of this information, the main results expected in the present study suggest that emotional eating style has an impact on purchasing behavior.

Moreover, evidence suggests that emotional eaters are more likely to develop eating disorders, such as binge eating [37] and anorexia nervosa [38]. These individuals have a high risk of weight gain as well as difficulty maintaining or losing weight [39]; they also present some personality traits such as neuroticism [29] and impulsivity [35]. Hence, since (a) food choices involve more than simple intake, (b) food purchase is the preliminary step in the decision-making process towards healthy or unhealthy intake [40], and (c) purchasing decisions share an impulsivity component with intake [41], it seems logical to suppose that, if emotional eating influences consumers' eating behavior in terms of intake, the same may occur when they are shopping and paying for food products.

In light of the above information, this study mainly aims to assess whether the emotional eating style has an influence (positive or negative) on consumer preferences for a food product (e.g., toasted bread) with nutritional claims (e.g., reduced-fat and low-salt). To achieve this objective, we conducted a real choice experiment (RCE), which is the most widely used stated preference multi-attribute method for valuing products or attributes. The results of the study contribute towards understanding individuals' psychological factors, which could be taken into consideration by stakeholders to improve the efficacy of public health policies or intervention programs.

2. Materials and Methods

2.1. Real Choice Experiment

This study follows the premise of non-hypothetical methods. An RCE, which is considered an incentive-compatible mechanism, was conducted with a real food product. The choice experiment (CE) is currently the most widely used stated preference method to evaluate consumer demand for market and nonmarket products. It is consistent with Lancaster's consumer theory [42] and random utility theory [43]. The main advantage of choice modeling against other valuation methods (e.g., contingent valuation, experimental auction) is the similarity of the choice task to actual consumer purchase situations, such as decisions made in supermarkets. For example, respondents are presented with several alternatives of a product, each with different attributes and levels in a series of multiple-choice tasks. In each task, they are required to choose which, if any, of the offered items they would purchase.

However, one main critique against this method is the existence of hypothetical bias, defined as the difference between the values obtained through hypothetical methods and those gathered through non-hypothetical approaches (or what an individual might actually pay for the good) [44,45]. Hence, participants may overstate their willingness to pay (WTP) in hypothetical settings and behave inconsistently when they do not have to support their choices through real financial commitments [46,47].

To mitigate this bias, the RCE was used. This is also its main advantage. The choice of methodology ensures that preferences are truthfully revealed and that estimated WTP values are real. Studies conducted by Chang et al. [48], Loomis et al., [49], Lusk and Shogren [50], Grebitus et al. [51], and de-Magistris et al. [52] confirmed that non-hypothetical choices are better approximations of true preferences than hypothetical ones based on comparisons of the hypothetical CE with the RCE and with actual market shares. Furthermore, since the final samples of the studies represent the population according to sociodemographic profiles, the estimated WTP can be considered representative of the Spanish population. This methodology allows estimation of consumer WTP for nutritional claims, which can be considered actual payments in the marketplace at the time of the experiment.

2.2. Recruitment and RCE Procedure

The study was conducted in March and April of 2015 in the capital town of Saragossa, Spain. As the city's size, sociodemographic characteristics, and level of income are representative of the country as a whole, the main results can be extrapolated to the Spanish population [53,54].

A total of 306 individuals were recruited for the experiment by a subcontracted professional market research agency using a stratified sampling procedure by gender, age, level of study, and BMI. The sample size resulted in a sample error of +/− 7% and a confidence level of 95.5% (K = 2) when estimating proportion (p = q = 0.5). The a priori power calculation analysis resulted in a medium effect size of d = 0.80 with $\alpha = 0.05$ and a power of 0.80, but the post-hoc analysis of the sample resulted in 0.995 [55]. The market research agency selected legal-age participants (18 years or older) who were primarily food buyers in households that consume toast. The experiment followed the guidelines in the Declaration of Helsinki, and the protocol was approved by the Ethics Committee of CITA (FP7-MC-CIG-332769).

Participants were divided into groups of 10 to 12. During the experiment, they were seated separately from one another to avoid any communication between them. Each session was conducted as follows. Before beginning, all the participants signed a consent form for their inclusion in the study. They were then informed that the experiment was structured in two parts, a questionnaire and a choice task. First, they filled in the questionnaire, which included the emotional eating scale (see Section 2.4.1 for more details) and items on their sociodemographic characteristics. After completing this, they were informed about the aim of the experiment, the methodology, and the food products used for the analysis. They were also notified that they would receive €10 in cash as a participation fee at the end of the session; this could be used to buy one of the products presented in the experiment.

The participants were allowed to inspect all the toast packages (which included the nutritional claims) on the designated shelf of the experiment room and which corresponded to products present in the Spanish market. During this inspection, they could find information related to the type of toast, nutritional claims, and price per package, but information about brands, ingredients, and manufacturing method was missing.

Afterwards, the researcher explained that the shelf contained 12 choice sets, which included three options: two packages of toast and a no-buy option. In addition, after checking each set, participants were required to write down their choices (one of the two items or no purchase) on a sheet given to them at the beginning of the session. Finally, the researcher instructed them to write a number from 1 to 12 (corresponding to each choice set) on another piece of paper, and she then selected one of them from an envelope to determine the binding choice set. If a participant chose any of the toast options, the researcher had to charge the indicated price (deducted from the €10 reimbursement) and give him or her the selected item. However, if the no-buy option was selected, the participant received the full €10.

2.3. Product and CE Design

A food product category representative of the Spanish diet was selected to analyze consumers' preferences in nutritional claims according to their eating styles. In particular, a package of 270 g of toast was selected, because it is essentially a cereal. Cereals are the principal food group that contributes to total energy intake (25%) in the Spanish diet [56], and bread is the main product in this category [56]. Moreover, toasted bread is one of the most commonly consumed types of bread in Spain (10%) [57].

Table 1 presents the selected attributes and levels used in the RCE. Three attributes were selected: price and two nutritional claims. To include an extended price range, four levels were considered (€0.70,

€1.15, €1.60, and €2.05); these were set based on the prices in the Spanish supermarket at the time of the experiment. A reduced-fat claim (FAT), a low-salt content claim (SALT), and an interaction between both claims (FSALT) were used to identify consumer preferences regarding nutritional information.

Two levels of nutritional claims were considered. The first, unlabeled condition represented the conventional toast option that did not carry any claims. The second level contained a nutritional claim. The FAT option indicated that the bread was produced with a 30% reduction in fat compared to traditional toast. Additionally, this claim was selected because consumers perceive bread to be a fattening food and tend to eliminate it from their diet [58]. On the other hand, it is well-known that high consumption of fat is related to the prevalence of non-communicable diseases such as cardiovascular illnesses and obesity [59,60]. Finally, the SALT claim indicated that the bread did not contain more than 0.03 g of salt per 100 g. This option was considered because, according to the WHO [61], consumption of salt is a risk factor in kidney disease, osteoporosis, and high blood pressure [62,63], and intake should be reduced to 5 g per day.

Table 1. Attributes and levels used in the choice experiment (CE) design.

Attributes	Levels
PRICE	€0.70 €1.15 €1.60 €2.05
Reduced-fat claim (FAT)	0 = No label 1 = The amount of fat is reduced by 30% compared to traditional toast
Low-salt content (SALT)	0= No label 1 = The amount of salt in the toast is no more than 0.03 g per 100 g of product

To reduce the hypothetical bias, a sequential Bayesian approach was used to minimize D-error. This design was selected because the parameters estimated (which measure taste preference for each nutritional claims separately) present the lowest possible standard error; consequently, the design can be considered efficient [64,65]. Following Scarpa and colleagues [65], three steps were performed to determine the choice task. First, a pilot study was conducted to specify the basic multinomial logit (MNL) model. Second, an orthogonal factorial design was created with the selected attributes and levels in the first step. Third, a database of the pilot study was used to establish a model whose estimated coefficients served as Bayesian priors. Finally, the last model indicated that a choice design with 12 tasks was needed to obtain an efficiency of 96.6%. Moreover, each choice task was required to contain three alternatives—two consisting of different options of toast and a no-buy option (Figure 2). Ngene software version 1.1.2 [66] was used to obtain the choice design.

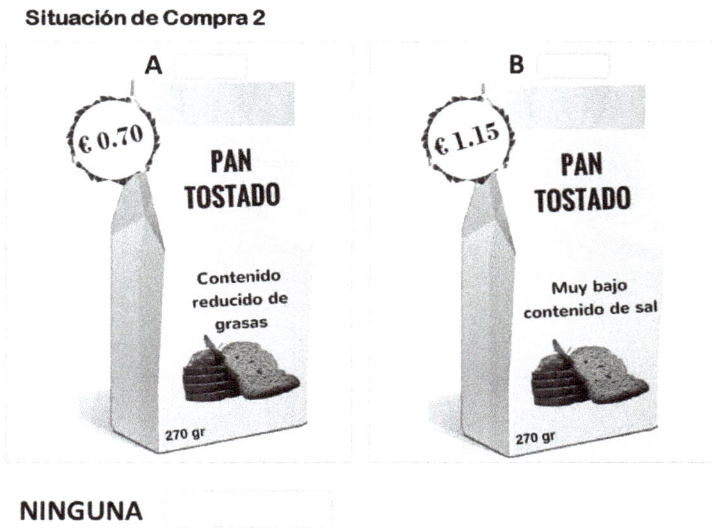

Figure 2. Example of a choice set.

2.4. Measures

2.4.1. Emotional Eater Questionnaire

Emotional eating was measured using the Spanish version of the emotional eater questionnaire (EEQ) proposed by Garaulet and colleagues [67]. This scale identifies the relationship between negative emotions and food intake in an individual. It is composed of ten items rated on a four-point system (never, sometimes, generally, and always). The total score is obtained by summing all the items; "never" corresponds to 1 point and "always" corresponds to a score of 4. Lower scores indicate that the individual exhibits a good relationship between emotions and food intake, therefore their eating behavior is healthy. Appendix A presents the English and Spanish version of the questionnaire.

Moreover, the EEQ enables classification of the sample into four groups based on total score. The first category is the non-emotional eaters (scores between 0 and 5). This group has stable eating behaviors; they decide to eat or not according to their hunger cues and not their emotions. The second group is the low emotional eaters (6 to 10); although their eating behavior is not based on emotions, they find certain foods irresistible. The third group is the emotional eaters (11 to 20); certain moods moderate how and how much they eat, but they are generally in control of their eating patterns. The last category, the very emotional eaters (scores between 21 and 30) present the least healthy eating behavior, where their moods moderate their food intake—these individuals are at high risk of developing eating disorders.

The Cronbach's alpha of the total scale was calculated to assess the adequacy of the EEQ in the sample. The result, 0.86, is a high score in this reliability index.

2.4.2. Model Specification

Choice experiments were developed following the Lancaster [42] and the random utility theories [43]. The former supposes that the utility of a food product can be broken down into the utility of its component attributes, while the latter suggests that its utility is clear to the individuals but not to the researcher. The researcher can only observe some attributes but not the components of the individual utility, which is treated as stochastic. Utility is considered a random variable obtained from the nth individual facing a choice among j alternatives in one of t choice occasions; it can be represented as follows:

$$U_{njt} = \beta x_{njt} + \varepsilon_{njt} \tag{1}$$

Here, β is a vector representing the estimated parameter, and ε_{njt} is a random term that follows an extreme value type I (Gumbel) distribution; the latter is also the independent identically distributed (iid) error term over time, people, and alternatives. Hence, given the attribute and the respective levels used in this study, the utility function becomes:

$$U_{njt} = \text{ASC} + \beta_1\, PRICE_{njt} + \beta_2\, FAT_{njt} + \beta_3\, SALT_{njt} + \beta_4\, FSALT_{njt} + \varepsilon_{njt} \tag{2}$$

ASC is an alternative-specific constant, coded as a dummy variable equal to 1 for the no-buy option and 0 otherwise. The price (PRICE) is represented in the model as a continuous variable. The claims *FAT*, *SALT*, and *FSALT* are coded as dummy variables where a value of 1 means that the corresponding claim is present on the toast package and 0 indicates that it is absent.

To assess preferences, various assumptions can be considered. First, consumers tend to choose the option that provides the highest utility from those available. Empirical evidence has demonstrated that consumer preferences are heterogeneous instead of homogeneous [15,68,69]. Hence, to obtain the mean and the standard deviation of each random parameter, it is necessary to consider a structured data panel to allow the next assumption that each individual made several choices, i.e., 12 in the case of the present study [70].

Consequently, a parameter logit model (RPL) could be established. However, this model assumes that the two toast options are correlated with each other and the no-buy option (ASC) is constant across the choice tasks. It supposes that the correlation between the two buy options would be higher than the no-buy option—in other words, the three alternatives share an extra error component. Thus, an error component random parameter logit (ECRPL) model may be used to relax this assumption and account for the extra error component missing in the utility function [64]. At this point, although emotional eating style can explain the heterogeneity in consumer preferences, it would be difficult to confirm that the former causes the latter.

Additionally, the RPL model also assumes that taste parameters are independent of one another; however, some attributes of the food product are likely interrelated. Thus, a Cholesky matrix could determine this interdependence in the taste parameter utilities of toast [71]. With this new assumption, an ECRPL model with emotional eating style interaction could be estimated.

As the aim of this study is to analyze whether emotional eating style explains heterogeneity in taste preferences regarding nutritional claims, EE was introduced in the model as a continuous variable. Following Bazzani et al. [72], before this introduction, the total score of an individual was mean-centered to allow for not only the marginal utility estimation of each nutritional claim but also the influence of emotional eating style on heterogeneity. In particular, the model reveals how variation from the mean value of the emotional eating score affects individual preferences for reduced-fat and low-salt content claims on toast.

The utility function is transformed to include the EE effects [Equation (3)]:

$$\begin{aligned} U_{njt} = \text{ASC} + \beta_1\, PRICE_{njt} + \beta_2\, FAT_{njt} + \beta_3\, SALT_{njt} + \beta_4\, FSALT_{njt} + \beta_5\, FAT_{jt} \times EE_n \\ + \beta_6\, SALT_{jt}\, \times EE_n + \beta_7\, FSALT_{jt}\, \times EE_n + 1_j \eta_{nt} + \varepsilon_{njt} \end{aligned} \tag{3}$$

In the above formula, β_5, β_6, and β_7 represent the respective coefficients of the interaction term between the attributes FAT, SALT, and FSALT with the emotional eating scale. As in Equation (2), the attributes are assumed to be random, while PRICE, ASC, and the interaction term variables are considered fixed. The $1_j \eta_{nt}$ term represents all the alternatives of the toast package design (1_j) as well as the error component associated with only the two alternatives of the product and not the no-buy option (η_{nt}) [73]

To assess the robustness and the consistency of the model and confirm that the estimated taste parameters were in line with actual purchasing behavior, changes in the Akaike information criterion (AIC) and the log likelihood (LL) were considered. Lower values of both indicate that the model is a better fit. Nlogit software version 6 was used to performer the statistical analysis and to estimate the final econometric model.

3. Results

3.1. Descriptive Analysis of Sociodemographic Characteristics

Three hundred and six individuals participated in the CE, and their sociodemographic characteristics are presented in Table 2. The majority of the sociodemographic characteristics of the sample did not show a significant statistical difference from the characteristics of the Spanish population. However, a slight majority were females (60%), who were overrepresented in comparison to the Spanish population. The average age of the sample was 45—two years older than the mean age of the nation. In this regard, 41% of the participants were middle-aged (35 to 55 years old). Most completed secondary studies (43%) and had household incomes between €1501 and €2500. The average BMI of the sample was 26, meaning than the participants were generally overweight, although the majority percentage of the sample were of normal weight (48%). The mean score on the emotional eating scale is 12, thus the sample can be classified as emotional eaters. In fact, 50% of the participants were rated to have an emotional eating style.

Table 2. Sample sociodemographic characteristics.

Definition Variable	Population	Sample/(n = 306)
Sex		
Male	42.7	40.5
Female	50.9	59.6 ***
Age; mean (standard deviation)	42.9	45.4 (16.6)
18 to 34 years	24.1	27.8
35 to 55 years	39.2	40.9
More than 55 years	36.7	31.4
Education Level		
Primary studies	17.0	19.9
Secondary studies	50.0	43.5
University studies	33.0	36.6
Household Income		
Below €1500	N/A	32.2
Between €1501 and €2500	N/A	38.8
More than €2500	N/A	29.0
Body Mass Index; mean (standard deviation)		26 (4.4)
Normal weight (Below 25 kg/m^2)	47.4	47.7
Overweight (Between 25 and 29.99 kg/m^2)	35.7	33.0
Obesity (More than 30 kg/m^2)	16.9	19.3
Emotional eating; mean (standard deviation)		12.6 (6)
Non-emotional eater *(score between 0–5)*		15.0
Low emotional eater *(score between 6–10)*		27.1
Emotional eater *(score between 11–20)*		50.0
Very emotional eater *(score between 21–30)*		7.8

N/A, Not Available; kg/m^2, kilograms/square meters; ***, **, * indicate significance at 1%, 5%, and 10% level.

3.2. Nutritional Claim Preferences Regarding Emotional Eating

Table 3 reports the estimated results for the models specified in Section 2.4.2. The first (model 1) is the basic RPL, which assumes heterogeneity in consumer preferences. Model 2 incorporates an error component into the utility function to account for correlation between the alternatives, and model 3 makes the same assumption but considers the interaction between utility and EE style.

For the estimation, 500 Halton draws were used instead of pseudo-random ones to provide a more accurate simulation for the econometric model specification.

The LL, the AIC, and the AIC/N of each model were compared to identify which of the three models presents the best fit. In this regard, the LL improved noticeably from models 1 to 2 and slightly more from 2 to 3. On the other hand, the AIC and the AIC/N decreased from models 1 to 2 and then stabilized for the subsequent models. This improvement in fit from model to model indicated not only the heterogeneity in consumer preferences but also a potential source thereof.

Most of the standard deviation estimates of the Cholesky matrix were statistically significant in model 3, confirming that emotional eating style can explain heterogeneous preferences for nutritional claims presented on toast packages.

The results of model 3 are displayed below. The alternative specific constant (ASC) or no-buy option was found to be negative and significant at 1%, indicating that consumers obtain more utility from either of the two toast options than no purchase. Following economic theory, PRICE coefficient was negative and statistically significant, which means that consumers lose utility in proportion to the price increase of the alternatives. All estimated coefficients related to nutritional claims were positive, but only FAT and SALT were non-zero at 1%, suggesting that consumers perceive more utility from reduced-fat and low-salt toast—in this order—than conventional toast. The non-statistical significance of the FSALT coefficient indicates that they do not perceive any utility in toast packages bearing both nutritional claims and hence prefer the original product.

The derived standard deviations for all three parameters were non-zero at 1% significance, confirming that utility function differs among individuals; therefore, consumer preferences regarding nutritional claims are heterogeneous. All interaction term coefficients were statistically significant at 1% when the nutritional claim interacted with emotional eating style. This suggests that emotional eating style is a source of the preference heterogeneity. In particular, the interactions between the reduced-fat and the low-salt content claims with EE style (FAT × EE and SALT × EE, respectively) were negative, indicating that utility to an emotional eater decreases when the toasted bread presents only one nutritional claim (either reduced-fat or low-salt content by equal proportions).

Table 3. Parameter estimates from the parameter logit model (RPL) and the parameter logit model with error content (RPL-EC) models.

	Model 1	Model 2	Model 3
Variable Coefficients	(RPL)	(RPL-EC)	(RPL-EC with EE interactions)
FAT	0.52 (0.18) ***	0.94 (0.22) ***	1.28 (0.22) ***
SALT	0.01 (0.16)	0.36 (0.21) *	0.77 (0.24) ***
FSALT	1.06 (0.21) ***	0.36 (0.27)	0.11 (0.25)
PRICE	−2.46 (0.13) ***	−3.11 (0.12) ***	−3.07 (0.12) ***
ASC	−4.14 (0.24) ***	−5.70 (0.31) ***	−6.02 (0.33) ***
Error component		3.68 (0.18) ***	3.69 (0.18) ***
Interaction terms with EE			
FAT × EE			−0.11 (0.22) ***
SALT × EE			−0.11 (0.03) **
FSALT × EE			0.10 (0.03) ***
No. of parameters	8	12	12
Log likelihood (LL)	−2536.30	−2231.15	−2222.53
AIC	5088.6	4486.3	4475.1
AIC/N	1.39	1.22	1.22

***, **, * indicate significance at 1%, 5%, and 10% level. FAT= reduced-fat; SALT= low-salt content; FSALT= an interaction between both reduced-fat and low-salt content; EE= emotional eating; ASC= alternative specific constant; AIC: Akaike information criterion.

Finally, following the procedure in Bazzani et al. [72], marginal WTP for reduced-fat and low-salt content claims at the mean values of the emotional eating style were calculated. Marginal WTP was obtained as a negative ratio of the mean value coefficient of the nutritional claims to the coefficient of the PRICE variable (FAT/PRICE, SALT/PRICE). The marginal WTP of the nutritional claim depended on the interaction with EE style; it was calculated as a negative ratio of the sum of the mean value coefficient of the nutritional claim and that of the interaction term divided by the PRICE coefficient (−(FAT + FAT × EE) ÷ PRICE).

Table 4 reports the calculated mean values of WTP for model 3. The WALD test revealed that only marginal WTP for reduced-fat and low-salt content was non-zero at 1% statistical significance. In particular, consumers were willing to pay €0.42 for reduced-fat toast and €0.25 for low-salt toast. However, when the WTP of the interaction term was compared, decreases of 9.52% and 16% on WTP for reduced-fat and low-salt claims, respectively, were observed. This result suggests that emotional eating style negatively impacts purchasing behavior.

Table 4. Marginal willingness to pay (WTP) for toasted bread with reduced-fat and low-salt content claims.

	Marginal WTP	Standard Error
FAT ***	€0.42	0.08
SALT ***	€0.25	0.08
FAT × EE ***	€0.38	0.08
SALT × EE ***	€0.21	0.08

***, **, * indicate significance at 1%, 5% and 10% level.

4. Discussion and Conclusions

Today, food choices remain a hot topic in research on both eating and purchasing behavior, especially in the context of poor public health and high prevalence of non-communicable diseases. However, despite all efforts from policymakers to help citizens improve their diet, results are yet to be seen. Several behavioral theories can explain consumer preferences for healthier food products (i.e., those with nutritional claims). Jacquier et al. [74] and Bublitz et al. [75] highlighted that, to improve the effectiveness of public health policies, it is important to include other factors such as emotions in the analysis of consumers' decision-making process. In this regard, the present study analyzes consumer preferences regarding toasted bread with nutritional claims and assesses if emotional eating style, as a coping strategy, influences purchase of this product.

The present results agree partially with previous empirical evidence. Cavaliere et al. [8], de-Magistris and López-Galán [10], Jurado and Gracia [12], and de-Magistris et al. [15] demonstrated that consumers show particular interest in reduced-fat claims. In fact, buyers perceive higher utility from reduced-fat toast than from other varieties. However, although preference for low-salt claims were positive, this effect is minor compared to reduced-fat claims; this finding supports Cavaliere et al. [8], who found that only 26% of their sample displayed strong interest in this nutritional claim. Conversely, de-Magistris and López-Galán [10] reported that Spanish consumers prefer conventional cheese over varieties with low-salt claims.

Additionally, de-Magistris et al. [15] noted heterogeneous preferences regarding low-salt potato chips. While pooled consumer samples appear to prefer low-salt claims, in fact, only obese individuals display a positive preference for this reformulated potato chip. When consumers consider the utility gained from toast carrying two nutritional claims, they do not show an interest in this type of toast. This finding contradicts de-Magistris et al. [10], who found that Spanish consumers prefer cheese bearing both reduced-fat and low salt claims. It also contrasts with de-Magistris et al.'s [15] findings that Spanish consumers perceive higher utility in potato chips bearing these two nutritional claims compared to the conventional variety. This discrepancy can be explained in part by the types of food product under investigation. Perhaps toasted bread is perceived by consumers to be healthier than cheese or potato chips. Another reason could be low market exposure to this type of nutritional claim, as only 6% of the food products on the Spanish market carry any nutritional claims related to salt content [76].

On the other hand, this study was the first to consider emotional eating as an influencing factor in the purchase of food products with nutritional claims. It identified whether emotions, manifesting as emotional eating, could represent an obstacle to effective public health policies (e.g., nutritional claims) and pose a challenge for policymakers in improving the food-related decisions of consumers.

In this respect, the mean EE scale rating and the frequency of the categories revealed that the participant sample presented an emotional eating style. Additionally, the analysis of consumer preferences suggested that emotional eating style negatively impacts purchasing behavior, as the WTP of the nutritional claims decreased between 9% and 16% for every unit increase in emotional eating. This finding supports the theory of emotional eating developed in the field of eating behavior, which states that negative emotions trigger high consumption of unhealthy foods such as energy-dense and sweet items [29–32,39].

In terms of the impact size, the psychological factor only explains between one-sixth and one-tenth of the heterogeneity in purchasing behavior seen in this study. This low explanatory power is due in part to the large percentage of participants who scored between 11 and 20 points on the EE scale (50%). These individuals know how to manage their emotions, especially negative ones. In some cases, they can be influenced by certain moods, but they have control over their eating behavior most of the time [67]. Thus, it is logical to assume that, when the harmful eating pattern (measured by the EE scale) is not clearly evident, the impact size on purchase decision will be minor.

While most studies have related emotional eating to consumption of energy-dense and sweet items, a salty food product was used in this analysis, which demonstrated the negative influence of this eating pattern on purchasing behavior. This result can be explained in part because bread is one of the first foods cut down when an individual begins a diet [58]. Furthermore, researchers such as Roberts [23] have found that some salty food products also motivate less healthy choices (e.g., emotional eating).

This study faced some limitations that offer insights for future research. First, the EEQ used considered only negative emotions as a stressor of emotional eating behavior. Hence, the impact of positive moods on purchasing behavior was not tested in the study. It would also be interesting to assess emotional influence on purchases conducted in a naturalistic environment, adopting other methodologies such as the ecological momentary assessment (EMA) to register behaviors, moods, and experiences of consumers in real time and in daily life. Second, because this study explored a non-typical energy-dense food product, future research could include a typical energy-dense food along with a neutral one to test if the product type moderates the impact of emotional eating style on purchasing behavior.

Finally, behavioral, political, and managerial implications may be derived from the present findings. First, the study highlighted that decision-making processes must be studied in a broader perspective, one that includes both rational and unconscious approaches—although this poses the challenge of developing new methodologies and measures for this purpose. Second, emotions—and the ability to manage them—should be considered by policymakers in designing public health policies that emphasize the pleasures associated with healthy eating habits. In this sense, outreach activities could be a key to reconditioning the public that sweets and other palatable foods are not the only ways to "comfort the soul". Third, companies (e.g., supermarkets and food producers) must be more ambitious in their commitment to proposing more reformulated product (e.g., those with nutritional claims). They may even encourage consumers to choose items bearing less preferred nutritional claims with some promotional strategies.

The solution to public health intervention is undoubtedly complex and requires the engagement of society, the public, and the private sector as a whole.

Author Contributions: B.L.-G. and T.d.-M. conceived the experiment, development of methods, data extraction, data analysis, interpreted the results, and wrote the paper.

Funding: This study has been funded by FP7-MC-CIG- 332769, Fighting against obesity in Europe: the role of health related claims in food products (OBESCLAIM)

Acknowledgments: We are grateful to Vincenzina Caputo and Riccardo Scarpa for their help on the Bayesian design of this study.

Conflicts of Interest: The authors declare no conflict of interest. The founding sponsors had no role in the design of the study; in the collection, analyzes, or interpretation of data; in the writing of the manuscript, and in the decision to publish the results.

Appendix A

Table A1. Cuestionario del Comerdor Emocional Garaulet (CCE).

¿La báscula, tiene un gran poder sobre ti? ¿Es capaz de cambiar tu estado de humor?			
□ Nunca	□ A veces	□ Generalmente	□ Siempre
¿Tienes antojos por ciertos alimentos específicos?			
□ Nunca	□ A veces	□ Generalmente	□ Siempre
¿Te cuesta parar de comer alimentos dulces, especialmente chocolate?			
□ Nunca	□ A veces	□ Generalmente	□ Siempre
¿Tienes problemas para controlar las cantidades de ciertos alimentos?			
□ Nunca	□ A veces	□ Generalmente	□ Siempre
¿Comes cuando estás estresado, enfadado o aburrido?			
□ Nunca	□ A veces	□ Generalmente	□ Siempre
¿Comes más de tus alimentos favoritos, y con más descontrol, cuando estás solo?			
□ Nunca	□ A veces	□ Generalmente	□ Siempre
¿Te sientes culpable cuando tomas alimentos "prohibidos", es decir, aquellos que crees que no deberías, como los dulces o snacks?			
□ Nunca	□ A veces	□ Generalmente	□ Siempre
Por la noche, cuando llegas a casa cansado de trabajar ¿es cuando menos control tienes con tu alimentación?			
□ Nunca	□ A veces	□ Generalmente	□ Siempre
Estás a dieta, y por alguna razón comes más de la cuenta, entonces piensas que no vale la pena y ¿comes de forma descontrolada aquellos alimentos que piensas que más te van a engordar?			
□ Nunca	□ A veces	□ Generalmente	□ Siempre
¿Cuántas veces sientes que la comida te controla a ti en vez de tú a ella?			
□ Nunca	□ A veces	□ Generalmente	□ Siempre

Table A2. Emotional Eater Questionnaire (EEQ) Garaulet.

Never	Sometimes	Generally	Always
Do the weight scales have a great power over you? Can they change your mood?			
□ Never	□ Sometimes	□ Generally	□ Always
Do you crave specific foods?			
□ Never	□ Sometimes	□ Generally	□ Always
Is it difficult for you to stop eating sweet things, especially chocolate?			
□ Never	□ Sometimes	□ Generally	□ Always
Do you have problems controlling the amount of certain types of food you eat?			
□ Never	□ Sometimes	□ Generally	□ Always
Do you eat when you are stressed, angry or bored?			
□ Never	□ Sometimes	□ Generally	□ Always
Do you eat more of your favorite food and with less control when are alone?			
□ Never	□ Sometimes	□ Generally	□ Always
Do you feel guilty when eat "forbidden" foods, like sweet or snacks?			
□ Never	□ Sometimes	□ Generally	□ Always
Do you feel less control over your diet when you are tired after work at night?			
□ Never	□ Sometimes	□ Generally	□ Always
When you overeat while on a diet, do you give up and start eating without control, particularly food that you think is fattening?			
□ Never	□ Sometimes	□ Generally	□ Always
How often do you fell that food controls you, rather than you controlling food?			
□ Never	□ Sometimes	□ Generally	□ Always

References

1. Popkin, B.M. Global nutrition dynamics: The world is shifting rapidly toward a diet linked with noncommunicable diseases. *Am. J. Clin. Nutr.* **2006**, *84*, 289–298. [CrossRef] [PubMed]
2. WHO. Noncommunicable disease. In *Health in 2015: From MDGs, Millennium Development Goals to SDGs, Sustainable Development Goals*; World Health Organization: Geneva, Switzerland, 2015; p. 204. ISBN 978-9-24-156511-0.
3. Von Lengerke, T.; Krauth, C. Economic costs of adult obesity: A review of recent European studies with a focus on subgroup-specific costs. *Maturitas* **2011**, *69*, 220–229. [CrossRef] [PubMed]
4. Dobbs, R.; Sawers, C.; Thompson, F.; Manyika, J.; Woetzel, J.; Child, P.; McKenna, S.; Spatharou, A. *Overcoming Obesity: An Initial Economic Analysis*; McKinsey Global Institute: New York, NY, USA, 2014.
5. Gupta, S.; Richard, L.; Forsythe, A. The humanistic and economic burden associated with increasing body mass index in the EU5. *Diabetes Metab. Syndr. Obes.* **2015**, *8*, 327–338. [CrossRef] [PubMed]
6. European Commission. *White Paper on A Strategy for Europe on Nutrition, Overweight and Obesity Related Health Issues*; The European Commission: Brussels, Belgium, 2007.
7. Comisión Europea Reglamento (CE). No. 1924/2006 del Parlamento Europeo y del Consejo, de 20 de diciembre de 2006, relativo a las declaraciones nutricionales y de propiedades saludables en los alimentos. *Diario Oficial de la Unión Europea* **2006**, *404*, 9–25.
8. Cavaliere, A.; Ricci, E.C.; Banterle, A. Nutrition and health claims: Who is interested? An empirical analysis of consumer preferences in Italy. *Food Qual. Prefer.* **2015**, *41*, 44–51. [CrossRef]

9. Miklavec, K.; Pravst, I.; Grunert, K.G.; Klopčič, M.; Pohar, J. The influence of health claims and nutritional composition on consumers' yoghurt preferences. *Food Qual. Prefer.* **2015**, *43*, 26–33. [CrossRef]
10. De-Magistris, T.; López-Galán, B. Consumers' willingness to pay for nutritional claims fighting the obesity epidemic: The case of reduced-fat and low salt cheese in Spain. *Public Health* **2016**, *135*, 83–90. [CrossRef]
11. De-Magistris, T.; Gracia, A. Consumers' willingness to pay for light, organic and PDO cheese: An experimental auction approach. *Br. Food J.* **2016**, *118*, 560–571. [CrossRef]
12. Jurado, F.; Gracia, A. Does the Valuation of Nutritional Claims Differ among Consumers? Insights from Spain. *Nutrients* **2017**, *9*, 132. [CrossRef]
13. Prieto-Castillo, L.; Royo-Bordonada, M.A.; Moya-Geromini, A. Information search behavior, understanding and use of nutrition labeling by residents of Madrid, Spain. *Public Health* **2015**, *129*, 226–236. [CrossRef]
14. Carrillo, E.; Varela, P.; Fiszman, S. Packaging information as a modulator of consumers' perception of enriched and reduced-calorie biscuits in tasting and non-tasting tests. *Food Qual. Prefer.* **2012**, *25*, 105–115. [CrossRef]
15. De-Magistris, T.; López-Galán, B.; Caputo, V. The impact of body image on the WTP values for reduced-fat and low-salt content potato chips among obese and non-obese consumers. *Nutrients* **2016**, *8*, 830. [CrossRef]
16. López-Galán, B.; De-Magistris, T.; Caputo, V. The impact of emotional intelligence of consumers when purchasing products with nutritional claims. *Econ. Agroaliment.* **2017**, *19*, 273–286. [CrossRef]
17. Gardner, M.P.; Wansink, B.; Kim, J.; Park, S.-B. Better moods for better eating? How mood influences food choice. *J. Consum. Psychol.* **2014**, *24*, 320–335. [CrossRef]
18. Köster, E.P.; Mojet, J. From mood to food and from food to mood: A psychological perspective on the measurement of food-related emotions in consumer research. *Food Res. Int.* **2015**, *76*, 180–191. [CrossRef]
19. Gibson, E.L. Emotional influences on food choice: Sensory, physiological and psychological pathways. *Physiol. Behave.* **2006**, *89*, 53–61. [CrossRef]
20. Macht, M. How emotions affect eating: A five-way model. *Appetite* **2008**, *50*, 1–11. [CrossRef]
21. Bongers, P.; Jansen, A. Emotional Eating Is Not What You Think It Is and Emotional Eating Scales Do Not Measure What You Think They Measure. *Front. Psychol.* **2016**, *7*, 1932. [CrossRef]
22. Macht, M.; Simons, G. Emotional Eating. In *Emotion Regulation and Well-Being*; Nyklíček, I., Vingerhoets, A., Zeelenberg, M., Eds.; Springer: New York, NY, USA, 2011; pp. 281–295. ISBN 978-1-44-196953-8.
23. Roberts, C.J. The effects of stress on food choice, mood and bodyweight in healthy women. *Nutr. Bull.* **2008**, *33*, 33–39. [CrossRef]
24. Bongers, P.; Jansen, A.; Havermans, R.; Roefs, A.; Nederkoorn, C. Happy eating: The underestimated role of overeating in a positive mood. *Appetite* **2013**, *67*, 74–80. [CrossRef]
25. Bennett, J.; Greene, G.; Schwartz-Barcott, D. Perceptions of emotional eating behavior. A qualitative study of college students. *Appetite* **2013**, *60*, 187–192. [CrossRef]
26. Tan, C.C.; Chow, C.M. Stress and emotional eating: The mediating role of eating dysregulation. *Pers. Ind. Diff.* **2014**, *66*, 1–4. [CrossRef]
27. Groesz, L.M.; McCoy, S.; Carl, J.; Saslow, L.; Stewart, J.; Adler, N.; Laraia, B.; Epel, E. What is eating you? Stress and the drive to eat. *Appetite* **2012**, *58*, 717–721. [CrossRef]
28. Raspopow, K.; Abizaid, A.; Matheson, K.; Anisman, H. Anticipation of a psychosocial stressor differentially influences ghrelin, cortisol and food intake among emotional and non-emotional eaters. *Appetite* **2014**, *74*, 35–43. [CrossRef]
29. Keller, C.; Siegrist, M. Does personality influence eating styles and food choices? Direct and indirect effects. *Appetite* **2015**, *84*, 128–138. [CrossRef]
30. Van Strien, T.; Herman, C.P.; Anschutz, D.J.; Engels, R.C.M.E.; de Weerth, C. Moderation of distress-induced eating by emotional eating scores. *Appetite* **2012**, *58*, 277–284. [CrossRef]
31. Fay, S.H.; Finlayson, G. Negative affect-induced food intake in non-dieting women is reward driven and associated with restrained–disinhibited eating subtype. *Appetite* **2011**, *56*, 682–688. [CrossRef]
32. Lazarevich, I.; Esther Irigoyen-Camacho, M.; del Consuelo Velazquez-Alva, M. Obesity, eating behavior and mental health among university students in Mexico city. *Nutr. Hosp.* **2013**, *28*, 1892–1899.
33. Van Strien, T.; Cebolla, A.; Etchemendy, E.; Gutiérrez-Maldonado, J.; Ferrer-García, M.; Botella, C.; Baños, R. Emotional eating and food intake after sadness and joy. *Appetite* **2013**, *66*, 20–25. [CrossRef]
34. Cardi, V.; Leppanen, J.; Treasure, J. The effects of negative and positive mood induction on eating behavior: A meta-analysis of laboratory studies in the healthy population and eating and weight disorders. *Neurosci. Biobehav. Rev.* **2015**, *57*, 299–309. [CrossRef]

35. Bongers, P.; van den Akker, K.; Havermans, R.; Jansen, A. Emotional eating and Pavlovian learning: Does negative mood facilitate appetitive conditioning? *Appetite* **2015**, *89*, 226–236. [CrossRef]
36. Evers, C.; Dingemans, A.; Junghans, A.F.; Boevé, A. Feeling bad or feeling good, does emotion affect your consumption of food? A meta-analysis of the experimental evidence. *Neurosci. Biobehav. Rev.* **2018**, *92*, 195–208. [CrossRef]
37. Leehr, E.J.; Krohmer, K.; Schag, K.; Dresler, T.; Zipfel, S.; Giel, K.E. Emotion regulation model in binge eating disorder and obesity—A systematic review. *Neurosci. Biobehav. Rev.* **2015**, *49*, 125–134. [CrossRef]
38. Haynos, A.F.; Fruzzetti, A.E. Anorexia Nervosa as a Disorder of Emotion Dysregulation: Evidence and Treatment Implications. *Clin. Psychol. Sci. Pract.* **2011**, *18*, 183–202. [CrossRef]
39. Van Strien, T.; Peter Herman, C.; Verheijden, M.W. Eating style, overeating and weight gain. A prospective 2-year follow-up study in a representative Dutch sample. *Appetite* **2012**, *59*, 782–789. [CrossRef]
40. Wansink, B.; Sobal, J. Mindless eating: The 200 daily food decisions we overlook. *Environ. Behav.* **2007**, *39*, 106–123. [CrossRef]
41. Kidwell, B.; Hardesty, D.M.; Childers, T.L. Emotional Calibration Effects on Consumer Choice. *J. Consum. Res.* **2008**, *35*, 611–621. [CrossRef]
42. Lancaster, K.J. A New Approach to Consumer Theory. *J. Political Econ.* **1966**, *74*, 132–157. [CrossRef]
43. McFadden, D. Conditional logit analysis of qualitative choice behavior. In *Frontiers in Econometrics*; Zarebka, P., Ed.; Academic Press: New York, NY, USA, 1973; pp. 105–142.
44. Carlsson, F.; Martinsson, P. Do Hypothetical and Actual Marginal Willingness to Pay Differ in Choice Experiments? *J. Environ. Econ. Manag.* **2001**, *41*, 179–192. [CrossRef]
45. Harrison, G.W.; List, J.A. Field Experiments. *J. Econ. Lit.* **2004**, *42*, 1009–1055. [CrossRef]
46. List, J.A.; Gallet, C.A. What Experimental Protocol Influence Disparities Between Actual and Hypothetical Stated Values? *Environ. Resour. Econ.* **2001**, *20*, 241–254. [CrossRef]
47. Murphy, J.J.; Allen, P.G.; Stevens, T.H.; Weatherhead, D. A Meta-analysis of Hypothetical Bias in Stated Preference Valuation. *Environ. Resour. Econ.* **2005**, *30*, 313–325. [CrossRef]
48. Chang, J.B.; Lusk, J.L.; Norwood, B.F. How closely do hypothetical surveys and laboratory experiments predict field behavior? *Am. J. Agric. Econ.* **2009**, *91*, 518–534. [CrossRef]
49. Loomis, J.; Bell, P.; Cooney, H.; Asmus, C. A Comparison of Actual and Hypothetical Willingness to Pay of Parents and Non-Parents for Protecting Infant Health: The Case of Nitrates in Drinking Water. *J. Agric. Appl. Econ.* **2009**, *41*, 698–712. [CrossRef]
50. Lusk, J.L.; Shogren, J.F. *Experimental Auctions: Methods and Applications in Economic and Marketing Research*; Cambridge University Press: Cambridge, UK, 2007; ISBN 978-0-51-161126-1.
51. Grebitus, C.; Lusk, J.L.; Nayga, R.M. Explaining differences in real and hypothetical experimental auctions and choice experiments with personality. *J. Econ. Psychol.* **2013**, *36*, 11–26. [CrossRef]
52. De-Magistris, T.; Gracia, A.; Nayga, R.M. On the use of honesty priming tasks to mitigate hypothetical bias in choice experiments. *Am. J. Agric. Econ.* **2013**, *95*, 1136–1154. [CrossRef]
53. Baba, Y.; Kallas, Z.; Realini, C.E.; Pérez-Juan, M.; Sañudo, C.; Albertí, P.; Insausti, K. Impacto de la experiencia sensorial y la información sobre las preferencias de los consumidores por la carne de vacuno enriquecida en omega-3 y ácido linoleico conjugado en tres ciudades españolas. *ITEA-Información Técnica Económica Agraria* **2017**, *113*, 192–210. [CrossRef]
54. Barreiro-Hurlé, J.; Gracia, A.; De-Magistris, T. Market implications of new regulations: Impact of health and nutrition information on consumer choice. *Span. J. Agric. Res.* **2009**, *7*, 257–268. [CrossRef]
55. Cohen, J. *Statistical Power Analysis for the Behavioral Sciences*; Lawrence Earlbaum Associates: Hillsdale, NJ, USA, 1988; ISBN 080-5-80283-5.
56. Varela-Moreiras, G.; Ruiz, E.; Valero, T.; Avila, J.M.; del Pozo, S. The Spanish diet: An update. *Nutr. Hosp.* **2013**, *28* (Suppl. 5), 13–20.
57. Mercasa La Alimentación en España 2018. *Producción, Industria, Distribución y Consumo*, 21st ed.; Mercasa-Distribución y Consumo: Madrid, Spain, 2019; ISBN 846-9-56-1715.
58. Navia Lombán, B.; Ortega Anta, R.M.; Perea Sánchez, J.M.; Apararicio Vizuete, A.; López Sobaler, A.; Andrés Carvajales, P.; González Rodríguez, L.; Cons Ferreiro, M. *Influencia del Consumo de Pan en la Calidad de la Dieta y Hábitos Alimentarios de Adultos Españoles. Percepciones y Conocimientos Existentes en Torno a Este Alimento*; Secretaría técnica y de Comunicación Campaña Pan cada día: Madrid, Spain, 2016; p. 14.

59. Fernández, M.G.; Basulto Marset, J.; Bretón Lesmes, I.; Quiles Izquierdo, J. Recomendaciones nutricionales basadas en la evidencia para la prevención y el tratamiento del sobrepeso y la obesidad en adultos. *Revista Española de Obesidad* **2011**, *9*, 6–78.
60. WHO. *Obesity and Overweight*; World Health Organization: Geneva, Switzerland, 2015.
61. WHO. *Diet, Nutrition and the Prevention of Chronic Diseases. Consultation, Report of a Joint WHO/DAO Expert*; World Health Organization: Geneva, Switzerland, 2003.
62. He, F.J.; MacGregor, G.A. A comprehensive review on salt and health and current experience of worldwide salt reduction programmes. *J. Hum. Hypertens.* **2009**, *23*, 363–384. [CrossRef]
63. Campbell, N.; Correa-Rotter, R.; Neal, B.; Cappuccio, F.P. New evidence relating to the health impact of reducing salt intake. *Nutr. Metab. Cardiovasc. Dis.* **2011**, *21*, 617–619. [CrossRef]
64. Scarpa, R.; Campbell, D.; Hutchinson, W.G. Benefit Estimates for Landscape Improvements: Sequential Bayesian Design and Respondents' Rationality in a Choice Experiment. *Land Econ.* **2007**, *83*, 617–634. [CrossRef]
65. Scarpa, R.; Zanoli, R.; Bruschi, V.; Naspetti, S. Inferred and stated attribute non-attendance in food choice experiments. *Am. J. Agric. Econ.* **2013**, *95*, 165–180. [CrossRef]
66. Metric, C. *Ngene 1.1.2. User Manual & Reference Guide. The Cutting Edge in Experimental Design*; Choice Metrics: Sydney, Australia, 2014; p. 241.
67. Garaulet, M.; Canteras, M.; Morales, E.; López, G.; Sánchez, D.; Corbalán, M.D. Validation of a questionnaire on emotional eating for use in cases of obesity; the Emotional Eater Questionnaire (EEQ). *Nutr. Hosp.* **2012**, *27*, 645–651.
68. López-Galán, B.; Gracia, A.; Barreiro-Hurle, J.; Barreiro-Hurlé, J. What comes first, origin or production method? An investigation into the relative importance of different attributes in the demand for eggs. *Span. J. Agric. Res.* **2013**, *11*, 305–315. [CrossRef]
69. Gracia, A.; Barreiro-Hurlé, J.; Galán, B.L. Are Local and Organic Claims Complements or Substitutes? A Consumer Preferences Study for Eggs. *J. Agric. Econ.* **2014**, *65*, 49–67. [CrossRef]
70. Train, K. *Discrete Choice Methods with Simulation*; Cambridge University Press: New York, NY, USA, 2002; ISBN 052-1-01-7157.
71. Scarpa, R.; Del Giudice, T. Market Segmentation via Mixed Logit: Extra-Virgin Olive Oil in Urban Italy. *J. Agric. Food Ind. Organ.* **2004**, *2*. [CrossRef]
72. Bazzani, C.; Caputo, V.; Nayga, R.M.; Canavari, M. Revisiting consumers' valuation for local versus organic food using a non-hypothetical choice experiment: Does personality matter? *Food Qual. Prefer.* **2017**, *62*, 144–154. [CrossRef]
73. Scarpa, R.; Philippidis, G.; Spalatro, F. Product-country images and preference heterogeneity for Mediterranean food products: A discrete choice framework. *Agribusiness* **2005**, *21*, 329–349. [CrossRef]
74. Jacquier, C.; Bonthoux, F.; Baciu, M.; Ruffieux, B. Improving the effectiveness of nutritional information policies: Assessment of unconscious pleasure mechanisms involved in food-choice decisions. *Nutr. Rev.* **2012**, *70*, 118–131. [CrossRef]
75. Bublitz, M.G.; Peracchio, L.A.; Andreasen, A.R.; Kees, J.; Kidwell, B.; Miller, E.G.; Motley, C.M.; Peter, P.C.; Rajagopal, P.; Scott, M.L.; et al. Promoting positive change: Advancing the food well-being paradigm. *J. Bus. Res.* **2013**, *66*, 1211–1218. [CrossRef]
76. López-Galán, B.; de-Magistris, T. Prevalencia de las declaraciones nutricionales en la prevención de la obesidad en el mercado español. *Nutr. Hosp.* **2017**, *34*, 154–164. [CrossRef]

nutrients

MDPI

Article

Qualified Health Claim Language affects Purchase Intentions for Green Tea Products in the United States

Amanda Berhaupt-Glickstein [1,*], Neal H. Hooker [2] and William K. Hallman [3]

1 Department of Nutritional Sciences, Rutgers, The State University of New Jersey, New Brunswick, NJ 08901-2882, USA
2 John Glenn College of Public Affairs, The Ohio State University, Columbus, OH 43210, USA; hooker.27@osu.edu
3 Department of Human Ecology, Rutgers, The State University of New Jersey, New Brunswick, NJ 08901-8520, USA; hallman@aesop.rutgers.edu
* Correspondence: Amanda.berhaupt@gmail.com; Tel.: +1-202-774-7286

Received: 1 April 2019; Accepted: 22 April 2019; Published: 24 April 2019

Abstract: Qualified health claims (QHC) describe diet–disease relationships and summarize the quality and strength of evidence for a claim. Companies assert that QHCs increase sales and take legal action to ensure claims reflect their interests. Yet, there is no empirical evidence that QHCs influence consumers. Using green tea as a case study, this study investigated the effects of QHCs on purchase intentions among adults 55 years and older living in the US. An online survey using a between-subjects design examined QHCs about the relationship between green tea and the reduced risk of breast and/or prostate cancer or yukichi fruit juice and the reduced risk of gastrocoridalis, a fictitious relationship. QHCs written by a green tea company generated greater perceptions of evidence for the relationship, greater confidence in green tea and cancer, and increased purchase intentions for green tea than other QHCs. Factors that mitigated the claim's effects on purchase intentions are: Race/ethnicity; age; importance of health claims; supplement use; health; worry about health/becoming sick with cancer; worry that led to dietary change; green tea consumption; and familiarity with the green tea–cancer. Consumers who made health-related dietary change in the past year and consider health claims important indicated greater purchase intentions than others.

Keywords: qualified health claim; older adult; green tea; cancer; purchase intentions

1. Introduction

The food and dietary supplement industries have a vested interest in qualified health claims (QHCs) [1,2] because they offer a way to communicate with consumers about a product's potential health benefit. QHCs describe the relationship between the consumption of a dietary substance and the reduced risk for a disease or health condition (i.e., diet–disease relationship). However, these claims are nuanced in that they also characterize the level of scientific support for the diet–disease relationship through a disclaimer [3]. One example of a QHC about calcium and colon or rectal polyps reads, "Very limited and preliminary evidence suggests that calcium supplements may reduce the risk of colon/rectal polyps. FDA concludes that there is little scientific evidence to support this claim" [4].

The level of evidence for a diet–disease relationship (and QHC) is determined by the US Food and Drug Administration (FDA). The FDA catalogues the scientific support for diet–disease relationships into four levels. A claim supported by strong scientific evidence (i.e., significant scientific agreement (SSA)) is described as level A. Such claims are straightforward and, importantly, do not require a disclaimer. An example of an A-level claim is: Adequate calcium throughout life, as part of a well-balanced diet, may reduce the risk of osteoporosis in later life [5]. By contrast, diet–disease relationships behind B, C, and D-level QHCs are supported by partial or incomplete evidence and require disclaimers that summarize this evidence. However, disclaimers do not indicate a level of evidence using letter-grades. Instead, the FDA prescribes the language in QHCs to accurately describe the level of scientific support [6].

Fundamentally, the value of a QHC for companies is the degree to which it persuades consumers to purchase their products [1,2]. Consequently, the disclaimer, which describes the evidence for the claimed relationship, is the crucial component of a QHC in determining product appeal to consumers. For examples of disclaimers, please see the QHCs composed by the US FDA and the court in Table 1. The disclaimer is the last sentence of the QHC.

1.1. Consumer Research about Qualified Health Claims

The food and supplement industries understand that consumers are motivated to buy products with health benefits and expect that QHCs will lead to increased purchases of their products. However, the expectation that QHCs can increase product purchases is based on research about the marketing impacts of A-level health claims that do not require a disclaimer. Such studies demonstrate a financial advantage [6], including greater purchase intentions [7] and increased market share [8,9]. Consumers also indicate a willingness to pay 50% to 200% more for products making health claims than for comparable products [10], including yogurt [11].

Anecdotally, companies have reported QHCs as valuable for increasing product sales [1]. Yet, only one study has examined the market potential for products with a QHC which demonstrated an increase in purchase intentions [12]. However, the experiment tested a graphic QHC, which is a format not permitted by the FDA. Therefore, there is no evidence that QHCs affect consumers' purchase intentions or hold commercial value.

Other research has explored consumers' understanding of the scientific support described in the QHC disclaimer. Most US consumers misunderstand QHCs as an indication of the quality of a product versus an indication of the level of evidence for a claim [13]. Further, consumers' understanding of claims largely stems from prior experience and knowledge of a diet–disease relationship rather than a claim itself [13]. Consumers require more information for lesser-known diet–disease relationships. One study found that more detailed messages had stronger impacts on consumers when the nutrient or diet–disease link was less familiar [14]. Research also indicates that US consumers have a limited ability to act on scientific information [15] and rarely read labels because of excessive information and a lack of time [16], and these reasons may be why QHCs are infrequently found on labels [17,18].

Nonetheless, the potential of QHCs to increase product sales has motivated companies to take legal action against the FDA about how to fairly and accurately characterize the level of evidence for diet–disease relationships [2]. This is richly illustrated by the history of lawsuits filed against the FDA by manufacturers and marketers, which have sought to alter the wording of QHC disclaimers [13].

1.2. The Evolving Qualified Health Claim for the Green Tea and Cancer Relationship

In 2004, the FDA assigned the green tea–cancer relationship the lowest level of evidence (i.e., D grade), yet there have been seven QHCs for this claimed relationship [4,19]. Six of the seven claims are no longer in use because they mischaracterized the scientific evidence for the relationship [20,21]. The seven claims are presented in Table 1 and identify the language contested by a green tea company (Fleminger, Inc. (i.e., Fleminger)) and the FDA. QHCs are referenced by the year in which they appeared (e.g., 2004).

Of the seven claims, three were written by Fleminger (Table 1, 2004; 2008; 2010) and serve to highlight the health value of their products [1,2]. However, Fleminger's claims were rejected by the FDA since they are scientifically inaccurate. While the three additional QHCs were written by the FDA (Table 1, 2005p; 2005b; 2011) and are scientifically accurate [22], the courts found them too technical (Table 1, 2005p; 2005b; 2011). The federal court wrote the seventh claim, which the FDA currently allows companies to use on green tea labels (Table 1, 2012) [21]. The court's QHC aims to strike a balance between the interests of stakeholders to communicate a health benefit of a product and to ensure the accuracy of scientific evidence to prevent consumers from being misled [2].

Table 1. Seven qualified health claims petitioned, unlawfully used by manufacturers, or prescribed by the US Food and Drug Administration or Federal Court, 2004–2012 (*n* = 1335).

Author	Year	Status	Qualified Health Claim	*n*
Fleminger	2004	• Scientifically inaccurate • Petitioned claim	Daily consumption of 40 ounces of typical green tea containing 170µg/mL of natural (-) epigallocatechin gallate (EGCG) may reduce the risk of certain forms of cancer. **There is scientific evidence supporting this health claim although the evidence is not conclusive**.	185
Fleminger	2008	• Scientifically inaccurate • Unlawfully used	Green tea may reduce the risk of cancer of the breast and the prostate. There is **credible evidence** supporting this claim although the evidence is limited.	176
Fleminger	2010	• Scientifically inaccurate • Illegally names FDA • Unlawfully used	Green tea may reduce the risk of breast and prostate cancers. The **FDA** has concluded that there is **credible evidence** supporting this claim although the evidence is limited.	185
FDA	2005p	• Scientifically accurate • Overly technical • No longer allowed	**One weak and limited study** does not show that drinking green tea reduces the risk of prostate cancer, but **another weak and limited study** suggests that drinking green tea may reduce this risk. Based on these studies, FDA concludes that it is **highly unlikely** that green tea reduces the risk of prostate cancer.	211
FDA	2005b	• Scientifically accurate • Overly technical • No longer allowed	**Two studies** do not show that drinking green tea reduces the risk of breast cancer in women, but **one weaker, more limited study** suggests that drinking green tea may reduce this risk. Based on these studies, FDA concludes that it is **highly unlikely** that green tea reduces the risk of breast cancer.	179
FDA	2011	• Scientifically accurate • Disclaimer negates claim • No longer allowed	Drinking green tea may reduce the risk of breast or prostate cancer. **FDA does not agree** that green tea may reduce that risk because there is very little scientific evidence for the claim.	206
Federal Court	2012	• Scientifically accurate • Technically appropriate • Allowed	Green tea may reduce the risk of breast or prostate cancer. FDA has concluded that there is very little scientific evidence for this claim.	193

Note: Bold words in qualified health claims (QHCs) indicate issues in the description of evidence.

The green tea QHC is an example that demonstrates the challenges to regulating QHCs. Using green tea as a case study, we explore the practical use of QHCs to influence consumer choice in the US. We examine the seven claims to understand their potential to influence the purchase intentions of older adults for green tea products and consider theory-based consumer and product-specific factors [23]. We also clarify whether the current QHC, permitted by the FDA, increases consumer purchase intentions for green tea products in comparison with claims no longer in use.

2. Materials and Methods

2.1. Sample

A sample of non-institutionalized, English-speaking adults aged 55 years and older, who reside in the United States (US), participated in the study. This segment was selected because older adults are affected by cancer and other chronic diseases outlined in QHCs [16,24,25]. They also represent a large portion of the population with robust purchasing power [26,27]. Research shows that older consumers are also more health-oriented than younger adults, more knowledgeable about diet and health, and adopt preventive behaviors [28], such as reading food labels [29,30] and taking dietary supplements [31,32]. One study in Finland identified health-seeking older consumers willing to purchase functional foods and believe they can influence their health through diet and lifestyle [33]. Functional foods are "foods or dietary components that may provide a health benefit beyond basic nutrition and may play a role in reducing or minimizing the risk of certain diseases and other health conditions" [34].

2.2. Study Design

An online survey was administered by GfK Custom Research, LLC (GfK) in January 2014. Probability-based recruitment was used to obtain a representative sample of older adults in the US. All subjects gave their informed consent. The study (protocol #12-454) was approved by the Rutgers, The State University of New Jersey Institutional Review Board.

Evidence indicates that experience with a product or prior knowledge about a diet–disease relationship motivates consumer behavior [35–38] and generates a greater perception of control over disease risk [39]. To mitigate the potential interaction of beliefs or behaviors associated with green tea and cancer, we employed a 2 × 7 between-subjects study design. This design allowed investigators to determine the effects of the QHC language. Participants were randomized into one of two conditions: (1) Green tea–breast/prostate cancer, or (2) yukichi fruit juice–gastrocoridalis, a fictitious but comparable diet–disease relationship. Yukichi fruit juice was described as a typical drink sold in stores and gastrocoridalis was introduced as a potentially painful and fatal disease. Participants were then randomized into one of seven groups so that each group viewed one QHC (see Table 1). The QHCs differed by condition in that (a) yukichi fruit juice substituted green tea as the dietary component, and (b) gastrocoridalis replaced breast and/or prostate cancer as the disease outcome.

This was a text-only study to examine consumer perceptions in response to the language in each claim, without distraction of label or package images. The QHC stimuli remained on screen for participants to refer to while responding to questions.

The survey began with, "This survey is designed to determine what Americans know and think about the health benefits of certain foods and dietary supplements. We are interested in what you currently know and feel, without using the internet or other resources to learn more about them. We are asking that you finish this survey in one sitting." The first section asked participants about their perceptions, beliefs, and behaviors related to food and health. This included questions about their consumption of green tea and yukichi fruit juice in the past year, and their reasons for consuming them. Participants also indicated the extent to which they worry about their health and whether their worry led them to change their diet. They also indicated the extent to which they worry about becoming sick with cancer or gastrocoridalis, and whether they had a past diagnosis of either disease. Section two presented one QHC and participants responded to questions about their perceptions, based on the QHC and its description of evidence for the claimed relationship.

2.3. Measures

Demographic data were collected prior to the survey as part of the GfK panelist database. The modifying factors identified in the health belief model (HBM) include: Age, sex, race/ethnicity, education, employment, and income, which were also included in analysis [12,23,38,40–44].

Dependent variables were identified based on stakeholder interests and from the Health Claims Framework (HCF), which postulates factors that predict purchase of products with health claims [37]. Purchase intentions for green tea or yukichi fruit juice were included since they are the primary outcome for the food and supplement industries [38,41] (Table 2).

The primary outcome for the FDA is consumer understanding of the level of scientific evidence [39] described in the QHC disclaimer. To test whether the QHC language led to different perceptions of evidence, participants responded a 13-point scale (Table 3) that provided a wide range of ordinal responses for the level of evidence. Further, during analysis, the scale could be separated into four evidence levels; 0 represented "No evidence"; 1–3 represented a "D"; 4–6 a "C"; 7–9 a "B"; and 10–12 an "A". The scale does not necessarily reflect FDA's rating of evidence; however, it is based on the Murphy (2005) approach to develop a scale that may correspond to the levels of evidence [45]. The green tea and yukichi fruit juice claims represent a D-level of scientific evidence and are based on the same evidence.

Two questions explored the perceived benefits of green tea and yukichi fruit juice. One question inquired about their perceived disease risk reduction and the second asked about confidence in the drink as a means to reduce their risk (Table 2). Participants also indicated the perceived dose needed to achieve the health benefit (Table 2), since it is dose-dependent and in the context of the overall diet [46].

Independent variables that included product-specific and consumer-specific measures were also obtained from the HBM and HCF. Familiarity lends itself to an increase in acceptance of functional foods that bear health claims [12,38,42,47–49]. Therefore, familiarity with the diet–disease relationship as well as with the QHC were measured. Similarly, existing behavior may also influence purchase intentions [50]. To account for past green tea or yukichi fruit juice consumption, participants indicated the frequency of their consumption in the past year. Taste preference was also measured, since it is a prominent consideration when selecting products [38,51,52] and, for some, outweighs the perceived health value of a product [53]. Participants also reported their health status, health worries, and their perceived susceptibility for cancer or gastrocoridalis. Further, they indicated whether their worries had prompted them to make a dietary change in the past year. Other questions measured personal relevance of cancer or gastrocoridalis, knowledge about diet and health, dietary supplement use, and perceived importance of health claims when making a purchase decision.

Table 2. Variable descriptions.

	Variable	Survey Question	Scale
DEPENDENT VARIABLES	**Purchase intention**	If priced the same as other GTs *(YFJ **) without this statement, how likely is it that you would purchase a bottle of GT (YFJ) with this statement?	1 = Not at all likely, 2 = Slightly likely, 3 = Somewhat likely, 4 = Fairly likely, 5 = Very likely, 6 = Extremely likely, 7 = Absolutely certain
	Evidence	Based on this statement, how much evidence is there that drinking GT (YFJ) may reduce the risk of Cancer (Gastrocordialis)?	See Table 3
	Confidence	Based on this statement, how confident are you that drinking this GT (YFJ) will reduce the risk of Cancer (Gastrocordialis)?	1 = Not at all confident, 2 = Slightly confident, 3 = Somewhat confident, 4 = Fairly confident, 5 = Very confident, 6 = Extremely confident, 7 = Absolutely confident
	Risk perception	Based on this statement, how much can drinking GT (YFJ) (as part of a regular diet) reduce the risk of Cancer (Gastrocordialis)?	1 = Not at all, 2 = Slight reduction, 3 = Some reduction, 4 = Modest reduction, 5 = Large reduction, 6 = Extreme reduction, 7 = Complete reduction
	Dose	How often would someone have to drink GT (YFJ) to reduce their risk of Cancer (Gastrocordialis)?	1 = Never, 2 = Less than once a month, 3 = Once a month, 4 = 2–3 times a month, 5 = At least once a week

Table 2. *Cont.*

	Variable	Survey Question	Scale
INDEPENDENT VARIABLES	**Reason for drinking GT (YFJ)?**	Why do you drink GT (YFJ)?	1 = To reduce the risk of X-type cancer, 2 = For another specific health reason, 3 = I enjoy the taste, 4 = For other reasons
	Familiarity—QHC	Have you ever seen this [QHC] on a: food or dietary supplement label/in an advertisement/on a website or in an article?	−1 = No, 0 = I don't know, 3 = Yes
	Familiarity—diet–disease relationship	Whether you have seen this statement or not, how familiar are you with the idea that drinking GT (YFJ) can reduce the risk of Cancer (Gastrocordialis)?	1 = Not all familiar, 2 = Somewhat, 3 = Fairly, 4 = Very, 5 = Extremely familiar
	Health Status	Would you say that in general your health is:	1 = Poor, 2 = Fair, 3 = Good, 4 = Very good, 5 = Excellent
	Cancer Worry	How worried are you about becoming ill with Cancer (Gastrocordialis)?	1 = Not at all, 2 = Somewhat, 3 = Moderately, 4 = Very, 5 = Extremely
	Diet Δ for Health Worry	How much has worrying about your health led you to change the way you ate in the past year?	1 = Not at all, 2 = A little, 3 = Somewhat, 4 = Quite a bit, 5 = All the time
	Health Worry	How often have you worried about your overall health in the past year?	1 = Not at all, 2 = A little, 3 = Somewhat, 4 = Quite a bit, 5 = All the time
	Cancer Diagnosis	Has a doctor ever told you that you had Cancer (Gastrocordialis)?	0 = No, 1 = Yes
	Perceived Nutrition Knowledge	How well informed are you about diet and health?	1 = Not informed, 2 = Somewhat, 3 = Fairly, 4 = Very, 5 = Extremely informed
	Existing behavior	Over the past 12 months, how often did you drink GT (YFJ)?	1 = Never, 2 = Less than once a month, 3 = Once a month, 4 = 2–3 times a month, 5 = At least once a week
	Importance of Health Claims on Supplement Labels	When you consider buying a new dietary supplement, how important are statements on the label that describe the product's health benefits?	1 = Not at all important, 2 = Somewhat, 3 = Important, 4 = Very, 5 = Absolutely essential
	Supplement Use	Over the past 12 months, how often did you take a dietary supplement? A dietary supplement could be a vitamin, or a mineral that is taken to supplement the diet (e.g., MVI).	1 = Never, 2 = Less than 1 day/month, 3 = 1–3 days/month, 4 = 1–3 days/week, 5 = 4–6 days/week, 6 = Everyday
	Importance of Health Claims on Food Labels	When you consider buying a new food product, how important are statements on the label that describe the product's health benefits?	1 = Not at all important, 2 = Somewhat, 3 = Important, 4 = Very, 5 = Absolutely essential

* GT: Green tea, ** YFJ: Yukichi fruit juice.

Table 3. Semantic differential scale of evidence and collapsed categories for data analysis.

Evidence Scale	None		Minimal				Some				Complete		
	o	o	o	o	o	o	o	o	o	o	o	o	o
Level of evidence *	N/A		D			C			B			A	
FDA claim category	N/A					Qualified Health Claim						Health Claim	

* A footnote in the 2009 Final Guidance indicated the health claim grading system is no longer in effect (Food and Drug Administration, 2003b; Food and Drug Administration, 2011). However, an implicit scale of evidence remains, as it is the fundamental difference between health claims and qualified health claims. For this reason, the current study used the 2003 scale of evidence to explore consumer perceptions of evidence. o = response scale points.

2.4. Data Analysis

Descriptive statistics characterized the study sample and Spearman rank correlation tests identified associations. Based on the HCF, the measures considered to influence purchase intentions are: Perceived evidence, risk reduction, and confidence in the claimed relationship. The 13-point scale of evidence was collapsed to correspond with the grades of evidence from the FDA ranking system, where 0 corresponded to "No evidence"; 1–3 a "D"; 4–6 a "C"; 7–9 a "B"; and 10–12 an "A" (Table 3) [45]. A multivariate analysis of variance test measured interactions between these dependent variables in the green tea/yukichi fruit juice conditions and QHC groups, followed by univariate tests. Independent t-tests and Chi-square tests examined group and condition differences in terms of: Health status, worry about health, worry about cancer, worry that led to dietary change, dietary supplement use, importance of health statements on food and dietary supplement package labels, previous diagnosis of cancer, and sociodemographics.

Hierarchical multiple linear regression was used to determine the predictive value of the independent measures on purchase intentions for green tea in the presence of a QHC. The analysis focused on the green tea condition since the yukichi fruit juice–gastrocoridalis condition was fictitious. Participants in the green tea condition reported their enjoyment (or not) of the taste of green tea, whether they drank green tea in the past year, and their familiarity with the green tea–cancer relationship and with the actual claim statement. One-way analysis of variance (ANOVA) was used to compare responses between QHC groups. The Welch's F-test results are reported when the assumption of homogeneity of variances is violated. For non-normal distributions, the potential bias was corrected with the resampling method of bootstrapping. To create a parsimonious predictive model for purchase intentions for green tea [54], nonsignificant variables were withheld from entry into the final regression model analysis, and significant predictors were added as blocks to examine their contribution to explained variance.

The HCF and HBM guided the addition of specific variables to the hierarchical model in discrete steps. At step one were the demographic, consumer-specific, and perceived susceptibility measures, step two was existing green tea consumption, step three was familiarity with the green tea–cancer relationship, and step four were the QHCs which were entered as a block of dummy-coded variables. This was done to understand how the claim statements may contribute to purchase intentions for green tea. A second regression model was created to isolate the effects of the dependent measures (confidence, risk reduction, evidence perceptions, purchase intentions) from the QHCs and included steps one through three with the removal of the dummy-coded QHCs. Adjusted R^2 is reported to account for the number of predictor variables, and change in R^2 is reported to indicate the contribution of each variable to the predictive model. $p < 0.05$ was considered statistically significant. Statistical analyses were performed using the Statistical Package for Social Sciences, version 22 (SPSS Inc., Chicago, IL, USA).

3. Results

A total of 1335 participants completed the online experiment, of the 2219 sampled, for a 60% completion rate. The experiment consisted of two conditions (green tea–cancer/yukichi fruit juice–gastrocordialis) and seven QHC groups. Each group viewed one QHC in one of two conditions; seven QHCs were about the green tea–cancer relationship (n = 669) and the remaining seven were identically worded QHCs about the yukichi fruit juice–gastrocoridalis relationship (n = 666).

The majority of participants were between the ages of 55 and 74 years old (n = 1123, 89.1%), White (n = 1060, 79.4%), who held a high school degree or more (n = 1233, 92.4%), with a household income under $100,000 ($n$ =1028, 77.0%). Given the random assignment of participants, there were no differences between conditions or across groups by race and ethnicity, age, education, employment, household income or incidence of breast or prostate cancer (Table 4).

Overall, the majority of participants reported that they were in good health (n = 1010, 76%) and had never received a cancer diagnosis from their doctor (n =1132, 84.8%). Participants were "a little" or "somewhat" worried about their health overall (n = 929, 69.9%) and fewer worried about becoming sick with cancer ("somewhat" n = 441, 39.2%; "not at all" n = 296, 26.3%). However, most had made a dietary change in the past year (n = 1,022, 76.8%) due to a health-related concern. Slightly more than half of the participants reported that they drink green tea (n = 691, 51.8%), and most claimed to drink it because they enjoy its taste (n = 215, 62.5%). Of the respondents who reportedly consumed green tea in the past year (n = 345), fewer than 10% drank it to reduce their risk of cancer (n = 32, 9.3%). The majority of the participants believe they are informed about diet and health (i.e., perceived nutrition knowledge) (n = 1298, 97.2%) and consider health claims important on food and dietary supplement product labels (n = 1168, 88.0%; n = 926, 91.5%, respectively). No differences were found between groups or conditions with respect to general health, perceived nutrition knowledge, green tea consumption in the past year, worry about cancer, or health-related dietary changes.

Table 4. Sample characteristics, n (%).

	Green Tea	Yukichi Fruit Juice	All Participants
Characteristics	(n = 666)	(n = 669)	(n = 1335)
Sex *			
Female	323 (46.9)	365 (53.0)	688 (51.5)
Male	343 (53.0)	304 (46.9)	647 (48.5)
Race/Ethnicity			
White	532 (39.9)	528 (39.6)	1060 (79.4)
Black	65 (4.9)	49 (3.7)	114 (8.5)
Hispanic	39 (3.0)	39 (3.0)	77 (6.0)
Other	20 (1.5)	15 (1.1)	35 (2.6)
2 or more races	26 (1.9)	23 (1.7)	49 (3.7)
Age			
55 to 64 years old	330 (24.7)	328 (24.6)	658 (49.3)
65 to 74 years old	229 (17.2)	236 (17.7)	465 (34.8)
75 or older	107 (8.0)	105 (7.9)	212 (15.9)
Education			
Less than high school	52 (7.8)	50 (7.5)	102 (7.6)
High school	244 (36.6)	209 (31.2)	453 (33.9)
Some college	195 (29.3)	197 (29.4)	392 (29.4)
Bachelor's degree or higher	175 (26.3)	213 (31.8)	388 (29.1)
Household Income			
$0–$49,999	283 (21.2)	309 (23.1)	592 (44.3)
$50,000–$99,999	229 (17.2)	207 (15.5)	436 (32.7)
$100,000–$149,999	98 (7.3)	107 (8.0)	205 (15.4)
$150,000 and above	52 (3.9)	50 (3.7)	102 (7.6)
Health Status			
Cancer (general)	104 (7.8)	99 (7.4)	203 (15.2)
Breast cancer	19 (1.4)	28 (2.1)	47 (3.5)
Prostate cancer	25 (1.9)	20 (1.5)	45 (3.4)
Gastrocoridalis	7 (0.5)	8 (0.6)	15 (1.1)

* $p < 0.05$.

Overall, participants rated the level of evidence as a "2" (i.e., D-grade) in both conditions (Table 3), and when the 13-point scale of evidence was collapsed into four grades of evidence, the results are consistent with a D level of evidence [53]. More than half of the participants were not at all confident in the ability of green tea/yukichi fruit juice to reduce the risk of cancer/gastrocoridalis (n = 709, 53.5%) (Mean (M)= 1.83, Standard Deviation (SD)= 1.10) but in both conditions, nearly two-thirds (n = 757, 57%) think there would be at least a slight risk reduction for cancer/gastrocoridalis (M = 1.98, SD = 1.10) if they drank the beverage. Most also think they would need to drink green tea/yukichi fruit juice once a week or more (n = 601, 80%) to reduce their risk. However, most would not buy green tea/yukichi fruit juice with the QHC (n = 755, 57.1%).

The majority of participants in the green tea condition reported that they were familiar with the green tea–cancer relationship (n = 343, 51.9%) but had never seen the QHC on a label of a green tea product (n = 480, 73.5%), in an advertisement or on a website or in an article (n = 453, 70.9%).

3.1. Purchase Intentions for Green Tea and Yukichi Fruit Juice

The diet–disease condition had a significant effect on perceptions of confidence, evidence, risk reduction, and purchase intentions (Pillai's Trace = 0.102, F = 36.896, df = 4, 1296, $p < 0.0001$).

A statistically significant effect was also found by QHC group on perceptions of confidence, evidence, risk reduction, and purchase intentions (Pillai's Trace = 0.138, F = 7.719, df = 24, 5196, $p < 0.0001$) as well as a significant interaction between the condition and group on the same variables (Pillai's Trace = 0.034, F = 1.834, df = 24, 5196, $p < 0.0001$).

However, these measures are strongly correlated (Table 5). A significant main effect of condition demonstrated that participants had greater purchase intentions for green tea than yukichi fruit juice $F(1, 1299) = 132.320$, $p < 0.0001$. There was also a significant main effect of QHC, such that claims written by Fleminger produced greater purchase intentions than claims written by the FDA, $F(6, 1299) = 8.047$, $p < 0.0001$. However, no interaction was found between condition and group, $F(6, 1299) = 1.713$, $p = 0.114$.

Table 5. Group means and standard deviations of ratings for evidence, cancer risk reduction, and confidence in the green tea–cancer relationship by QHC. Correlations (rs) between group perceptions of evidence, risk reduction, and confidence in the green tea–cancer relationship.

							Risk Reduction	**Confidence**
Measure	QHC	*n*	*Min*	*Max*	*M*	*SD*	r_s, *sig*	
Evidence	2004	95	1	10	4.18	2.67	0.536, p < 0.01	0.523, p < 0.01
	2008	93	1	13	3.91	2.69		
	2010	102	1	13	4.70	2.95		
	2005P	92	1	11	2.43	1.98		
	2005B	88	1	12	2.72	2.18		
	2011	96	1	8	2.38	1.77		
	2012	94	1	12	3.23	2.62		
Risk reduction	2004	95	1	5	2.65	1.07		0.804, p < 0.01
	2008	95	1	6	2.35	1.12		
	2010	102	1	5	2.59	1.18		
	2005P	93	1	5	1.81	1.09		
	2005B	90	1	5	1.73	0.88		
	2011	96	1	4	1.98	0.93		
	2012	94	1	5	2.06	1.09		
Confidence	2004	95	1	6	2.47	1.17		
	2008	95	1	5	2.22	1.09		
	2010	102	1	6	2.55	1.25		
	2005P	93	1	7	1.71	1.09		
	2005B	90	1	7	1.73	1.12		
	2011	96	1	5	1.86	1.00		
	2012	94	1	5	2.07	1.23		

Min = Minimum rating; Max = Maximum rating; M = mean; SD = standard deviation.

Significant differences were found between condition, with participants perceiving greater evidence for the green tea–cancer relationship than for the yukichi fruit juice–gastrocordialis relationship, $F(1, 1299) = 25.407$, $p < 0.0001$. There was also a significant main effect of the QHC group, with greater perceptions of evidence among claims written by Fleminger than those written by the FDA, $F(6, 1{,}299) = 25.491$, $p < 0.0001$. Irrespective of the QHC, participants reported greater evidence for the green tea–cancer relationship than for yukichi fruit juice and gastrocoridalis, suggesting there are other factors, in addition to the measures in the current study, that contribute to their perceptions. There was also a significant interaction between condition and QHC group on evidence ratings, meaning that participants who viewed a yukichi fruit juice–gastrocoridalis QHC rated the evidence lower than those who viewed a green tea–cancer QHC, regardless of the author, $F(6, 1299) = 2.804$, $p = 0.10$.

Similarly, there were main effects of both condition and QHC group on perceptions of disease reduction. Participants perceived greater reductions in cancer risk resulting from drinking green tea than gastrocoridalis from drinking yukichi fruit juice, $F(1, 1299) = 36.369$, $p < 0.0001$. Groups that viewed a Fleminger QHC also perceived greater reductions in disease risk than those who viewed a QHC written by others, $F(6, 1299) = 19.263$, $p < 0.0001$. A statistically significant interaction was also found between conditions and QHC groups on perceived risk reduction, F(6, 1299) = 3.085, $p = 0.005$. Participants' perceptions of gastrocoridalis risk did not differ among yukichi fruit juice QHCs, while the perceived cancer risk was different among green tea QHCs.

Finally, participants in the green tea condition were significantly more confident in the claimed relationship than those in the yukichi fruit juice condition, F(1, 1299) = 78.922, $p < 0.0001$. A main effect was also identified between QHC groups. Participants who viewed a Fleminger claim reported greater confidence in the claimed relationship than in other groups, F(6, 1,299) = 16.919, $p < 0.0001$. There was also a statistically significant interaction between QHC group and condition. Participants' confidence in the green tea condition was greater regardless of the QHC, whereas in the yukichi fruit juice–gastrocoridalis condition, participants were overall less confident, F(6, 1299) = 2.917, $p = 0.008$.

There were no statistically significant differences between the green tea and yukichi fruit juice conditions for dietary supplement use ($p = 0.993$), reported importance of health claims on dietary supplement or food labels ($p =.505$; $p = 0.138$, respectively), perceived nutrition knowledge ($p = 0.837$), worry about overall health ($p = 0.979$) that led to a dietary change in the past year ($p = 0.360$), worry about becoming sick with gastrocoridalis ($p = 0.240$) or cancer ($p = 0.066$), or a previous diagnosis of gastrocoridalis ($p = 0.800$) or cancer ($p = 0.677$).

3.2. Predictors of Purchase Intentions for Green Tea

To determine the factors' predictive value for purchase intentions of green tea, consumer-specific and sociodemographic variables were individually tested for the green tea condition only ($n = 666$). The significant sociodemographic predictors are age and race/ethnicity and account for 3.5% of the variation in purchase intentions for green tea, $F(5, 656) = 5.857$, $p < 0.0001$. Participants reported lower purchase intentions with each incremental increase in age, $b = -0.090$, $t(661) = -2.334$, $p < 0.05$. Black ($b = 0.126$, $t(656) = 3.274$, $p = 0.001$). Hispanic ($b = 0.141$, $t(656) = 3.670$, $p < 0.0001$) participants reported statistically significantly greater purchase intentions for green tea than White participants.

The predictive consumer specific variables accounted for 4.1% of the variance in purchase intentions and included: Dietary supplement use and the perceived importance of health claims on food products and on dietary supplement products, $F(3, 501) = 8.162$, $p < 0.0001$. For each incremental increase in dietary supplement use in the past year, participants reported that they are *less* likely to intend to purchase green tea, ($b = -0.114$, $t(503) = -2.582$, $p = 0.01$). Yet, the more important participants consider health statements on dietary supplement products, the greater their purchase intentions ($b = 0.135$, $t(501) = 2.209$, $p < 0.05$).

Measures of perceived susceptibility that significantly predict green tea purchase intentions are: General health, worry about overall health, worry about becoming ill with cancer, and worry that led to a dietary change in the past year, $F(4, 548) = 8.876$, $p < 0.0001$. The better participants rate their general health, the greater their purchase intentions for green tea, $b = 0.126$, $t(548) = 2.722$, $p < 0.0001$. Similarly, participants who are more worried about their overall health ($b = 0.105$, $t(550) = 2.228$) or about becoming sick with cancer ($b = 0.103$, $t(549) = 2.262$) demonstrated greater purchase intentions, $p < 0.05$. Participants have greater purchase intentions with each incremental increase in dietary changes made in response to a health worry in the past year, $b = 0.212$, $t(548) = 4.423$, $p < 0.0001$. Non-explanatory predictors were removed from further analysis; sex, education, employment, income, perceived nutrition knowledge, and past cancer diagnosis.

3.3. Model 1

A block of the modifying sociodemographic, consumer-specific, and perceived susceptibility predictors was entered as independent variables into a regression model to predict green tea purchase intentions. These variables included: Race/ethnicity, age, perceived importance of health claims on food products and on dietary supplement products, dietary supplement use, and self-reported health status, worry about overall health, worry about becoming ill with cancer, and worry that led to a dietary change in the past year. The model significantly predicted purchase intentions for green tea and accounted for 9.3% of the variance, $F(12, 406) = 4.581$, $p < 0.0001$ (Table 6).

Table 6. Hierarchical regression analysis for variables predicting purchase intentions for green tea.

Variable	*Model 1*					*Model 2*				
				95% CI					95% CI	
	B	*SE_B*	*Beta*	*Lower*	*Upper*	*B*	*SE_B*	*Beta*	*Lower*	*Upper*
Age	−0.029	0.091	−0.014	−0.207	0.150	0.017	0.081	0.009	−0.142	0.176
Race/Ethnicity										
Black	0.154	0.265	0.027	−0.367	0.676	−0.002	0.236	0.000	−0.467	0.463
Hispanic	0.487	0.254	0.087	−0.012	0.985	0.284	0.224	0.051	−0.155	0.723
Other, non-Hispanic	0.307	0.379	0.037	−0.437	1.051	0.271	0.335	0.033	−0.388	0.930
2+, non-Hispanic	0.281	0.342	0.037	−3.90	0.953	0.258	0.302	0.034	−0.335	0.852
DS Use	−0.163	0.055	−0.133 **	−0.272	−0.054	−0.121	0.049	−0.099 *	−0.218	−0.024
Importance of Health Claims										
Food Labels	−0.053	0.082	−0.042	−0.215	0.109	−0.081	0.073	−0.064	−0.224	0.062
DS Labels	0.169	0.077	0.138 *	0.017	0.321	0.131	0.068	0.107	−0.003	0.266
Health Status	0.057	0.088	0.034	−0.115	0.230	0.026	0.077	0.016	−0.126	0.179
Health Worry	−0.034	0.089	−0.022	−0.209	0.141	−0.014	0.079	−0.010	−0.169	0.140
Cancer Worry	0.062	0.072	0.044	−0.080	0.205	−0.024	0.064	−0.017	−0.149	0.101
Worry Diet Δ	0.155	0.075	0.115 *	0.008	0.301	0.142	0.065	0.106 *	0.014	0.271
Green Tea Consumption	0.200	0.045	0.213 **	0.111	0.288	0.171	0.040	0.182 **	0.093	0.250
Familiarity	0.272	0.066	0.200 **	0.143	0.402	0.048	0.061	0.035	−0.072	0.168
QHC Group										
QHC 2004	0.617	0.246	0.150 *	0.133	1.101					
QHC 2008	0.609	0.250	0.140 *	0.117	1.101					
QHC 2010	0.654	0.237	0.164 **	0.188	1.121					
QHC 2005b	0.126	0.2407	0.031	−0.346	0.599					
QHC 2011	0.162	0.246	0.039	−0.320	0.645					
QHC 2012	0.358	0.241	0.088	−0.115	0.832					
Evidence						0.130	0.026	0.238 **	0.079	0.181
Confidence						0.369	0.060	0.314 **	0.250	0.487
Adjusted R^2		0.202					0.372			
F for Δ in R^2		2.535*					60.891 **			

CI = confidence interval; *B* = unstandardized beta; SE_B = standard error for the unstandardized beta; *Beta* = standardized beta; * $p < 0.05$ ** $p < 0.01$; Reference categories: Race/ethnicity = White; QHC Group = 2005 p.

Participants who drank green tea in the past year indicated significantly greater purchase intentions ($M = 2.73$, $SD = 1.49$) than those who did not, *Welch's* $F(1, 640.890) = 87.810$, $p < 0.001$. By contrast, participants who had not consumed green tea in the past year indicated that they are only "slightly likely" ($M = 1.76$, $SD = 1.17$) to buy it in the future, despite having seen a QHC about the green tea–cancer relationship. As a result, current consumption was added to the model at step 2, with a resulting significant R^2 change = 0.052, $F(13, 403) = 6.404$, $p < 0.0001$, adj. $R^2 = 0.144$ (Table 6).

There were no differences between consumers who drank green tea in the past year and those who did not in terms of whether they had seen the claim on a label, *Welch's* $F(2,114.967) = 2.826$, $p = 0.063$, website, advertisement, or article, *Welch's* $F(2, 148.997) = 2.270$, $p = 0.107$. Yet those who drank green tea are significantly more familiar with the green tea–cancer relationship ($M = 2.11$, $SD = 1.10$) than those who did not ($M = 1.61$, $SD = 0.87$), *Welch's* $F(1, 642.179) = 42.389$, $p < 0.0001$. Accordingly, familiarity with the green tea–cancer relationship was added to the model at step 3. Results showed that familiarity with the green tea and cancer is a significant predictor of purchase intentions for green tea, $F(14, 401) = 7.685$, $p < 0.0001$, and accounted for 18.4% of the variance in the model, R^2 change = 0.041 (Table 6). At step 4, we then entered the QHCs, which were significant predictors of future behavior, $F(20, 395) = 6.263$, $p < 0.0001$, though they only accounted for a small amount of variance in the model, adj. $R^2 = 0.202$, R^2 change = 0.029 (Table 6). In the full model, the significant predictors of purchase intentions for green tea are: The perceived importance of health statements on dietary supplements, worry about health that led to dietary change in the past year, having consumed green tea in the past year, familiarity with the diet–disease relationship, and exposure to a QHC written by Fleminger. Dietary supplement use is negatively related to purchase intentions for green tea.

3.4. Model 2

To understand if perceptions of evidence for the green tea–cancer relationship, risk reduction for cancer, and confidence in the claimed relationship predict purchase intentions for green tea, three additional steps were added to the model: Ratings for evidence, risk reduction, and confidence in the green tea–cancer relationship. Next, a second model was created to isolate the effects of these variables from the QHCs and included steps 1 (i.e., sociodemographics, consumer-specific, and perceived susceptibility variables), 2 (i.e., past behavior), and 3 (i.e., familiarity), and the dummy-coded QHCs were removed. The following analysis represents steps 4, 5, and 6 for the second model.

Perception of evidence was found to be a strong and significant predictor of purchase intentions for green tea, $F(15, 399) = 19.781$, $p < 0.0001$, adj. $R^2 = 0.315$, R^2 change $= 0.128$, $p < 0.0001$. Confidence in the green tea–cancer relationship is also a significantly strong predictor of future intentions to purchase green tea, $F(16, 398) = 16.345$, $p < 0.0001$, adj. $R^2 = 0.372$, R^2 change $= 0.056$, $p < 0.0001$. Entry of the third variable, perception of cancer risk reduction, showed that the overall model was significant for predicting purchase intentions for green tea, $F(1, 400) = 18.736$, $p < 0.0001$, although the variable did not add any more to the model, adj. $R^2 = 0.371$, R^2 change $= 0.000$, $p = 0.896$ (Table 6).

However, evidence, risk, and confidence are multicollinear and demonstrate strong positive correlations (Table 5), which likely explains the weak predictive value of perceived risk reduction on purchase intentions. Therefore, a backwards stepwise regression model tested the three predictors in a block. The model removed risk reduction and maintained evidence perceptions and confidence in the green tea–cancer relationship, indicating that perceived disease risk reduction did not account for any additional variance and so was removed.

The two significant predictors were re-entered into the model (i.e., perceived evidence, confidence) at Step 4. As expected, this block was a strong predictor of purchase intentions, $F(16, 398) = 16.345$, $p < 0.0001$, adj. $R^2 = 0.372$, R^2 change $= 0.185$ (Table 6). For every incremental increase in perceived evidence, there was a 0.130 increase in purchase intention for green tea, and for every incremental increase in confidence in the green tea–cancer relationship, there was a 0.369 increase in purchase intention. While dietary supplement use remained negatively related to purchase intentions, the importance of health statements on supplements, worry that led to dietary change, past green tea consumption, and familiarity with the claimed relationship were predictive of purchase intention (Table 6).

Neither dose nor taste were tested in a predictive model, since no association was detected between perceived dose necessary to achieve risk reductions and purchase intentions, $\chi^2(24) = 29.845$, $p = 0.190$ or with the taste of green tea, $\chi^2(6) = 7.284$, $p = 0.295$.

To summarize, the significant individual predictors for purchase intentions were entered into the models at step (1) and include: Sociodemographics (race/ethnicity, age); consumer-specific variables (dietary supplement use, importance of health claims on food and supplement labels); and perceived susceptibility (general health, worry about health, worry about cancer, worry that led a dietary change). Model 1 significantly predicted purchase intentions and included steps (2) green tea consumption, (3) familiarity with green tea and cancer relationship, and (4) the seven QHCs, $F(16, 400) = 7.841$, $p < 0.0001$, adj. $R^2 = 0.208$, R^2 change $= 0.028$. Model 2 included steps (1–3), removed step 4 (i.e., QHCs from Model 1) and replaced it with (4) evidence perceptions and (5) confidence in the green tea–cancer relationship, which led to a statistically significant increase in R^2 of 0.185, $F(13, 401) = 20.225$, $p < 0.0001$, adj. $R^2 = 0.376$.

4. Discussion

With respect to their future intentions to purchase green tea, the data suggest that older consumers are influenced by their past behaviors and preferences. However, even after controlling for those preferences and experiences, exposure to the claims written by Fleminger in 2004, 2008, and 2010 demonstrated greater intentions to purchase green tea than other claims. Our results also demonstrated that some QHCs increase consumer perceptions of evidence for the diet–disease relationship, which affects their confidence in the relationship and purchase intentions for green tea.

Research has identified several sociodemographic characteristics that are associated with greater acceptance of functional foods with health claims. We expected some overlap between these same characteristics of consumers who are receptive to health claims, and QHCs.

In the current study, race or ethnicity was a strong predictor of purchase intentions for green tea when entered at step one in the regression model. Black and Hispanic consumers intended to purchase green tea more than White consumers. This finding is in contrast to previous research, which has pointed to White adults as more accepting of functional food products [40,42–44,55] although this difference may be explained in part by the presence of a QHC which is unique to the current study. However, race or ethnicity did not remain a significant predictor of purchase intentions for green tea in the full regression model(s).

Other research has indicated that women are more receptive to functional foods and to nutrition information in health claims [37,42]. However, the current study did not find sex to significantly predict purchase intentions for green tea. Further, no other sociodemographic variables demonstrated predictive value including education, employment, or income.

Previous research also demonstrates mixed results for age as a predictor of functional food acceptance with health claims [40,42–44]. For example, one study found that functional foods were accepted by women 35–54 years old [42], whereas another study identified a greater range of accepting adults, 30–70 years old [41]. Still, two other studies did not find any relationship between age and functional food acceptance [42,44].

Our study initially indicated age as a strong predictor for green tea purchase intentions in adults aged 55–64 but not for those over 65 years old. However, in the full model, age was not a significant predictor of purchase intentions. More than half of our sample already drinks green tea for the taste, not to reduce their risk of cancer. While there is strong evidence that taste is a predictor for purchase intentions and behavior related to health claims [38,52], it was not correlated with purchase intentions for green tea after viewing a QHC.

Purchase intentions for yukichi fruit juice were significantly lower than for green tea. Participants could not have had prior experience with yukichi fruit juice or gastrocordialis due to their fictitious or novel nature. Therefore, consumers considered other information when responding to questions. In addition, existing green tea consumption and familiarity with the green tea–cancer relationship were both significant predictors of behavioral intentions. These findings are consistent with previous research. Familiarity and experience with a product and/or a health condition have been found to significantly predict consumer behaviors [12,38,42,47–49], including purchase intentions for functional foods.

Dietary supplementation was negatively associated with the intention to purchase green tea, after viewing a QHC. This is in contrast to other research that indicates consumers who regularly take supplements are more accepting of functional foods [55], such as green tea [56].

Claim language is regarded as a predictor of purchase intentions for functional products with health claims [38], and our results support this theory. While the results of the current study cannot isolate the language or wording in QHCs that may have led to greater purchase intentions for green tea, the results show that QHCs written by Fleminger led to greater purchase intentions compared with other QHCs. However, the FDA concluded that the claims written by Fleminger overstate the level of scientific support for the green tea–cancer relationship [19,21]. Our results provide support for the agency's conclusion, demonstrating that these same QHCs produced greater perceptions of evidence and greater intentions to purchase green tea, compared with the other claims. The 2004 Fleminger QHC indicates, "There is scientific evidence supporting this health claim although the evidence is not conclusive", while the 2008 and 2010 Fleminger claims suggest the evidence is "credible but limited". Further, the 2010 claim identifies the FDA as the reviewer of evidence which, altogether, aimed to increase the trustworthiness of the claim. The 2005p and 2005b QHCs reference the number and quality of studies behind the claimed relationship which is similar to other diet–disease claims that the court found to be too technical for consumers to understand [20]. Although technically accurate, the claims are not consumer-friendly. Since QHCs were required by case law, it is possible that the agency sought to protect the consumer from being misled about the health value of a product by drafting overly technical claims [20,21]. The 2011 claim, which states that the FDA does not agree with the claimed relationship, provides insight into the agency's perspective [21]. The 2012 FDA QHC suggested by the federal court as a compromise between companies' abilities to market the health benefit of green tea and the FDA's mission to prevent consumer confusion about scientific certainty for the relationship [21] led to greater perceptions of evidence than the FDA QHCs in 2005p/b and 2011. The 2012 QHC also produced greater purchase intentions for green tea than the FDA QHCs, but the results were not statistically significant, which suggests the current QHC may strike a balance between interests of making profit for companies and the interest of the FDA in preventing consumer confusion.

The reported confidence in the diet–disease relationship was a very strong predictor of purchase intention. On its own, the perceived extent of the reduction in the risk of cancer was associated with intentions to purchase green tea. However, it did not contribute additional predictive value to the model of purchase intentions after accounting for the measure of confidence in the green tea–cancer relationship. Our results support the hypothesis that specific claims differentially affect the perceived health benefits of green tea products since perceptions of evidence and confidence in the green tea–cancer relationship were associated with variations in the QHC.

The overall proportion of variance in purchase intentions that can be explained by the independent variables was nearly 40% (Table 6). There are several additional measures in the HCF and HBM that were not included in the current study and so, it is possible that the addition of these measures could lead to models with greater predictive power of purchase intentions for green tea with a QHC.

There is considerable potential for QHCs in the market, since most US consumers believe that functional foods can reduce the risk of becoming sick with a specific disease and believe they have some control over their personal health [57]. Participants in our sample of older consumers had greater intentions to purchase green tea if in the past year, they worried about their health and if their worry led to a dietary change. Indeed, more than three-quarters of our sample changed their diet to address health concerns. Since US consumers perceive cancer as a greater risk than other diseases, including heart disease, type 2 diabetes, and stroke [58], there is likely a considerable marketing opportunity for products associated with reducing the risks of cancer. Future research could examine the complexities of QHCs through a more interdisciplinary lens. More specifically, collaborating with market researchers could expand the application of our results to general consumer behavior and consider its place within the theory of consumption.

However, it should be noted that the current study examines QHCs for the green tea–cancer relationship, a D-grade claim, which meets the lowest level of evidence for a diet–disease relationship [19]. Future study might examine science communication in B and C grade QHCs to gain greater understanding about existing claims as well as how federal lawsuits may impact the language of QHCs, which ultimately affects consumers. Although claims do not necessarily distinguish categorically healthy versus unhealthy foods [59], the perceived health quality of products by consumers may yield different perceptions of evidence and purchase intentions [60] with QHCs.

Strengths and Limitations

Few recent studies have examined how consumer perceptions change in response to different QHCs. This study is unique in its contribution to the understanding of the influence of QHCs on purchase intentions. Only one other study has explored purchase intentions with respect to product labels with QHCs but was limited in that the format tested has never been allowed for use on labels [12]. A major strength of this study is that it included QHCs written by different stakeholders, including Fleminger, the federal court, and the FDA, many of which have been used on labels or on the Internet.

One potential limitation of the study is that it did not include a control group that did not view any QHC. However, the existence of QHCs associated with green tea and a significant degree of prior familiarity with the green tea–cancer relationship among the participants made inclusion of a control group problematic. Instead, prior familiarity with the green tea–cancer relationship and with the QHC seen by the participant was used as a within-subjects control variable, creating a more ecologically valid design. That is, those who were unfamiliar with the relationship and had no prior familiarity with the QHC to which they were exposed serve as comparisons for those who were familiar with the relationship and/or the QHC. In addition, the study aimed to evaluate the differences in consumer perceptions among *actual* QHCs proposed or used by stakeholders and to test the assertions made by the various stakeholders about those claims. Unfortunately, members of a traditional control group would be unable to respond to the key questions about the level of perceived evidence and confidence or cancer risk reduction, based on a QHC. However, future research could test the differences between a control group and the current green tea QHC that is enforced by the FDA.

This study used a 2 × 7 between-subjects design. An alternative within-subjects study design was tested in which participants viewed several QHCs and responded to the same battery of questions. However, the time to complete the survey was considerably lengthened, threatened primacy effects, and increased participant burden, and so this design was rejected as impractical.

Finally, the study results are based on self-reported data which, in general, have a low to moderate correlation with observed behaviors [61,62]. There are likely differences between self-reported intentions and actual purchases of green tea with a QHC.

5. Conclusions

Overall, the QHCs impacted consumers' perceptions of evidence for a diet–disease relationship which impacted the health perceptions for that product, as well as purchase intentions. There is a segment of older, US consumers that positively respond to green tea QHCs. Consumers who have made a dietary change to address a health concern and consider health claims on supplements to be important reported greater purchase intentions than others. Consumers who perceived there to be greater evidence for the green tea–cancer relationship and reported greater confidence in the claimed relationship also intended to purchase green tea more than others. The three claims written by Fleminger, the green tea manufacturer, led to perceptions of greater evidence, perceived cancer risk reduction, and confidence in the claimed relationship, which also produced greater purchase intentions for green tea when compared with QHCs written and permitted by the FDA or the Court.

Author Contributions: Conceptualization, A.B.-G., W.K.H., and N.H.H.; Data curation, A.B.-G. and W.K.H.; Formal analysis, A.B.-G. and W.K.H.; Funding acquisition, W.K.H. and N.H.H.; Investigation, A.B.-G., W.K.H. and N.H.H.; Methodology, A.B.-G., W.K.H. and N.H.H.; Project administration, W.K.H.; Resources, W.K.H. and N.H.H.; Software, W.K.H.; Supervision, W.K.H.; Writing—original draft, A.B.-G.; Writing—review and editing, A.B.-G., W.K.H., and N.H.H.

Funding: The research was funded by the Research funded by Robert Wood Johnson Foundation Center for Health Policy (66487) and the Robert T. Rosen Scholarship from Rutgers, The State University of New Jersey.

Conflicts of Interest: The authors declare no conflict of interest.

References

1. Emord, J.W.; Schwitters, B. Do qualified health claims deceive when they are not misleading? Perspectives from the European Union and United States. *Food Drug Policy Forum* **2012**, *2*, 1–11.
2. *Pearson v. Shalala*, 164 F. 3d 650. 1999.
3. U.S. Food and Drug Administration. *Claims That Can Be Made for Conventional Foods and Dietary Supplements*. Available online: http://www.fda.gov/food/labelingnutrition/labelclaims/ucm111447.htm (accessed on 20 November 2011).
4. Berhaupt-Glickstein, A.; Hallman, W.K. Communicating scientific evidence in qualified health claims. *Crit. Rev. Food Sci. Nutr.* **2017**, *57*, 2811–2824. [CrossRef] [PubMed]

5. U.S. Food and Drug Administration. *Guidance for Industry: Health Claims on Calcium and Osteoporosis; and Calcium, Vitamin D, and Osteoporosis*. Available online: https://www.fda.gov/Food/GuidanceRegulation/GuidanceDocumentsRegulatoryInformation/LabelingNutrition/ucm152626.htm (accessed on 20 January 2018).
6. Williams, P. Consumer understanding and use of health claims for foods. *Nutr. Res.* **2005**, *63*, 256–264. [CrossRef]
7. Roe, B.; Levy, A.; Derby, B.M. The impact of health claims on consumer search and evaluation outcomes: Results from FDA experimental data. *J. Public Policy Mark* **1999**, *18*, 89–105. [CrossRef]
8. Freimuth, V.S.; Hammond, S.L.; Stein, J.A. Health advertising: Prevention for profit. *Am. J. Public Health* **1988**, *78*, 557–561. [CrossRef]
9. Levy, A.; Stokes, R. Effects of a health promotion advertising campaign on sales of ready-to-eat cereals. *Public Health Rep.* **1987**, *102*, 398–403.
10. Nakaweesa Munene, C. Analysis of Consumer Attitudes and Their Willingness to Pay for Functional Foods. Master's Thesis, Louisiana State University and Agricultural and Mechanical College, Baton Rouge, LA, USA, 2006.
11. Vecchio, R.; Van Loo, E.J.; Annunziata, A. Consumers' willingness to pay for conventional, organic and functional yogurt: Evidence from experimental auctions. *Int. J. Consum. Stud.* **2016**, *40*, 368–377. [CrossRef]
12. Reinhardt-Kapsak, W.; Schmidt, D.; Childs, N.M.; Meunier, J.; White, C. Consumer perceptions of graded, graphic and text label presentations for qualified health claims. *Crit. Rev. Food Sci. Nutr.* **2008**, *48*, 248–256. [CrossRef] [PubMed]
13. Berhaupt-Glickstein, A.; Nucci, M.L.; Hooker, N.H.; Hallman, W.K. The evolution of language complexity in qualified health claims. *Food Policy* **2014**, *47*, 62–70. [CrossRef]
14. Lin, J.C. How do consumers interpret health messages on food labels? *Nutrition Today* **2008**, *43*, 267–272. [CrossRef]
15. Crowell, A.; Schunn, C. Scientifically literate action: Key barriers and facilitators across context and content. *Public Underst. Sci.* **2014**, *23*, 718–733. [CrossRef]
16. Moreira, M.J.; Garcia Diez, J.; M de Almeida, M.M.; Saraiva, C. Evaluation of Food Labelling Usefulness for Consumers. *Int J Consum Stud.* **2019**. [CrossRef]
17. Hieke, S.; Kuljanic, N.; Pravst, I.; Miklavec, K.; Kaur, A.; Brown, K.A.; Egan, B.M.; Pfeifer, K.; Gracia, A.; Rayner, M. Prevalence of nutrition and health-related claims on pre-packaged foods: A five-country study in Europe. *Nutrients* **2016**, *8*, 137. [CrossRef]
18. Government Accountability Office. *Food Labeling: FDA Needs to Reassess Its Approach to Protecting Consumers from False or Misleading Claims*. Available online: http://www.gao.gov/products/GAO-11-102 (accessed on 9 February 2012).
19. U.S. Food and Drug Administration. *Letter Updating The Green Tea and Risk of Breast Cancer and Prostate Cancer Health Claim April 17, 2012*. Available online: http://www.fda.gov/Food/LabelingNutrition/LabelClaims/QualifiedHealthClaims/ucm301644.htm (accessed on 10 November 2012).
20. *Alliance for natural health U.S. v. Sebelius*, 714 F.supp.2d 48. 2010.
21. Fleminger, Inc. v. US Department of Health and Human Services. 2012. Available online: https://www.hhs.gov/blog/authors/hhs (accessed on 20 January 2018).
22. U.S. Food and Drug Administration. *Consumer Health Information for Better Nutrition Initiative—Attachment D Consumer Studies Research Agenda*. Available online: http://www.fda.gov/Food/IngredientsPackagingLabeling/LabelingNutrition/ucm096079.htm (accessed on 9 June 2013).
23. European Food Information Council. *Can Health Claims and Symbols Lead to Healthier Eating Habits?* Available online: http://www.eufic.org/jpage/en/page/PRESS/fftid/clymbol-healthclaims-symbols/ (accessed on 15 January 2015).
24. American Cancer Society. *Cancer Facts and Figures 2015*. Available online: http://www.cancer.org/research/cancerfactsstatistics/cancerfactsfigures2015/index (accessed on 1 July 2015).
25. King, D.E.; Matheson, E.; Chirina, S.; Shankar, A.; Broman-Fulks, J. Research letter: The status of baby boomers' health in the United States: The healthiest generation? *JAMA Intern. Med.* **2013**, *173*, 385–386. [CrossRef]
26. Colby, S.L.; Ortman, J.M. *The Baby Boom Cohort in the United States: 2012 to 2060. Population Estimates and Projections*. Available online: https://www.census.gov/prod/2014pubs/p25-1141.pdf (accessed on 31 March 2017).

27. Nielsen. *Introducing Boomers. Marketing's Most Valuable Generation*. Available online: http://www.nielsen.com/content/dam/corporate/us/en/reports-downloads/2012-Reports/nielsen-boomers-report-082912.pdf (accessed on 31 March 2017).
28. Nocella, G.; Kennedy, O. Food health claims—What consumers understand. *Food Policy* **2012**, *37*, 571–580. [CrossRef]
29. Academy of Nutrition and Dietetics. *Nutrition and You: Trends 2011 Public Opinion on Food and Nutrition: 20 Years of Insights*. Available online: http://38.98.131.45/nutritiontrends/#.WsnTyhNubUI (accessed on 20 November 2011).
30. Govindasamy, R.; Italia, J. *Evaluating Consumer Usage of Nutritional Labeling: The Influence of Socio-Economic Characteristics*. Available online: https://ageconsearch.umn.edu/bitstream/36734/2/pa990199.pdf (accessed on 28 April 2017).
31. Balluz, L.S.; Kieszak, S.M.; Philen, R.M.; Mulinare, J. Vitamin and mineral supplement use in the United States. Results from the third national health and nutrition examination survey. *Arch. Fam Med.* **2000**, *9*, 258–262. [CrossRef]
32. Gray, S.L.; Hanlon, J.T.; Fillenbaum, G.G.; Wall, W.E., Jr.; Bales, C. Predictors of nutritional supplement use by the elderly. *Pharmacotherapy* **1996**, *16*, 715–720. [PubMed]
33. Puhakka, R.; Valve, R.; Sinkkonen, A. Older consumers' perceptions of functional foods and non-edible health-enhancing innovations. *Int. J. Consum. Stud.* **2017**, *42*, 111–119. [CrossRef]
34. Crowe, K.M.; Francis, C. Position of the academy of nutrition and dietetics: Functional foods. *J Acad. Nutr. Diet.* **2013**, *113*, 1096–1103. [CrossRef] [PubMed]
35. Dobrenova, F.; Terlutter, R. The effect of familiarity on consumer ratings of the scientific support of health claims on food. *J. Food Prod. Market.* **2015**, *21*, 569–587. [CrossRef]
36. Hatch, C.D.; Buttrick, H. The impact of qualified health claims on advertising evaluations: The cases of POM wonderful and minute maid. *J. Consumer Policy* **2019**. [CrossRef]
37. Lahteenmaki, L. Claiming health in food products. *Food Qual. Prefer.* **2013**, *27*, 196–201. [CrossRef]
38. Wills, J.M.; Storcksdieck genannt Bonsmann, S.; Kolka, M.; Grunert, K.G. Symposium 2: Nutrition and health claims: Help or hindrance European consumers and health claims: Attitudes, understanding and purchasing behavior. *Proc. Nutr. Soc.* **2012**, *71*, 229–236. [CrossRef] [PubMed]
39. Contento, I.R. *Nutrition Education: Linking Research, Theory, and Practice*, 2nd ed.; Jones and Bartlett Publishers: Sudbury, MA, USA, 2007.
40. Gilbert, L. Marketing functional foods: How to reach your target audience. *AgBioForum* **2000**, *3*, 20–38.
41. Glanz, K.; Rimer, B.K.; Viswanath, K. *Health Behavior and Health Education: Theory, Research, and Practice*, 4th ed.; Jossey-Bass: San Francisco, CA, USA, 2008.
42. Pothoulaki, M.; Chryssochoidis, G. Health claims: Consumers' matters. *J. Funct. Foods* **2009**, *1*, 222–228. [CrossRef]
43. Cox, D.N.; Evans, G.; Lease, H.J. The influence of product attributes, consumer attitudes and characteristics on the acceptance of: (1) novel bread and milk, and dietary supplements and (2) fish and novel meats as dietary vehicles of long chain omega 3 fatty acids. *Food Qual. Prefer.* **2011**, *22*, 205–212. [CrossRef]
44. Contini, C.; Casini, L.; Stefan, V.; Romano, C.; Jørn Juhl, H.; Lähteenmäki, L.; Grunert, K.G. Some like it healthy: Can socio-demographic characteristics serve as predictors for a healthy food choice? *Food Qual. Prefer.* **2015**, *46*, 103–112. [CrossRef]
45. Murphy, D. *Working Paper No. 277: Consumer Perceptions of Qualified Health Claims in Advertising*. Available online: https://www.ftc.gov/sites/default/files/documents/reports/consumer-perceptions-qualified-health-claims-advertising/wp277_0.pdf (accessed on 10 April 2019).

46. U.S. Food and Drug Administration. Experimental study of qualified health claims: Consumer inferences about monounsaturated fatty acids from olive oil, EPA and DHA omega-3 fatty acids, and green tea. OMB no. 0910-0952. Available online: http://www.fda.gov/Food/IngredientsPackagingLabeling/LabelingNutrition/ucm207549.htm (accessed on 27 April 2017).
47. Dean, M.; Lampilaa, P.; Shepherd, R.; Arvola, A.; Saba, A.; Vassallo, M.; Claupein, E. Perceived relevance and foods with health-related claims. *Food Qual. Prefer.* **2012**, *24*, 129–135. [CrossRef]
48. Hasler, C.M. Health claims in the United States: An aid to the public or a source of confusion? *J. Nutr.* **2008**, *138*, S1216–S1220. [CrossRef]
49. Saldanha, L.G. The qualified health claims quandry. *Nutrition Today* **2006**, *41*, 54–55. [CrossRef]
50. Mazis, M.B.; Raymond, M.A. Consumer perceptions of health claims in advertisements and on food labels. *J. Consumer Aff.* **1997**, *31*, 10–26. [CrossRef]
51. International Food Information Council. *2011 Functional Foods/Foods for Health Consumer Trending Survey*. Available online: http://www.foodinsight.org/Resources/Detail.aspx?topic=2011_Functional_Foods_Foods_For_Health_Consumer_Trending_Survey (accessed on 20 November 2013).
52. Nielsen, & National Marketing Institute. *Health and Wellness in America, August 2014*. Available online: http://www.nielsen.com/us/en/insights/reports/2014/health-and-wellness-in-america-the-consumer-perspective.html (accessed on 4 August 2017).
53. Anders, S.; Schroeter, C. Estimating the effects of nutrition label use on Canadian consumer diet-health concerns using propensity score matching. *Int. J. Consum. Stud.* **2017**, *41*, 534–544. [CrossRef]
54. Field, A. *Discovering Statistics Using IBM SPSS Statistics*; Sage: Los Angeles, CA, USA, 2013.
55. Reinhardt-Kapsak, W.; Rahavi, E.B.; Childs, N.M.; White, C. Functional foods: Consumer attitudes, perceptions, and behaviors in a growing market. *J. Am. Diet. Assoc.* **2011**, *111*, 804–810. [CrossRef] [PubMed]
56. Berhaupt-Glickstein, A.; Hallman, W.K. A profile of older green tea consumers in the USA. *Br. Food J.* **2017**, *119*, 2932–2944. [CrossRef]
57. International Food Information Council. *2013 Food and Health Survey*. Available online: http://www.foodinsight.org/LinkClick.aspx?fileticket=34E2C49BOcc%3d&tabid=1480 (accessed on 1 June 2013).
58. Wang, C.; O'Neill, S.M.; Rothrock, N.; Gramling, R.; Sen, A.; Acheson, L.S.; Rubenstein, W.S.; Nease, D.E., Jr.; Ruffin, M.T., 4th. Family Healthware Impact Trial (FHITr) group. Comparison of risk perceptions and beliefs across common chronic diseases. *Prev. Med.* **2009**, *48*, 197–202. [CrossRef]
59. Schaefer, D.; Hooker, N.H.; Stanton, J.L. Are Front of Pack Claims Indicators of Nutrition Quality? Evidence from 2 Product Categories. *J. Food Sci.* **2016**, *81*, H223–H234. [CrossRef]
60. Talati, Z.; Pettigrew, S.; Dixon, H.; Neal, B.; Ball, K.; Hughes, C. Do health claims and front-of-pack labels lead to a positivity bias in unhealthy foods? *Nutrients* **2016**, *8*, 787. [CrossRef]
61. Kendall, P.A.; Elsbernd, A.; Sinclair, M.; Schroeder, M.; Chen, G.; Bergmann, V.; Hillers, V.N.; Medeiros, L.C. Observation versus self-report: Validation of a consumer food behavior questionnaire. *J. Food Prot.* **2004**, *67*, 2578–2586. [CrossRef]
62. Prince, S.A.; Adamo, K.B.; Hamel, M.E.; Hardt, J.; Connor Gorber, S.; Tremblay, M. A comparison of direct versus self-report measures for assessing physicaly activity in adults: A systematic review. *IJBNPA* **2008**, *5*, 1–24.

nutrients

MDPI

Article

Perceived Correspondence of Health Effects as a New Determinant Influencing Purchase Intention for Functional Food

Ágoston Temesi [1,*], Ágnes Bacsó [1], Klaus G. Grunert [2] and Zoltán Lakner [1]

1 Department of Food Economics, Faculty of Food Science, Szent István University, Villányi str. 29-43, 1118 Budapest, Hungary; agi.bacso@gmail.com (Á.B.); lakner.zoltan@etk.szie.hu (Z.L.)

2 MAPP Centre, Department of Management, Aarhus BSS, Aarhus University, Fuglesangsalle 4, DK-8210 Aarhus V, Denmark; klg@mgmt.au.dk

* Correspondence: temesi.agoston@etk.szie.hu; Tel.: +36-1-305-7100 (ext. 6178)

Received: 8 February 2019; Accepted: 28 March 2019; Published: 29 March 2019

Abstract: This study has revealed the role of a new factor, perceived correspondence of health effects, in consumer acceptance of functional foods. Using a web survey of 1016 people, we hypothesized and verified the following: when an ingredient does not occur naturally in the carrier but the consumer assigns the same health effect to it as to the carrier, the product's acceptance will be more positive than it would be if an identical health effect was not associated with the carrier and the functional ingredient. Factors influencing consumer acceptance were examined via binary logistic regression models. According to the results, if a functional food developer fortifies the carrier with an ingredient that does not occur naturally in the carrier, the product can expect higher acceptance if the health effects perceived by consumers are properly matched. In general, it has been found that expected taste and awareness of the product were decisive in all demographic and income groups, whereas perceived correspondence of health effects had a lesser, but still positive influence on acceptance.

Keywords: consumer research; carrier foods; functional ingredients; loglinear analysis

1. Introduction

Two decades ago Childs [1] claimed that the success of functional foods depends on their consumer acceptance. In spite of comprehensive research on consumer attitudes to functional foods, a decade later Siró et al. [2] (p. 456) still claimed that "the development and commerce of these products is rather complex, expensive and risky, as special requirements should be answered". Nowadays food fortification is still a significant trend all around the world, which is important for developing countries and for vulnerable consumer groups as well e.g., [3–5]. This increases the need for revealing new functional food product development opportunities that keep consumer behavior in mind.

It quickly became obvious in former research that one of the most important aspects of consumer acceptance is taste or expected taste of the product [6–10]. In fact, several studies have concluded that the consumer is not likely to compromise on the product's taste in exchange for its beneficial health effects [11–16]. As Lalor et al. [15] (p. 57) express: "participants were not prepared to purchase foodstuffs if they did not "taste good", irrespective of health or any other issue". Although Urala and Lähteenmäki [17] remark that some types of food might have such a strong effect on health that the consumer may renounce their requirements regarding taste, Verbeke's [18] cross-sectional research confirmed the views of the majority of the scholars, namely that the customer group that accepts an unpleasant taste of a functional food in exchange for its health-promoting effects is rather limited. Verbeke [18] (pp. 130–131) states: "hoping for consumer willingness to compromise on the taste of functional foods for health is highly speculative, risky, and deemed to yield a niche market strategy".

Several studies have examined how perception of the healthiness status of a carrier influences the acceptance of the functional food. Most of the scholars conclude that consumers more readily accept functional food based on carriers considered healthy [19–25]. Only Bech-Larsen and Grunert [26] argue with this opinion. In their research, consumers thought fortification of foods considered unhealthy was more reasonable than that of healthy products.

As for the choice of functional ingredient, studies have shown that consumers appreciate if the ingredient used in the enrichment can also be found naturally in the carrier food (e.g., adding calcium to milk, or adding fiber to rye-bread) [7,21,27]. Finally, familiarity with a given functional ingredient significantly influences the acceptance of the food [27]. Urala and Lähteenmäki [8], Messina et al. [28] as well as Krutulyte et al. [29] do not separate familiarity with the new ingredient from familiarity with the whole food, and emphasize the importance of familiarity with the specific functional food.

Based on this earlier work, Krutulyte et al. [29] developed a conceptual framework for the acceptance of functional foods where the perceived fit of carrier and ingredient is a major determinant of purchase intention, together with health concern and attitude to functional foods. The connection between attitude to fortification and purchase intention has been confirmed recently by e.g., Jahn et al. [30] and Bromage et al. [3], while the connection of perceived fit and purchase intention has been confirmed by Lu [31] and Bruschi et al. [25]. This perceived fit in turn is affected by the expected taste of the functional food, the health image of the carrier, the natural presence of the selected ingredient in the carrier, and finally, familiarity with the product/ingredient.

2. Conceptual Model and Aim of Research

The purpose of the present research was to investigate a new determinant that can influence consumers' intention to buy functional foods, thus extending the elements of the conceptual model of Krutulyte et al. [29]. The new determinant, called "perceived correspondence of health effects", is an extension of previous research on perceived fit of ingredient and carrier. Even though an ingredient does not naturally occur in the carrier, the health effect of the two (ingredient and carrier) can be perceived as identical. According to our hypothesis, consumers prefer ingredient/carrier combinations where the ingredient added to the carrier possesses the same perceived health effect as the carrier itself, compared to those where the perceived health effects of carrier and ingredient are different. The research of Krutulyte et al. [29] indicated a connection between purchase intention and perceived fit of carrier–ingredient combination, suggesting that a "correspondence of health effects" would influence the purchase intention for functional foods indirectly.

The hypotheses of the research are visualized in Figure 1.

Figure 1 shows our conceptual framework. It is based on the original work by Krutulyte et al. [29] extended with our hypothesis and indicating the references which confirm the effects of the different influencing factors. We hypothesize that the perceived match of the health effects of carrier and ingredient, together with the naturalness of the ingredient in the carrier, affects the perceived overall health effect of carrier/ingredient combination. This is based on the argument that an increased health effect of the product can be obtained in two ways: by increasing the amount of the healthy ingredient naturally contained in the product (enrichment), or by adding a new component which has the same perceived health effect as the product (fortification). Our main hypothesis is that the perceived correspondence of health effects in a carrier/ingredient combination directly increases the perceived fit of carrier/ingredient combination.

The aim of research has been the empirical testing of this conceptual model and notably our main hypothesis. The ultimate aim of research has been to empirically test a new approach to functional products.

If this hypothesis was supported, it would expand the options of product developers, since not only enriching the product with ingredients which naturally occur in the carrier will be appropriate, but also all ingredients to which consumers assign the same health effect as to the carrier.

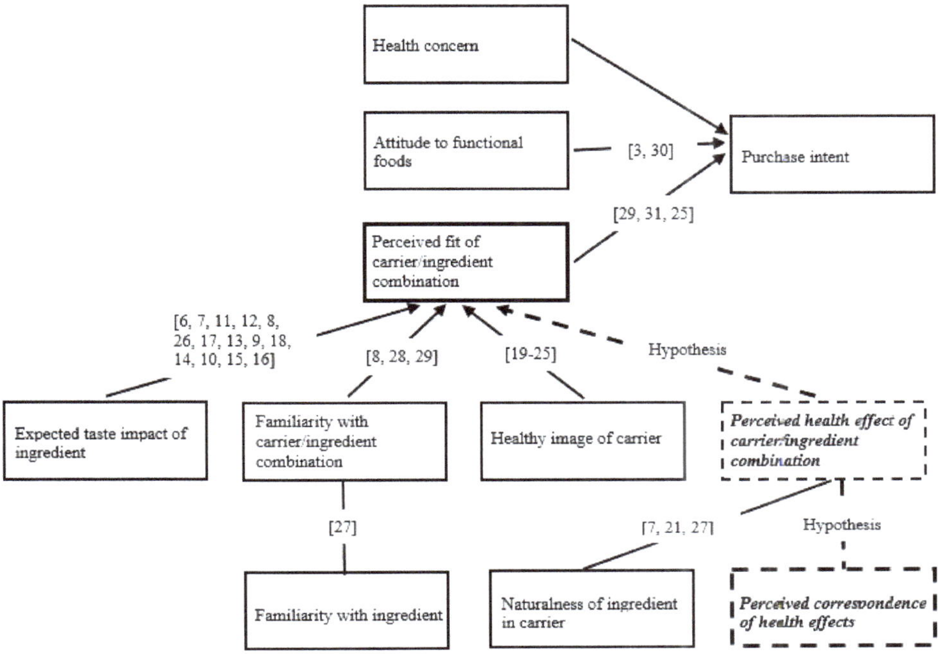

Figure 1. Determinants (confirmed and hypothesized) influencing purchase intent for functional foods.

3. Materials and Methods

A non-representative consumer survey was carried out to investigate the research hypothesis. Respondents were asked to complete a 10-min online questionnaire in the first two weeks of October 2015. We shared the Hungarian language survey in social media using a snowball sampling method started with Hungarians.

Respondents were informed that the aim of the questionnaire was to measure factors that might influence their purchase intent for functional foods. All respondents filled out anonymous questionnaires and they could quit without submitting it at any point of the questionnaire. Respondents filled out the questionnaire as volunteers and did not receive any incentives.

3.1. Characteristics of the Sample

1016 people provided valid answers. The composition of the sample is shown in Table 1.

Participants in the online survey were mostly women, between the ages of 18 and 29, and with relatively high education (secondary or tertiary); facts that can be attributed to the methodology of the survey and to these groups' interest in the issue as we have used the social media to recruit respondents. Recent findings however show that functional components are more important for women and better educated people [32].

3.2. Composition of the Questionnaire

Five carriers have been selected for the product combinations (100% pure orange juice, muesli bar from oat-flakes, natural yogurt, dark chocolate with at least 70% cocoa content, rye-bread), which were combined with 11 types of functional ingredients (vitamins A, E, C, and D, calcium, magnesium, probiotics, linseed, oat-flakes, ginseng, caffeine). Of the 55 possible carrier/ingredient combinations, only those were retained where the ingredient does not naturally occur in the carrier or occurs only in traces according to nutritional experts we have consulted. This resulted in 28 remaining combinations as shown in Table 2.

Table 1. Socio-demographic characteristics of the sample.

Socio-Demographic Characteristics	%
Gender	
female	88.7
male	11.3
Age group	
18–29	51.4
30–44	20.7
45 or older	27.9
Residence	
village	15.7
town/city	43.1
capital	41.1
Education	
primary	6.3
secondary	49.5
tertiary	44.2
Perceived income	
below average	17.2
average	67.1
above average	15.7

Table 2. Carrier foods and the ingredients combined with them.

Carrier Foods	100% Pure Orange Juice	Muesli Bar from Oat-Flakes	Natural Yogurt	Chocolate with at Least 70% Cocoa Content	Rye-Bread
Functional ingredients	Vitamin A Vitamin D Vitamin E Calcium Ginseng	Vitamin C Vitamin D Magnesium Probiotics Linseed	Vitamin E Oat-flakes Linseed Ginseng	Vitamin C Vitamin D Calcium Magnesium Probiotics Ginseng Caffeine	Vitamin C Vitamin D Calcium Oat-flakes Linseed Probiotics Ginseng

The questionnaire contained measures corresponding to the constructs in Figure 1. The questions of the questionnaire were developed by the authors.

The perceived healthiness of the carrier food was measured on a semantic differential scale from 1 (very unhealthy) to 5 (very healthy). In a similar way, the expected taste of product combinations was measured on a semantic differential scale from 1 (having a very bad taste) to 5 (very tasty). Familiarity was measured by asking whether respondents had seen the product combination before and, if yes, if they had tasted it.

In order to measure the perceived health effects of the 5 carriers and 11 functional ingredients, five statements about possible health effects were formulated after consultation with nutritional experts. From these respondents could choose the ones they associated with the ingredients and carriers. The statements were the following: "strengthens the immune system and adds to general well-being", "helps digestion and gastrointestinal functions", "increases mental performance", "supports the formation of healthy bone weight and the health of bones", and finally "can reduce the risk of heart and other cardiovascular diseases, reduces the cholesterol level of the blood". Respondents could also choose "I do not know" answer. Based on these data the variable "perceived correspondence of health effects" was formed. It had a score of 1 when the respondent marked at least one common health effect for both the carrier of the combination and the functional ingredient and a score of 0 otherwise.

In the next part of the questionnaire, respondents were asked to evaluate the 28 product combinations on the extent to which the carrier and the ingredient fit. Again, a scale from 1 (does not think the carrier and ingredient fit) to 5 (believe that the carrier and ingredient fit appropriately) was applied.

Before the beginning of the web-based consumer survey, the questionnaire was pre-tested in individual interviews to check the wording and clarity of the questions.

3.3. Data Analysis

Preliminary analyses of the scales used for the determinants indicated that they have a skewed distribution, with answers concentrated at the higher or lower scale values. For further analysis, the scale values were therefore transformed to binary variables. For the perceived healthiness of the carrier food, the results were transformed such that values of 4 and 5 were converted into a value of 1, indicating a positive consumer judgement on healthiness, while smaller values were converted into 0, indicating a low level of perceived healthiness. Likewise, for expected taste, scores 4 and 5, showing that respondents believe the product combination has a good taste, were converted into 1, while lower values were converted into 0. Also, for ratings of perceived fit between carrier product and ingredient, values 4 and 5 (indicating good fit) were transformed to 1, whereas lower values (indicating little perceived fit) were converted to 0.

Familiarity was coded based on previous exposure to the combination. Having seen but not tasted the product and having seen and tasted the product were recoded into 1 and named as awareness, and having neither seen nor tasted the product was recoded into 0.

A new variable "correspondence of health effects" was formed which was assigned the value 1 if the respondent marked at least one common health effect for both the carrier of the combination and the functional ingredient, and 0 otherwise.

A pooled set of data was formed. The original table recorded 1016 rows of observations. The analysed four determinants and the purchase intention for 28 product combinations as well as the demographic variables of respondents were listed. We "unstacked" the data into stacked form. In this way, all of the data for a determinant (e.g., perceived correspondence of health effects) have been grouped together in a single column and finally, we had 28448 (1016 × 28) records. This data structure allowed us to analyse the effect of different factors.

Determinants of perceived fit between carrier product and ingredient were examined by binary logistic regression models. Because of the large number of predictors, and to reduce the distorting effect caused by the relations among them, binary logistic regression with stepwise backward elimination method was applied. The removal testing has been based on the probability of the likelihood-ratio statistic ($p \leq 0.05$), based on the maximum partial likelihood estimates. The analysis was carried out individually for all 28 carrier/ingredient combination and for the pooled data. After assessing the influencing factors for the whole sample, we repeated the analysis for different demographic groups. The effects of determinants on acceptance of product combinations in different socio-economic groups were also estimated based on the pooled data. After the linear regression analysis, we simply counted and summarized the number of significant differences between various factors. All variables were also checked for relationships with demographic characteristics. All statistical analyses were done at the 5% significance level ($p \leq 0.05$).

4. Results and Discussion

4.1. Role of Determinants in Perceived Carrier/Ingredient Fit

The main aim of our research was to test whether correspondence of health effects has an impact on perceived fit between carrier and ingredient, and to compare the size of the effect with other determinants that in previous research have been shown to have an effect on perceived fit. Results of the binary logistic regressions are in Table 3.

Table 3. The role of the influencing factors in the fit perception of 28 functional food carrier/ingredient combinations (n = 1016, 28 answers per respondent).

	Perceived Correspondence of Health Effects (expB)	Health Image of Carrier (expB)	Expected Taste of Combination (expB)	Awareness of the Combination (expB)
Orange juice + vitamin A	1.329 *	2.200 **	2.928 **	1.800 **
Orange juice + vitamin E	1.449 **	2.007 **	2.723 **	1.482 *
Orange juice + ginseng			3.493 **	
Orange juice + calcium		2.273 **	2.728 **	2.138 **
Orange juice + vitamin D		1.979 **	2.626 **	2.317 **
Muesli bar + probiotics	1.668 **		2.917 **	1.480 *
Muesli bar + linseed	2.160 **		5.546 **	1.604 **
Muesli bar + vitamin C	1.422 *		3.706 **	1.691 **
Muesli bar + magnesium	1.700 **	1.379 *	3.360 **	1.417 *
Muesli bar + vitamin D		1.361 *	3.967 *	
Yogurt + linseed	1.625 **		5.033 **	2.161 **
Yogurt + oat-flakes	1.887 **	1.740 *	5.040 **	4.589 **
Yogurt + vitamin E	1.606 **	2.480 **	2.810 **	1.620 **
Yogurt + ginseng		1.807 *	4.328 **	2.632 **
Dark chocolate + caffeine			3.565 **	2.642 **
Dark chocolate + ginseng		1.586 **	3.764 **	
Dark chocolate + calcium		1.565 **	2.770 **	1.428 *
Dark chocolate + magnesium		1.305 *	2.980 **	
Dark chocolate + vitamin C		1.458 **	3.608 **	2.189 **
Dark chocolate + probiotics		1.483 **	3.375 **	
Dark chocolate + vitamin D			3.912 **	2.083 *
Rye-bread + linseed	1.880 **	1.531 *	4.859 **	2.734 **
Rye-bread + oat-flakes	2.041 **	1.527 *	4.439 **	2.924 **
Rye-bread + calcium	1.550 *	1.604 **	2.866 **	2.567 **
Rye-bread + vitamin C		1.537 *	3.728 **	2.412 **
Rye-bread + probiotics	1.475 *		3.634 **	
Rye-bread + ginseng		1.577 **	5.199 **	2.480 **
Rye-bread + vitamin D	1.440 *	1.613 **	3.179 **	2.021 *
All carrier/ingredient combinations	1.330 **	1.425 **	3.606 **	2.556 **

* $p \leq 0.05$; ** $p \leq 0.01$.

The influence of the new determinant, perceived correspondence of health effects, on consumer perception of carrier/ingredient fit is different both by product group and by each individual combination. The strongest effect can be found in the following combinations, in descending order: muesli-bar–linseed, rye-bread–oat-flakes, yogurt–oat-flakes, and rye-bread–linseed. Consumer opinion least depends on perceived correspondence of health effects in the case of the combinations of dark chocolate (independently of the applied bioactive component) and of products containing ginseng.

The role of correspondence of health effects could not be shown in any of the products fortified with ginseng, which is probably due to the fact that this herb is not well- known in Hungary, so its effects also remain hidden from the public. As for dark chocolate, Balasubramanian and Cole [19] provide an explanation. Their research found that for products belonging to the category "fun foods", nutritional information is less important than the need for a good taste, so the researchers do not recommend their use as carriers. This is in line with the results of the present research.

The results support the importance of the health image of the carrier food in consumer perception of functional foods, a determinant that has been researched by many scholars. An effect is found in several combinations of all five food groups; at the same time, independently of the combinations, a significant link has been demonstrated between considering a carrier food healthy and attitudes towards its carrier/ingredient combination. Based on our results, we share the opinion of other scholars [19–24] that the perception and market success of a functional food is enhanced if the fortified carrier is well-known for being healthy.

Expected taste turns out to be the most important determinant for all examined combinations and also in general. The results are in accordance with the findings of previous research e.g., [6–14,18].

Among the carriers, the role of taste is most prominent in the examined combinations of yogurt, where the chance that the consumer perceives a fit between carrier and ingredient is 4.3 times higher if the resulting taste is expected to be pleasant. This value is significantly higher than the one for the examined combinations of orange juice, where an expected good taste results in a 3-fold increase in perceived fit. In the combinations of the muesli bar, taste plays the most important role, too; nonetheless, awareness of the combinations was also very important, and for some combinations (linseed, magnesium, probiotics), the perceived correspondence of health effects also had a high chance quotient.

Awareness is a determining factor mostly in the case of product concepts containing yogurt, rye-bread, and oat-flakes, or linseed. The results concerning awareness of the combinations confirm the findings of Grunert et al. [27], according to which consumer attitude (in the researched combinations and in general as well) is more favourable if the respondent has already met the carrier–ingredient combination earlier. This underlines the importance of marketing communication or any other activity that focuses on introducing the product to consumers or informing the public about the product.

4.2. Role of Determinants in Perceived Carrier/Ingredient Fit in Case of Socio-Demographic Groups

The factors influencing the perception of the 28 carrier/ingredient combinations have been examined according to the respondents' gender, age group, residence, schooling and income perception. Table 4 shows how many of the 28 examined product combinations were significantly influenced by the 4 predictors within the different demographic subgroups.

Table 4. The number of functional food combinations (out of 28) where perception of carrier/ingredient fit was significantly affected by predictors within different demographic groups.

	Perceived Correspondence of Health Effects	Health Image of the Carrier	Expected Taste of the Combination	Awareness of the Combination
male	0	2	26	6
female	14	17	28	22
ages 18–29	9	5	28	15
ages 30–44	5	15	27	11
45 or older	9	13	28	8
village	4	8	25	6
town/city	7	15	28	16
capital	6	7	28	15
primary education	2	5	25	2
secondary education	13	7	28	22
tertiary education	6	10	28	10
income below average	1	3	28	8
average income	11	13	28	17
income above average	3	11	26	9

* $p \leq 0.05$; ** $p \leq 0.01$.

Results show that in almost all of the examined 28 product combinations, expected taste was significantly related to perceived fit in all respondent groups. It is also obvious that for men, for those from villages, for people with an education below secondary level, and for people with below average income, this is the only factor that influences their perceived fit of a carrier/ingredient combination. In contrast, women are influenced by more factors when judging the fit of a product concept.

For respondents between the ages of 18–29 and for those who live in the capital, awareness of the product is important in several cases. Beside awareness of the product combination, the health image of the carrier is also of great importance for those older than 30, for people who live in a town/city, for those who have tertiary education, and for women.

Perceived correspondence of health effects influenced perceived fit of a carrier/ingredient combination mostly among women, for those with secondary education, and for people with an average income. With the exception of male respondents, perceived correspondence of health effects had a favourable influence on perceived fit for at least some of the 28 product combinations.

Table 5 shows the same analysis for the data pooled across all carrier/ingredient combinations.

Table 5. The influence of individual factors on the perceived fit of carrier/ingredient combinations in different demographic groups.

	Perceived Correspondence of Health Effects (expB)	Health Image of the Carrier (expB)	Expected Taste of the Combination (expB)	Awareness of the Combination (expB)
male	1.23 *	1.29 **	3.59 **	2.15 **
female	1.33 **	1.43 **	3.63 **	2.63 **
Age 18–29	1.29 **	1.19 **	3.16 **	2.96 **
Age 30–44	1.34 **	1.87 **	4.12 **	2.59 **
45 or older	1.36 **	1.88 **	4.11 **	2.03 **
village	1.27 **	1.95 **	2.96 **	2.18 **
Town/city	1.41 **	1.56 **	3.40 **	2.50 **
Capital	1.26**	1.17 **	4.14 **	2.84 **
Primary education	1.68 **	2.32 **	5.18 **	1.62 **
Secondary education	1.34 **	1.41 **	3.18 **	2.66 **
Tertiary education	1.28 **	1.36 **	3.96 **	2.62 **
Income below average	1.39 **	1.41 **	5.05 **	2.37 **
Average income	1.36 **	1.35 **	3.38 **	2.54 **
Income above average	1.15 *	1.81 **	3.36 **	2.94 **

* $p \leq 0.05$; ** $p \leq 0.01$.

The results indicate that the examined determinants have a positive influence on the attitudes towards a carrier–ingredient combination in all socio-demographic and perceived income groups. In general, the expected taste of the combination had a great importance. On average, expected good taste resulted in a 3.6 times bigger chance of perceived fit, with some cases reaching a 5-fold increase. This factor was especially important for those with primary education, people with below average income, those older than 30, and for people from the capital. Another aspect of great importance was awareness of the combination for all respondent groups, but especially for women, young people and for those from the capital. Health image of the carrier, verified in previous research, and perceived correspondence of health effects, verified in the present study, both positively influence perceived fit of carrier/ingredient combination in each respondent group, although to a lesser extent than the two other determining factors. Specifically, perceived correspondence of health effects led to a 30% bigger chance of perceived fit.

4.3. Managerial Implications

"Ensure healthy lives and promote well-being for all at all ages" is considered as a priority, being part of the Sustainable Development Goals (Agenda 2030) of the United Nations [33]. The majority of leading food producers are committed to enhancing the portfolio of functional products.

Due to these, it is a very important question, how to motivate the consumers to choose the alternatives that are beneficial for their health. The understanding of interrelationships between functional ingredients, carriers, and consumers' preference system can be an important aid in the phase of conceptualisation, development, and promotion of novel food products. These pieces of knowledge can be important in decreasing the failure rate of the launching of new products and could contribute to increasing the efficiency of the utilisation of R+D funds.

4.4. Limitations

As a limitation of the research, we have to add that it has been carried out in Hungary using a web survey for the non-representative, convenient sampling of Hungarian respondents, which might cause the sample to be biased. This should be kept in mind when considering the results. We have tested our hypothesis for only 28 product combinations. International researches with increased number of possible carrier ingredient combinations could improve the generalizability of our findings. More types of demographic variables and representative sampling could help finding the target group where perceived correspondence of health effects has the strongest effect on purchase intention.

5. Conclusions

We have examined the effect of different determinants on the perceived fit of 28 carrier/ingredient combinations created from 5 carriers and 11 ingredients. In the case of all individual products and also in general, it was found that expected taste had the greatest influence on the perceived fit of a carrier–ingredient combination. At the same time, awareness of the combination, the health image of the carrier and perceived correspondence of health effects also had an influence. Nonetheless, the role of these determinants is different for each carrier food. The largest variances between the various combinations were found in connection with perceived correspondence of health effects. This factor has the highest influence on the perceived fit of muesli-, rye-bread-, and yogurt-based product combinations, while in the case of dark chocolate no such effect was found.

As a summary, it can be stated that for numerous products, enhancing an existing and well-known health effect of the carrier increases positive perceived fit even if the ingredient is not originally present in the carrier.

In the case of almost all product combinations in all respondent groups, a substantial influence of the expected taste was found. Perceived correspondence of health effects positively influenced the perceived fit of the carrier/ingredient combinations among women, among those with secondary education and among people with average income.

When generalising over the individual acceptance of products, it was found that beside taste, awareness of the product (that is whether the respondent has heard of the given product or not) is also of great importance when forming an opinion about it. At the same time, perceived correspondence of health effects was proven to be a positive influencing factor in all respondent groups.

It can be stated that in addition to the influencing factors researched in previous studies, perceived correspondence of health effects also plays an important role in whether consumers perceive a good fit between a carrier product and a functional ingredient. Nonetheless, the effect of individual determinants differ significantly in each food category and combination.

Author Contributions: Á.T. and Á.B. worked out the concept. All the authors took part in developing the methodology and writing the paper. Investigation, Á.B., Supervision, K.G.G. and Z.L.

Funding: The Project is supported by the European Union and co-financed by the European Social Fund (grant agreement no. EFOP-3.6.3-VEKOP-16-2017-00005).

Acknowledgments: The authors would like to express their acknowledgements for the assistance they received from Dr. Tibor Deutsch in the field of statistical analysis and Anna Parditka-Temesi in the field of high standard database management.

Conflicts of Interest: The authors declare no conflict of interest.

References

1. Childs, N.M. Foods that help prevent disease: Consumer attitudes and public policy implications. *J. Consum. Market* **1997**, *14*, 433–447. [CrossRef]
2. Siró, I.; Kápolna, E.; Kápolna, B.; Lugasi, A. Functional food. Product development, marketing and consumer acceptance—A review. *Appetite* **2008**, *51*, 456–467. [CrossRef] [PubMed]

3. Bromage, S.; Gonchigsumlaa, E.; Traeger, M.; Magsar, B.; Wang, Q.; Bater, J.; Li, H.; Ganmaa, D. Awareness and attitudes regarding industrial food fortification in Mongolia and Harbin. *Nutrients* **2019**, *11*, 201. [CrossRef]
4. Guo, J.; Lovegrove, J.; Givens, D. A Narrative review of the role of foods as dietary sources of vitamin D of ethnic minority populations with darker skin: The underestimated challenge. *Nutrients* **2019**, *11*, 81. [CrossRef]
5. Engle-Stone, R.; Nankap, M.; Ndjebayi, A.; Gimou, M.M.; Friedman, A.; Haskell, M.J.; Tarini, A.; Brown, K.H. Vitamin A status of women and children in Yaoundé and Douala, Cameroon, is unchanged one year after initiation of a national vitamin A oil fortification program. *Nutrients* **2017**, *9*, 522. [CrossRef] [PubMed]
6. Jonas, M.S.; Beckmann, S.C. *Functional Foods: Consumer Perceptions in Denmark and England*; MAPP: Aarhus, Denmark, 1998. Available online: https://pure.au.dk/ws/files/91/wp55.pdf (accessed on 29 March 2019).
7. Poulsen, J.B. *Danish Consumers' Attitudes towards Functional Foods*; MAPP: Aarhus, Denmark, 1999. Available online: https://pure.au.dk/ws/files/32297714/wp62.pdf (accessed on 29 March 2019).
8. Urala, N.; Lähteenmäki, L. Reasons behind consumers' functional food choices. *Food Sci. Nutr.* **2003**, *33*, 148–158. [CrossRef]
9. Urala, N. *Functional foods in Finland: Consumers' Views, Attitudes and Willingness to Use*; VTT Publications: Helsinki, Finland, 2005. Available online: https://www.vtt.fi/inf/pdf/publications/2005/P581.pdf (accessed on 29 March 2019).
10. Krutulyte, R.; Scholderer, J.; Hagemann, K.S.; Grunert, K.G.; Elgaard, P.; Nielsen, B.; Graverholt, J.P. Motivational factors for consuming omega-3 PUFAs: An exploratory study with Danish consumers. *Appetite* **2008**, *51*, 137–147. [CrossRef] [PubMed]
11. Gilbert, L. The functional food trend: what's next and what Americans think about eggs. *J. Am. Coll. Nutr.* **2000**, *19*, S507–S512. [CrossRef]
12. Tuorila, H.; Cardello, A.V. Consumer response to an off-flavour in juice in the presence of specific health claims. *Food Qual. Prefer.* **2002**, *13*, 561–569. [CrossRef]
13. Cox, D.N.; Koster, A.; Russel, C.G. Predicting intentions to consume functional foods and supplements to offset memory loss using an adaptation of protection motivation theory. *Appetite* **2004**, *33*, 55–64. [CrossRef] [PubMed]
14. Lyly, M.; Roininen, K.; Honkapää, K.; Poutanen, K.; Lähteenmäki, L. Factors influencing consumers' willingness to use beverages and ready-to-eat frozen soups containing oat β-glucan in Finland, France and Sweden. *Food Qual. Prefer.* **2007**, *18*, 242–255. [CrossRef]
15. Lalor, F.; Madden, C.; McKenzie, K.; Wall, P.G. Health claims on foodstuffs: A focus group study of consumer attitudes. *J. Funct. Foods* **2011**, *3*, 56–59. [CrossRef]
16. Kraus, A. Factors influencing the decisions to buy and consume functional food. *Brit. Food J.* **2015**, *117*, 1622–1636. [CrossRef]
17. Urala, N.; Lähteenmäki, L. Attitudes behind consumers' willingness to use functional foods. *Food Qual. Prefer.* **2004**, *15*, 793–803. [CrossRef]
18. Verbeke, W. Functional foods: Consumer willingness to compromise on taste for health? *Food Qual. Prefer.* **2006**, *17*, 126–131. [CrossRef]
19. Balasubramian, S.K.; Cole, C. Consumers' search and use of nutrition information: The challenge and promise of the nutrition labelling and education act. *J. Mark.* **2002**, *66*, 112–127. [CrossRef]
20. Van Kleef, E.; Van Trijp, H.C.M.; Luning, P. Functional foods: Health claim-food product compatibility and the impact of health claim framing on consumer evaluation. *Appetite* **2005**, *44*, 299–308. [CrossRef]
21. Ares, G.; Gámbaro, A. Influence of gender, age and motives underlying food choice on perceived healthiness and willingness to try functional foods. *Appetite* **2007**, *49*, 148–158. [CrossRef]
22. Siegrist, M.; Stempfli, N.; Kastenholz, H. Consumers' willingness to buy functional foods. The influence of carrier, benefit and trust. *Appetite* **2008**, *51*, 526–529. [CrossRef]
23. Annunziata, A.; Vecchio, R. Functional foods development in the European market: A consumer perspective. *J. Funct. Foods* **2011**, *3*, 223–228. [CrossRef]
24. Annunziata, A.; Vecchio, R. Consumer perception of functional foods: A conjoint analysis with probiotics. *Food Qual. Prefer.* **2013**, *28*, 348–355. [CrossRef]
25. Bruschi, V.; Teuber, R.; Dolgopolova, I. Acceptance and willingness to pay for health-enhancing bakery products—Empirical evidence for young urban Russian consumers. *Food Qual. Prefer.* **2015**, *46*, 79–91. [CrossRef]

26. Bech-Larsen, T.; Grunert, K.G. The perceived healthiness of functional foods—A conjoint study of Danish, Finnish and American consumers' perception of functional foods. *Appetite* **2003**, *40*, 9–14. [CrossRef]
27. Grunert, K.G.; Lähteenmäki, L.; Boztug, Y.; Martinsdóttir, E.; Ueland, Ø.; Åström, A.; Lampila, P. Perception of health claims among Nordic consumers. *J. Consumer Policy* **2009**, *32*, 269–287. [CrossRef]
28. Messina, F.; Saba, A.; Turrini, A.; Raats, M.; Lumbers, M.; Food in Later Life Team. Older people's perceptions towards conventional and functional yoghurts through the repertory grid method: A cross-country study. *Brit. Food J.* **2008**, *110*, 790–804. [CrossRef]
29. Krutulyte, R.; Grunert, K.G.; Scholderer, J.; Lähteenmäki, L.; Hagemann, K.S.; Elgaard, P.; Nielsen, B.; Graverholt, J.P. Perceived fit of different combinations of carriers and functional ingredients and its effect on purchase intention. *Food Qual. Prefer.* **2011**, *22*, 11–16. [CrossRef]
30. Jahn, S.; Tsalis, G.; Lähteenmäki, L. How attitude towards food fortification can lead to purchase intention. *Appetite* **2019**, *133*, 370–377. [CrossRef]
31. Lu, J. The effect of perceived carrier-ingredient fit on purchase intention of functional food moderated by nutrition knowledge and health claim. *Brit. Food J.* **2015**, *117*, 1872–1885. [CrossRef]
32. Kraus, A.; Annunziata, A.; Vecchio, R. Sociodemographic factors differentiating the consumer and the motivations for functional food consumption. *J. Am. Coll. Nutr.* **2017**, *36*, 116–126. [CrossRef]
33. United Nations. *Transforming Our World: The 2030 Agenda for Sustainable Development (A/70/L.1)*; United Nations: New York, NY, USA, 2015. Available online: http://www.un.org/ga/search/view_doc.asp?symbol=A/70/L.1&referer=/english/&Lang=E (accessed on 19 March 2019).

MDPI
St. Alban-Anlage 66
4052 Basel
Switzerland
Tel. +41 61 683 77 34
Fax +41 61 302 89 18
www.mdpi.com

Nutrients Editorial Office
E-mail: nutrients@mdpi.com
www.mdpi.com/journal/nutrients

www.ingramcontent.com/pod-product-compliance
Lightning Source LLC
Chambersburg PA
CBHW061601220926
51850CB00005B/387